Brain Function and Oscillations

Springer
Berlin
Heidelberg
New York
Barcelona
Hong Kong
London
Milan
Paris
Singapore
Tokyo

Springer Series in Synergetics

Editor: Hermann Haken

An ever increasing number of scientific disciplines deal with complex systems. These are systems that are composed of many parts which interact with one another in a more or less complicated manner. One of the most striking features of many such systems is their ability to spontaneously form spatial or temporal structures. A great variety of these structures are found, in both the inanimate and the living world. In the inanimate world of physics and chemistry, examples include the growth of crystals, coherent oscillations of laser light, and the spiral structures formed in fluids and chemical reactions. In biology we encounter the growth of plants and animals (morphogenesis) and the evolution of species. In medicine we observe, for instance, the electromagnetic activity of the brain with its pronounced spatio-temporal structures. Psychology deals with characteristic features of human behavior ranging from simple pattern recognition tasks to complex patterns of social behavior. Examples from sociology include the formation of public opinion and cooperation or competition between social groups.

In recent decades, it has become increasingly evident that all these seemingly quite different kinds of structure formation have a number of important features in common. The task of studying analogies as well as differences between structure formation in these different fields has proved to be an ambitious but highly rewarding endeavor. The Springer Series in Synergetics provides a forum for interdisciplinary research and discussions on this fascinating new scientific challenge. It deals with both experimental and theoretical aspects. The scientific community and the interested layman are becoming ever more conscious of concepts such as self-organization, instabilities, deterministic chaos, nonlinearity, dynamical systems, stochastic processes, and complexity. All of these concepts are facets of a field that tackles complex systems, namely synergetics. Students, research workers, university teachers, and interested laymen can find the details and latest developments in the Springer Series in Synergetics, which publishes textbooks, monographs and, occasionally, proceedings. As witnessed by the previously published volumes, this series has always been at the forefront of modern research in the above mentioned fields. It includes textbooks on all aspects of this rapidly growing field, books which provide a sound basis for the study of complex systems.

A selection of volumes in the Springer Series in Synergetics:

Synergetics An Introduction
3rd Edition By H. Haken

Chemical Oscillations, Waves and Turbulence
By Y. Kuramoto

Synergetics of the Brain
Editors: E. Başar, H. Flohr, H. Haken, A. J. Mandell

Self-Organization Autowaves and Structures Far from Equilibrium Editor: V. I. Krinsky

Temporal Disorder in Human Oscillatory Systems Editors: L. Rensing, U. an der Heiden, M. C. Mackey

Computational Systems – Natural and Artificial Editor: H. Haken

From Chemical to Biological Organization
Editors: M. Markus, S. C. Müller, G. Nicolis

Propagation in Systems Far from Equilibrium Editors: J. E. Wesfreid, H. R. Brand, P. Manneville, G. Albinet, N. Boccara

Neural and Synergetic Computers
Editor: H. Haken

Synergetics of Cognition
Editors: H. Haken, M. Stadler

Theories of Immune Networks
Editors: H. Atlan, I. R. Cohen

Neuronal Cooperativity Editor: J. Krüger

Synergetic Computers and Cognition
A Top-Down Approach to Neural Nets
By H. Haken

Rhythms in Physiological Systems
Editors: H. Haken, H. P. Koepchen

Self-organization and Clinical Psychology
Empirical Approaches to Synergetics in Psychology Editors: W. Tschacher, G. Schiepek, E.J. Brunner

Inside Versus Outside Endo- and Exo-Concepts of Observation and Knowledge in Physics, Philosophy and Cognitive Science
Editors: H. Atmanspacher, G. J. Dalenoort

Ambiguity in Mind and Nature
Multistable Cognitive Phenomena
Editors: P. Kruse, M. Stadler

Modelling the Dynamics of Biological Systems
Editors: E. Mosekilde, O. G. Mouritsen

Principles of Brain Functioning A Synergetic Approach to Brain Activity, Behavior and Cognition By H. Haken

Erol Başar

Brain Function and Oscillations

Volume II: Integrative Brain Function. Neurophysiology and Cognitive Processes

With 198 Figures

Springer

Professor Dr. Erol Başar
Institute of Physiology
Medical University Lübeck
D-23538 Lübeck, Germany
e-mail: ebasar@physio.mu-luebeck.de
and
Brain Dynamics Research Center
TÜBITAK
Research Council of Turkey
06100 Ankara, Turkey

Series Editor:

Professor Dr. Dr. h.c.mult. Hermann Haken
Institut für Theoretische Physik und Synergetik der Universität Stuttgart
D-70550 Stuttgart, Germany
and
Center for Complex Systems, Florida Atlantic University
Boca Raton, FL 33431, USA

Library of Congress Cataloging-in-Publication Data
Başar, Erol. Brain function and oscillations / Erol Basar. p. cm. -- (Springer series in synergetics, ISSN 0172-7389) Includes bibliographical references and index. Contents: v. 1. Principles and approaches -- v. 2. Integrative brain function. ISBN 3-540-64345-1 (v. 1 : hardcover : alk. paper) 1. Electroencephalography. 2. Evoked potentials (Electrophysiology) I. Title. II. Series [DNLM: 1. Electroencephalography. 2. Magnetoencephalography. 3. Brain--physiology. 4. Evoked Potentials--physiology. WL 150 B297b 1998] QP376.5.B376 1998
616.8 ' 047547--dc21 DNLM/DLC for Library of Congress 98-3612

ISSN 0172-7389

ISBN 3-540-64345-1 Springer-Verlag Berlin Heidelberg New York

Typesetting: Data conversion by Protago, Berlin
Cover design: *design & production*, Heidelberg
SPIN 10552473 55/3144 - 5 4 3 2 1 0 - Printed on acid-free paper

In memory of
Hans Berger

Years ago, as a young scientist, I hope some day to contribute to the dream of Hans Berger, that human behavior and thought could be understood with the help of EEG. Indeed, EEG oscillations (reflected also in MEG) may have a great and fundamental role in brain functioning, possibly more than we can today imagine.

Erol Başar

List of Co-Authors

Ahmet Ademoğlu, Ph.D., Department of Biomedical Engineering, Bogazici University, Istanbul, Turkey

Prof. Dr. Erol Başar, Institute of Physiology, Medical University Lübeck, Lübeck, Germany; TÜBITAK Brain Dynamics Research Unit, Ankara, Turkey

Prof. Dr. Canan Başar-Eroğlu, Institute of Physiology, Medical University Lübeck, Lübeck, Germany; Institute of Psychology and Cognition Research, University of Bremen, Germany

Prof. Dr. Theodore H. Bullock, Department of Neurosciences, Scripps Institute, La Jolla (CA), USA

Prof. Dr. Tamer Demiralp, Electro-Neuro-Physiology Research and Application Center, Istanbul, Turkey; TÜBITAK Brain Dynamics Research Unit, Ankara, Turkey; Institute of Physiology, Medical University Lübeck, Lübeck, Germany

Prof. Dr. Sirel Karakaş, Institute of Experimental Psychology, Hacettepe University, Ankara, Turkey; TÜBITAK Brain Dynamics Research Unit, Ankara, Turkey

Assoc. Prof. Dr. Vasil Kolev, Institute of Physiology, Bulgarian Academy of Sciences, Sofia, Bulgaria; Institute of Physiology, Medical University Lübeck, Lübeck, Germany

Dr. med. Ralph Parnefjord, Institute of Physiology, Medical University Lübeck, Lübeck, Germany; Department of Psychiatry, Technical University Aachen, Aachen, Germany

Lic. Rodrigo Quian Quiroga, Department of Neurophysiology and Epilepsy Program, Institute of Neurological Investigations Raúl Carrea (FLENI); Institute of Physiology, Medical University Lübeck, Lübeck, Germany

Dr. med. Elke Rahn, Institute of Physiology, Medical University Lübeck, Lübeck, Germany; Department of Psychiatry, Hospital Neustadt, Neustadt (Holstein), Germany

Cand. med. Oliver Sakowitz, Institute of Physiology, Medical University Lübeck, Lübeck, Germany

Priv.-Doz. Dr. med. Martin Schürmann, Institute of Physiology, Medical University Lübeck, Lübeck, Germany

Dr. rer. nat. Atsuko Schütt, Institute of Physiology, Medical University Lübeck, Lübeck, Germany

Prof. Dr. phil. Michael Stadler, Institute of Psychology and Cognition Research, University of Bremen, Germany

Assoc. Prof. Dr. Juliana Yordanova, Institute of Physiology, Bulgarian Academy of Sciences, Sofia, Bulgaria; Institute of Physiology, Medical University Lübeck, Lübeck, Germany

Foreword

by W. J. Freeman

These two volumes on "**Brain Oscillations**" appear at a most opportune time. As the "**Decade of the Brain**" draws to its close, brain science is coming to terms with its ultimate problem: understanding the mechanisms by which the immense number of neurons in the human brain interact to produce the higher cognitive functions. The ideas, concepts, methods, interpretations and examples, which are presented here in voluminous detail by a world-class authority in electrophysiology, summarize the intellectual equipment that will be required to construct satisfactory solutions to the problem.

Neuroscience is ripe for change. The last revolution of ideas took place in the middle of the century now ending, when the field took a sharp turn into a novel direction. During the preceding five decades the prevailing view, carried forward from the 19th century, was that neurons are the carriers of nerve energy, either in chemical or electrical forms (Freeman, 1995). That point of view was enormously productive in terms of coming to understand the chemical basis for synaptic transmission, the electrochemistry of the action potential, the ionic mechanisms of membrane currents and gates, the functional neuroanatomy that underlies the hierarchy of reflexes, and the neural fields and their resonances that support Gestalt phenomena. No better testimony can be given of the power of the applications of this approach than to point out that it provides the scientific basis for contemporary neurology, neuropsychiatry, and brain imaging.

With the development of the first generations of analog and digital computers at the dawn of the age of information, the energy metaphor came to be seen as inadequate. Brains were to be understood not in terms of channeling and expending energy but instead as processing information, for purposes of communication, cooperation, and control. (Shaw and Palm (1988) have collected and reprinted the crucial articles of this historical development). The focus of this revolution was the interpretation given by Warren McCulloch and Walter Pitts of the action potential. Heretofore it was viewed as an electrical wave of energy in networks comparable to those for the telegraph and the telephone. In their hands it became a binary digit, on-off, 0-1, enabling neural networks to do Boolean algebra and other logical functions. In other words, the neuron came to be viewed not as a dynamic element in a metabolic machine but as a symbol generator, capable of "representation" of information in the brain.

Previously, when the firing of a single neuron in a sensory cortex first became observable with a microelectrode, neurobiologists gave a neutral description of its "receptor field", meaning the spatiotemporal stimulus configuration that served to excite or inhibit the firing maximally. Jerry Lettvin, Horace Barlow, and others, however, transformed and replaced the interpretation by specifying a "feature" of an object, such as a line, color or tone, which they concluded was "represented" by the firing of the neuron. There was some initial reluctance by neurophysiologists to accept this shift in viewpoint, but resistance rapidly disappeared, for two reasons. One is that the concept of "mental representation" is deeply embedded in the philosophical literature deriving from Descartes and Kant, so that it could be easily extended to networks, assemblies, and populations of neurons, for maintaining memory banks for stored images, cognitive maps, personal histories, and world views. The other reason is the emergence of a new discipline, fueled by the development of digital computers: cognitive science and its close relative, robotics.

Cognitive science has been highly productive of information processing systems and devices in a wide range of commercial, scientific, and military applications. The entire digital computer industry can be viewed as founded on a misconception of how neurons work. Yet cognitivism has not fared well as a descriptor of brain function. Rule-driven symbol manipulation is increasingly being seen as excessively rigid, incapable of expressing or incorporating meaning, and lacking in common sense understanding of real world situations. In neurobiology it has led to the intractable "binding problem": how are representations of features combined so as to form representations of objects, and how are the representations compared with retrieved memories in the process of identification? Various solutions to the binding problem have been proposed, such as through quantum coherence, reentrant signaling, synchronization of oscillations through resonance in dendritic networks, but none is generally accepted.

The representationalist view is seductive, because it seems to be supported by data from neurobiology. However, cognitivists seem to be unaware of a subtle circularity in their appeal to empirical evidence.. About 50 years ago, with great developments in electronics and computer science, there began an invasion of researchers and ideas from the physical, engineering, and cognitive sciences, which grew to a flood that transformed neurobiology. Experimental designs and the interpretations of data were reformulated in terms of information, memory storage, analog comparators, networks, filters, integrators, logical gates, etc. In other words, to the extent that neurobiology is identified with computational neuroscience, it becomes indistinguishable from artificial intelligence.

Physicists, philosophers, molecular biologists, and immunologists coming to this recent literature cannot see that its current contents have already been formulated in terms of the concepts for which they then claim to find evidence. That is, what they are looking for in their research and how they interpret their findings have already been determined by these concepts of information, feature detection and representation. The perspective needed to see this circularity can only be gained through a detailed understanding of the neurobiological literature of the preceding half century, which they do not have, nor does the younger generation of neurobiologists.

Fortunately, there is an alternative research approach that is not beholden to the information processing view and which, therefore, is not confronted with the intractable binding problem. This approach is using nonlinear dynamics to describe the multilevel organization of neurons through their capacity of generating and sharing oscillations. It is not representational because, on this account, a stimulus input acts to trigger the interactions of masses of neurons, whose interconnections have been determined by previous experience through the mechanisms of learning. The brain then constructs the significance of the input for the behavior of an organism, rather than merely representing the features of an object that is the source of the stimulation. The key new concept that is needed is the hierarchical organization of neurons with each other to form assemblies, then of assemblies to form brains, then of brains to cause muscles to move the organism into the surrounding environment, thereby, through controlling the relations of the sensory receptors to the world, enabling brains to select their own input and adapt it to their own purposes.

The mechanisms of self-organization through the genesis of oscillations through various kinds of interaction in physical, chemical, biological, psychological, and social systems have been most deeply explored in recent decades by Aharon Katzir-Katchalsky (1974) in the nonequilibrium thermodynamics of brain cells; by Nobelist Ilya Prigogine (1980) in studies of dissipative structures and chaotic state transitions; by Hermann Haken (1983) in the "**synergetics**" of lasers, with circular causality between macroscopic and microscopic phenomena; and by Michel Foucault (1976) in his descriptions of the "power-knowledge duo" in social systems. Recent applications of dynamical systems theory to enactive robotics have been reviewed by Clark, and the potential of dynamics to depict the developmental processes in infants and children have been demonstrated by Thelen and Smith (1994). Models of the relations afforded by reciprocal interactions are implemented with the tools of nonlinear dynamics, and the approach has been popularized in theories of chaos and complexity. The present two volumes are most appropriate for neuroscientists, because they are focused on the techniques, problems and results of observation, measurement, and analysis of macroscopic oscillations in human and animal brains.

Applications to brains of these theories of dynamical interactions are at four hierarchical levels. First, complex biochemical feedback pathways within cells support the emergence of oscillations at cycle durations of minutes, hours, and days, and they underlie the recurrence patterns of normal cyclical behavior, as well as the epilepsies, mood disorders, and other pathologies.

Second, large numbers of neurons interacting through innumerable synapses under the influence of external and internal stimuli and of endogenous neurohormones form macroscopic populations. These are not genetically preformed netlets that are selected by Darwinian mechanisms. They correspond more closely to the "**nerve cell assemblies**" conceived by Donald Hebb (1949), which are formed and modulated through experience by changing the strengths of connections, which have come to be called "Hebbian synapses". In these assemblies, relationships of the neurons to the mass are explained by Haken's synergetic theory, whereby the microscopic neurons contribute to the macroscopic order and then are "enslaved" by that order, similarly to particles in lasers and soap bubbles.

At the third level, neuron populations interact with each other across extended regions of the brain by large bundles and tracts of axons. Each part of cortex and basal ganglia maintains its own "soap bubble" dynamics, with specializations based in its history and input, and it is pushed by these interactions into creating new patterns within itself that reflect and contribute to an ever-shifting global pattern involving the entire forebrain. These are not reentrant "mappings" that correspond to transfers of information in computational neural networks. They are dynamical flows with continuous distributions and trajectories, comparable to hurricanes and tornadoes. The mathematics needed to describe them has undergone striking development in recent years with the aid of computer graphics and digital computers, particularly as adapted by Abraham and co-workers (1990) for non-specialists. This is the level of predominant concern in the present two volumes.

At the fourth level, the integration of psychological phenomena with the dynamics of brains is undertaken. Perception is viewed as an active process that begins with an emergent pattern of activity in the forebrain. From that pattern, firings go into the motor systems that induce search movements. Firings from that same pattern also go as "corollary discharges" to all of the sensory cortices, to prepare them for the consequences of the intended actions, and to specify the classes of stimuli that are sought. This aspect, also called "reafference", was discovered by Helmholtz in the 1870's in his studies of patients with paralysis of the muscles controlling the position of the eyes. When asked to look in the direction that they could not, the patients reported that the world seemed to move in the opposite direction. Helmholtz called this the manifestation of the "effort of will". The closure that is required for interaction between brain and environment comes with the arrival of the stimulus and the resulting perturbation of the central structures, to which the stimulus-evoked activity is transmitted. The dynamical interplay of motor output and corollary discharge with proprioceptive and exteroceptive feedback, and with repeated update of the hippocampal cognitive map are required for orientation of action in time and space.

The internal updating and restructuring of its past, as the basis for constructing each next step into its future, is the essence of the function of each brain. The availability of that structuring for the guidance of actions by each individual, uniquely expressed in an evolutionary unfolding, has its subjective aspect in the experience of that individual, the conscious awareness of his or her unique personal history. All available parts of the forebrain participate, and the entire body of past experience, in the form of synaptic modifications and neurohormonal modulations, is brought to bear in varying degree at each moment, waking or sleeping. This is the process that is revealed by EEG and MEG analyses of the oscillatory space-time patterns that support subjective experience and objective behavior. Through the accounts provided by nonlinear dynamics, we get a form of explanation that is apart from top-down representational causality and bottom-up physicochemical causality. It corresponds to phenomenologists' descriptions of how we experience everyday meaningful activity. No other existing approach can give that explanatory power. Yet its full utility cannot be realized by dynamicists and mathematicians unfamiliar with the nervous system, nor by neural computationalists unfamiliar with the state variables of neural activity that are continuous in time and space. These are the properties of brains, essential for realizing the opportunities of this approach, that are offered by the materials in these volumes.

Citations

Abraham FD, Abraham RH, Shaw CD, Garfinkel A (1990): A Visual Introduction to Dynamical Systems Theory for Psychology. Santa Cruz CA: Aerial Press.
Clark A (1996): Being There. Cambridge MA: MIT Press
Freeman WJ (1995) Societies of Brains. Mahwah NJ: Lawrence Erlbaum Associates.
Foucault M (1976): The History of Sexuality: Vol. 1. An Introduction (R Hurley, Trans.). New York: Random House (1980).
Haken H (1983) : Synergetics: An Introduction. Berlin: Springer.
Hebb DO (1949) : The Organization of Behavior. New York: Wiley.
Katchalsky A, Rowland V, Blumenthal R (eds) (1974): Dynamic patterns of brain cell assemblies. Neuroscience Research Program Bulletin 12: 3-87.
Prigogine I (1980): From Being to Becoming: Time and Complexity in the Physical Sciences. San Francisco: WH Freeman.
Shaw GL, Palm G (1988): Brain Theory. Reprint Volume. Singapore: Worlds Scientific Press.
Thelen E, Smith LB (1994): A Dynamic Systems Approach to the Development of Cognition and Action. Cambridge MA: MIT Press.

Department of Molecular & Cell Biology — Walter J. Freeman
Division of Neurobiology, LSA129
University of California
Berkeley CA 94710-3200 USA
tel 510-64204220 fax 510-643-6791
wfreeman@socrates.berkeley.edu

Preface to Volume II

This book proposes a new "*neurons–brain doctrine*" and aims to develop a framework for an "*integrative neurophysiology.*" These goals will be approached by bridging a large spectrum of physiological and cognitive results manifested with "*EEG-oscillations.*"

This is also the first, and yet the only, book on oscillatory brain phenomena that tries to establish a general experimental approach to integrative brain functions in a wide frequency range between 1 and 1000 Hz.

The integrative approach is based on both human and animal experiments. Sensory, cognitive, and motor processes are analyzed and compared by applications of a strategy called the *brain dynamics research program*, explained in detail in the companion volume. The author and his colleagues have undertaken a *voyage* of discovery by measuring electrical activity from several types of brains, ranging from invertebrates to the human brain, by employing all the tools of this program and considering sensory and cognitive responses of the brain under a common aspect: observation of diverse oscillations, or changes of their amplitudes, frequencies, enhancements, and phases.

The material has been collected in various laboratories in collaboration with several colleagues from San Diego, Copenhagen, Vancouver, New York, Ankara, Istanbul, Sofia, Perth, Buenos Aires, Moscow, Stockholm, Shanghai. Approximately 20 scientists besides my coworkers in Lübeck have been involved in the experiments and their evaluation.

By working in this way and aiming so broadly, I have gained much multidisciplinary experience, which I would like to communicate to other scientists and graduate students working towards the understanding of brains. Thus, I do not consider the material in these two volumes as a textbook, but as an *atelier* that can will give us an opportunity to learn more together as we approach the year 2000.

All my books have been ateliers, or workshops, revolving around a simple idea: *resonant oscillations*, first treated by Galileo Galilei in the 17th century. These books have had a steady evolution. *Dynamics* was the first key word, followed by *not noise*, *chaos*, *induced rhythms*, *functional alphas*, *brain oscillations*, and finally *integrative brain activity*. Accordingly, the voyage will continue. A number of chapters can never be perfectly finished. It is my wish that this *large atelier* will be used by many young scientists who are willing to enter it. Our books are never finished. We begin a new voyage, hoping that many young people will join in this new endeavor for new discoveries. Time and patience will open new and unexpected vistas.

Some of the key points are mentioned briefly in the following:

- An important statement is that the most general dynamics in the brain are governed by the brain's natural oscillations (alpha, theta, delta, gamma, etc.). It is hypothesized that these oscillations provide basic links to brain functions, especially for *communication* and *associative* functions. The author's team are among the first groups of scientists who published, almost 30 years ago, the first reports on gamma-band responses. However, it is the major aim of this book to show that other basic frequencies are as important as the oscillations in the gamma band. *Multiple oscillatory responses provide the integrative brain functions.*
- The following working hypothesis is strongly emphasized: *Selectively distributed alpha, theta, delta, and gamma oscillatory systems* control the integrative brain functions at all sensory and cognitive levels. As a possible approach (strongly based on data obtained from a diversity of experiments) we suggest to consider the EEG-frequencies as a type of *alphabet of brain function.*
- Our results show that brain oscillations are correlated with multiple functions depending on task, sensation, and structures. At least 50 different functional correlates of brain oscillations are analyzed and summarized. These descriptions include *sensory registration*, *perception*, *movement*, and *cognitive processes* related to *attention*, *learning*, and *memory.*
- A tentative synopsis of functions related to the frequencies strongly indicates new horizons in this emerging field. A large ensemble of functions related to different frequencies or to the superposition of different frequencies is described. This step provides a tentative start to formulating an *integrative neurophysiology* based on EEG-oscillations.
- *Complex functions* are, as a rule, manifested by several *superimposed oscillations* with various degrees of amplitudes, durations, and delays.
- Based on our results, the last chapter proposes a new *neurons–brain doctrine (i.e, neuron assemblies doctrine)* incorporating the concepts of Sherrington, Berger, Adrian, Barlow, Mountcastle and Freeman, and considering new trends in this field. This is an ambitious goal; however, its seems feasible: A link will be established between conventional *EEG frequencies* (or the *brain's natural frequencies*) and the *integrative brain function* consisting of interwoven sensory and cognitive processes.

During the last two decades I have had the opportunity to collaborate at various levels or exchange ideas with several outstanding neuroscientists. Among them are T. H. Bullock, R. Galambos, W. J. Freeman, H. Haken, F.,H. Lopes da Silva, R. Hari, H. Petsche, G. Pfurtscheller, D. Sheer, R. Adey, W. Klimesch, H. Weinberg, G. Roth, M. Stadler, and H. Stampfer.

Further, the four Conferences on Brain Dynamics and Oscillatory Phenomena that I organized (Berlin (1985), Berlin (1987), New York (1990), and Lübeck (1994)) and editing the corresponding books gave me a unique opportunity to discuss closely and to correspond intensively with a large number of leading neuroscientists. Unfortunately, it is not possible to cite all of them here. These interactions provided a high-level learning effect and contributed enormously to enriching my horizons in this important emerging branch of the neurosciences.

Lübeck, January 1998 Erol Başar

Acknowledgments

I set out to write a long book at the beginning of the 1990s. I was hoping to finish it in two or three years. However, the writing was interrupted several times due to new extensions of results such as parallel processing and the expansion of our group since 1993 with the establishment of the TÜBITAK Brain Dynamics Research Center in Ankara. The interactions of our institute in Lübeck with the groups in Istanbul and Ankara demanded a considerable commitment of time. Moreover, new results from TÜBITAK joint research caused an expansion of data.

Parallel to the writing of this manuscript I have edited two more books: (1) *Induced Oscillations in the Brain*, together with T. H. Bullock, in San Diego, and (2) *Alpha Processes in the Brain*, together with R. Hari, F. Lopes da Silva, and M. Schürmann. The editing of both books and organization of the conferences in New York and Lübeck took considerable time. Additionally, after the editing of these books, more material on oscillatory phenomena in the brain has been accumulated, and my horizons have been widened due to outstanding new results of my colleagues, who have submitted chapters. Finally, when the book was almost finished, in 1995, Springer-Verlag proposed the publication of two books instead of one to enable readers to select topics according to their particular interests.

A great number of colleagues, coworkers, and friends have assisted in the preparation of manuscripts that have been in a continuous evolving state, due to the reasons cited above. An outstanding leading neuroscientist, Prof. Dr. Theodore Holmes Bullock, in San Diego, has observed and surveyed my work over the past 15 years, enriching my knowledge and scope. I have enormously profited from his advice and constructive criticism, from our joint experiments, and from the editing of our two books. He has been for me an important teacher. He always encouraged me to write a book with an integrative approach.

Four of my companions and coworkers made the most essential contributions to both volumes: Their Herculean labors will be described in chronological order.

(1) Prof. Dr. Canan Başar Eroğlu, my wife, played a key role in the evolution of these books throught her numerous contributions to the experimental work in both volumes. For 15 years she has conducted a wide variety of in-

vestigations comprising animal experiments, clinical studies, pharmacological studies, research into children's EEG and high cognitive processes, including the difficult problem of bistable perception. Thus, she has clearly played an essential role in the evolution of these books. For the past two years she has been working at the University of Bremen, and despite the heavy demands on her time, her help to my scientific work has been considerable.

(2) Prof. Dr. Sirel Karakaş, my former graduate student and colleague for 25 years, is not only the coauthor of several chapters, but she also played an outstanding and essential role in assisting me to integrate various concepts and chapters. She developed several constructive ideas, from selecting the titles to conceiving the integrative chapters. She proposed extremely useful approaches for splitting the initial manuscript into two separate books. She provided untiring questioning and criticism and suggested bridges between chapters. She came to work with me during a very difficult period in the preparation of the book by restoring and correcting about half of the chapters. She made many extensions and changes by her incessant stream of questions and proposals in the last years of our work in Lübeck, Ankara, Istanbul, and Zürich. Accordingly, she played an essential role in reshaping the book, impossible to describe in detail here. Sirel Karakaş was another invisible key person, editing my earlier book EEG-Brain Dynamics: She undertook the responsibility of the final organization of the manuscript in Ankara while I was spending a year as Richard Merton Professor at the University of Kiel, in Germany, 20 years ago, and she has provided magnificent assistance. We have had great profit from her early experience. Further, I am honored by the consistent support of Sirel Karakaş over the years, who provided the scientific bridge between Ankara and Lübeck and who contributed greatly to the development of the Brain Dynamics Research Center, in Ankara. The interaction with these laboratories has provided an essential contribution to these volumes, and she has been the untiring architect of this new and fruitful international collaboration.

(3) Priv.-Doz. Dr. med. Martin Schürmann has been from the beginning my most important colleague in the creation of these books. He is not only the coauthor of several chapters, but he has been a real architect in structuring the books. Dr. Schürmann has written with me the most fundamental chapters. My research group has been of crucial importance in the practical realization of these books. Dr. Schürmann who is not only a neurophysiologist but also has an M.S. degree in informatics, substantially contributed to finishing the manuscripts by guiding our team during my long periods at absence in recent years. My gratitude to Dr. Schürmann is based mostly on his modesty; he has done everything in a simple way to stress the importance of the scientific investigations summed up in these books. Working with him has been of enormous scientific value.

(4) Mrs. Heidi Wolfframm, my former secretary, was responsible for the preparation of the manuscripts between 1992 and 1995, the period during which 80% of the core material of the books was written. Not only did she

provide efficient secretarial help, but she also performed organizational and editorial work of high quality. She kept the manuscripts consistently in a very ordered shape. She was able to process my difficult handwriting, and was able to take dictation over two to three hours that she then quickly corrected and transcripted. Her devotion to the writing of these books has been simply great. Her ability to find the true path between disordered paragraphs made it possible that the core chapters of the books could be speedily finished without long interruptions, enabling the author to review everything within a very short time. I am very much indebted to Mrs. Wolfframm for her devotion to these books and her enthusiasm in participating in such a project. Without her enormous contribution it would have been difficult to begin and improve the books.

A special vote of thanks is due to Assoc. Prof. Dr. med. Dr. rer. nat. J. Röschke, University of Mainz, again one of my earlier graduate students and longstanding coworkers and friends. Dr. Röschke helped enormously to develop and enrich our scientific work during his stay in Lübeck. He has successfully worked on a variety of problems ranging from analysis of chaos (in the first days of the chaos area) up to the construction of the experimental setup to record invertebrate ganglia activity. Curiously enough, during his research period in my laboratories he completed two doctoral dissertations, one in physics, the other in medicine.

Assoc. Prof. Dr. Tamer Demiralp made important scientific contributions to the development of our data, especially on wavelet analysis and behavior experiments. He also played a major role in creating bridges between Lübeck and Istanbul and between his group and San Diego. His contribution to the material of these books has been very important.

Several other people have made important contributions. Dr. A. Schütt developed several experimental approaches and discussed with me several issues related to the books and other publications. Assoc. Prof. Dr. V. Kolev and Dr. J. Yordanova, also coauthors of several chapters, helped me tremendously with essential and substantial criticism in all steps of preparing the manuscripts. In the last four years, both scientists provided me essential support concerning the methodology and reevaluation of my concepts.

Dipl. Ing. F. Greitschus has developed over the years the data-processing systems for all the experiments and has performed excellent work for the organization and functioning of the laboratories. Dipl. Ing. M. Gehrmann has worked on difficult engineering problems in our EEG laboratory and in our computer network during the writing of the book.

Our technical assistants, Mrs. B. Stier, Mrs. K. Leffler, and Mrs. G. Huck, for years have worked heavily on data processing. Mrs. Stier performed excellent assistance during animal experiments. Mrs. R. Garnath and Mrs. G. Fletschinger carefully prepared illustrations in collaboration with our technical assistants. Mrs. B. Ranwig was responsible for language and spelling in approximately half of the chapters: she accomplished very useful work.

Mrs. Beate Nürnberg joined our staff at the last stages of the book. She adapted very quickly to my working style and provided magnificent and untiring help by text processing and all types of related organizational tasks. Due to her energy and punctiliousness the final manuscript could be realized. I am indebted to her for of her excellent approach to all problems related to the neurophysiology research group.

Finally, but by no means least, Dr. D. Strüber, University of Bremen, has very carefully read the final manuscripts and offered most valuable corrections and criticisms.

Several institutions have supported our research during the last fifteen years: TÜBITAK, DFG, BMBF, Deutsche EEG-Gesellschaft, VW-Stiftung. Without the financial support of these institutions, especially for international scientific exchange programs, these books could not have been realized to this extent.

Abbreviations and Glossary

1 Anatomical Abbreviations

- **CE:** *Cerebellum*
- **CA3:** *CA3 layer in hippocampus*
- **GEA:** *Gyrus ectosylvian anterior, auditory cortex*
- **MG:** *Medial geniculate nucleus*
- **LG:** *Laterale geniculate nucleus*
- **IC:** *Inferior colliculus*
- **HI:** *Hippocampus*
- **OC:** *Occipital cortex, area 17*
- **RF:** *Reticular formation*
- **SC:** *Superior colliculus*

2 Glossary and Abbreviations

- **AFC:** Amplitude–frequency characteristics. The spectra of the evoked responses in the frequency domain. In the present books AFCs are obtained by the application of Fourier transform to the transient evoked potentials (see Chap. 4, Volume I)
- **ALPHAS** is an expression characterizing the ensemble of diverse 10 Hz oscillations in the brain (see Chap. 24, Volume II).
- **Alpha response:** Oscillatory component of an evoked potential in approximately 8–13 Hz frequency range (see Chap. 24, Volume II)
- **Alpha system (selectively distributed):** see selectively distributed oscillatory systems (also Chap. 24, Volume II)
- **AEP:** auditory evoked potential
- **CAP:** Combined analysis procedure of EEG and evoked potentials
- **BDRP:** Brain Dynamics Research Program
- **EEG:** Electroencephalogram
- **EHF (enhancement factor):** In a given experimental record of EEG-EP epochs, the enhancement factor EHF is the ratio of the maximal time-locked response amplitude (max) to the rms value of the spontaneous activity just prior to the stimulus, with both signals (spontaneous and evoked

activities) being filtered within the same frequency pass bands (see Chap. 4, Volume I)

$$\text{EHF} = \frac{\max}{2\sqrt{2}\,\text{rms}}$$

- **ERP:** Event-related potential
- **EP:** Evoked potential
- **Evoked oscillations:** See Chap. 8 (Volume I)
- **Event-related oscillation:** It includes also evoked oscillations and induced rhyhthms (see Chap. 8, Volume I)
- **Evoked frequency response:** Dominant maximum in AFC
- **FFT:** Fast Fourier transform
- **Delta response:** Oscillatory component of an evoked potential in approximately 0.5–3.5 Hz frequency range (see Chaps. 20, 21, Volume II)
- **Gamma response:** Oscillatory component of an evoked potential in the approximately 30–60 Hz frequency range (see Chaps. 23, 26, Volume II)
- **Gamma system (selectively distributed):** see selectively distributed oscillatory systems (also Chaps. 3, 26, Volume II)
- **Induced oscillations:** See Chap. 8, Volume I
- **Internal EPs:** The "*rule of excitability*" is formulated as follows: "If a brain structure has spontaneous rhythmic activity in a given frequency channel, then this structure is tuned to the same frequency and is producing *internal evoked potentials* to internal afferent impulses originating in the CNS, or it is responding in the form of *evoked potentials* to external sensory stimuli with patterns similar to those of internal evoked potentials."
- **MEG:** Magnetoencephalogram
- **MEF:** Magnetic evoked field
- **MOR (Major operating rhythms):** Experiments have shown that in several areas of the brain some rhythms are more distinguished and dominant in comparision to others. Example: the posterior 10 Hz and frontal theta (Chap. 12, Volume I)
- **Phase-locked and non-phase-locked activity:** Non-phase-locked activities contain evoked oscillations that are not rigidly time-locked to the moment of stimulus delivery. These are, for example, induced alpha, beta, gamma, etc. oscillations that may relate to specific aspects of information processing. In the framework of the additive model of evoked potentials, non-phase-locked activity includes the background EEG. For analysis of only non-phase-locked or both phase-locked and non-locked EEG responses, specific approaches have been used. Phase-locked activity is suggested to include all types of event-related brain potentials. For quantification of the phase-locked activity, an averaging procedure is usually applied whereby the phase-locked responses are enhanced and the non-phase-locked ones are attenuated.
- **Resonance:** Resonance is the response that may be expected of underdamped systems when a periodic signal of a characteristic frequency is

applied to the system. The response is characterized by a "surprisingly" large output amplitude for relatively small input amplitude (i.e., the gain is large). A translation of these comments by illustration is afforded by the annoying vibrations developed in a house when certain periodic stimuli.

- **RMS:** Root-mean-square
- **REM:** Rapid eye movements
- **Selectively distributed oscillatory systems in the brain:** By means of the application of combined analysis procedure of EEG and EPs we recently emphasized the *functional importance of oscillatory responses* (in the framework of brain dynamics) related to association and ("long-distance") communication in the brain. We assumed that *alpha networks, theta networks, and gamma networks (or systems)* are selectively distributed in the brain (for the delta, theta, and alpha ranges see Chaps. 24, 25, 26 in Volume II). We also have tentatively assigned *functional properties*, namely sensory–cognitive functions, to alpha, theta, delta, and gamma resonant responses. According to this theory a sensory stimulation evokes 10 Hz enhancements in several structures of the brain, both cortical (primary auditory cortex, primary visual cortex) and subcortical (hippocampus). The selectively distributed oscillators systems in the brain are treated in detail in Volume II.
  The synchronous occurrence of such responses in multiple brain areas hints at the existence of distributed oscillatory systems and parallel processing in the brain. Such diffuse networks would facilitate the information transfer in the brain according to the general theory of resonance phenomena. Although alpha responses are observable in multiple brain areas, they are markedly dependent on the site of recording. The dependence of the alpha response on whether or not the stimulus is adequate for the brain area under study thus hints at a special functional role of alpha responses in primary sensory processing.
  The term *diffuse* was used in order to describe the distributed nature of the frequency response in the brain. It is not yet possible to define connections between the elements of these systems neuron by neuron, or to define the directions of signal flow and exact boundaries of neuronal populations involved. However, this description is necessary to emphasize that rhythmic phenomena in these frequency ranges are not unique features of the observed single subsystem of the brain and that their simultaneous existence in distant brain structures may be a relevant and important point in the description of an integrative neurophysiology.
- **SSWI:** A method of single sweep oscillatory analysis (see Chap. 6, Volume I)
- **SWS:** Slow wave sleep
- **Theta response:** Oscillatory component of an evoked potential in approximately 4–8 Hz frequency range (see Chap. 25 and Sect. 18.4, Volume II)
- **Theta system (selectively distributed):** See selectively distributed systems, Chap. 25 and Sect. 18.4, Volume II)

- **TRFC method:** A Fourier method that enables one to obtain the frequency characteristics from the transient response (Chap. 4, Volume I)
- **VEP:** Visual evoked potential
- **Wavelet analysis:** Method of time-frequency analysis (for mathematical details and a discussion of properties, see Chap. 5, Volume I). This method can be used to search and find repeatable and phase-locked signals in a given frequency window (details demonstrated in Chap. 21, Volume II)

Table of Contents

Prologue: The Voyage in EEG ... 1

Part I. Dynamics of Electrical Signals in the Animal Brain

1. Dynamics of Potentials in the Visual and Auditory Pathway, Hippocampus, and Reticular Formation of the Cat Brain . 19

1.1 Surgery, Experimental Conditions, and Raw EEG ... 19
1.2 Sensory Pathways in the Cat Brain ... 20
1.3 Evoked Potentials to Auditory Stimulation in the Cat Brain – Time Domain ... 22
1.4 Evoked Potentials to Visual Stimulation in the Cat Brain – Time Domain ... 23
1.5 Amplitude–Frequency Characteristics Obtained with Auditory Stimulation ... 23
1.5.1 Auditory Cortex ... 24
1.5.2 Medial Geniculate Nucleus ... 25
1.5.3 Mesencephalic Reticular Formation ... 26
1.5.4 Inferior Colliculus ... 26
1.5.5 Hippocampus ... 26
1.5.6 Cerebellar Cortex ... 26
1.6 Amplitude–Frequency Charateristics: Visual Stimulation ... 28
1.6.1 "Filtered Potentials" ... 29
1.7 Coherence Functions Between All Possible Pairings of Recording Electrodes–Auditory Stimulation ... 29
1.8 Phase Synchronization Demonstrated by Phase Spectra – Auditory Stimulation ... 30
1.9 Coherence Functions Between All Possible Pairings of Recording Electrodes–Visual Stimulation ... 32
1.10 Phase Synchronization Demonstrated by Phase Spectra – Visual Stimulation ... 34

2. Cross-Modality Experiments on the Cat Brain
by Erol Başar, Tamer Demiralp, Martin Schürmann, and
Canan Başar-Eroğlu 37
2.1 Introduction 37
2.2 What Are Multimodal Recognition and Cross-Modality Attention? View of Hartline 37
2.3 The Present Chapter Combines Cross-Modality Experiments, Frequency Analysis, and Wavelet Transform Approaches 39
2.4 Results 40
2.4.1 Averaged EPs (Single Animal, Grand Average) 40
2.4.2 Amplitude–Frequency Characteristics 42
2.4.3 Results of Digital Filtering 43
2.4.4 Results of Wavelet Analysis of EPs 46
2.4.5 Statistical Comparison of Results of Wavelet and Frequency Analysis 49
2.5 Single-Trial Analysis of EPs 49
2.5.1 Example of Single-Trial Analysis 49
2.5.2 Wavelet Analysis of Single-Trials 51
2.6 Physiological Implications of Cross-Modality Experiments ... 51
2.6.1 Hippocampus Is a Supramodal Center 54
2.6.2 Possible Functional Roles of Evoked Alpha Oscillations 54
2.7 EP/ERP Frequency Components – "Real Components" Related to Psychophysiological Functions 55
2.8 Monomodal vs. Bimodal Stimulation 56

3. Selectively Distributed Gamma-Band Responses Studied in Cortex, Reticular Formation, Hippocampus, and Cerebellum
by Martin Schürmann, Tamer Demiralp, Canan Başar-Eroğlu,
and Erol Başar 61
3.1 Gamma Responses to Auditory Stimuli Recorded from Various Structures 62
3.2 Gamma Responses to Visual Stimuli Recorded from Various Structures 65
3.3 Gamma Responses – Multiple Functional Correlates 67

4. Highest Frequency Range in Reticular Formation and Inferior Colliculus (100–1000 Hz) 69
4.1 Introduction 69
4.2 Selectively Averaged Transient Evoked Potentials 70
4.3 Amplitude–Frequency Characteristics 70
4.4 Consistent Selectivities in the Highest Frequency? 72
4.5 Very High Frequency Responses in the Human Brain 73

5. **The Brain of the Sleeping Cat: Dynamics of Electrical Signals** . . . 75
5.1 Some Sleep Stages of the Cat . . . 75
5.1.1 Spindle Sleep Stage . . . 75
5.1.2 Slow Wave Sleep Stage . . . 77
5.2 Evoked Potentials During Sleep Stages . . . 78
5.3 Amplitude–Frequency Characteristics During Sleep Stages . . . 79
5.3.1 Spindle Sleep (SS) Stage . . . 79
5.3.2 Slow Wave Sleep (SWS) Stage . . . 81
5.4 Application of Combined Analysis Procedure to the Spontaneous and Evoked Activities . . . 83
5.4.1 Simultaneously Recorded and Filtered EEG-EP Epochs (1–45 Hz) . . . 83
5.4.2 The Coherence Functions Between All Possible Pairings of Recording Electrodes . . . 85
5.5 Further Comments on the Component Analysis and the Real Responses in Evoked Potentials . . . 87
5.6 Interpretation of Results on Stereodynamics in the Auditory Pathway During the Slow Wave Sleep Stage . . . 88
5.6.1 Synchronization and Coupling of Resonances in the Responses of Various Brain Centers in Alpha and Beta Frequency Ranges . . . 88
5.7 Human Frequency Responses During SWS Sleep . . . 89

6. **Dynamics of Potentials from Invertebrate Brains** by Atsuko Schütt, Theodore H. Bullock, and Erol Başar . . . 91
6.1 Introduction . . . 91
6.2 Anatomy and Physiology of the Invertebrate (Gastropods) Nervous System . . . 92
6.2.1 The Abdominal Ganglia Complex . . . 93
6.2.2 The Pedal and Buccal Ganglia . . . 94
6.2.3 Microscopic Anatomy . . . 94
6.3 Materials and Methods . . . 94
6.4 Results . . . 95
6.4.1 Ongoing Compound Field Potentials . . . 95
6.4.2 Spikes . . . 96
6.4.3 Relationship Between EEG of Vertebrates and Field Potential Fluctuations of Invertebrates . . . 97
6.5 Potentials Evoked by Means of Electrical Stimulation . . . 101
6.5.1 *Aplysia* . . . 101
6.5.2 *Helix Pomatia* . . . 101
6.6 Gamma (30–50 Hz) Activity . . . 104
6.7 Neurochemical Modulation . . . 105
6.8 Unsolved Problems . . . 108

7. Dynamics of Potentials from the Brain of Anamniotes (Vertebrates)
by Erol Başar, Atsuko Schütt, and Theodore H. Bullock 109
7.1 Introduction 109
7.2 Methods 109
7.2.1 Ray 109
7.2.2 Goldfish 110
7.3 Results 110
7.3.1 Ray 110
7.3.2 Goldfish 111
7.4 The Reasons for Neuroethological Comparison 114
7.5 Similarities and Differences 115
7.5.1 Unsolved Questions 116

8. Frequency Response of the Cat Brain Is Influenced by Pharmacological Agents
by Canan Başar-Eroğlu, and Erol Başar 117
8.1 Effects of Ceruletide in the Brain 117
8.2 Methodological Remarks on Experiments with Pharmacological Agents (Haloperidol, Neostigmine, Acetylcholine) 118
8.2.1 Experimental Procedure and Data Analysis 118
8.3 Auditory EPs (AEPs) upon Application of Cerulein, Haloperidol, and Neostigmine 119
8.4 Amplitude–Frequency Characteristics 120
8.5 Interpretation of Pharmacologically Induced Changes by Application of Cerulein, Neostigmine, and Haloperidol 121
8.6 The Utility of Frequency Analysis to Neuropharmacological research 124

Part II. The Human Brain: Dynamics of EEG, Evoked Potentials, and MEG

9. Evoked Alpha and Theta Responses in Humans to Auditory and Visual Stimuli
by Martin Schürmann and Erol Başar 129
9.1 Subjects, Methods, Environment 130
9.1.1 Evoked Potentials: Auditory and Visual Stimuli 130
9.1.2 Frequency-Domain Approach to Evoked Potentials 131
9.1.3 Component Analysis by Means of Digital Band-Pass Filtering 136
9.2 Brain Resonance Phenomena and their Manifestation in Evoked Potentials 139

9.3 Single EEG-EP Epochs, Averaged EPs, and AFCs for the Study of Brain Resonance Phenomena 141
9.4 Functional Correlates of Theta and Alpha EP Components in Responses to Inadequate and Adequate Stimuli 141
9.5 Prospective and Future Research 142
9.6 Conclusions 142

10. "Cross-Modality" Experiments in Humans
by Erol Başar and Martin Schürmann 143
10.1 Analysis of Evoked Potentials and Their Frequency Characteristics: Auditory and Visual Stimuli 144
10.2 Filtered Evoked Potentials 147
10.3 Cross-Modality Responses Analyzed with Single EEG-EP Sweeps 148
10.4 Immediate Interpretation of Cross-Modality Experiments 149
10.5 Cat Intracranial Recordings Support the Results from Human Data 150
10.6 Physiological Implications of "Cross-Modality" Experiments: Possible Functional Roles of Induced Rhythmicities 151
10.7 Responses to Adequate and Inadequate Stimuli in MEG Recordings in Human Subjects 152
10.8 Further Thoughts Concerning Functional Correlates of Theta and Alpha Responses 152

11. The Bisensory Evoked Theta Response – A Correlate of Supramodal Association?
by Oliver W. Sakowitz and Erol Başar 155

12. Evoked Delta Oscillations on the Hearing Threshold
by Ralph Parnefjord and Erol Başar 161
12.1 Slow Wave Oscillations at Hearing Level: An Individual Experiment 161
12.2 AEP Investigations at the Threshold Level 162
12.3 Experimental Procedure 163
12.3.1 The Threshold Experiment: 164
12.4 Brain Response to Auditory Stimuli with Different Intensities 164
12.4.1 Time Domain Averages 164
12.4.2 Digitally Filtered AEPs 165
12.4.3 Grand Average Amplitude–Frequency Analysis 167
12.4.4 Selectively Filtered Auditory EPs 169
12.4.5 Frequency Distribution in Single Subjects 170
12.5 Has the Frequency Shift a Sensory–Cognitive Interpretation? . 172
12.5.1 Possible Origin of the Delta Response 173
12.5.2 The Decision-Memory System 174

13. Evoked Oscillations in Magnetoencephalography 177
13.1 Technical Remarks and Advantages of MEG 177
13.2 Neural Currents Underlying the ECD 179
13.3 The Electric and Magnetic Alpha: A Comparative Study of Auditory and Visual Evoked Fields 180
13.4 Evoked Fields to Sensory Stimulation: Alpha Response 181
13.4.1 Methods 181
13.5 Human MEG Responses – Temporoparietal Versus Occipital Alpha and Delta–Theta Responses 181
13.6 Evidence of 10 Hz and 5 Hz Evoked Magnetic Rhythm 186

Part III. Cognitive Processes

14. Selective Attention and Memory: Neurophysiology and Cognitive Psychology 193
14.1 Background and Perspective 193
14.2 Comparative Studies 195
14.3 Concept of Selective Attention and P300 196
14.4 Visual Selective Attention 197
14.4.1 Selective Attention: Experiments with Monkeys 199
14.5 Stages of Memory Processing: Encoding, Storage, and Retrieval 202
14.6 Encoding and Sensory Register 202
14.7 Memory 203
14.7.1 Short-Term Memory 204
14.7.2 Long-Term Memory 204
14.8 Pattern Recognition 205

15. Memory Templates in Event-Related Oscillations, P300, MMN 207
15.1 Remarks on Family of P300 Responses: ERPs 207
15.2 Experimental Setup and Paradigms 208
15.2.1 Paradigm 1 – Oddball 209
15.2.2 Paradigm 2 – Oddball with Increased Certainty of Alternating Targets 210
15.3 Frequency Analysis of ERPs: Preliminary Results........... 211
15.3.1 Comparative Analysis of Poststimulus Frequency Changes Under Different Experimental Conditions and Their Contribution to Different Latency Peaks ... 213
15.3.2 Formation of Peaks 215
15.3.3 Comparison of ERP Responses to Regular and Random Infrequent Target Stimuli.................. 216
15.4 Orientation Reaction and Learning During Repetitive Stimulation 218

15.5 Analysis of Pre- and Poststimulus Activity in Single Sweeps: "Preparation Rhythms" ... 219
15.6 Event-Related Theta Oscillations ... 220
15.7 Event-Related 10 Hz Oscillations ... 221
15.8 The Modulation of P300 Activity by Preparation Rhythms .. 222
15.9 P300, Prestimulus EEG Activity and Their Relation to Short-Term Memory: Memory Templates ... 223
15.10 Theta and Alpha Oscillations in Klimesch's Memory Model . 224
15.11 Habituation ... 224
15.12 Appendix: Frequency Analysis of MMN ... 225
15.12.1 MMN Formation of Peaks ... 226

16. Dynamics of Compound Potentials (P300) in Freely Moving Cats
by Canan Başar-Eroğlu and Erol Başar ... 229
16.1 Introduction ... 229
16.2 Methods and Paradigms Utilized for Obtaining P300 from Freely Moving Cats ... 229
16.3 Systematic Analysis of the Effect of Omission Rate on ERPs Recorded from the Cat Hippocampus ... 231
16.4 Utility of Analysis in the Frequency Domain ... 234
16.5 Multiple Electrodes in the Hippocampus ... 235
16.6 Hippocampal P300 and Its Cognitive Correlates: The Theta Component in the CA3 Layer ... 240

17. The Compound P300–40 Hz Response of the Cat Hippocampus
by Canan Başar-Eroğlu and Erol Başar ... 243
17.1 The P30–40 Hz Compound Potential ... 243
17.2 Gamma Activity in Earlier Studies ... 247

18. Event-Related Potentials During States of High Expectancy: Results on the Cat Hippocampus, Cortex, and Reticular Formation ... 249
18.1 Neuronal Activity of the Hippocampus During Learning, Searching, and Decision Making ... 249
18.1.1 Unit Activity and Behavior ... 250
18.1.2 Multiple Sensory–Behavioral Correlates in Single Neurons – Theta Cells in the Hippocampus: View of Ranck ... 250
18.1.3 Training-Induced Increase in Hippocampal Unit Activity: View of Thompson ... 251
18.1.4 Signal Detection in the Hippocampus ... 251
18.2 Event-Related Potentials in Cortex and Hippocampus in a P300-like Paradigm ... 254

18.3 Frequency Responses During States of "High Expectancy" ... 255
18.3.1 Time-Domain Analysis of the Responses to the 1st, 2nd, 3rd, and 4th Stimuli Preceding the Omitted Stimulus ... 255
18.3.2 Frequency-Domain Analysis by Means of the AFCs ... 256
18.3.3 The Differences Between AFCs of Responses to 1st, 2nd, and 4th Stimuli Preceding the Omitted Stimulus ... 257
18.3.4 Adaptive Digital Filtering of the Responses and Statistical Testing of the Results ... 258
18.4 Selectively Distributed Theta System of the Brain: The Limbic, Frontal, and Parietal Areas Are Mainly Involved 262
18.4.1 Frequency Selectivity of the Amplitude Enhancements in Hippocampus ... 262
18.4.2 Comments on the Anatomical and Physiological Links Between the Hippocampal Formation and the Association Areas of the Neocortex ... 263
18.4.3 The Integrative Analysis of the Increased Theta Response in the Brain: Diffuse Theta Response System in the Brain ... 264
18.5 Interpretation of Changes in ERPs ... 265
18.6 Why We Compare EP Results with Conventional Experiments on Hippocampus ... 267

19. Event-Related Potentials During States of High Expectancy and Attention in Human Subjects
by Tamer Demiralp and Erol Başar ... 269
19.1 Selective Theta Distribution ... 269
19.2 Experimental Paradigm ... 270
19.3 ERPs to Repetitive Stimuli ... 270
19.3.1 Averaged Responses ... 272
19.3.2 Adaptive Filtering of Respectively Applied Evoked Responses ... 275
19.4 Increase in Theta Components Is Highest in Frontal Recordings ... 281
19.5 In Visual Modality the Secondary Dominant Theta Increase Occurs in the Parietal Recordings ... 281
19.6 The Cognitive Theta Components of ERPs as a Sign of Hippocampocortical Interaction ... 282
19.7 Concluding Remarks ... 282

20. Topological Distribution of Oddball "P300" Responses
by Erol Başar, Tamer Demiralp and Canan Başar-Eroğlu 283
20.1 Experimental Paradigm 283
20.2 Topological Differences Between AFCs of AEPs 283
20.3 Differences Between Time-Domain Grand Averages and AFCs of Responses in the Three Paradigms 284
20.4 Adaptive Filtering of the Responses 285
20.5 The "Selectively Distributed Theta-Response System in the Brain" and the Corticohippocampal Interaction 289
20.6 Paradigms Used 289
20.6.1 Oddball Paradigm 289
20.6.2 Paradigm with Omitted Fourth Signal and to-Be-Attended 3rd Signal 290

21. Wavelet Analysis of Oddball P300
by Tamer Demiralp, Ahmet Ademoglu, and Erol Başar 293
21.1 Results 293
21.2 The P300 Wave Can Be Detected in the Single-Trial ERPs by the Spline Wavelet Coefficients in the Delta Frequency Range 299
21.3 The Response-Based Classification of the ERP Trials Yields Enhanced P300 Amplitudes Compared with the Average of the Target Responses 300
21.4 A Functional Interpretation 301
21.5 The Number of Sweeps Containing a P300 Wave May Be Used as an Additional Measure in ERP Analysis 302

22. Dynamic Memory Manifested by Induced Alpha 303
22.1 Why Look for Internal Event-Related Oscillations? 303
22.2 Coherent and Ordered States of EEG due to Cognitive Tasks 303
22.2.1 Preliminary Experiments and Method 303
22.2.2 Preliminary Results in Special Cases 304
22.2.3 Global Trends of Pretarget Event-Related Rhythms. Expectation and Reaction of Subjects; Subject Variability 305
22.3 Paradigms with Increasing Probability of Occurrence 306
22.4 Experiments with Light Stimulation 310
22.4.1 Examples of Experiments with Varied Probabilities of Stimulus Occurrence 310
22.5 Long-Standing Experiments with Subject A.F. 317
22.6 Quasi-deterministic EEG, Cognitive States, Dynamic Memory 318
22.6.1 What Is New in the "Dynamics of Time-Locked EEG Patterns" 320
22.7 Appendix 320

23. Event-Related Oscillations as a Strategy in Cognition Research ... 323
23.1 Generalization of Cognitive Responses: Advantages of the Brain Dynamics Research Program and the Concept of Event-Related Oscillations ... 323
23.2 Component Analysis Towards Functional Understanding During Cognitive Processes ... 325

Part IV. Integrative Systems in Brain Function

24. Functional Alphas Selectively Distributed in the Brain – A Theory ... 331
24.1 The EEG 10 Hz Band Rhythms Classified ... 331
24.2 Why the Expression "Alphas"? ... 332
24.3 Alphas and Alpha Activity Revisited ... 333
24.4 Some Physiologically Based Theories on the Generation of Alpha Rhythms ... 334
24.4.1 The Facultative Pacemaker Theory ... 334
24.4.2 The Scope of Lopes da Silva and Coworkers ... 335
24.4.3 Survey by Andersen and Andersson ... 336
24.5 Multifunctional and Selectively Distributed 10 Hz Oscillations – A New Survey ... 338
24.6 Secondary Alpha Response or Alpha Response with Delay ... 345
24.7 Synopsis on Multiple Functions of "Alphas" ... 348
24.7.1 Memory Mechanisms and Alpha ... 348
24.7.2 Alpha as Sensory Response ... 349
24.7.3 Alpha and Motor Processes ... 349
24.7.4 Association Mechanisms and Attention ... 349
24.8 "Alphas" Selectively Distributed in the Brain ... 350
24.9 An Integrative Theory of Alphas ... 351

25. Theta Rhythms in Integrative Brain Function ... 353
25.1 Functional Importance of Theta Rhythms ... 353
25.1.1 A Summary of Theta Rhythms in the Limbic System ... 353
25.2 Earlier Experiments on Induced or Evoked Theta Oscillations 356
25.3 Correlating with Orienting: Review by Miller ... 357
25.4 Theta Activity in the Prefrontal Cortex ... 358
25.5 Miller's Description of the Relation of Intracellular Potentials to EEG Activity in the Theta EEG Activity ... 359
25.6 Selectively Distributed and Induced Theta Oscillations in the Brain; A Theory ... 365

26. Gamma-Band Responses in the Brain: Functional Significance
by Canan Başar-Eroğlu, Erol Başar, Martin Schürmann, Atsuko Schütt, Daniel -Strüber, Michael Stadler, and Sirel Karakaş 367
26.1 Historical Note: Four Phases of Pioneering Studies Related to the Gamma Band ... 368
26.2 A Classification of Gamma-Band Activities by Galambos ... 369
26.3 40 Hz Responses at the Cellular Level ... 369
26.4 40 Hz Responses in Field-Potential Recordings: Sensory and/or Cognitive Processes? ... 370
26.4.1 Measurements in Animals ... 370
26.4.2 Measurements in Humans ... 373
26.5 Functional Interpretation of 40 Hz Responses in Light of Comparative Data ... 377
26.5.1 The Binding Problem: Gamma-Band-Induced Rhythm as a Mechanism of Feature-Linking in the Visual Cortex ... 377
26.5.2 The Diffuse and Selectively Distributed Gamma System of the Brain ... 378
26.5.3 Conclusion ... 380

27. Structures, Brain Waves, and Their Functions ... 381
27.1 Parallel Processing – A Principle of Brain Function Accessible to Investigation by Means of Field Potentials and EEG ... 381
27.2 The Basic Cortical Circuit and Cortical Oscillatory Responses 382
27.3 Thalamus: Sensory Gate for the Alpha Response ... 385
27.3.1 Classical Thalamocortical Projection ... 385
27.4 Hippocampus ... 387
27.4.1 Hippocampus as "Supramodal Structure" ... 387
27.4.2 Cross-Modality Experiments ... 388
27.5 Frontal Cortex ... 389
27.6 Cerebellum ... 390

28. Brain Functioning: Integrative Models
by Erol Başar and Martin Schürmann ... 393
28.1 EEG Frequencies as General Operators ... 393
28.2 Do EEG Frequencies Reflect Repertoires of Higher Brain Function? ... 394
28.3 From "Functional EEG Modules" to "Selectively Distributed Frequency Systems" in the Brain ... 395
28.3.1 Does Cortico-Cortical communication between EEG Modules in the Distant Parts of the Cortex Exist? ... 398
28.4 Tentative Conclusions ... 398
28.4.1 Activation of Alpha System with Light ... 400

28.4.2 Activation of the Alpha System with Auditory Stimulation ... 402
28.4.3 Activation of Theta and Delta Systems ... 402
28.4.4 Experiments with Focused Attention ... 402

29. EEG and Event-Related Oscillations as Brain Alphabet ... 407
29.1 The Integrative Character of the Compound Potential EP ... 407
29.2 Compound P300 Potential ... 408
29.3 A Cognitive Input Reduces the Compound Potentials to Almost Homogeneous Oscillatory Responses ... 409
29.4 How Is a Compound EP Almost Reduced to Homogeneous Oscillatory Response Potential in the Delta Frequency Range? 410
29.5 Event-Related Rhythms in 5 Hz and 10 Hz: Reduction of the Compound Potential by Topological Differentiation ... 410
29.6 Brain Codes: Brain Alphabet EEG? ... 411
29.7 Examples of the Brain Alphabet EEG ... 413
29.8 The Concept of "EEG Codes" as an Important Step Towards the New Integrative Neurophysiology ... 413
29.9 Thoughts Concerning the So-Called Grandmother Cell ... 414
29.10 Possible Operator Properties of EEG Frequencies ... 415

30. Event-Related Oscillations in Brain Function ... 417
30.1 Selectively Distributed Theta Oscillations: Properties, Functions, and Hypotheses ... 417
30.1.1 Properties ... 417
30.1.2 Functions ... 418
30.1.3 Hypotheses ... 419
30.2 Selectively Distributed Alpha Oscillations: Properties, Functions, and Hypotheses ... 420
30.2.1 Properties and Functions ... 420
30.2.2 Hypotheses ... 422
30.3 Functions and Hyphotheses Related to the Selectively Distributed Gamma Oscillations ... 423
30.3.1 Hypotheses ... 424
30.4 Selectively Distributed Delta Oscillations: Functions and Hyphotheses ... 426
30.5 Conclusion: Multiple Functions ... 426

Part V. Conclusion

31. An Integrative Neurophysiology Based on Brain Oscillations
by Erol Başar and Sirel Karakaş 429
31.1 Oscillations Govern the General Transfer Functions in Neural Tissues of the Brain 429
31.2 Brain Oscillatory Theory and Functional Interpretations 431
31.2.1 Spontaneous Oscillations 431
31.2.2 The Origin of Event-Related Oscillations 431
31.2.3 Functional Interpretation 432
31.2.4 The Approaches of Relevance 434
31.2.5 Final Conclusion 434

32. A "Neurons–Brain" Doctrine: New Thoughts 435

33. Epilogue: EEG Oscillations in Integrative–Cognitive Neurophysiology 437

References 439

Author Index 469

Subject Index 473

Prologue: The Voyage in EEG

Difficult problems like nature of thinking and memory are approached from two directions, through neurophysiological methods studying the physics, chemistry and biology of the nerve systems, and through psychological methods studying the observed phenomena in the brain. This research can be thought of as boring a tunnel from two sides: The two approaches have not yet met but hopefully some day they will.

B. Weiskopf

Weißt du, wie es ward? Weißt du, wie das wird? *(Do you know how it was? Do you know what will be?)*

R. Wagner, Götterdämmerung, Act III

For building a bridge between the present volume and the volume I we have to recall what happened in the foregoing description. The above title was chosen for this reason. When we are aware of the past we can be prepared for things to come. Accordingly, the companion volume began with the basic description of *"dynamics"* going back to the time of Newton and Galileo. The theoretical framework for the present volume has already been prepared in Vol. I, where the following mote of thought was introduced:

The domain of mechanics, which describes the motion of bodies without reference to the causes of motion, is called *kinematics*, whereas the domain that studies the resulting motions is called *kinetics*. These two domains are combined under the expression *"dynamics,"* and it is this general topic that is to be treated in this book. In other words, the dynamics of the brain also contain a relationship of mutual influences between bodies as reflected by kinetics. The trajectories reflecting the activity of neuronal populations can also be described as somewhat similar to the analysis of motion. Accordingly, by using the expression *"brain dynamics,"* we intend to elucidate the causes, or mechanisms, that give rise to the trajectories manifested as electrical signals from the brain.

Newton was interested not only in describing motions of planets, but he also wanted to find the mechanism of gravitation between the planets. Galileo did not only observe the oscillations of clocks, but he also wanted to learn about their machinery. Einstein was interested in describing tracks of molecules as in the case of Brownian motion, but he also analyzed the causes of Brownian motion. Furthermore, Einstein was searching for causes of gravitation, but he also wished to understand the causes of dissipating energy. To establish what has happened in the cosmos, he predicted the existence of *"black holes"* not only by using the description of astrophysical events, but also by combining all the data registered concerning the motion of stars and the given laws of physics. He tried to describe the nature of stars, including those that are not visible with conventional means, i.e., black holes.

At the beginning of the twentieth century *"Newtonian dynamics"* was extended with the theory of relativity and quantum mechanics. In this new era the orbits of electrons in an atom could not be perfectly described. They were described only in a *"cloudy way."* The configuration of atoms and molecules and also the structure of several planets could be determined using *atomic spectra.* These data helped to explain the global dynamics at the atomic and also at the galactic level. Later on, the properties and physical interactions of *elementary particles* helped enormously in increasing our understanding of fundamental processes. However, the properties of most physical systems could be successfully examined without knowing the exact position and exact motion of every emitted electron or interacting photon in nature.

In this volume we try to elucidate neurophysiological phenomena by utilizing tools and concepts of brain dynamics related to EEG. We will not only describe oscillations of the EEG, or time courses, of evoked potentials (EPs), but we will attempt to discover the underlying functional mechanisms, or causes, giving rise to these signals. In order to do this, knowledge of the anatomy of the various brain structures, the composition of cells, their function and the different signals from different neural populations, is essential. A description of the EEG with some analytic functions would be *"EEG-Brain Kinematics."* By looking for causal factors for this kinematic description we are dealing with *"EEG-Brain Dynamics."*

This approach can be compared to the efforts of physicists who did not try to learn the trajectory of every electron in nature but started with global observations of systems. Similarly, we will not attempt to learn about the firing of every single neuron upon sensory–cognitive inputs. Processes at the single-cell level are the most important fundamental neural ones. Accordingly, we will try to correlate the global behavior of neural dynamics of populations with firing properties of neurons. We will describe the relevant experiments, correlating EEG and neuronal firing in detail. To give an example, let us consider *alpha activity* of the brain. In clinical studies, this signal is mostly used to describe state changes related to waking and sleep stages or to epileptic seizures. In contrast, we will try to locate function-related alpha activities. However, we take into consideration not only anatomical and physiological evidence but also a rich ensemble of mathematical tools described in Vol. I.

Some New Trends in Neuroscience

Within the period from 1980 to 1990 five trends in neuroscience have, in my opinion, presented fruitful developments toward the understanding of brain dynamics and function:

(1) The discovery of *oscillatory phenomena at the cellular level*, such as the 40 Hz studies by the groups of Singer and Eckhorn (Gray and Singer 1987; Eckhorn et al. 1988; Gray et al. 1989) and measurements of 10 Hz and 5 Hz oscillatory behavior as well at the membrane level by Dinse and coworkers (Dinse et al. 1997), Llinás and coworkers (Llinás 1988).
(2) The application of *chaos theory* to EEG signals to demonstrate that the EEG is not just a noise signal (see Chapt. 10 in Vol. I).
(3) Important developments based on the acceptance of *cognitive function analysis* by the use of EEG and event-related potentials (ERPs).
(4) The use of *magnetoencephalography* as a complementary tool to EEG.
(5) The development of *fast laboratory computers* and the availability of *sophisticated neurocomputing software* have considerably accelerated all the above-mentioned fields of research. The laboratory computer has replaced the important revolutionary role of galvanometers and cathode ray tubes in use at the beginning of this century, as reflected in the first step towards integrative neurophysiology in the work of Sherrington (1948).

"EEG is not Noise": The Working Hypothesis

In the late 1970s or early 1980s the EEG was considered a *"background noise"* or *"idling of the brain."* The research on EPs thus appeared to be more promising than that on EEG signals, which was considered to be a simple "noise" to be rejected from the recordings. However, I myself considered the EEG signal to be one of the most important signals of the brain, to be taken into serious consideration for understanding brain function. According to the developments of the last ten years, I have seen a notable turnaround in this field. I regard the turning point to be at the beginning of the 1980s.

If one has decided to force a turning point, one is very aware of the difficulties and dangers therein. By considering the EEG to be one of the most important reflections of the brain's behavior, I was aware of the difficulties to come, but when I had written the last page of *"EEG Brain Dynamics"* in 1980, my concept was a combination of experiments and data. At the time I preferred to interpret the data, which I was able to collect by means of the best computers available, with on-line processing. The introduction of the concept of chaos in the analysis of brain functions was one of the best supports for interpreting the EEG as functional signal.

Studies of ERPs, especially cognitive or endogenous potentials under aspects of abnormal conditions, have been advanced to a great degree and have increased our knowledge. Studies on *focused attention and memory deficits* found a new basis by employing tools of EEG and evoked potentials.

What Is This Book About?

Oscillatory Brain Theory Seeks Functional Analysis

The general framework established with EEG dynamics, stating the importance of resonance in the alpha and gamma bands, is now the focus of a broad spectrum of neurophysiological work treating sensory and cognitive processes. Accordingly, I have seen the need to integrate these emerging ideas with my long-standing experimental work.

Studies of brain oscillatory activity have now been extended to include many neural systems in many species, in a number of different behavioral states. The general proposition driving the field is that the stimulus-induced slow wave oscillations are related to those higher-order neural operations intercalated between initial central sensory processing and such complex brain functions as *perception*, execution of *movements*, and storage in *memory*, in short, those functions whose study makes up a large part of what is now called by the inclusive term *cognitive neurosciences.*

Resonant Parallel Systems in the Brain Merit More Consideration for Functional Analysis

According to the results presented in Vol. I, our brain is tuned in such a way that signals coming *from our sensory organs* or signals arriving at our central nervous system *from endogenous and/or hidden sources* create reactions or responses in the brain by enhancing the natural frequencies of the brain, namely of EEG.

The sensory and cognitive functions analyzed and described in various chapters of this book lead to the concepts of parallel processing and diffusely distributed systems, which I called the *"selectively distributed alpha system," "selectively distributed theta system,"* and *"selectively distributed gamma system"* of the brain. The activation of these systems governs the modulation of event-related potentials, the spontaneous EEG, and functionally related induced rhythms. The search for the function of induced or evoked rhythms enables various sensory and cognitive processes to be described by a common language.

It is not possible to describe a given phenomenon with a given rhythm or with a particular tuning frequency. In any given operation of our brain, several oscillations function together. By studying dynamics or changes in the composition of tuned frequencies I have tried to find an approach for the description of integrative neurophysiology as yet impossible to reach by analysis of firing single neurons. One can also find similar resonance phenomena between ensembles of brain rhythms from neural populations, starting with invertebrate ganglia up to the human brain; one can also find similar resonance phenomena between these if one makes the assumption that common modes of operations within brain structures exist. This assumption constitutes the critical step from *"brain dynamics"* to *"integrative neurophysiology."*

Newton, Galileo, and Einstein turned to special planets or stars in studying dynamics in parts of the galaxy. These scientists profited by learning from general rules of dynamics and took examples from special stars to learn about the entire galaxy.

The step I have undertaken in this book is difficult. But I hope to establish the general importance of the EEG frequencies as constituting *common modes* and *functional building blocks* in the brain. There are several examples and chapters, especially in the last part of this book, that are dedicated to integrative neurophysiology.

My hope for the future is to realize that some more people will pursue this direction, expanding the number of examples and creating a new trend in brain research. Since 1980 an unbelievable turnaround has been made by people working with the tools of *EEG and oscillatory brain dynamics.* I think that this domain of the neurosciences will be highly important and fruitful as we approach the year 2000.

Brain Oscillations Are Building Blocks of Integrative Brain Functions

Several results on *evoked* and *induced oscillations* and changes in the frequency of these oscillations have been described. The studies cover recordings from the human brain, several intracranial structures of the cat brain, and furthermore, a description of the oscillations (ongoing, induced, and evoked) from various parts of the brain of low vertebrates and ganglia of invertebrates.

In Part I, containing the core material of this book, it will be shown that EPs recorded from the cat brain are *ensembles of evoked and/or induced rhythms* in the *alpha, theta, beta, and gamma frequency ranges.* The evoked and/or induced rhythms are subject to different oscillatory behavior depending on the experimental conditions: For example, if ample light stimulation is used, the induced alpha rhythms are within the 12 Hz frequency range in the occipital cortex; however, if sound stimulation is used, the induced rhythms are within the 9 Hz frequency range in the vertex or temporal recordings. Light stimulation induces 12 Hz rhythms in the occipital cortex (in the visual cortex), whereas sound stimulation does not induce large alpha enhancements in the visual cortex. If sound stimulation is used and the recordings are made within the primary auditory area of the cat, we can observe induced rhythms in a lower frequency range of 10 Hz. The 40 Hz induced rhythms were measured in various structures of the cat brain and also in the occipital and frontal areas of the human brain (Chap. 3). Depending on the kind of experiments, and especially in magnetoencephalographic (MEG) recordings, sometimes one can detect induced rhythms as *pure 10 Hz* or *pure theta rhythms* (such recordings are presented in Part II). In other words, one can register single EEG–EP recordings that are oscillations with a homogeneous frequency or induced rhythms "par excellence." Part I further deals with the results from lower brains and invertebrates that have amplitude–frequency characteristics with similar frequency response and induced rhythms by application of pharmacological agents. For example, by application of the neuropeptide cerulein, the frequency contents of evoked rhythms are altered, thus showing dominant responses in the slower theta frequency range (see Chap. 8); the unfiltered EPs resemble an oscillatory waveform in the theta frequency range.

In Part II, similar experiments performed with scalp electrodes in humans will be presented. These experiments also show that the evoked and/or induced rhythms are dependent on the experimental conditions: For example, *cross modality experiments* were performed where responses to auditory and visual stimuli were compared. Recordings from an occipital electrode show ample 10 Hz responses only when the stimulation is a visual one, i.e., an adequate stimulus for the visual cortex underlying the occipital electrode. Theta responses, however, are less dependent on whether or not the stimulus is adequate. Some particular stimulation paradigms (near-threshold stimuli, bimodal stimuli) shed light on processes beyond primary sensory processing. MEG recordings provide examples of evoked rhythms restricted to a narrow frequency band ("pure alpha responses," etc.).

In Part III, we will make use of the strategy of Fourier analysis to obtain the amplitude–frequency characteristics, this time applied to ERPs elicited with cognitive tasks. The strategy was the following one: After recording control EPs from the human and the cat brain, the subjects or animals were loaded with cognitive tasks, and in this way one could see the changes in the frequency domain.

- By using P300 paradigms (or related paradigms requiring the subject's attention and expectancy) it could be observed that induced rhythms *"switched over"* to lower frequencies in the delta and theta frequency range. Now, we call this switch of frequency tentatively *"regime changes"* (see Chap. 14, 21, and 25). If the cognitive task includes *"a surprise," "decision making," or "a new discovery,"* then induced rhythms are centered in the delta frequency range.
- Furthermore, it will be shown that during cognitive tasks the so-called *induced alpha oscillations* could be prolonged, and we have been able to show short phase shifts or delays in some of the paradigms. These results will be interpreted in relation to the cross-modality experiments presented in Parts I and II.

- An important step in the description of evoked rhythms were the experiments showing that *"omitted stimuli"* could elicit *"a 40 Hz oscillatory response"* in the hippocampus of the cat brain and also in cortex recordings of the human brain (Chap. 17).

In Part IV, all these types of sensory evoked and/or induced rhythms modulated by cognitive tasks are discussed by emphasizing functional correlates of *event-related oscillations.* This approach is based on the concept that EPs are ensembles of oscillations and on experiments showing that (1) these induced oscillations are distributed throughout the brain and (2) these rhythms can be either delayed or prolonged, or one rhythm can become much more dominant than the other ones due to cognitive states. Part IV serves to propose a synthesis on functional correlates for a new trend that we call *"EEG-related neurophysiology"* on the level of integrative functions:

- Chaps. 27 and 28 treat some core problems and provide keys to the analysis of brain functions *using the language of oscillations.*
- In Chap. 29 we will show that the compound potential, which is an ensemble of event-related oscillations, is usually related to *complex functions.* Therefore, we will discuss the concept of the *"compound potential"* with several examples. We will argue that 10 Hz rhythms, theta rhythms, and 40 Hz rhythms are probably generated from distributed networks, and they can be correlated to various functions. A particular combination of several rhythms observed in a specific structure might be related to a certain complex function.

In this part, we will proceed step by step, covering different frequency windows. We will start with the 10 Hz frequency window, since the author believes that this is one of the most important signals in the brain (see Başar et al. 1997).

These last-mentioned chapters contain as yet unrefined results and hypotheses. The extension of such views, however, could gain important utility in the near future. Accordingly, they serve not only to close the book, but they also prepare new experiments and new interpretations that should be extended.

Toward an Integrative Neurophysiology Based on Concepts of Galilei, Darwin, Berger, and Adrian

The first convergence between the fields of biology and psychology dates back to the work of Charles Darwin in the nineteenth century. Darwin's discovery showed behavioral continuity between animals. A woman suggested to him that the study of human behavior may profit from the examination of its simpler analogues in lower forms of animals. Kandel's (1976) fascinating description of the notion initiated by Darwin stimulated the comparative study of behavior and led to the development of animal models designed to relate the nervous system to behavior.

The reason I mention the pioneering work of a physicist such as Galileo Galilei in the context of modern neurophysiology stems from the fact that in a similar way we probably need a breakthrough now, not only combining the biological side with physical concepts, but also with psychological tools, which gains great importance, since the methods of neurophysiology are successfully applied by psychologists.

I try to look through the general window of the development of *"dynamics"* in the work of Galileo and Newton and later on of Einstein, thus guiding my thoughts regarding the global properties of the brain. But this is not enough. Darwin taught us to look to the realm of comparative neurophysiology in order to judge the ability and performance of the human brain.

Hans Berger measured the EEG activity, and Lord Adrian established the link between EEG and integrative neurophysiology. When we travel from the analysis of isolated ganglia of invertebrates such as aplysia or helix pomatia up to the human brain, we enrich our knowledge in a similar fashion by moving from the scale of atoms to the level of the domains of planets and the universe. The ganglia of invertebrates have only a limited number of neurons, in the range of a few thousands. This is a very small pool of neurons in comparison with the billions of neurons in the mammalian brain. The idea of learning about brains of lower vertebrates and invertebrates to understand the principle of the *"higher"* brain was given to me by working with T.H. Bullock, from whom I also inherited the expression *"compound potentials."*

The fact that properties such as the 10 Hz or 40 Hz oscillations in invertebrate ganglia are similar to the oscillations in the mammalian brain argues that we consider the common resonance properties to be universal building blocks. I believe that in the new integrative neurophysiology the building block EEG will play one of the most important roles. The ideas of Galileo, Darwin, Berger, and Adrian should be applied further.

The Voyage

From 1980 until the present day, we have brought about a cooperative effort among 15 institutions, including my research team at the laboratory in Lübeck. I have had the opportunity to analyze the behavior of various brains by applying concepts and experimental abilities of other groups. I have searched and averaged various different opinions, concepts as well as experimental paradigms, for the purpose of finding common rules with which to understand brain functions. It is certainly not a one-man job to analyze simultaneously and in a parallel manner chaos and snail ganglia, cognitive processes and brains under pathological conditions. Accordingly, I expect some gaps in the experiments we are going to present here. The experiments were not necessarily designed in the best possible way. However, I have had the unique opportunity of establishing this trend of combining several perspectives.

Now I mention again Darwin and Darwinism. The approach of Darwin was one of the most ingenious contributions to the development of human biology, and probably the most efficient one. He was far ahead of his time, leaving an extraordinary scientific legacy, as influential to modern thought as Newton or Einstein.

In his young years Charles Darwin was invited to sail around the world on HMS *Beagle* as a naturalist. Five years later, Darwin returned from that famous *voyage* with some astonishing new theories about the origin of species. Today his writings are regarded as seminal works for modern physics, astronomy, and, above all, biology. Especially his thinking on topics such as *natural selection*, evolution, and genetics are of a revolutionary nature.

Darwin used all the modern technologies for measurements and observations, probably as several other scientists did. His essential step was accomplished by comparing diverse systems, diverse species, and by sailing around the world by going out of the observed systems. He created an integrative *biology* of evolution and drew conclusions about the origin of species. The voyage was the key for his breakthrough; the voyage gave him the possibility to compare and to integrate.

Today's brain research is far from reaching such an integrative view despite modern technologies such as MRI, MEG, and arrays of microelectrodes. Unfortunately, most scientists prefer to stay in closed societies and insist on explaining the brain by looking through specific windows. It is not possible to understand the frontal cortex by means of microelectrodes in the frontal cortex, not possible to find the origin of alpha generators by recording only in thalamus and cortex. Only a voyage in the brain with global recordings might elucidate this point. The same is true for the gamma activity. Furthermore, only a chain of cognitive paradigms anchored with physiological knowledge and anatomical descriptions can permit us to understand cognitive phenomena. In order to understand the origin of brain rhythms it is not sufficient to study one specific brain. The voyage should include the comparison of several types of brains.

This is the methodological concept of *"going out of the system"* that I proposed in 1976. I had learned from the cardiovascular system the efficiency of resonating circuits. This view has been applied, consequently, in my research laboratories in the last twenty years or so. Accordingly, I felt as if I was on a long voyage in the vocabulary of brains. Wherever I landed I was confronted with alphas, gammas, resonant systems, with common components like *"letters of an alphabet."* These findings gave me the courage to throw a first stone to establish an *"EEG-related integrative neurophysiology."* I am convinced that the appropriate analysis of the brain's natural frequencies, rather than single-unit recordings, will open new avenues.

The title of this prologue should not give the impression that my goal is as ambitious as Darwin's work. My step is a very modest and yet uncompleted one. I attack only a new frontier and try to explain to other brain scientists the attractions of such a voyage that I have started. This is also an *"invitation au voyage,"* as beautifully expressed by Charles Baudelaire.

By working in this way and aiming so broadly I have gained much multidisciplinary experience. I would like to communicate to other scientists and graduate students working toward an understanding of brains. Thus, I do not consider the material I present here as a teaching textbook, but as a medium that will give us an opportunity to learn more together toward the year 2000.

Suggested Rules About Brain Dynamics (1980) and Brain Oscillations (1997)

The Working Hypothesis: "EEG Is not Noise – It Is the Most Important Signal of the CNS":

In the years since the publication of *EEG-Brain Dynamics* (1980), in which the dynamic features of EEG and EPs in normal and abnormal conditions of health and behavior were comparatively studied, EEG and EPs have gained new importance. Indeed, studies of event-related potentials, especially cognitive or endogenous potentials under aspects of abnormal conditions, have advanced research to a great degree and increased our knowledge. Studies on focused attention and memory deficits founded a new basis by employing tools of EEG and evoked potentials.

In the late 1970s and early 1980s the EEG was considered a *"background noise"* or *"idling of the brain."* The research on EPs thus appeared to be more promising than on the EEG signal, which was considered to be a simple noise to be rejected from the recordings. However, our group considered the EEG signal to be one of the most important signals of the brain, to be taken into serious consideration in order to understand brain function. In the late 1980s, new findings about gamma oscillations at the cellular level as candidate correlates of feature integration were among the first signs of a renewed interest in brain oscillations (Gray and Singer 1987; Gray et al. 1989; Eckhorn et al. 1988). Since then, numerous articles have appeared, extending the field to include alpha, theta, and delta oscillations (see Mulholland 1995; Shaw 1996; Klimesch 1996). Two special issues of the *International Journal of Psychophysiology* have recently summarized these developments (Gruzelier 1996; Başar et al. 1997).

Some Suggested Rules for the Analysis and Understanding of Evoked Potentials and Event-Related Oscillations

1. The averaged EPs are compound potentials that are generated by superposition of responses of homogeneous neural groups. The elementary EPs

that give rise to the form of compound averaged EPs usually have oscillatory character. These can be damped oscillations as well as oscillatory patterns having an amplitude-modulated waveform (see also Başar and Ungan 1973; Başar, 1976).

2. The existence of a number of peaks (waves) with different latencies in the time course of an EP does not necessarily indicate the existence of different functional structures or neural groups giving rise to these waves. For example, in Fig. P.1 it is clearly seen that the 30–55 Hz component (gamma response) of the cat hippocampus alone has influence on different waves in the time course of the EP. It does not influence a unique wave. In other words, in order to resolve a particular activity, it is not sufficient and justifiable to examine a single peak. The whole response must be frequency analyzed in order to avoid possible deceptive interpretations.
3. If the time course of an EP has a particular wave in a certain behavioral state and if this wave disappears during other behavioral or sleep stages, the disappearance of this wave does not necessarily mean that a neural group ceases its activity during this stage. On the contrary, another neural group can cause the disappearance of the wave under question. In other words, the disappearance of a peak in the time course of averaged EPs can be due to an active phenomenon.
4. As a consequence of the previous principles (1) and (2), it can be stated that if an averaged EP does not have a large number of waves, this fact does not mean that the averaged EP has a simple functional structure. Different activities can compensate each other in such a way that these activities do not result in significant waves in the time course of the averaged EP.
5. Slope and slope changes and the entire time course of evoked potentials carry the whole information concerning different activities and frequency selectivities of brain structures rather than the number and latencies of the waves. Therefore, existence of different mechanisms can be confirmed only by systems-theoretical criteria (such as frequency characteristics or filtering results) and not by simply counting the number or measuring the latencies of waves.
6. The existence of only a few resonant-frequency systems (with different attenuations or gains) in substructures of the brain can give rise to a number of varieties of transient evoked responses.
7. There exists an important relationship between spontaneous activity and EPs. This relation is characterized by resonances and evoked frequency stabilization, which are described in Chap. 11 of Vol. I.
8. *The background activity:* From the results presented and discussions in the preceding, it emerges that in the understanding of EPs, the *spontaneous activity should not be considered as a background activity* that is to be eliminated by an averaging procedure. The spontaneous activity prior to stimulus is a part of the evoked potential. It is not possible to understand the real components in the EP without a knowledge of the EEG prior to stimulus.

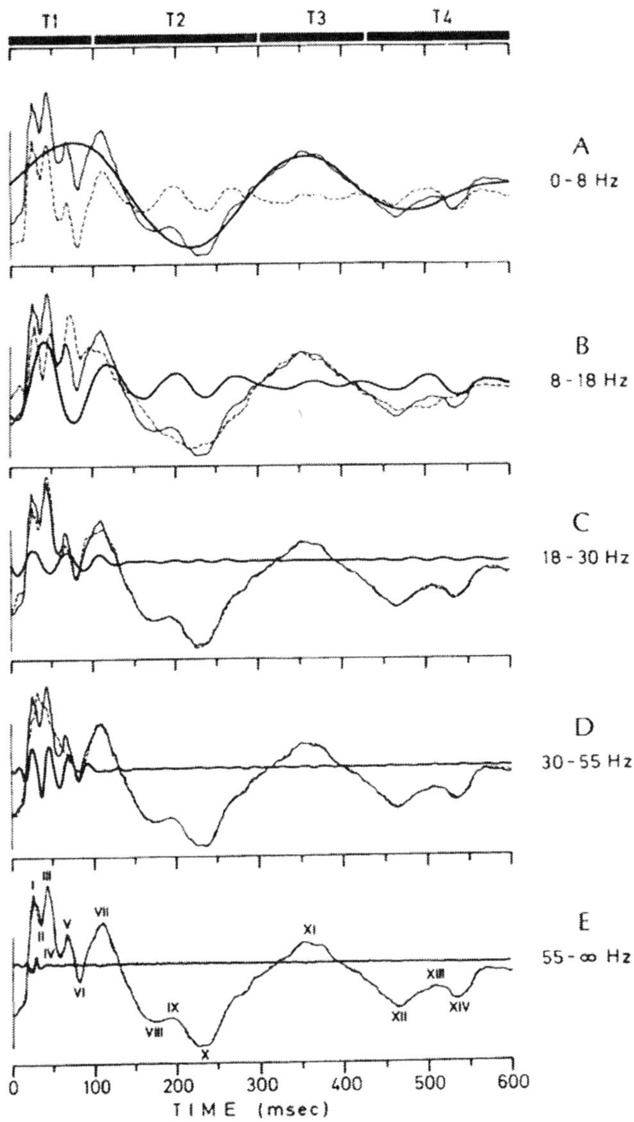

Fig. P.1. Filtering of a hippocampal selectively averaged EP with different stop-band and pass-band filters. *Solid curves* are filtered averaged EPs obtained with application of pass-band filters. *Dashed curves* are filtered averaged EPs obtained with application of stop-band filters. The band limits (shown on the *right side* of the averaged EPs) of the applied filters are chosen according to the TRFC method. The original selectively averaged EP is shown for comparison with all the filtered averaged EPs. Time sections T1–T4 are shown at the *top* of the illustration. (From Başar E. *EEG Brain Dynamics*. Amsterdam: Elsevier 1980.)

9. *Internal evoked potentials:* The findings on stereodynamics have shown that the evoked responses in all the nuclei and in all the frequencies are strongly dependent on the spontaneous activities just prior to stimulus. There are cases in which the filtered EEG–EP epochs already depict, in the EEG portion, ample potentials similar to the filtered EP signals taking place immediately after stimulation. The resemblance in the shapes of EPs and such EEG bursts leads us to use the expression *"internal evoked potentials"* for the description of large-amplitude and synchronized EEG recordings. When the relevant internal evoked potentials are recorded before stimulation, usually the EPs induced by stimulation do not have large amplitudes. Accordingly, we assume that analysis of single-trial EPs will also contribute to an understanding of EEG population dynamics.

The General Principles Obtained from the Study of Stereodynamics in the Brain

1. *Synchronization of resonance phenomena:* The resonance phenomena in alpha and beta frequency ranges occur synchronously in the auditory and visual pathways of the brain. The responsiveness in the studied brain nuclei shows that upon stimulation, they act as coupled oscillators without phase lag in alpha and beta frequencies.
2. It is demonstrated that there is no phase shift between well-developed single EP responses of all measured nuclei in a frequency range up to 25 Hz. (This information could not have been obtained with the information based only on averaged evoked potential measurements; see also Chaps. 1 and 2.)
3. The combined study of the dynamics of single EEG–EP recordings serves an understanding of *coherency* before and after stimulation: The important coherency increase in EPs can be interpreted, at first glance, as a stimulus-induced regulation of energy increase in the form of accumulation in certain discrete frequency channels, accompanied by the phase stabilization effect of the stimulation.
4. Since the coherency values presented reflect the common and synchronized power measured from pairings of analyzed brain structures, one can conclude that upon stimulation, a significant fraction of the common energy of all the studied brain centers is accumulated in a sharp frequency region between 10 Hz and 13 Hz during the waking stage. Within the alpha–beta frequency range, the energy stabilization following the stimulation shifts to slightly higher frequencies during SWS.
5. In the study of frequency characteristics obtained from averaged EPs recorded from the cat brain, we have been able to show prominent peaks in the alpha frequency range (for all nuclei), in the 40 Hz range (GEA, HI), in the 50–60 Hz range (MG and RF), and in the 70 Hz range (IC). The mutual dependence of these components could be shown, however,

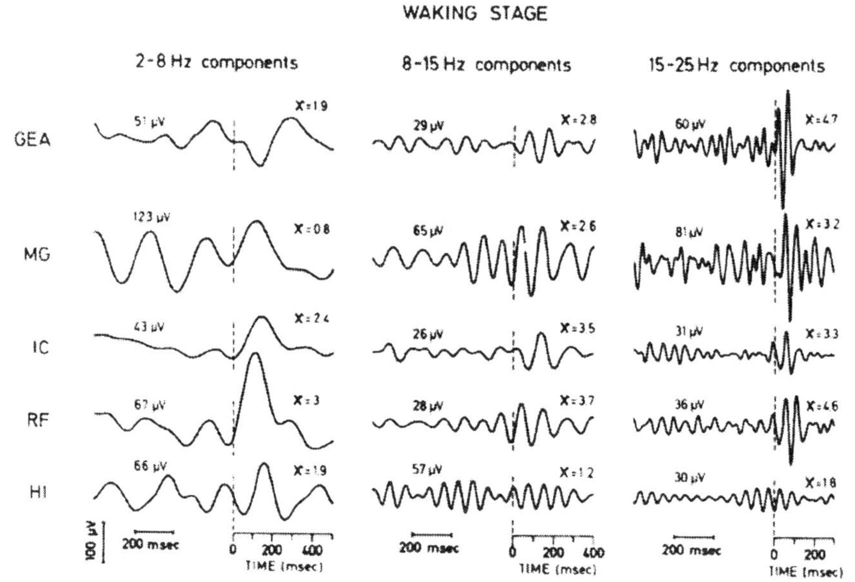

Fig. P.2. Three components of a typical set of single EEG-EP epochs (auditory stimulation) with relatively high enhancement factors. These epochs were simultaneously recorded from various brain structures of the cat during waking stage, and were filtered without phase shift. The pass-bands of the theoretical filters applied were chosen according to the shared band limits of the selectivities in the respective amplitude–frequency characteristics (not shown). The exact frequency limits of the pass-bands, which are approximately quoted at the *top*, are as follows for different nuclei: GEA (1–6 Hz, 7–17 Hz, 20–30 Hz); MG (1–6 Hz, 8–18 Hz, 18–25 Hz); IC (2–8 Hz, 8–13 Hz, 15–26 Hz); RF (1–8 Hz, 8–17 Hz, 17–26 Hz); HI (2–7 Hz, 8–14 Hz, 15–26 Hz). The rms values of the spontaneous parts and the enhancement factors are also given *above* each filtered EEG-EP epoch. (From Başar E. et al. (1979) *Biological Cybernetics* 34: 1–19.)

only with the studies of single EEG–EP recordings that were recorded simultaneously from these nuclei.

6. *Principle of synchronized selectivities:* As revealed by the alpha and beta components of simultaneously recorded EPs, as well as by the coherence functions computed for all possible pairings of the recording electrodes, the selectivities in the frequency range of 8–25 Hz (alpha–beta range), which are displayed by the frequency characteristics in the form of prominent and common-amplitude maxima, are synchronized selectivities of the studied nuclei. A synchronization between these nuclei is observed in the time and frequency domains regarding both the simultaneously recorded in-phase EP components (Figs. P.2, P.3) and the coupling and harmony in a proportional increase or decrease of magnitudes in successive single EP recordings.

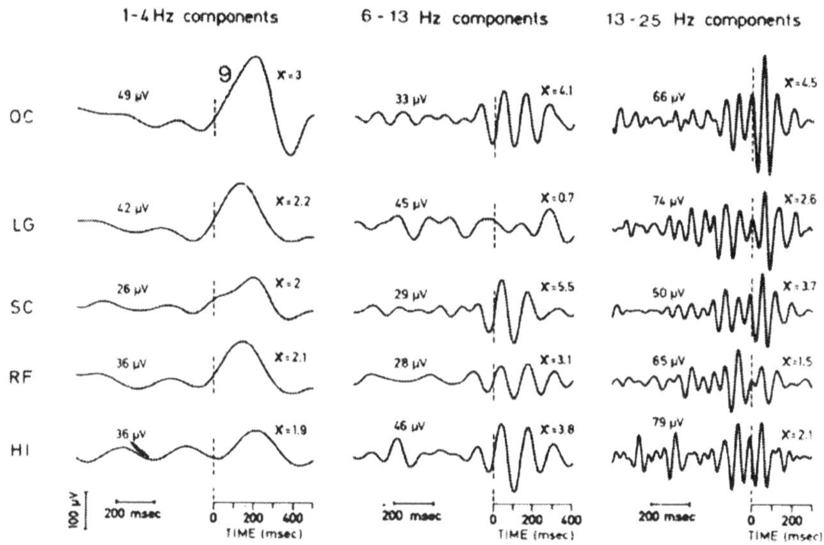

Fig. P.3. Three components of a typical set of single EEG-EP epochs (visual stimulation) with relatively high enhancement factors. These epochs were simultaneously recorded from various brain structures of the cat during waking stage, and were filtered without phase shift. The pass-bands of the theoretical filters applied were chosen according to the shared band limits of the selectivities in the respective amplitude–frequency characteristics (not shown). The exact frequency limits of the pass-bands, which are approximately quoted at the top, are as follows for different nuclei: OC (1–3 Hz, 5–13 Hz, 13–25 Hz); LG (1–7 Hz, 8–15 Hz, 15–27 Hz); SC (1–6 Hz, 7–14 Hz, 14–30 Hz); RF (1–5 Hz, 7–13 Hz, 13–20 Hz); HI (1–3 Hz, 6–12 Hz, 12–23 Hz). The rms values of the spontaneous parts and the enhancement factors are also given above each filtered EEG-EP epoch. (From Başar E. et al. (1979) *Biological Cybernetics* 34: 1–19.)

Distributed Systems and Functional Selectivity

Mesulam (1990, 1994) described in a very useful manner the neurocognitive networks and related *"selectively distributed processing."* The synchrony of selectivities earlier described by our group could have a conceptual parallelism with the selectively distributed processing. In Mesulam's neurological model of cognition, the unimodal areas of cortex provide the most precise building blocks of experience (for functional anatomy see also Chap. 3 in Vol. I). Transmodal nodes bind information in a way that introduces temporal and contextual coherence. The formation of specific templates belonging to objects and memories occurs in distributed form but with considerable specialization. This arrangement leads to a highly flexible and powerful computational system that underlies the selectively distributed processing. In our earlier work we often used the expression or concept of distributed oscillatory systems and their resonance as selective activities. According to Mesulam, functional selectivities exist in distributed functions that are based on anatomy. The electrophysiological activity of selectively distributed systems must be also of selective behavior. Accordingly, oscillatory response susceptibility of the sensory cortices, of the hippocampus, thalamus, or cerebellum, should also be differentiated, depicting selective behavior to stimulation from the milieu interieur or exterieur.

This means that the oscillatory 10 Hz response or delta response should also show selective behavior in various anatomical structures distinguished by their physiological functions. This is, indeed, the case, as Chaps. 15, 20, 24, 25, and 26 will show. Accordingly, in this book we will modify the expression "distributed oscillatory systems" in favor of "selectively distributed oscillatory systems."

Part I

Dynamics of Electrical Signals in the Animal Brain

1. Dynamics of Potentials in the Visual and Auditory Pathway, Hippocampus, and Reticular Formation of the Cat Brain

In all chapters of Part I the amplitude–frequency characteristics (AFCs) and filtered EPs will again play a major role in establishing a framework for the understanding of brain dynamics in cortical and subcortical structures. As it was argued in Chaps. 4 and 18 of Vol. I, the analysis of frequency characteristics gives global, important information concerning the real EP frequency components. The filtered responses open better windows to localizing the response at the time axis, and the response amplitudes can be described more exactly in comparison with AFCs. In using either of these methods, statements concerning minor peaking have to be avoided.

The combined analysis procedure explained in Chap. 4 in Vol. I has been used by Başar and coworkers in several neurophysiological studies since 1972. The prediction of the so-called *gamma response* in the cat brain (Başar and Özesmi 1972, Başar and Ungan 1973) and later the measurement of the human *gamma response* (Başar et al. 1976b) were not only confirmed by several authors, but the gamma band studies are nowadays one of the most sought-after in neurophysiological and cognitive function analysis. Again, the measurements of Dudkin et al. (1978), Dinse et al. (1997), and Silva et al. (1991; see Chaps. 8 and 9 in Vol. I for details) demonstrate that the so-called *alpha response*, or *evoked alpha*, will probably provide an interesting research area in brain function. As we have cited at the beginning of the chapter, the first discovery was probably made by Lord Adrian, who measured induced and/or evoked alpha and theta.

1.1 Surgery, Experimental Conditions, and Raw EEG

Measurements of brain potentials, which will be described in the following sections of this chapter, were performed in cats with chronically implanted electrodes. These electrodes were implanted – under Nembutal anesthesia – in several structures of the auditory pathway (gyrus ectosylvianus anterior, GEA; medial geniculate nucleus, MG; inferior colliculus, IC) and of the visual pathway (occipital cortex, OC; lateral geniculate gyrus, LG; superior colliculus, SC). Further electrodes were implanted in centers such as the mesencephalic reticular formation (RF), the hippocampus (HI), and the cerebellum

Table 1.1. Coordinates (in mm) of implanted electrodes (for the abbreviations see text)

	GEA	MG	IC	OC	LG	SC	HI	RF	CE
ANTERIOR/POSTERIOR	15.5	3.5	−2.5	−4.0	4.0	0.5	3.5	3.0	−1.30
LATERAL	12.5	9.0	5.0	2.0	−8.7	1.5	6.5	4.0	0.05
HORIZONTAL	10.0	1.5	3.5	*	5.5	4.4	8.8	−1.0	0.73

*2 mm below dura level

(CE), all of which are indirectly related to the auditory and visual pathways. The electrodes were placed in these nuclei according to the stereotaxic atlas by Snider and Niemer (1964) with the coordinates given in Table 1.1.

The electrodes were stainless steel and 0.1 mm in diameter. The derivations were against a common reference, which consisted of three stainless steel screws in different regions of the skull. A David Kopf 1760 instrument was used for stereotaxic surgery. Experimental sessions started approximately two weeks after surgery. During the experiments, the cats could move freely, resting or sleeping in an echo-free and soundproof room.

1.2 Sensory Pathways in the Cat Brain

Figure 1.1 represents the afferent auditory pathways with systems theory configuration. The system's input is to the ear. Different recordings (output) can be defined or chosen; when the investigator locates his electrodes in the cortex, the output will be measured in the cortex; if he locates the electrodes in the medial geniculate nucleus, the output signal will be the potential changes in the medial geniculate nucleus. Signal transport through the auditory pathway occurs, as this schema illustrates, in the major brain centers, such as the midbrain, brain stem reticular formation, thalamus, cerebellum, and cortex. In other words, the whole brain is involved with the signal transmission of signals elicited by auditory stimulation. Therefore, it is to be expected that the transfer function of signal transmission through the auditory pathway is very complicated.

It would be very difficult (or almost impossible) to consider the auditory pathway as a fully isolated system in the brain. Although the hippocampus is not an auditory nucleus, it has direct connections with nuclei of the thalamus (see Chap. 2 in Vol. I). On the other hand, all these nuclei are interconnected through the reticular formation. Therefore, when measuring outputs from the auditory pathway, one can neither exclude influences of the limbic system nuclei nor those of the visual pathway (collaterals between superior and inferior colliculi). Similarly, measurements in different parts of the cortex cannot be considered to be outputs of morphologically or functionally isolated structures, since various cortical areas are interconnected through association areas (for example, the acoustical and motor cortices; see Thompson 1967).

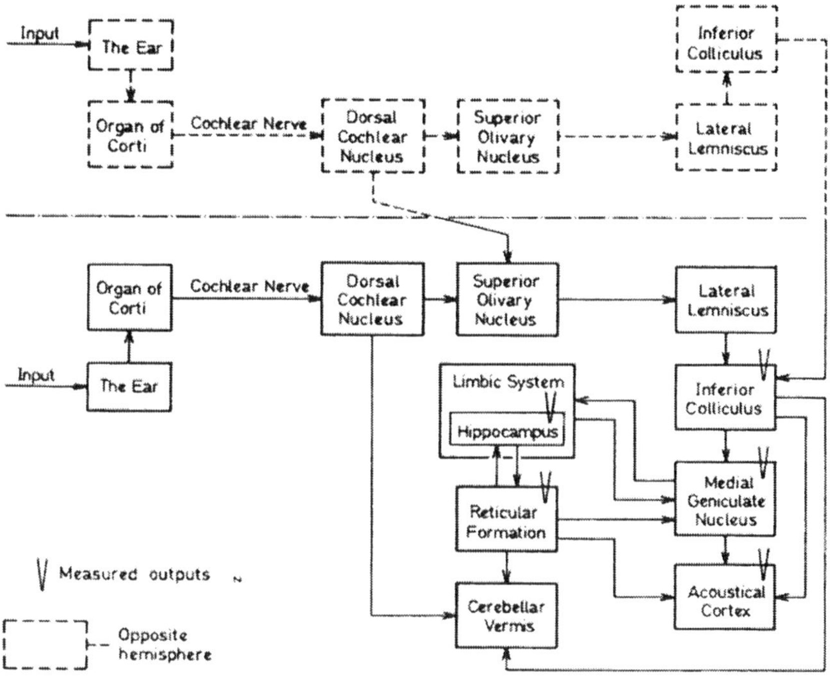

Fig. 1.1. A simplified systems theoretical illustration of the afferent auditory pathways and some related nuclei. Note that only the cross connections from one hemisphere to the other are presented. (From Başar E. et al. (1975) *Biological Cybernetics* 20: 145–160.)

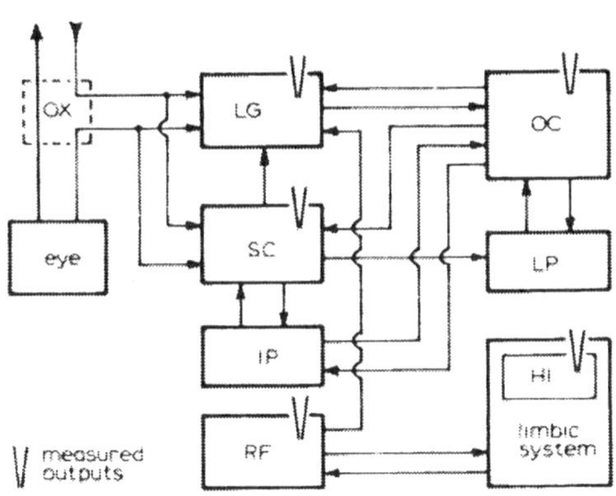

Fig. 1.2. A simplified systems theoretical illustration of the visual pathways and some related nuclei. (OC) occipital cortex; (LP) lateral posterior nucleus; (HI) hippocampus; (LG) lateral geniculate nucleus; (SC) superior colliculus; (IP) inferior pulvinar; (RF) reticular formation; (OX) optic chiasm. (From Ungan P. (1979) thesis: Hacettepe University, Ankara.)

Figure 1.2 represents an oversimplified block schema of the visual pathway and related centers with systems-theory configuration (compare the description of the visual system in Chap. 2 in Vol. I). The system's input is to the eye; on the other hand, different outputs can be chosen or defined. Depending on the location of the recording electrodes, the output may be the potential changes in the superior colliculus, lateral geniculate body, occipital cortex, or other structures.

1.3 Evoked Potentials to Auditory Stimulation in the Cat Brain – Time Domain

Figure 1.3 illustrates typical examples of the selectively averaged auditory evoked potentials in all the nuclei under study during the waking stage. The stimulation consisted of tone bursts of 2000 Hz and 80 dB SPL (auditory step functions).

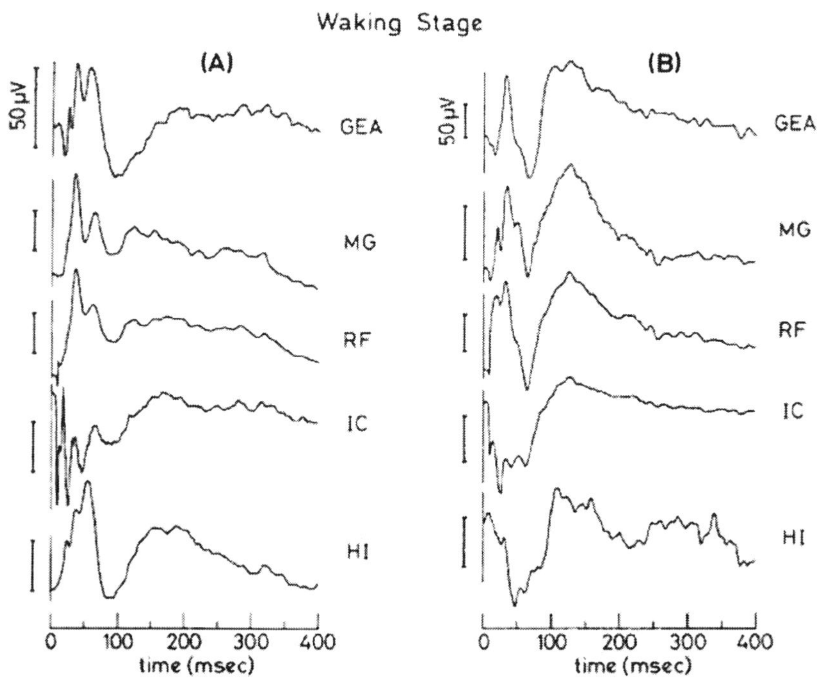

Fig. 1.3. Two typical sets of simultaneously recorded and selectively averaged evoked potentials in different brain nuclei of chronically implanted cats, elicited during the waking stage by auditory stimuli in the form of a step function. Direct computer plottings. Negativity *upwards*. (From Başar E. et al. (1975) *Biological Cybernetics* 20: 145–160.)

During the waking stage, transient EPs (selectively averaged EPs) usually depict three different kinds of waves or peaks.

(1) Short latency waves (3–12 msec latency) with fast slope changes and small amplitudes.
(2) Middle-range latency waves (20–50 msec).
(3) Long-latency waves (80–300 msec) with relatively small slope changes and large amplitudes.

The EPs of the reticular formation and of the inferior colliculus contain a significant number of short-latency waves, whereas auditory cortex (gyrus, ectosylvianus anterior), marginal gyrus, and hippocampus have fewer short-latency components.

We wish to emphasize that the EPs of Fig. 1.3 are simultaneously recorded and selectively averaged EPs and that for the reader interested in empirical findings, the EPs in these illustrations might be useful for comparative studies.

1.4 Evoked Potentials to Visual Stimulation in the Cat Brain – Time Domain

Figures 1.4a and 1.4b illustrate typical examples of selectively averaged EPs recorded from the nuclei under study in response to light stimuli during the waking stage. We wish to emphasize that the recordings of Figs. 1.4a and 1.4b are simultaneously recorded and selectively averaged EPs and that the reader interested in the study of transient EPs might find these data useful for comparative studies.

1.5 Amplitude–Frequency Characteristics Obtained with Auditory Stimulation

Figure 1.5 shows typical amplitude–frequency characteristics in all the brain nuclei under study. These amplitude–frequency characteristics were computed using selectively averaged transient responses similar to those of Fig. 1.3 with the help of the mathematical TRFC-method presented in Chap. 4 in Vol. I. The study of frequency characteristics of different brain nuclei is nothing more than considering them as a filter system or a device impeding or activating the transmission of electrical signals. Withing the framework of systems theory, we will describe these frequency characteristics using the terminology of communication networks and refer to the expression "filtering properties."

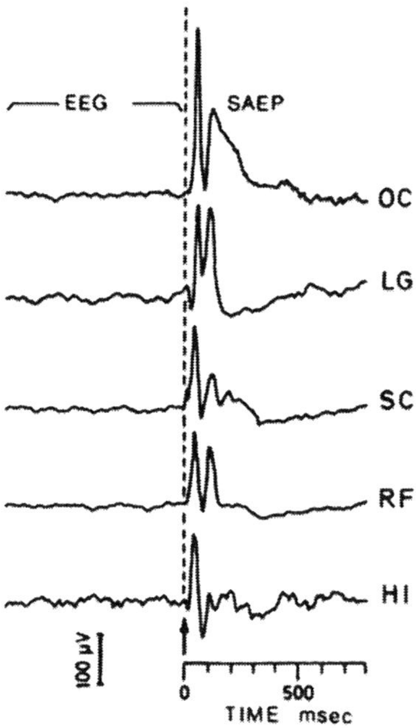

Fig. 1.4a. A typical set of simultaneously recorded and selectively averaged EPs in different brain nuclei of chronically implanted cap, elicited by visual stimuli in the form of a step function. Direct computer plottings. Negativity *upwards*. (From Başar E. *EEG Brain Dynamics*. Amsterdam: Elsevier 1980.)

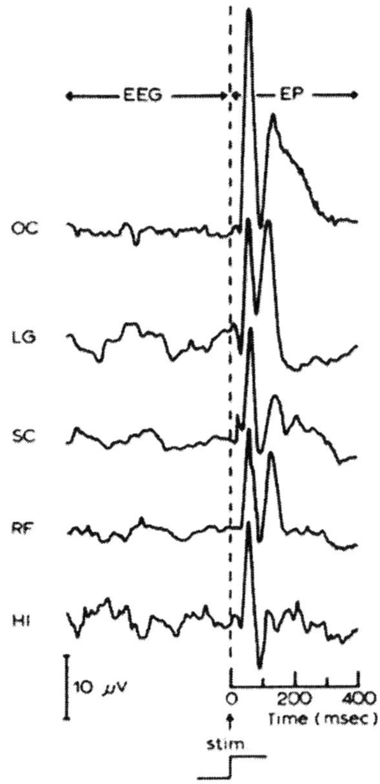

Fig. 1.4b. A typical set of simultaneously recorded and selectively averaged evoked potentials in different brain nuclei of chronically implanted cats, elicited by visual stimuli in the form of a step function. Direct computer plottings. Negativity *upwards*. (From Başar E. et al. (1975) *Biological Cybernetics* 20: 145–160.)

1.5.1 Auditory Cortex

Figure 1.5a presents two typical amplitude characteristics computed from the selectively averaged EPs of the cat acoustical cortex (gyrus ectosylvian anterior). Amplitude maxima in the frequency ranges of 8–15 Hz, 40 Hz, and 80 Hz were most commonly observed (upper curve). The amplitude maximum in the 8–15 Hz range sometimes shifted to higher frequencies of 14–20 Hz, as was observed in 35% of the curves (lower curve). The maximum around 8–15 Hz (i.e., alpha frequency range) was the most prominent maximum in almost all cases.

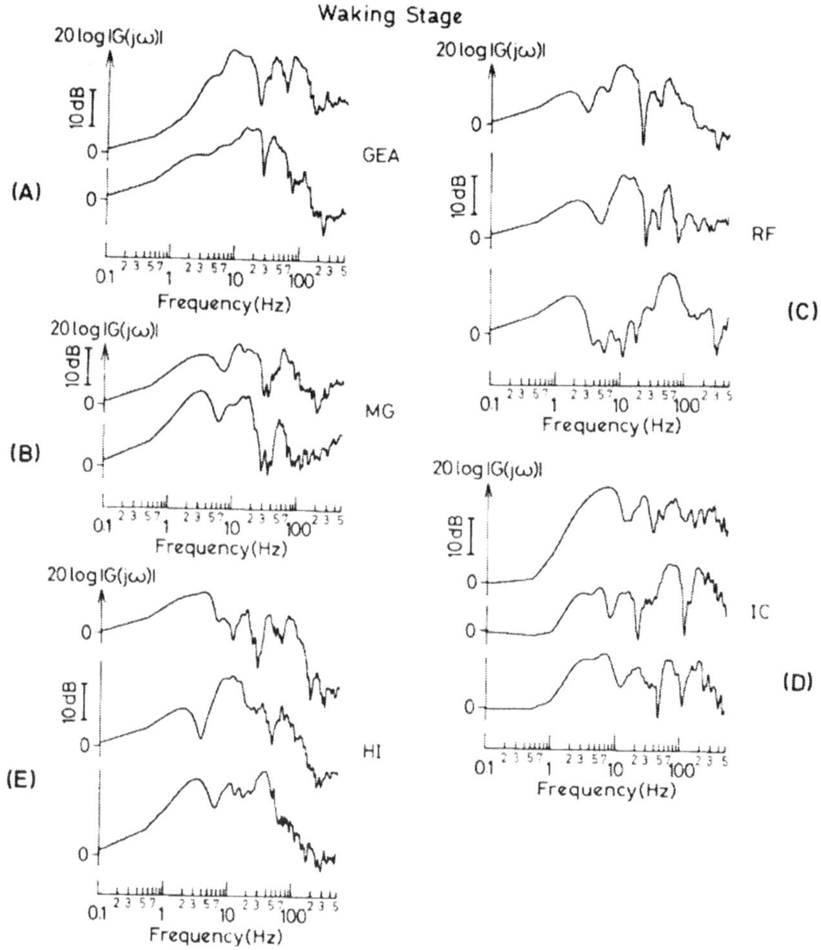

Fig. 1.5. Examples of typical amplitude–frequency characteristics of different brain nuclei of the cat determined by the TRFC method and using transient evoked responses similar to those shown in Fig. 1.4. Direct computer plottings. Along the abscissa is the input frequency in logarithmic scale, along the ordinate the potential amplitude, $|G(j\omega)|$, in decibels. The curves are normalized in such a way that the amplitude at 0 Hz is equal to 1 (or $20 \log 1 = 0$). (From Başar E. et al. (1975) *Biological Cybernetics* 20: 145–160.)

1.5.2 Medial Geniculate Nucleus

Amplitude maxima in the frequency ranges of 3–4 Hz (theta range), 8–15 Hz (alpha range), and around 50 Hz characterized the dynamics of the medial geniculate nucleus. The maximum around 8–15 Hz sometimes had narrow filter properties with sharp edges. The dominant maximum was either in the alpha frequency range (75% of the curves) or in the theta frequency range (25% of the curves). We draw attention here to the spontaneous activity of the medial geniculate nucleus during the waking stage, which depicted regular theta waves as in the case of the hippocampus (Fig. 1.5b).

1.5.3 Mesencephalic Reticular Formation

The most prominent amplitude maxima in the frequency characteristics of the reticular formation were at approx. 50 Hz and in a frequency range between 100 and 500 Hz. The 50 Hz maximum was the dominant maximum in almost all the curves. A maximum at approx. 30 Hz was also found to exist in all curves. Amplitude maxima in the alpha frequency range were also recorded. However, the alpha maximum was sometimes missing (30% of the curves) or reduced to a low magnitude (lowest curve; Fig. 1.5c).

1.5.4 Inferior Colliculus

Inferior colliculus amplitude–frequency characteristics exhibited the most prominent maxima in higher frequency ranges, similar to the characteristics of the reticular formation. Amplitude maxima around 80 Hz and 180 Hz were, in most of the curves (66% of the curves), the maxima with the largest magnitude. Maxima between 3 and 8 Hz were seen in all curves, while a maximum around 10–12 Hz was seen in only 40% of the experiments. When the maximum in the alpha frequency range was missing, a maximum in the 20–25 Hz (beta range) was observed (Fig. 1.5d).

1.5.5 Hippocampus

The hippocampus is a structure that shows less consistent dynamic characteristics. The amplitude–frequency characteristics may show variable maxima even during the waking stage, without the cat going through transition and sleep stages. Figure 1.5e illustrates some typical hippocampal amplitude characteristics during the waking stage. Sometimes the maximum in the theta frequency range was the most prominent one (the upper and the lowest curve). Dominant amplitude maxima in the alpha, beta, and in the 40 Hz frequency ranges were also observed. For an overview of extended experiments in the hippocampus, see Chap. 18 in Vol. I.

1.5.6 Cerebellar Cortex

Durusan (1976) determined the amplitude–frequency characteristics of the cat cerebellar cortex upon acoustical stimuli. This investigator also used the averaged EPs from the cerebellum upon application of a tone burst of

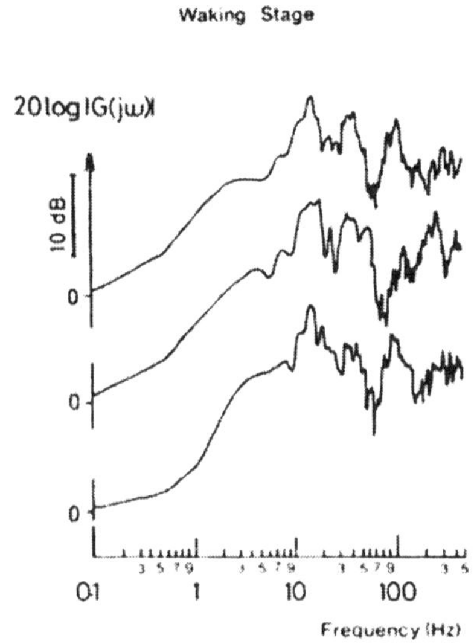

Fig. 1.6. Examples of typical amplitude–frequency characteristics of the cat cerebellum determined by the TRFC method. Direct computer plottings. Along the abscissa is the input frequency in logarithmic scale, along the ordinate is the potential amplitude, $|G(j\omega)|$, in decibels. The curves are normalized in such a way that the amplitude at 0 Hz is equal to 1 (or $20 \log 1 = 0$). (From Durusan R. (1976) thesis: Hacettepe University, Ankara.)

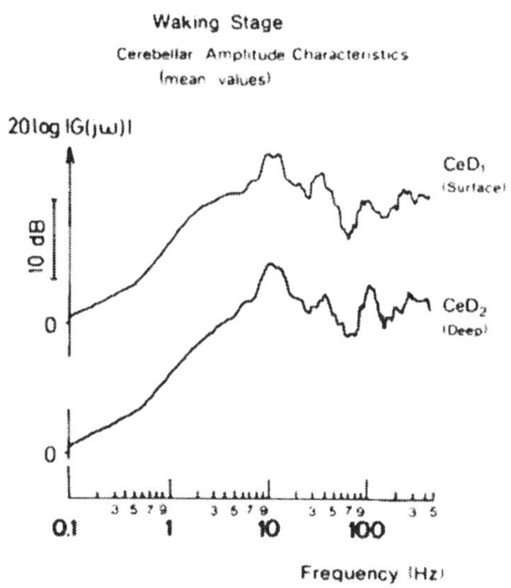

Fig. 1.7. Mean-value curves of cerebellar amplitude–frequency characteristics obtained from 12 experiments on 12 different cats. (The two curves are obtained from two different levels, one deeper (D2) than the other (D1), of the acoustical vermis. (From Durusan R. (1976) thesis: Hacettepe University, Ankara.)

2000 Hz. The TRFC-method described in Chap. 4 in Vol. I was used in order to obtain the frequency characteristics. Figure 1.6 illustrates three typical amplitude characteristics obtained from three different cats. The most prominent maximum was centered at around 12 Hz; minor peakings around 20 Hz were also observed. There are also relevant selectivities at 30–50 Hz, 100 Hz and at 250 Hz. In Fig. 1.7 the mean-value amplitude–frequency characteristics from twelve experiments are presented. Curves D1 and D2 show the mean-value curves obtained from two different levels, one deeper (D2) than the other (D1), of the acoustical vermis of the cat cerebellum.

1.6 Amplitude–Frequency Charateristics: Visual Stimulation

Figure 1.8 shows simultaneously recorded amplitude–frequency characteristics in the cat's visual pathway. For the sake of simplicity we first confine our attention to the responses lower than 30 Hz. The amplitude–frequency characteristics are presented here for the occipital cortex, lateral geniculate nucleus, reticular formation, superior colliculus, and hippocampus of the cat brain. Structures such as occipital cortex, lateral geniculate nucleus, and superior colliculus are mostly involved with visual signal processing, whereas hippocampus and reticular formations are polysensory and multimodal centers (see Chap. 3 in Vol. I). At first glance it is to be seen that the most dom-

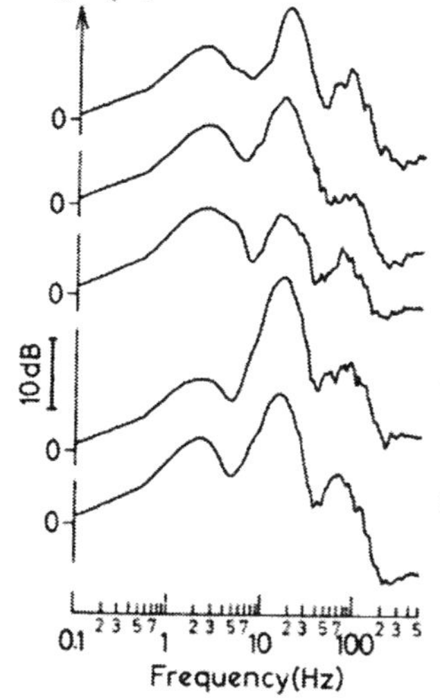

Fig. 1.8. Examples of typical amplitude characteristics of different brain nuclei of the cat determined by the TRFC method and using transient evoked responses similar to those shown in Fig. 1.5. Direct computer plottings. Along the *abscissa* is the input frequency in logarithmic scale, along the *ordinate* the potential amplitude, $|G(j\omega)|$, in decibels. The curves are normalized in such a way that the amplitude at 0 Hz is equal to 1 (or $20 \log 1 = 0$). (From Başar E. et al. (1975) *Biological Cybernetics* 20: 145–160.)

inant peak in all these simultaneously obtained amplitude–frequency characteristics are centered at a frequency between 12 and 15 Hz. A maximum at around 2–3 Hz is also detected in all the curves.

At this point, the reader may be interested in a comparison of amplitude–frequency characteristics for auditory vs. visual stimulation. Such a comparison is presented in Chap. 2. In brief, the dominant alpha response is centered between 7 and 10 Hz for auditory stimulation but shifted to higher frequencies for visual stimulation. It is remarkable that this frequency shift is obtained not only in the cortex but in all the other subcortical structures under study.

1.6.1 "Filtered Potentials"

Filtered curves obtained by the adaptive filtering procedure (see Chap. 4 in Vol. I) reveal further response properties of the brain structures studied. In particular, they show differences for auditory vs. visual stimulation in primary sensory areas of the cat brain. These results will be presented in Chap. 2. For the gamma-band response in the brain, see Chap. 3, and for response components in the high-frequency range between 100 Hz and 1000 Hz see Chap. 4.

1.7 Coherence Functions Between All Possible Pairings of Recording Electrodes –Auditory Stimulation

The coupling (and/or synchronization) of resonant responses from various nuclei in the alpha and beta frequency ranges can also be demonstrated using the coherence functions between all possible pairings of spontaneous and evoked activities in the brain structures studied. In Fig. 1.9, which presents results of a typical experiment, the coherencies in a frequency range of 3–60 Hz for spontaneous activities and evoked potentials of all possible pairings of the studied brain structures are illustrated. Due to the reasonably large number of single sweeps included in the computation of averages and due to the spectral window used, the auto- and cross-spectral amplitudes have been adequately smoothed, and a significance value of 0.2 has been attained for all the curves. Therefore, the area under the coherence function is darkened only if the curve surpasses this value, in order to give emphasis to those parts of the curves above the significance level. One recognizes immediately that in the alpha (8–14 Hz) and beta (14–25 Hz) frequency ranges the coherency usually has high values between 0.5 and 0.9 for evoked responses. However, the coherency between spontaneous activities of the same pairings of nuclei has definitely lower values. In the alpha and beta frequency ranges, the coherence of the spontaneous activity barely reaches 0.3 in a few cases. In other words, there exists an important coherency increase upon stimulation.

Table 1.2 represents averaged coherencies for alpha and beta frequencies calculated from 11 experiments on 11 cats. Averaged coherency is the mean

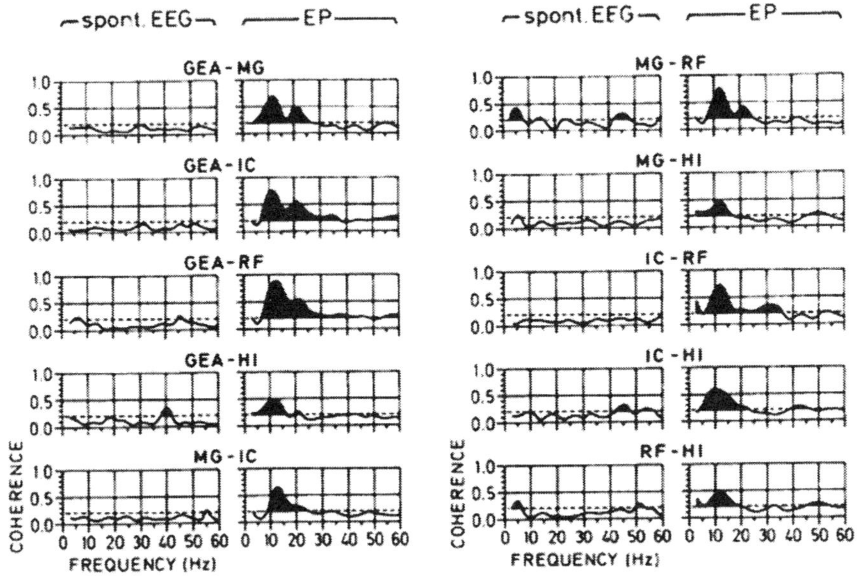

Fig. 1.9. A typical set of coherence functions computed from the spontaneous activities and EPs of all possible pairings of the studied brain structures during the waking stage. The scale is indicated at the bottom. *Along the abscissa* is the frequency from 0 Hz to 60 Hz, *along the ordinate* is the coherence between 0 and 1. The *horizontal broken lines* indicate the significance level, which is 0.2 for all the plots. The area under the coherence function is *darkened* only if the curve surpasses this level. In order to facilitate a comparison between the coherence values computed from spontaneous and evoked parts of the EEG-EPepochs, the respective coherence functions are presented adjacently as couples for all the pairings of recording electrodes. (From Başar E. (1980) *EEG Brain Dynamics*. Amsterdam: Elsevier 1980.)

value of the respective coherencies obtained from the results of 11 experimental sessions. We also calculated a mean value from coherencies between all possible pairings of nuclei. We call this value *overall coherency* of all the studied nuclei. In the alpha frequency range, the overall coherency increases from 0.08 to 0.50. In the beta band, the *overall coherency* increases from 0.08 to 0.35.

1.8 Phase Synchronization Demonstrated by Phase Spectra – Auditory Stimulation

A typical set of phase spectra is given in Fig. 1.10, representing the general features of all the results obtained from 11 cats. In order to avoid presentation of redundant information, only four couples of phase spectra, which are computed with reference to the activity of acoustical cortex, are given in this

Table 1.2. Averaged magnitudes of coherences in parings of spontaneous EEG and envoked potentials recorded from the respective brain structure of 11 cats during the waking stage. The mean values of the averaged coherences in all possible pairings of nuclei are also given, in the alpha and beta frequency ranges, as the overall coherency, which may represent the coherence state of the brain as a whole. Note that the overall coherency is significantly increased, in these frequency channels, upon auditory stimulation. (From Başar E. et al. (1979) *Biological Cybernetics* 34: 1–19)

	8–15 Hz (Alpha channel)		18–25 Hz (Beta channel)	
Pairings of Brain nuclei	EEG	EP	EEG	EP
GEA–MG	0.07	0.55	0.05	0.44
GEA–IC	0.08	0.58	0.07	0.49
GEA–RF	0.08	0.70	0.08	0.56
GEA–HI	0.05	0.41	0.05	0.30
MG–IC	0.05	0.50	0.10	0.39
MG–RF	0.09	0.59	0.09	0.43
MG–HI	0.06	0.32	0.05	0.17
IC–RF	0.08	0.46	0.12	0.37
IC–HI	0.08	0.46	0.10	0.16
RF–HI	0.10	0.41	0.12	0.19
Overall coherency	0.075	0.50	0.08	0.35
Factor of overall coherency increase	6.75		4.38	

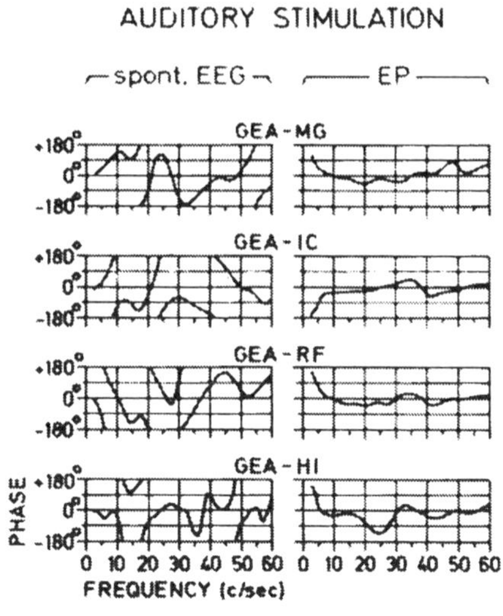

Fig. 1.10. With reference to the electrodes at the auditory cortex (GEA), a typical set of cross-phase specrtra computed from the spontaneous activities and EPs of other studied brain structures of the cat during the waking stage. The scale is indicated at the bottom. *Along the abscissa* is the frequency from 0 Hz to 60 Hz; *along the ordinate* is the phase angle between $-180°$ and $+180°$. The spectra computed from spontaneous and evoked parts of the recorded EEG-EPepochs are presented adjacently as couples for the same pairings of recording electrodes. (From Başar E. (1980) *EEG Brain Dynamics*. Amsterdam: Elsevier 1980.)

figure. By comparing the phase spectra in each couple, it is clearly seen that the phase values display an almost random variation when computed from the spontaneous EEG pairings, whereas the phase spectra of the EP pairings show a stable course around zero, at least within a certain frequency band covering the alpha–beta range. In other words, in the alpha–beta range, the responses of all the studied brain centers are almost in phase, although between the spontaneous activities of these centers there is no preferred phase angle, which remains significantly stable within the same frequency band.

1.9 Coherence Functions Between All Possible Pairings of Recording Electrodes –Visual Stimulation

Of the coherence functions computed from the simultaneous EEG-EPepochs, which were recorded from nine cats using visual stimuli, a typical set of curves is presented in Fig. 1.11. An immediate comparison of the curves

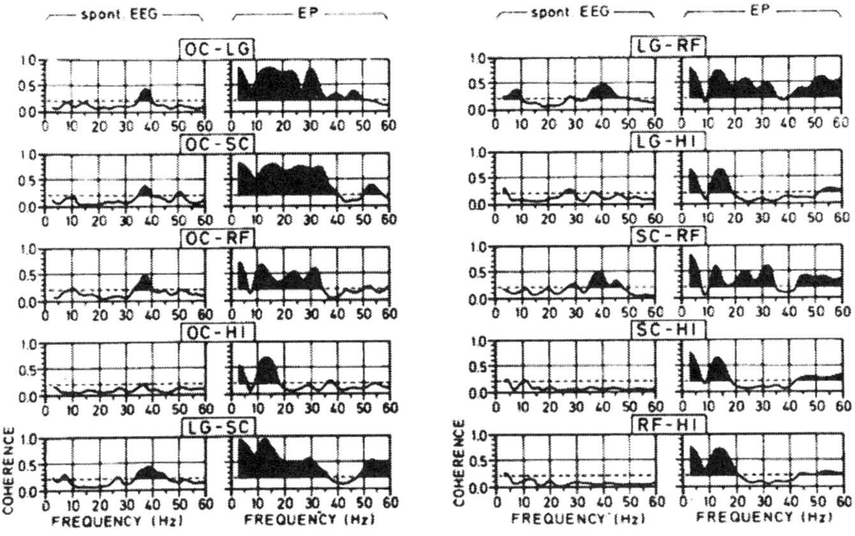

Fig. 1.11. A typical set of coherence functions computed from the spontaneous and visual evoked potentials of all possible pairings of the studied brain structures. The scale is indicated at the bottom. *Along the abscissa* is the frequency from 0 to 60 Hz; *along the ordinate* is the coherency between 0 and 1. The *horizontal broken lines* indicate the significance level, which is 0.2 for all the plots. The area under the coherence function is darkened only if the curve surpasses this level. In order to facilitate a comparison between the coherence values computed from spontaneous and evoked parts of the EEG-EPepochs, the respective coherence functions are presented adjacently as couples for all pairings of recording electrodes. (From Ungan P. (1979) thesis: Hacettepe University, Ankara.)

Table 1.3. Averaged magnitudes of coherences in parings of spontaneous EEG and visual EPs recorded from the respective brain structure of 9 cats. The standand deviations around the averaged magnitudes are also given. The mean values of the averaged coherences in all possible pairings of nuclei are given for delta (3 HZ) and alpha–beta (10–25 HZ) channels as the overall coherency. Note that upon visual stimulation, the overall coherency is significantly increased not onla in the frequency range of 1–25 Hz, but also at 3 Hz. (From Ungan P. (1979) thesis. Hacettepe University, Ankara)

	3 c/sec (Delta channel)				10–25 c/sec (Alpha–beta channel)			
	EEG		EP		EEG		EP	
Pairings of Brain Nuclei	Aver.	s.d.	Aver.	s.d.	Aver.	s.d.	Aver.	s.d.
OC-LC	*0.09*	0.06	*0.60*	0.09	*0.15*	0.08	*0.77*	0.13
OC-SC	*0.27*	0.15	*0.68*	0.22	*0.24*	0.10	*0.80*	0.23
OC-RF	*0.02*	0.01	*0.80*	0.15	*0.16*	0.09	*0.73*	0.21
OC-HI	*0.21*	0.17	*0.56*	0.13	*0.20*	0.06	*0.74*	0.15
LG-SC	*0.09*	0.02	*0.54*	0.15	*0.15*	0.07	*0.80*	0.14
LG-RF	*0.20*	0.13	*0.64*	0.12	*0.16*	0.09	*0.77*	0.14
LG-HI	*0.28*	0.11	*0.41*	0.16	*0.19*	0.10	*0.79*	0.18
SC-RF	*0.11*	0.11	*0.51*	0.17	*0.25*	0.10	*0.77*	0.19
SC-HI	*0.15*	0.13	*0.42*	0.10	*0.17*	0.08	*0.75*	0.20
RF-HI	*0.37*	0.13	*0.46*	0.14	*0.16*	0.06	*0.69*	0.15
Overall coherency	*0.18*		*0.56*		*0.18*		*0.76*	
Factor of overall coherency increase		*3.11*				*4.22*		

clearly shows that the coherence values computed from the evoked responses are much higher than those computed from the spontaneous activity preceding the stimulation. The evoked internuclei coherency can often reach values very close to 1 (the theoretical upper limit), whereas the spontaneous coherency can hardly exceed the significance level in a few cases. Furthermore, evoked coherence values are quite high, especially within the frequency range of 8–25 Hz, corresponding to the alpha–beta band, where the spontaneous coherences are not significant. It should be noted here that significant coherence values are also recorded in the delta frequency range (3 Hz) upon visual stimulation.

In order to find the mean values, we have computed the averaged magnitudes of coherences in pairings of spontaneous EEG and visual EPs recorded from the same brain structures of all the animals studied. These results are presented in Table 1.3. In this table, the values for 3 Hz are also given with those in the 10–25 Hz frequency channel, these frequency positions being the most important ranges where significant internuclei coherence is observed in the respective groups of brain centers with adequate stimulation modalities. For these frequency ranges, the mean value of the averaged coherences in all pairings of nuclei is also given as the overall coherency, which may represent the coherence state of the brain as a whole.

1.10 Phase Synchronization Demonstrated by Phase Spectra – Visual Stimulation

The phase spectra were computed from the same pairings of simultaneous recordings evaluated for coherence functions. In order to avoid presenting redundant information, however, the visual cortex (OC) is chosen as reference for the phase angles of activities recorded upon visual stimulation from the other centers. Hence, for the visual stimulus, only the four couples of phase spectra, which were computed from the pairings with these references, are included in Fig. 1.12. In this figure, a typical set of phase spectra is given, representing the general features of all the results obtained with visual stimuli. The presentation of these curves is made with the same format used for the coherence functions. This is also done for an easy comparison of the curves computed from the spontaneous and evoked parts of the combined EEG-EPepochs. Along the abscissa is the frequency from 0 Hz to 60 Hz, and along the ordinate is the phase angle between −180° and +180°. Both scales are linear. Since the same sweeps used for the coherence functions were included in the frequency domain averaging, the presented phase spectra are adequately smoothed.

An immediate comparison of the phase spectra in each couple shows that the phase values computed from the spontaneous pairings display an almost

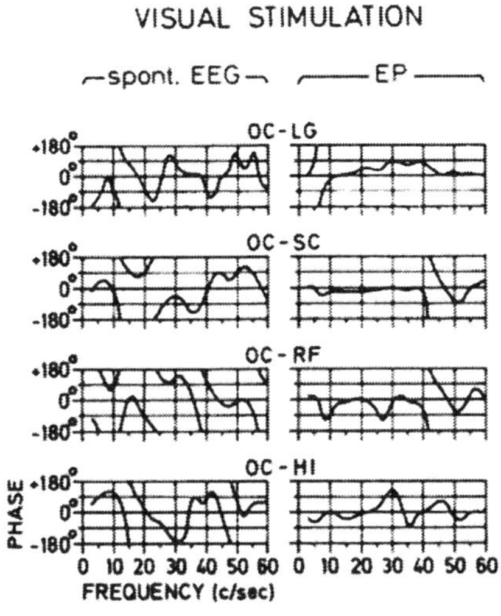

Fig. 1.12. With reference to the electrodes at the occipital cortex (OC), a typical set of cross-phase spectra computed from the spontaneous activities and EPs of other studied brain structures of the cat. The scale is indicated at the *bottom*. *Along the abscissa* is the frequency from 0 to 60 Hz; *along the ordinate* is the phase angle between −180° and +180°. The spectra computed from spontaneous and evoked parts of the recorded EEG-EPograms are presented adjacently as couples for the same pairings of recording electrodes. (From Başar E. (1980) *EEG Brain Dynamics*. Amsterdam: Elsevier 1980.)

random variation, whereas the phase spectra of the EP pairings show a rather stable course around zero, at least within a certain frequency band covering the alpha–beta range. This means that in the alpha–beta range, the responses of all the studied brain centers are almost in phase. On the other hand, there is no preferred phase angle that remains significantly stable within the same frequency band between the spontaneous activities of these centers. The reader should compare the given results with the phase spectra obtained with auditory stimulation (as given earlier in this chapter).

2. Cross-Modality Experiments on the Cat Brain

Erol Başar, Tamer Demiralp, Martin Schürmann, and Canan Başar-Eroğlu

2.1 Introduction

In Chap. 1, EPs and amplitude–frequency characteristics from various cortical and subcortical structures of the cat brain were described. For some of these structures, in particular the primary sensory areas, the stimuli may be adequate or inadequate; e.g., a visual stimulus is adequate for the visual cortex, but inadequate for the auditory cortex. It is noteworthy that even inadequate stimuli do evoke changes of the EEG in the respective structure (although responses to adequate stimulation are usually higher in amplitude). These types of changes already described by John et al. (1989) and Galambos (1989) are important for an understanding of integrative functions of the brain. In the present chapter we will analyze cross-modality responses in the cat brain with the strategies of the combined analysis procedure. We will start with a section describing the view of Hartline about the concept of cross-modality.

2.2 What Are Multimodal Recognition and Cross-Modality Attention? View of Hartline

Hartline (1987) summarized essential features of **multisensory convergence**, or **multimodal integration**, in a way sufficient for consideration in our book. In the following we will extract the information most important to our studies from Hartline's text.

> "**Classification of multisensory neurons:** Among invertebrates and vertebrates, there are many examples of neurons that are excited or modulated by input from more than one sensory system. The most common type of multimodal neuron, termed OR neuron, responds reliably to stimuli that activate any of two or more senses (e.g. visual or infrared senses in pit vipers, visual or auditory modalities in mammals, visual (oscella) or wind hair in insects). [...]
> **Functional roles of multisensory neurons:** OR neurons would be appropriate for a sensory substitution function – one in

which input via any active modality can provide the information needed for downstream processing. For example, locust ocellar and wind hair sensor inputs converge on interneurons that help stabilize the insect's flight. In many vertebrate species, OR neurons in the optic tectum (superior colliculus in mammals) probably allow an object to trigger an orientation movement via several modalities. An example of higher-order cells that manifest OR-type sensory integration is provided by hypothalamic neurons, recorded in primates, that respond to sight or taste of food. [...]

The intermodality facilitation of ENHANCED AND neurons could be a basis for several candidate functions: (1) multimodal recognition (e.g., snake infrared–visual AND neurons might participate in recognition of edible warm-blooded prey), (2) cross-modality attention (in which one sense primes appropriate neurons for vigorous response to a different sensory input), and (3) noise reduction (as in OR neurons). This would be advantageous for multimodal detection or localization of objects that generate near-threshold stimulation in more than one modality (ENHANCEMENT tends to be numerically more impressive as the stimuli approach threshold; furthermore, two subthreshold stimuli can interact to evoke a clear response).

The intermodality interactions of DEPRESSED neurons could participate in multimodal recognition by suppressing the response to objects having a particular multisensory signature. Depression might also occur as a byproduct of other mechanisms; e.g., it might reflect inhibition obtained by stimulating an antagonistic surround of an ENHANCED cell. ENHANCED and DEPRESSED neurons as well as OR neurons have been found in the optic tectum, in thalamic nuclei, and in several regions of frontal, parietal, and temporal association cortex. [...]

Examples [of multisensory representation of space] are found in (1) the optic tectum, where convergence among visual, somatic, infrared, auditory, and electrosensory modalities occurs in diverse species from elasmobranchs through primate, and (2) the visual cortex, where visual, auditory, and probably tactile senses converge. [...]

In each of several distinct regions of visual cortex, visual inputs are organized according to a spatiotopic map, as in the optic tectum. The visual cortex is thought to participate in processes subservient to recognition of objects or patterns. *Though this part of the brain is considered to be involved exclusively with vision, in cats about one-third of the neurons in areas 17 (striate cortex), 18, and 19 are reported to be responsive to sound as well as to retinal input, and some are reported to be modulated by somatosensory input.* The auditory receptive fields of these neurons are well-defined in the azimuthal dimension and are roughly congruent to the visual receptive fields of the same and nearby neurons, indicating that some spatial features are conserved across modalities. The function of auditory and somatosensory input to visual cortex is not known.

Nonspecific interactions: *[The nonspecific interactions in the brain merits important consideration in the present chapter because of the cognitive processes in which the CNS is involved.]* Many animals have quiescent or inattentive states. In vertebrates, a neural system associated with arousal to the attentive state (part of the reticular activating system) receives inputs from many senses. The arousal system influences neurons in almost all brain nuclei that deal with specific sensory information. Thus *midbrain, thalamic*, and *cortical* neurons primarily responsive to visual stimuli may be activated or modulated by auditory or somesthetic stimuli via nonspecific arousal mechanisms [see also Chap. 2 in Vol. I]."

In the present chapter we will analyse cross-modality responses in nuclei of the auditory and visual pathways and discuss them in the light of Hartline's outstanding and important view.

Our results in these cross-modality experiments demonstrate that the differences between responses to adequate stimulation and responses to inadequate stimulation ("adequate/inadequate differences") were marked in the alpha frequency range and less marked in the theta range. This hints at a possible special role of the alpha response in primary sensory processing.

In contrast, EP theta components appear to be less dependent on whether the stimulus is adequate or not, which is compatible with a predominantly associative–cognitive role of the theta response as hypothesized above and as published earlier (Başar et al. 1991, Demiralp and Başar 1992, Başar-Eroğlu et al. 1992; see also Chaps. 19, 16, 9, 10).

2.3 The Present Chapter Combines Cross-Modality Experiments, Frequency Analysis, and Wavelet Transform Approaches

The study was performed by using intracranial electrodes in cats (see Chap. 1 for details on electrode positions). Auditory stimuli were tone bursts of 1000 ms duration, 2000 Hz frequency, and 80 dB sound pressure level. The rise time was 0.5 ms. Auditory stimuli were presented binaurally via loudspeakers. The light stimulator was a 20 W fluorescent bulb that was electrically triggered. The duration of the light step was also 1000 ms. These stimulations permitted cross-modality experiments comprising adequate and inadequate stimulation of the auditory and visual cortex. Procedures for data acquisition and evaluation were as given in Chap. 4 in Vol. I with two main steps:

1. Computation of the amplitude–frequency characteristics (AFCs) by transforming the averaged EP to the frequency domain.

2. Digital filtering with filter limits chosen adequately according to the amplitude–frequency characteristics.

In the present study, additionally, the wavelet transform (see Chap. 5 in Vol. I) was applied to the same data in order to compare the results of both conventional and novel techniques. Wavelet analysis is, as yet, seldom applied in brain research, although it is now very popular in mathematical and engineering sciences. We used quadratic B-Spline wavelets for their suboptimal time–frequency localization and their linear phase. For the sake of comparison between EP/ERP frequency analysis and wavelet analysis, the frequency limits were chosen as follows: 0.5–4 Hz ("delta"), 4–8 Hz ("theta"), 8–16 Hz ("alpha"); 16–32 Hz ("beta"); 32–64 Hz ("gamma").

2.4 Results

2.4.1 Averaged EPs (Single Animal, Grand Average)

The time series (wide-band filtered, 0.3–70 Hz) of EPs recorded from the auditory cortex (gyrus ectosylvianus anterior, GEA), occipital cortex (OC), and hippocampus (HI) are given in Fig. 2.1. The figure shows results in a typical animal. In GEA and OC, marked modality-dependent differences are visible. With respect to GEA, the response to auditory stimulation has a

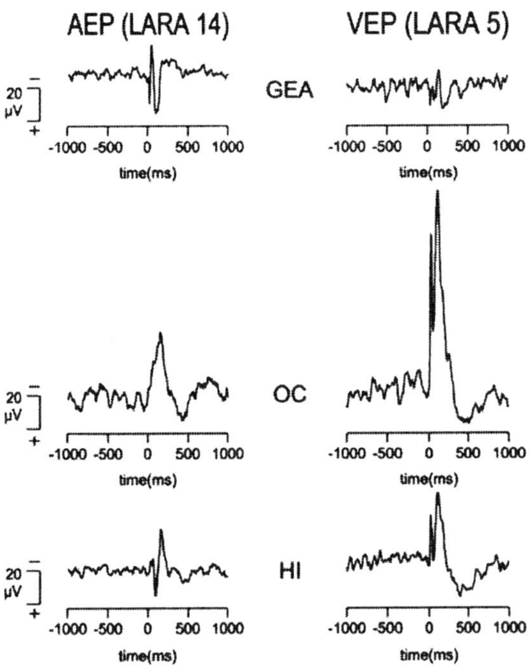

Fig. 2.1. *Averaged ERPs* in a typical animal. *Left column*: auditory stimulation; *right column*: visual stimulation. Recordings from auditory cortex (GEA), visual cortex (OC), and hippocampus (HI). Along the *x-axis*, time in ms; along the *y-axis*, amplitude in μV (negativity *upwards*).

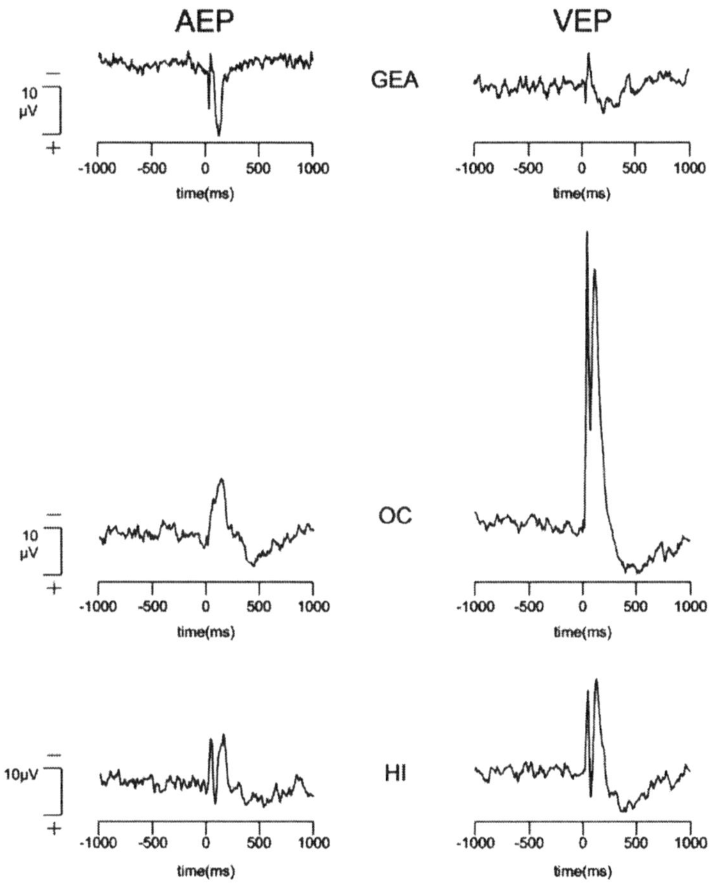

Fig. 2.2. *Grand average ERPs* ($N = 6$). *Left column*: auditory stimulation; *right column*: visual stimulation. Recordings from auditory cortex (GEA), visual cortex (OC), and hippocampus (HI). *Along the x-axis*: time in ms; *along the y-axis*: amplitude in µV (negativity *upwards*).

maximal poststimulus amplitude of approximately 80 µV. In contrast, the response to visual stimulation has an amplitude of approximately 50 µV. In OC, the amplitude of the auditory EP is approximately 100 µV, whereas the amplitude of the visual EP is nearly 300 µV (note the marked differences with respect to the overall waveshapes). Hippocampal (HI) response amplitudes are approximately 80 µV for auditory stimulation and approximately 120 µV for visual stimulation.

The respective grand average curves are given in Fig. 2.2. They are in accordance with the results for the typical animal. Distinct modality differences are observed for all locations, with an auditory EP of approximately 35 µV maximal poststimulus amplitude and a visual EP of approximately 25 µV

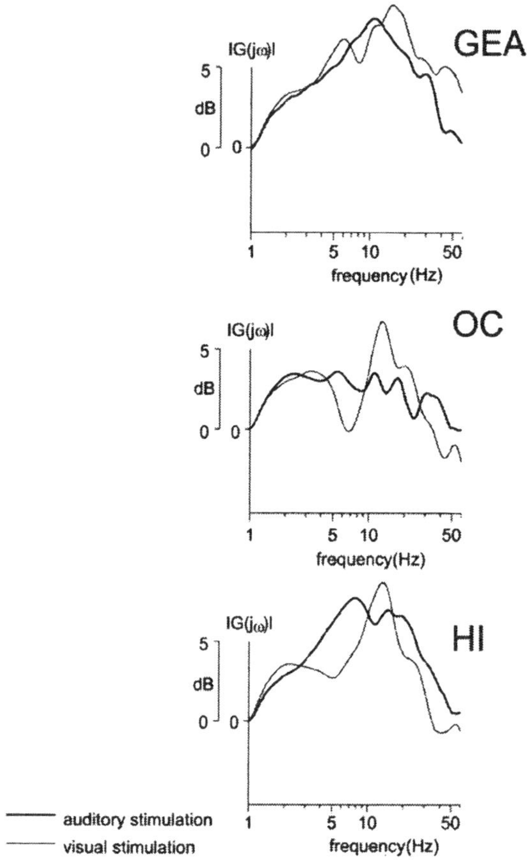

Fig. 2.3. *Amplitude frequency characteristics,* grand averages ($N = 6$). *Solid lines;* auditory stimulation; *dashed lines;* visual stimulation. Data from auditory cortex (GEA), visual cortex (OC), and hippocampus (HI). *Along the x-axis:* frequency in logarithmic scale. *Along the y-axis:* relative amplitude in decibels (dB). The amplitudes are normalized in such a way that the amplitude at 1 Hz is equal to 0. (Frequency domain averages.)

amplitude in GEA. With respect to OC, the auditory EP has an amplitude of approximately 30 μV, whereas the visual EP amplitude is approximately 150 μV. Hippocampal (HI) response amplitudes are approximately 30 μV for the auditory ERP and approximately 60 μV for the visual ERP.

2.4.2 Amplitude–Frequency Characteristics

Amplitude–frequency characteristics were computed from averaged EPs for the electrode locations mentioned above. Figure 2.3 shows grand averages for auditory cortex (gyrus ectosylvianus anterior, GEA), occipital cortex (OC), and hippocampus (HI).[1]

Figure 2.3 (GEA) shows amplitude–frequency characteristics for the auditory cortex with auditory vs. visual stimulation. *The center frequency is*

[1] Note that averaging across individual cats was performed in the frequency domain.

around 9–10 Hz for auditory stimulation ("adequate"). The visual stimulation does not elicit a relevant 10 Hz response in the auditory cortex.

The situation is completely different in the recordings from the occipital cortex. Figure 2.3 (OC) shows amplitude–frequency characteristics for the visual cortex with auditory vs. visual stimulation. For auditory stimulation ("inadequate"), no marked maxima in the alpha range are visible. However, responses to visual stimuli show a dominant peaking in the 10 Hz range, with a *peak frequency of 12–13 Hz.*

Figure 2.3 (HI) shows amplitude–frequency characteristics for hippocampal recordings. For *auditory stimulation* a response in the 10 Hz range is visible; the grand average of amplitude–frequency characteristics reveals a frequency responsiveness in the *8–10 Hz* range. *Visual stimulation* also elicits maxima in the alpha range. In the grand average, the maximum is centered at *12 Hz.* This means that the response is shifted to higher frequencies in comparison to auditory stimuli.

The difference in the evoked frequency response of the hippocampus in comparison to pure sensory centers such as cortex and thalamus is the following: In the 10 Hz frequency range, brain structures, e.g., thalamus and cortex, do not react with 10 Hz resonances or 10 Hz enhancements if the stimulation is not an *adequate* one. The visual cortex and the lateral geniculate nucleus do not show this all-around flexibility of the hippocampus. The hippocampus resonates and/or enhances the auditory and visual stimulation in 10 Hz frequency ranges, but the response has slightly different frequency values depending on the stimulation modality. The thalamus *blocks*, or *filters*, out the inadequate stimulation in the 10 Hz frequency range. The sensory cortices react in the same way.

2.4.3 Results of Digital Filtering

Typical Animal. The results obtained in the experimental animal "lara14/5" are shown in Fig. 2.4 (*left*, responses to auditory stimuli; *right*, responses to visual stimuli). For each type of stimulus, three electrode locations are shown. The uppermost row shows wide-band-filtered responses; the filter limits for the other curves are as given in the figure (gamma, beta, alpha, theta, delta). The figure permits a comparison of responses to *inadequate* stimuli (visual EP in GEA and auditory EP in OC) vs. responses to *adequate* (auditory EP in GEA and visual EP in OC) stimuli: It can be seen that the alpha range shows the largest differences, with marked alpha responses to adequate stimuli, whereas inadequate stimuli do not elicit alpha responses (no enhancement). Large inadequate vs. adequate differences are also observed in the beta and delta components of the visual EP. In all other recordings, this inadequate vs. adequate difference is smaller. Note that although the visual EP in GEA ("inadequate" stimulation) is barely visible in wide-band filtered curves, there is still a theta response (residual response).

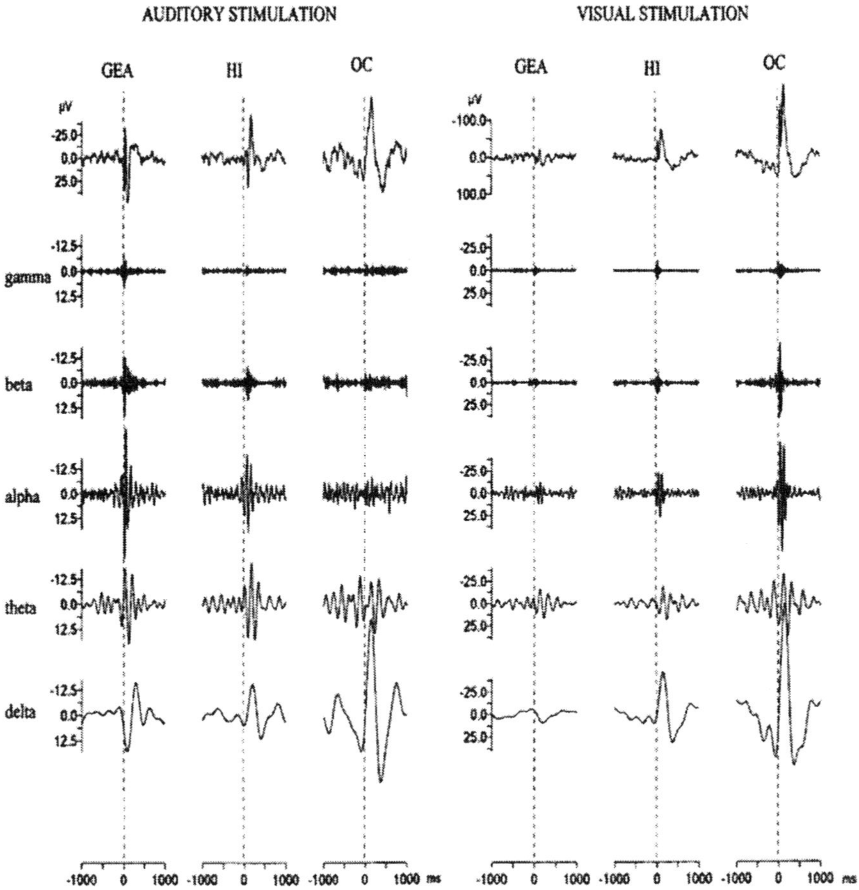

Fig. 2.4. Results of *band-pass filtering* in a typical animal. *Left:* auditory stimulation; *right:* visual stimulation. Each *column* refers to an electrode site (auditory cortex: GEA; visual cortex: OC; hippocampus: HI). The *uppermost row* shows the wide-band filtered curve. The *further rows* show the frequency components gamma (32–64 Hz), beta (16–32 Hz), alpha (8–16 Hz), theta (4–8 Hz), and delta (0.5–4 Hz).

Interim Summary: It may be pointed out that high-amplitude alpha responses were characteristic of responses to adequate stimuli in primary sensory areas. Adequate vs. inadequate differences were larger for alpha responses than for theta responses. This supports the hypothesis that alpha responses are predominantly associated with primary sensory processing.

Statistical Analysis. The effects of stimulus modality on the frequency content of the EPs of primary sensory cortices and the hippocampus were subjected to statistical analysis, which yielded several significant results as outlined below. The following preliminary remark has to be made: It is well known that the primary visual and auditory areas display strong EPs to

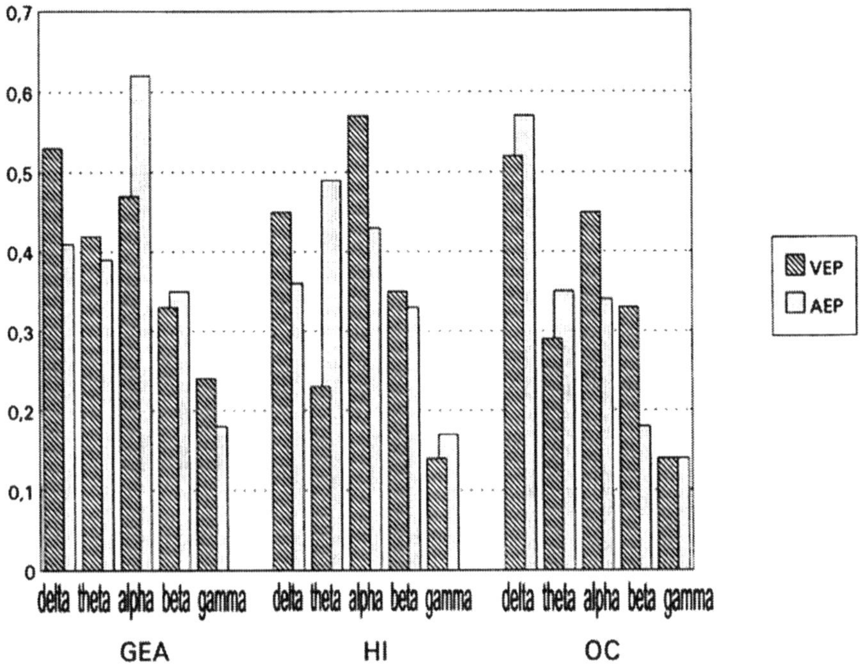

Fig. 2.5. Amplitudes of EP frequency components obtained by digital filtering. *Along the y-axis*, normalized amplitude (see text).

adequate stimuli but not to inadequate stimuli. An important question arising from this finding is whether this difference is reflected in the whole frequency content of the signals or whether it is the result of a dominant change in certain frequency bands. In order to identify the *relative contribution* of the frequency ranges to the compound EP, we used the following normalization procedure: To take into account interindividual differences in EP amplitude, we computed relative amplitudes (amplitude of the respective EP component divided by amplitude of the wide-band filtered EP). The results are shown in Fig. 2.5.

The results were further evaluated by means of a repeated measures ANOVA test, the results of which are given in Table 2.1. For auditory vs. visual stimuli, differences were largest in the alpha range, with higher alpha response amplitudes for adequate stimuli (auditory EP in GEA, visual EP in OC) than for inadequate stimuli (acoustical stimuli in OC, visual stimuli in GEA). This result was significant at the 0.05 level. Differences for all other frequency ranges were smaller (except beta in OC). It is noteworthy that theta responses showed an opposite behavior, with slight, yet significant (0.01 level), increases in amplitude for inadequate stimuli in comparison to adequate stimuli.

Table 2.1. Repeated measures ANOVA statistics of the effects of the stimulation modality, topography (lead) and modality x topography (lead) on the relative amplitudes of the **band-pass filtered frequency components**. Relative amplitudes were computed by dividing the peak-to-peak amplitude of the frequency component in the poststimulus period to the peak-to-peak amplitude of the unfiltered response (GEA, HI, OC).

	Modality		Lead		Modality × Lead	
	F	p	F	p	F	p
Delta	0.74	N.S.	1.33	N.S.	0.62	N.S.
Theta	5.80	0.04	2.14	N.S.	7.58	0.004
Alpha	0.50	N.S.	11.57	0.001	5.49	0.014
Beta	5.98	0.037	6.67	0.007	5.49	0.014
Gamma	0.17	N.S.	4.98	0.019	3.00	N.S.

2.4.4 Results of Wavelet Analysis of EPs

Typical Animal. The analysis of auditory and visual EPs in all three structures (GEA, HI, OC) by means of the wavelet method shows distinct response components in delta, theta, alpha, beta, and gamma frequency ranges, which are differently weighted according to the investigated brain structure and the stimulus modality. Figure 2.6 shows the results of the 5-octave wavelet analysis of the AEP and VEP recorded from the GEA, HI, and OC leads of a representative cat (lara 14/5, the same as in Fig. 2.4; note that the uppermost rows in Figs. 2.4 and 2.6 are identical and show wide-band filtered EPs). The basic observations on the averaged EPs were in general in accordance with the results shown in Fig. 2.4 (obtained by Fourier transform and adaptive digital band-pass filtering). However, especially in higher frequency bands the time-localization capability of the wavelet analysis was significantly better compared with conventional band-pass filters. The ringing effects that occur with the conventional band-pass filtering techniques, which lead to oscillations before the stimulation time point, were absent or clearly smaller in alpha, beta, and gamma frequency ranges. The results support our earlier studies. Here we mention two remarkable items:

1. The dominant frequency components of the visual EPs recorded in the OC area were in the delta and alpha ranges with comparable weights, whereas the auditory EPs (recorded in GEA) had a dominant component in the alpha range.
2. Alpha response components were most pronounced in the responses of OC and GEA to adequate stimuli and in HI in visual modality, whereas their amplitudes decreased extremely in responses of both GEA and OC to inadequate stimuli.

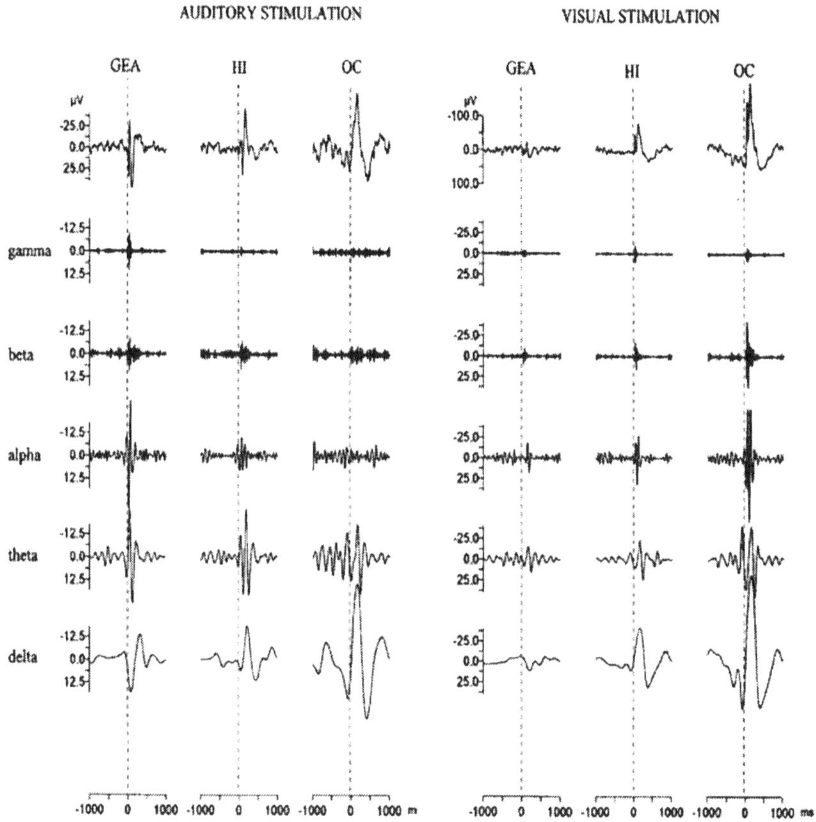

Fig. 2.6. Results of *wavelet decompostion* in a typical animal (same as in Fig. 2.4). *Left:* auditory stimulation; *right:* visual stimulation. Each *column* refers to an electrode site (auditory cortex: GEA; visual cortex: OC; hippocampus: HI). The *uppermost row* shows the wide-band filtered curve. The *further rows* show the frequency components gamma (32–64 Hz), beta (16–32 Hz), alpha (8–16 Hz), theta (4–8 Hz), and delta (0.5–4 Hz).

Statistical Analysis Concerning Cross-Modality Experiments. Amplitudes (normalized for the same reasons and in the same way as stated for digital filtering) are given in Fig. 2.7.

1. The dominant frequency components of the visual EPs recorded in the OC area were in the delta and alpha ranges with comparable weights, whereas the auditory EPs of the auditory cortex (GEA) had a dominant component in the alpha range. This topography effect (factor "lead") on the delta response components can be seen in Table 2.2. A similar effect of the topography with lower significance was also observed for the alpha component.

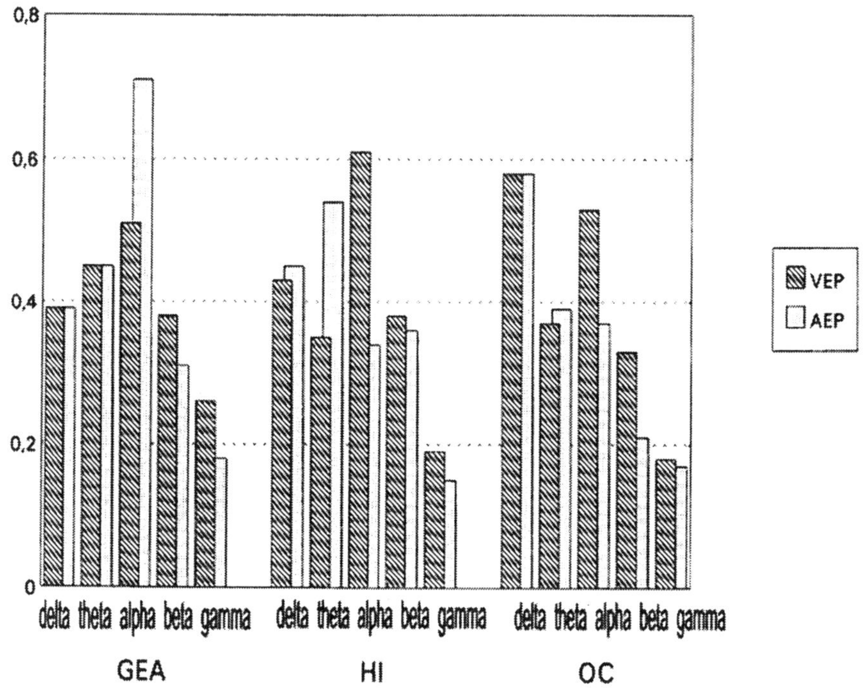

Fig. 2.7. Amplitudes of wavelet components. *Along the y-axis*, normalized amplitude (see text).

2. Gamma responses were obtained in all three structures in both stimulus modalities. There are no significant differences between the gamma responses obtained in different structures, which is in accordance with our hypothesis of a selectively distributed gamma response system (see Chaps. 3 and 26). However, the effects *modality* and *modality x lead* were significant on the gamma response components.
3. Alpha response components were most pronounced in the responses of the OC and GEA to adequate stimuli and in hippocampus in visual modality, whereas their amplitudes decreased extremely in responses of both GEA and OC to inadequate stimuli. The only significant effect of the factor *modality x lead*, which corresponds to the cross-modality effect, was on the alpha response component. It means that a significant topographical difference due to the application of stimuli of different sensory modalities was obtained only in the alpha frequency band (Table 2.2). A finer analysis of this result shows that there are significant increases in the contribution of the alpha components to the evoked responses of the GEA and OC, if the stimulation is an adequate one.
4. Hippocampus responded with significantly higher alpha component to visual stimuli and with significantly higher theta amplitude to auditory

Table 2.2. Repeated measures ANOVA statistics on the effects of stimulation modality, topography (lead) and modality x topography (lead) on the relative amplitudes of the **wavelet components**. Relative amplitudes were computed by dividing the peak-to-peak amplitude of the frequency component in the poststimulus period to the peak-to-peak amplitude of the unfiltered response (GEA, HI, OC).

	Modality		Lead		Modality × Lead	
	F	p	F	p	F	p
Delta	0.65	N.S.	12.7	0.000	0.06	N.S.
Theta	14.5	0.004.	0.89	N.S.	3.41	N.S.
Alpha	2.48	N.S.	5.46	0.02	21.0	0.000
Beta	17.9	0.002	2.89	N.S.	1.14	N.S.
Gamma	0.43	N.S.	1.61	N.S.	0.30	N.S.

stimuli. At this point, it should be mentioned that the limits of the theta and alpha frequency ranges show some changes in auditory and visual modalities. Therefore, the effect we obtain in the theta range, defined in the wavelet analysis as 4–8 Hz, might be partly due to the low alpha components. For a precise judgment of this possibility, wavelet decomposition could be applied in suboctaves of the theta range.

2.4.5 Statistical Comparison of Results of Wavelet and Frequency Analysis

The analysis of the frequency components of auditory and visual evoked responses recorded from the auditory cortex (GEA), hippocampus (HI), and occipital cortex (OC) of the cat by using chronically implanted electrodes revealed comparable results by the application of both adaptive digital band-pass filtering and wavelet decomposition techniques. The effect of cross-modal stimulation could also be detected by both methods.

2.5 Single-Trial Analysis of EPs

2.5.1 Example of Single-Trial Analysis

Figure 2.8 shows examples of single sweeps recorded in the primary visual cortex of the cat. Single sweeps recorded with auditory stimulation are in the left column; those with visual stimulation are in the right column. The filter limits are 8–15 Hz. Visual stimulation elicits distinct alpha responses, i.e., amplitude enhancement and phase-locking of the EEG. In contrast, these phenomena do not occur with auditory stimulation. It is only in the case of visual stimulation that a large number of sweeps show a high degree of similarity to the filtered averaged EP (Fig. 2.8b). For visual stimulation,

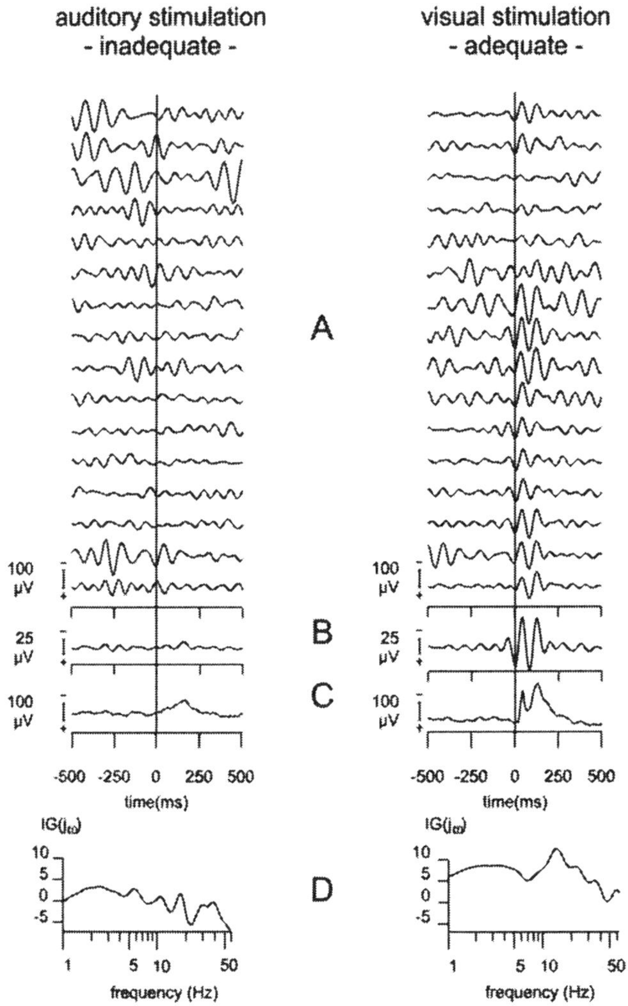

Fig. 2.8. Example of *ERP single trials* in a typical animal (visual cortex recordings). *Left column:* auditory stimulation ("inadequate"); *right column:* visual stimulation ("adequate", see text). (**a**) single trials, filtered with limits 8–15 Hz. (**b**) averaged ERP, filtered with limits 8–15 Hz. (**c**) averaged ERP, wide-band filtered. (**d**) amplitude–frequency characteristics.

the unfiltered averaged EP (Fig. 2.8c) shows a waveform that is in good accordance with the filtered averaged EP of Fig. 2.8b. Amplitude–frequency characteristics are displayed in Fig. 2.8d. For visual stimulation, they show a resonance maximum in the alpha range – in accordance with the distinct responses seen in single-trial EPs filtered in the alpha range (8–15 Hz).

2.5.2 Wavelet Analysis of Single-Trials

In Fig. 2.9, a further improvement in the analysis of EPs by using the wavelet transform is shown. In this example, the alpha component of the EP was used to classify different types of single sweeps. Because the wavelet transform represents the time course of different frequency components of the signal by distinct coefficients with a relevant time resolution for that frequency range, it yields a significant data reduction, which allows a simpler analysis of the frequency components of the response. In this example, we classified the sweeps, showing (i) early alpha enhancement and (ii) prolonged or late alpha enhancement. The averages of these subgroups showed significantly different shapes. Additionally, even by superimposing the broad-band filtered single sweeps of each group, the early and late or *prolonged alpha phase locking* can be clearly observed. In the early-alpha-response sweeps, a phase-locking effect can be obtained only in the first cycle of the poststimulus alpha activity, whereas in the other subgroup phase locking is visible during two alpha cycles.

2.6 Physiological Implications of Cross-Modality Experiments

The main goal of this chapter is to show the relevance of the frequency domain and time-scale plane analysis methods for the analysis of event-related oscillations and/or resonance phenomena in the brain. On the basis of auditory and visual evoked potential measurements in cats we will first discuss the physiological interpretation of the results of the frequency domain analysis. Second, the question of whether the frequency components obtained by frequency-domain analysis are "real components" will be briefly addressed (see Chap. 18 in Vol. I).

In particular, this last question is closely connected with the physiological basis of the signals. Considering different specific resonant frequencies and activation time courses of various brain structures involved with the investigated brain process, a time–frequency approach is important in the analysis of the compound response to be able to identify functionally significant signal components.

A variety of experimental paradigms has been used to investigate possible psychophysiological correlates of these evoked EEG oscillations. Converging results support the view that evoked EEG oscillations are correlated with – at least partly – separable functions of information processing in the brain. Some examples are as follow (see below for details):

1. The importance of *theta resonance phenomena* in relation to cognitive behavior has been pointed out in a number of recent reports (Buzsáki 1992, Miller 1991, Lopes da Silva 1992, Demiralp and Başar 1992; Başar-Eroğlu et al. 1992).

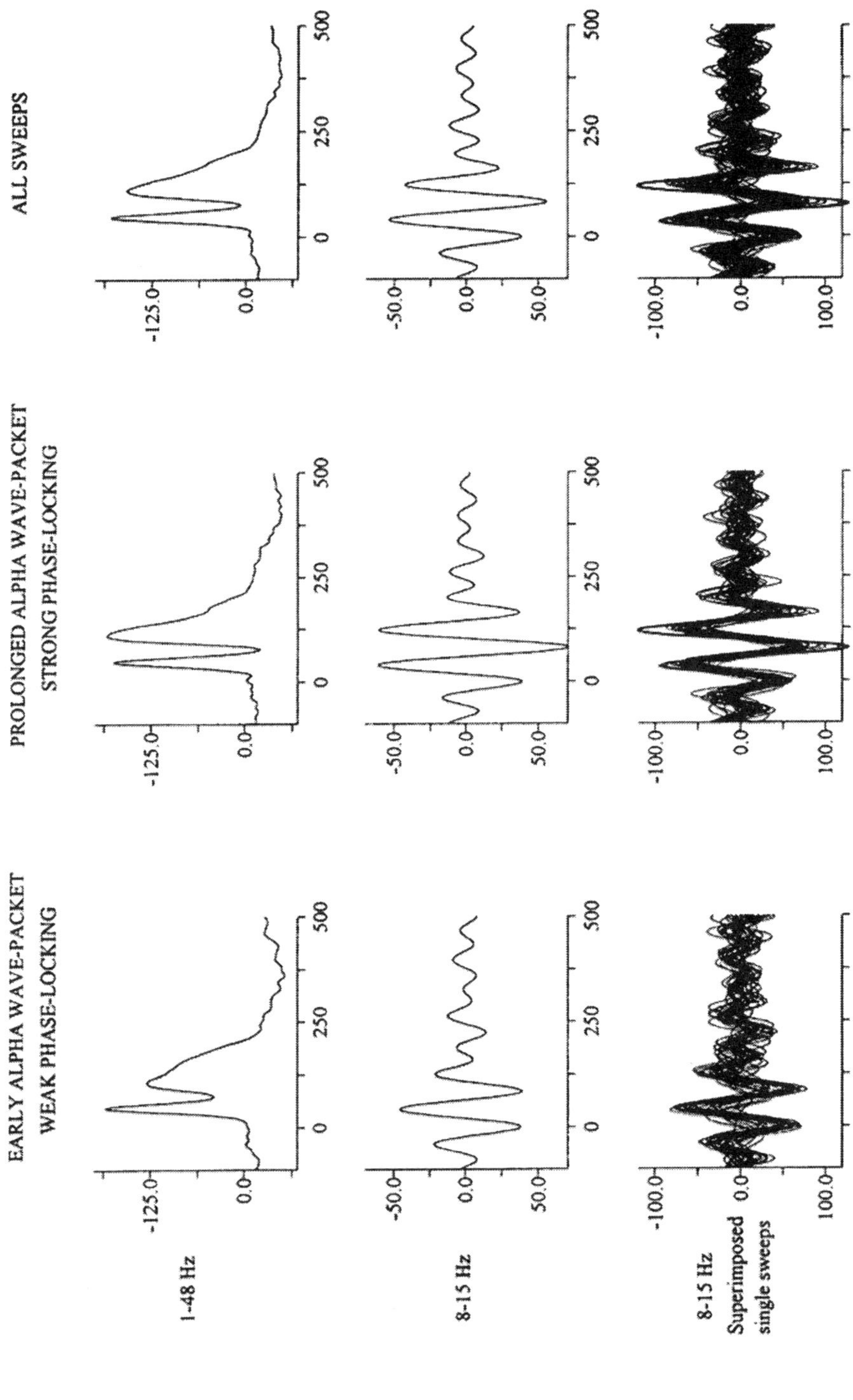
EARLY ALPHA WAVE-PACKET
WEAK PHASE-LOCKING
PROLONGED ALPHA WAVE-PACKET
STRONG PHASE-LOCKING
ALL SWEEPS
1-48 Hz
8-15 Hz
8-15 Hz
Superimposed
single sweeps
-125.0
0.0
-50.0
0.0
50.0
-100.0
0.0
100.0
0
250
500

2. The role of *8–12 Hz (alpha) oscillations* in information processing in the cat visual cortex has been pointed out by Başar et al. (Başar 1972, 1980; Başar et al. 1975a,b; Başar and Schürmann 1996) at the systems level and by Dinse et al. (1990, 1997; Kopecz et al. 1993) and Schreiner (1990) at the cellular level.
3. EP analysis based on systems theory rules and application of digital filters also showed that a marked *40 Hz component* follows stimulation (Başar et al. 1987). For further studies of evoked electric and magnetic 40 Hz activity see, e.g., Galambos et al. (1981), Pantev et al. (1991), Ribary et al. (1992). A large number of studies at the single-cell level complement these results (Llinás 1988; Gray and Singer 1987; Gray et al. 1989; Eckhorn et al. 1988).

The concept of *evoked and induced rhythms* (Başar and Bullock 1992; Bullock 1992) is a further approach parallel to frequency analysis and resonance phenomena. Results of several investigators working now with the concept of evoked rhythmicities and induced rhythmicities have recently been reviewed by Başar (1992; Başar et al., in press b), who emphasized that the complex dynamics of compound potentials and the resonance phenomena might play one of the most important roles in brain organization (we refer to Chap. 8 in Vol. I for a justification of the expression "event-related oscillation").

The present chapter is based mainly on intracranial recordings in cats. We demonstrated high-amplitude responses recorded from primary sensory areas responding to adequate stimuli. The lack of such responses to inadequate stimuli is another fact refuting the objection that alpha-response-like waveforms are a mere result of digital filtering. The good spatial resolution of such recordings may compensate for the lack of spatial resolution in human scalp measurements. (see also Chap. 18 in Vol. I).

One of the most important demonstrations concerning the physiological correlates of frequency components is obtained by the application of *cross-modality experiments*. This being the core topic of the current study, we add only that the results of measurements with scalp electrodes in human subjects (Başar and Schürmann 1994) are consistent with the data obtained in cats: 10 Hz enhancements in the filtered EPs or the dominant frequency peaks in the alpha frequency range in amplitude–frequency characteristics can be obtained if the physiological sensory stimulation is an adequate one. In the visual cortex only visual stimulation can give rise to highly significant

Fig. 2.9. *Selective averaging* according to results of wavelet decomposition (see text for details). The *left column* refers to single trials with early alpha (8–15 Hz) wave packet, the *middle column* to single trials with prolonged alpha wave packet, and the right column to all single trials. The *rows* show the wide-band filtered averaged ERP, the averaged ERP filtered in the alpha (8–15 Hz) range, and the single trials filtered in the alpha (8–15 Hz) range.

increases in the theta frequency range, whereas the theta response is not affected so much by inadequate stimulation.

If the recorded frequency response of all components by frequency analysis were obligatory, then we would see all these changes together. Here again a component is dominating if the sensory stimulation is adequate for the analyzed structure. This is an excellent example that the dissection of frequency components in frequency bands reveals functional relations.

2.6.1 Hippocampus Is a Supramodal Center

The hippocampus is a polysensory and supramodal center (for an overview and references, see Chap. 2 in Vol. I). In this structure *all types of sensory stimuli* give rise to high 10 Hz enhancements. On the other hand, thalamic or cortical structures can react with large alpha enhancements *only to adequate stimuli.* Again, we can see here that the 10 Hz response compound of the EPs is not obligatory and it is completely different depending on stimulus modality and structure studied.

Marked differences for adequate vs. inadequate stimulation were also observed in human occipital EPs (Başar and Schürmann 1994, see Chap. 10). The occipital electrode being adjacent to the visual cortex, we refer to these measurements as "cross-modality" experiments. Acoustical stimulation elicits no alpha enhancement in left-occipital recordings.

2.6.2 Possible Functional Roles of Evoked Alpha Oscillations

As to different functional roles of EP frequency components, the hypothesis underlying this approach was as follows: Alpha responses appear to be related to primary sensory processing, whereas theta and/or slower responses appear to be involved mainly in associative and cognitive processes.

We repeat here Hartline's statement (1987) about the visual cortex, which is considered to be involved exclusively with vision: "In cats about one-third of the neurons in area 17 (striate cortex), 18, and 19 are reported to be responsive to sound as well as to visual input [...]."

Furthermore, "a large number of neurophysiological studies showed that primary sensory stimuli elicit impulses or volleys converging over thalamic centers to primary sensory areas. On the other hand, the "sensory stimulation of second order" usually reaches the cortex over association areas (see, for example, Shepherd (1988)). Due to this consideration it is conceivable that the responses in the lower frequency ranges (theta, delta) might reflect the responsiveness of various brain areas in cases of association processes involved in global associative cognitive performance" (Başar et al. 1991).

On the basis of cognitive ERP measurements, Başar-Eroğlu et al. (1992) assumed that the brain's alpha, theta, and delta responsiveness is functionally related to cognitive processes such as selective attention, learning, and decision-making (see Chaps. 16, 17 and 18). The cross-modality responses of this study support this hypothesis. We assume – according to Hartline (1987) – that the occipital response is a cross-modality response that is probably not elicited directly by visual stimulation; it is evoked or induced through cognitive mechanisms acting mostly in association areas of the brain including the forebrain, parietal areas, and the limbic system. This last aspect provides further justification for regarding the theta response as a predominantly cognitive component.

This interpretation is also supported by results obtained with a different paradigm: It is mainly slow frequencies that contribute to differences between EPs obtained in an omitted stimulus paradigm and EPs recorded in a session without cognitive load. In a time-prediction task, selective averaging of responses to the last stimulus before omission showed increased delta–theta amplitudes (Demiralp and Başar 1992). This is noteworthy because the increase was most prominent in frontal and parietal electrodes that are closely related to association areas of the brain (for a comparable study on cats, see Demiralp et al. 1994 and Chap. 18).

Furthermore, the interpretation is supported by EP measurements in a group of multiple sclerosis (MS) patients (Başar-Eroğlu et al. 1993b; Schürmann et al. 1993). We regard this demyelinating disease as a model of impaired sensory input with consequences for primary sensory processing rather than associative–cognitive processing in the brain. We recorded visual and auditory EPs in an MS group in comparison with a control group. It was found that alpha responses were markedly reduced in the MS group, whereas theta responses were unaltered. This is in agreement with the above-mentioned hypothesis that alpha responses mainly reflect primary sensory processing and theta responses mainly associative–cognitive processing (see also Chap. 16 in Vol. I, which describes the "break of the alpha response").

2.7 EP/ERP Frequency Components – "Real Components" Related to Psychophysiological Functions

The EP frequency components (obtained by adaptive digital filtering) and also the amplitude–frequency characteristics reflect a *global* and *general* frequency behavior (or frequency contents) of the compound potentials, which we call EPs. As outlined by Başar (1980: chapters on the neural correlates of EEG and EPs), the compound potentials and the filtered responses give only a *general idea* about the frequency distribution of the real neuronal activity. As indicated in Başar's (1980) chapter on resonance phenomena in physics, we have here an analogy with quantum mechanics in which the position, exact location, or exact energy of a given elementary particle can be described only with a probability wave. In other words, "wave packets" in modern physics have been used in order to give a *cloudy description* of the observed particles and not a very exact physical entity. By the analysis of frequency responses of the compound potentials, we aim also to obtain *cloudy (or global) information about the frequency–amplitude characteristics.* Accordingly, minor changes of filtered responses or minor peakings in the amplitude–frequency characteristics should not to strongly be taken into account. Only *major* and *dominant changes* in the filtered EPs or *drastic changes* in the amplitude–frequency characteristics can be analyzed when we are trying to find psychophysiological correlates of EPs.

In our analysis we have indicated only such major changes in the EPs. They were usually due to anatomical differences of the structures studied or dependent on the type of experiments performed. Pathological changes also gave rise to such results. Chapter 18 in Vol. I gives further examples demonstrating that *we are not performing an arbitrary analysis of frequency components.* The dissection of the EP into frequency components can give us a very good idea as to the real psychophysiological contents of the potential when the concept is carefully applied.

2.8 Monomodal vs. Bimodal Stimulation

The analysis of dynamics described in Chap. 1 consisted only in the application of simple stimuli as tone or light signals. It was emphasized that the sensory cortices and hippocampus showed completely different behavior with regard to enhancements in the alpha frequency band immediately following the stimulation. As demonstrated in the present chapter, the auditory and visual cortices did not show enhancements in the alpha frequency band upon cross-modality stimulation (see Figs. 2.4 and 2.6 above).

The behavior of the hippocampus was completely different, as will be extensively discussed in Chap. 27. The hippocampus reacted with major enhancements to both auditory and visual stimulation (see Figs. 2.4 and 2.6 above). This means that in the hippocampus a manifold of neural population reacting upon visual and auditory stimuli in the 10 Hz frequency range should exist. We also use the expression *"supramodal association structure"* in describing the hippocampus.

The question now arising is, how will the auditory and visual cortex and hippocampus behave upon a complex stimulation consisting of visual and auditory stimulation? Therefore, the stimuli described above were applied in a combined way to cats with electrodes implanted in hippocampus and cortical structures. Chapter 11 is devoted to a comparison of EPs and amplitude–frequency characteristics for auditory stimulation, visual stimulation, and a combined auditory–visual stimulation. Some examples are presented here.

Figure 2.10 shows auditory EPs, visual EPs, and "double stimulation" EPs (auditory and visual stimulation applied simultaneously) recorded from the auditory cortex (GEA). The EPs have an amplitude of approx. 70 μV.

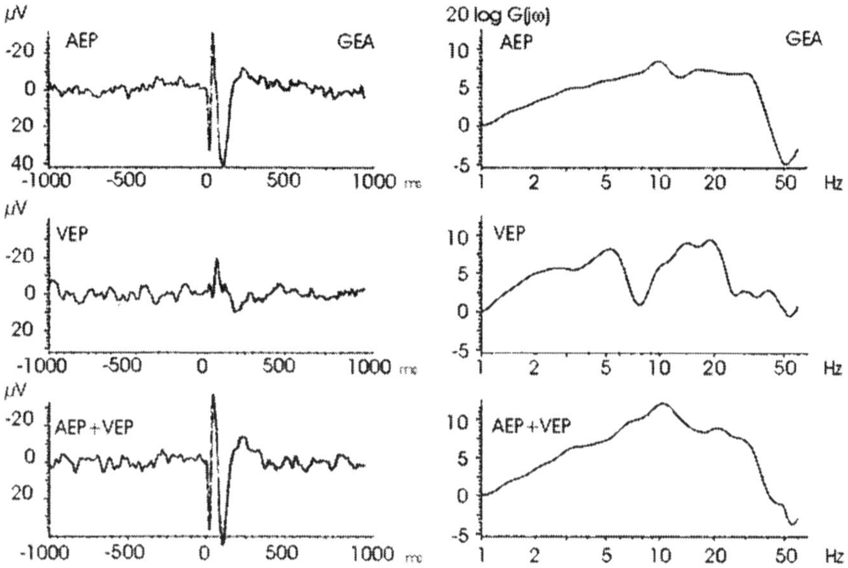

Fig. 2.10. Grand average EPs ($n = 4$ cats) recorded from GEA with auditory stimulation (AEP), visual stimulation (VEP), and bimodal stimulation (AEP+VEP). *Left column:*, wide-band filtered EPs; *right column:* amplitude–frequency characteristics computed from the EPs.

The visual EPs (inadequate stimulation) show amplitudes of only 20 μV. The evoked potentials obtained with double stimulation have very large amplitudes, up to 80 μV. The auditory evoked potentials have an amplitude maximum around 10 Hz, although the visual evoked potentials have no dominant alpha response. The evoked potentials resulting from double stimulation also have a dominant 10 Hz maximum.

Figure 2.11 shows corresponding EPs recorded from the hippocampus (HI). They have a completely different time course in comparison to the cortical recordings. The adequate and inadequate stimulation both evoke large evoked potentials with amplitudes in the range of 50 μV. The double stimulation also evokes very large evoked potentials. In all cases dominant theta responses were observed in the amplitude–frequency characteristics.

Figure 2.12 illustrates filtered EPs in the theta and alpha frequency ranges. It is important to note that in the auditory cortex the visual stimulation does not evoke any significant alpha responses.

On the other hand, in the hippocampus (Fig. 2.13) both 10 Hz responses and theta responses are recorded. The double stimulation evokes much larger theta components and also ample alpha components.

How can we interpret results when comparing the different types of stimulation described above? It seems that signals arriving at the hippocampus can excite neural populations sensitive to both visual and auditory stimu-

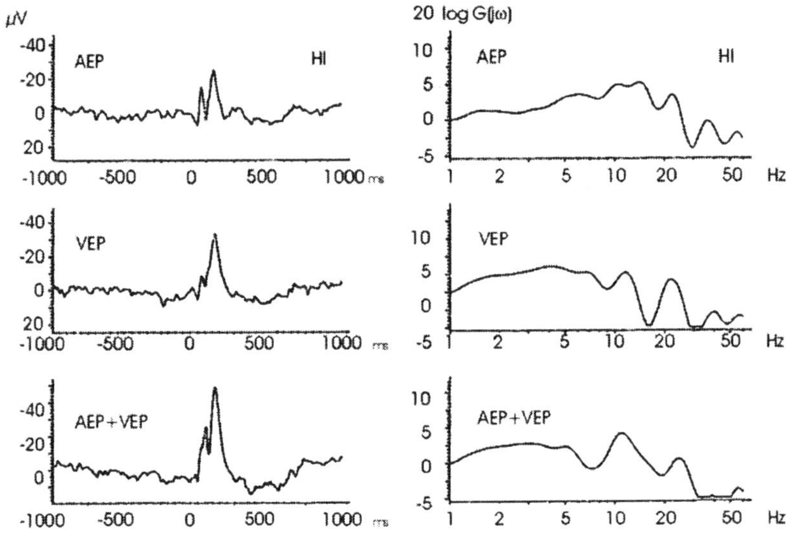

Fig. 2.11. Grand average EPs ($n = 3$ cats) recorded from HI with auditory stimulation (AEP), visual stimulation (VEP), and bimodal stimulation (AEP+VEP). *Left column:*, wide-band filtered EPs; *right column:* amplitude–frequency characteristics computed from the EPs.

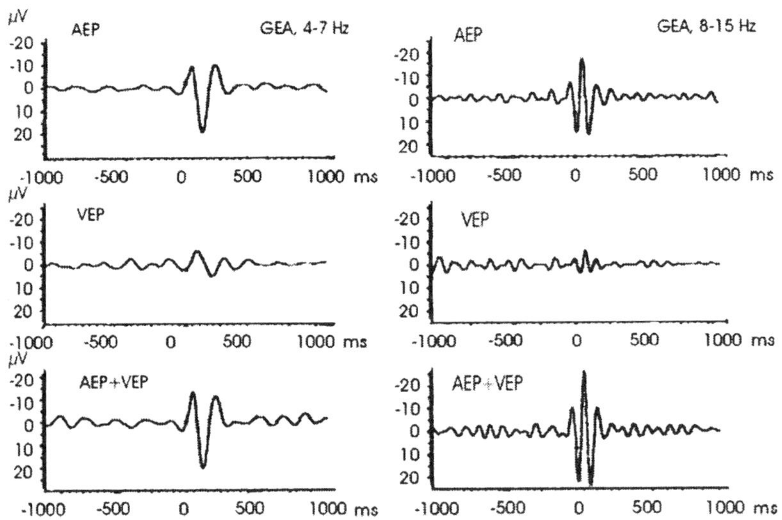

Fig. 2.12. Grand average EPs ($n = 4$ cats) recorded from GEA with auditory stimulation (AEP), visual stimulation (VEP), and bimodal stimulation (AEP+VEP). *Left column:* EPs filtered in the 4–7 Hz range; *right column:* EPs filtered in the 8–15 Hz range.

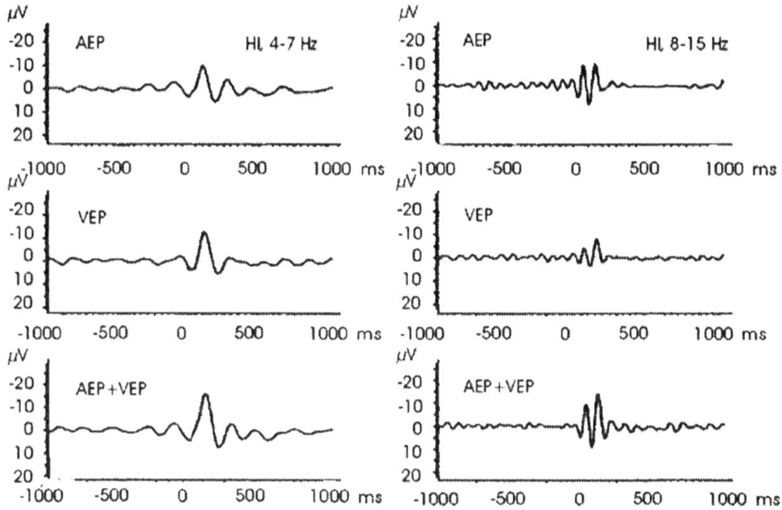

Fig. 2.13. Grand average EPs ($n = 3$ cats) recorded from HI with auditory stimulation (AEP), visual stimulation (VEP), and bimodal stimulation (AEP+VEP). *Left column:* EPs filtered in the 4–7 Hz range; *right column:* EPs filtered in the 8–15 Hz range.

lation. The sensory cortices, however, cannot react in the 10 Hz frequency range. The small-amplitude evoked potentials have a weak theta response, as was the case with inadequate stimulation.

These examples again demonstrate that several structures in the brain do not process electrical information in the same manner. The selectivity of the response can be understood better by means of frequency analysis. Upon application of double (bimodal) stimulation, the 10 Hz response of the auditory cortex is not augmented. The response of these structures in the 10 Hz range is almost the same. In the hippocampus the situation is completely different, further demonstrating the assumption of functional interpretation of theta and alpha responses described in previous sections. The reader is referred to Chap. 11 for a discussion and analysis of the theta response increase in the human brain.

3. Selectively Distributed Gamma-Band Responses Studied in Cortex, Reticular Formation, Hippocampus, and Cerebellum

Martin Schürmann, Tamer Demiralp, Canan Başar-Eroğlu, and Erol Başar

Although the amplitude–frequency characteristics upon auditory and visual stimuli were analyzed in detail in Chap. 2, and it was mentioned that these frequency responses contained marked 40 Hz components, in this chapter we will lay special emphasis on the gamma frequency response. We have undertaken this step because in the last decade several approaches have been tried, giving special functional emphasis to the gamma frequency band. Our aim will be to demonstrate that the brain has a distributed neural population system of generators that are excitable in the 40 Hz frequency range. In this chapter, firstly, an account on pioneering work in this rapidly developing branch of electrophysiology is given. Our scope concerning the selectively distributed gamma response system of the brain will be explained extensively in Chap. 26.

Various rhythms in the EEG and their possible functional meanings have been investigated in a number of studies during the past 70 years. Rhythms in the frequency ranges delta (0–4 Hz), theta (4–8 Hz), alpha (8–13 Hz), and beta (13–30 Hz) spontaneously exist in the EEG and have been found to correlate with various normal and pathological brain states. In connection with these studies, some authors have reported evoked, induced, or emitted rhythms in the EEGs (e.g., ERPs), which are correlated with external and/or internal events (Brazier 1960; Adrian 1942; Freeman 1975; Gray and Singer 1987, 1989; Eckhorn et al. 1988, 1989a,b; Galambos and Makeig 1988; Başar 1980, 1983, 1992; Bullock et al. 1990, 1992). In this class of brain electromagnetic rhythms, the 40 Hz, or gamma, rhythm (Freeman 1975; Gray and Singer 1988, 1989a,b; Eckhorn et al. 1988, 1989a,b; Galambos, 1988; Başar 1980, 1992; Bressler 1990) has gained importance during the past years. We will use the term "gamma rhythm" or "gamma-band activity" because according to the studies performed to date, activities with similar characteristics can be shown in the range 25 to 110 Hz, and this activity has not to be strictly tuned to 40 Hz (Eckhorn 1988; Galambos 1992).

3.1 Gamma Responses to Auditory Stimuli Recorded from Various Structures

The gamma-band limits were selected after the inspection of the AFCs of each animal in each electrode location (not shown). The center frequencies of the gamma peaks were varied between 30 and 70 Hz in different cats and in different localizations. Therefore, a broad gamma-band filtering between 30 and 70 Hz was carried out on the data to be able to include all these peaks in further analysis and comparisons and to avoid possible artifacts induced by narrow-band filtering.

Figure 3.1 shows the power spectra of the grand average prestimulus EEGs (left column) and the AFCs of the grand average EPs (right column) of 7 cats in 4 brain structures to give a general idea of the frequency-domain transformation of the data. Even though the gamma-band peaks occurred in different cats with different center frequencies, both the grand average power spectra of the prestimulus EEGs and the grand average AFCs of the EPs show globally the mean effect of these gamma peakings.

Figure 3.2 shows the averaged evoked responses of 7 cats in 4 brain structures – gyrus ectosylvianus anterior (GEA, auditory cortex), hippocampus (HI), formatio reticularis (RF), and acoustical vermis of the cerebellum (CE) – filtered in the gamma frequency range. All of the averaged responses demonstrated an enhancement in the gamma-band activity following the stimulation, which shows the time-locked nature of the gamma response. The grand averages of all the experiments filtered in the gamma range are presented at the bottom of the illustration.

In Fig. 3.3 the superimposed gamma-band filtered single sweeps of one cat in 4 structures is shown. An evident time locking of the gamma responses

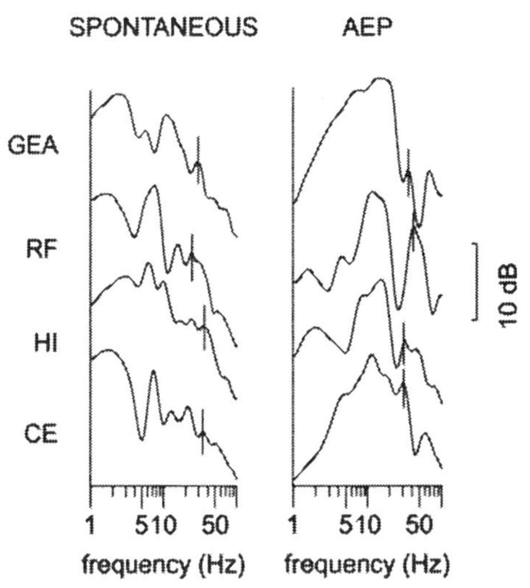

Fig. 3.1. The power spectra computed from the grand average of the prestimulus EEG periods *(left column)* and the amplitude–frequency characteristics of the grand average auditory EPs *(right column)* of Gyrus Ectosylvianus Anterior (GEA), formatio reticularis mesencephali (RF), hippocampus (HI), and cerebellum (CE) of 7 cats. The peaks in the gamma frequency range (30–70 Hz) are indicated (|). (From Başar E. et al. (1995) *IEEE Engineering in Biology and Medicine* 14: 400–410.)

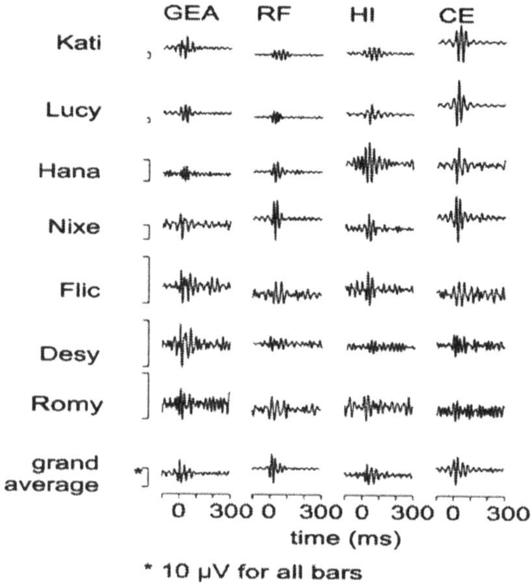

Fig. 3.2. Responses to auditory stimuli in single cats, filtered in gamma band (30–70 Hz) filtered auditory responses of single cats in GEA, RF, HI, and CE. The gamma-band filtered grand averages of responses are shown in the *bottom row*. (From Başar E. et al. (1995) *IEEE Engineering in Biology and Medicine* 14: 400–410.)

in single sweeps to the stimulation time point can be observed also by superimposition of single sweeps.

Figure 3.4 shows the superimposed broadband filtered (1–70 Hz) and gamma-band filtered (30–70 Hz) averaged responses of one cat in 4 brain structures. The gamma-band components of the evoked responses are also clearly visible in the broadband filtered signals, which is evidence that we are not dealing with filter artifacts.

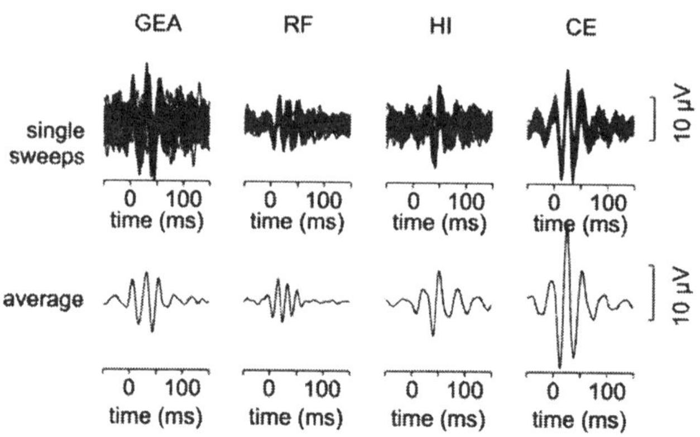

Fig. 3.3. The superimposed gamma band (30–70 Hz) filtered single sweeps of a AEP recording of one cat *(upper row)* and the gamma-band filtered average AEPs of the same cat *(bottom row)* in different brain structures. (From Başar E. et al. (1995) *IEEE Engineering in Biology and Medicine* 14: 400–410.)

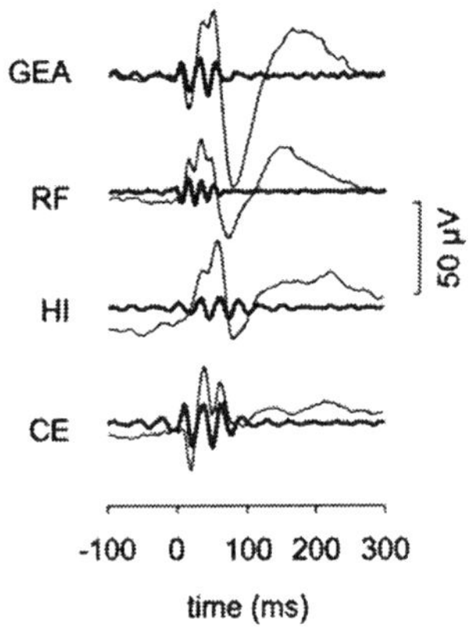

Fig. 3.4. The broadband (1–70 Hz, *thin line*) and gamma-band filtered (30–70 Hz, *thick line*) averaged auditory EPs of a representative cat in four brain structures. An extremely prominent gamma-band response can be observed in cerebellum. (From Başar E. et al. (1995) *IEEE Engineering in Biology and Medicine* 14: 400–410.)

For a robust *statistical test* of the gamma-band enhancements due to stimulation, the maximum peak-to-peak amplitudes in the 1000 ms prestimulus time window (between −1000 and 0 ms) have been compared with those in the 200 ms poststimulus time window (between 0 and 200 ms). Although the enhancement factor calculation defined as the ratio of the peak-to-peak amplitude of the evoked response to the RMS amplitude of the prestimulus period by Başar (1980) would give a physically more meaningful measure of the gamma-band response, we used the peak-to-peak amplitudes also in measuring the prestimulus activity to enhance the confidence range of the statistics. In this way any transient waveform occurring in the averaged prestimulus period EEG through phase locking by chance is also considered.

The differences between the pre- and poststimulus period maximum gamma amplitudes of 4 leads were tested by a 2-way ANOVA test (stimulation x lead; see Table 3.1). The main effect of the factor "stimulation" (pre- and poststimulus gamma-band amplitudes) was significant [$F(1,6) > 8.62$, $p < 0.03$], whereas the effect of the factor "lead" was not significant [$F(3,18) > 1.08$, $p = 0.38$]. Also, the interaction for the maximal amplitude measurements between "stimulation" and "lead" was insignificant [$F(3,18) > 1.01$, $p = 0.41$]. The mean values of the maximum gamma-band amplitudes in pre- and poststimulus periods are given in Fig. 3.5 in histogram form with mean percentage increase values.

Table 3.1. Results from 2-way analysis of variance (stimulation x lead) for the maximal amplitudes of the gamma band activities. (From Başar E. et al. (1995) *IEEE Engineering in Medicine and Biology* 14: 400–410)

Factor (df)	F	p
Stimulation (1,6)	8.62	0.03
Lead (3,18)	1.08	0.38
S × L (3,18)	1.01	0.41

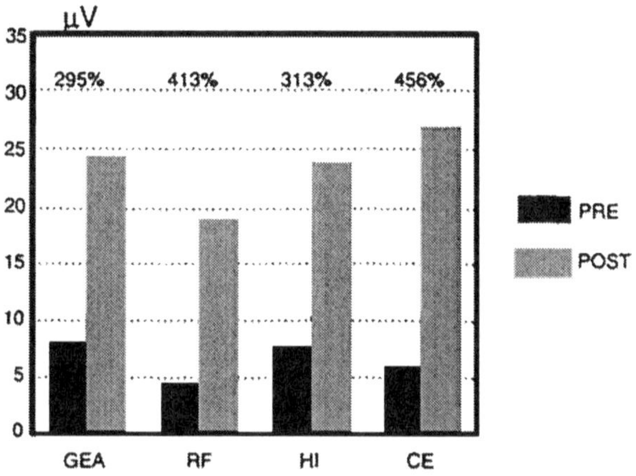

Fig. 3.5. The mean values of the maximum peak-to-peak amplitudes of the gamma-band activity in pre- and poststimulus periods of averaged auditory responses recorded in GEA, RF, HI, and CE. The increases of the amplitudes in the poststimulus period are given as the percent of the prestimulus amplitudes *on top* of the bars. The amplitude enhancement due to the stimulation is significant over all leads ($p < 0.03$). The differences between the gamma-amplitude enhancements in different locations are insignificant. (From Başar E. et al. (1995) *IEEE Engineering in Biology and Medicine* 14: 400–410.)

3.2 Gamma Responses to Visual Stimuli Recorded from Various Structures

For measurements with visual stimuli, recordings were made from chronically implanted electrodes in the occipital cortex (area 17, referred to as OC), lateral geniculate nucleus (LG), superior colliculus (SC), dorsal hippocampus (HI), mesencephalic reticular formation (RF), and auditory cortex (gyrus ectosylvianus anterior, GEA). EPs were elicited by means of intense visual step function stimuli ($N = 100$) delivered via a fluorescent bulb (with pseudo-random interstimulus intervals of 2.5 to 3.5 s, mean: 3 s).

Figure 3.6 shows the *averaged evoked responses* in 9 cats filtered in the gamma frequency range. An enhancement in the gamma-band activity following the stimulation is visible in all structures, thus showing the time-locked nature of the gamma response.

In Fig. 3.7 the superimposed gamma-band filtered *single trials* of one cat in 4 structures is shown. An evident time locking of the gamma responses in single trials to the stimulation time point can be observed.

Figure 3.8 illustrates the stimulus-evoked gamma enhancement. A significant difference between pre- and poststimulus gamma amplitudes was observed ($p < 0.01$, Wilcoxon, for OC, HI, RF; $p < 0.05$ for SC). Note that no significant difference was observed for the auditory cortex (GEA), and that

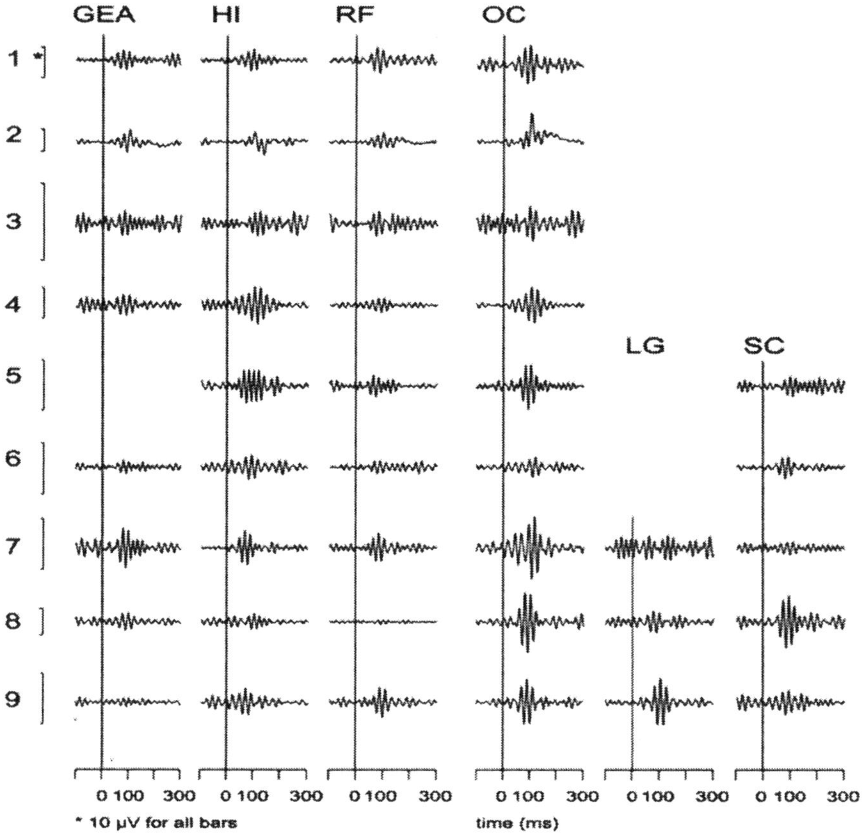

Fig. 3.6. Responses to visual stimuli (gamma-band filtered, 30–48 Hz) in nine single cats. Recordings from GEA (gyrus ectosylvianus anterior, auditory cortex), RF (reticular formation), HI (hippocampus), OC (occipital, visual cortex), LG (lateral geniculate nucleus), SC (superior colliculus). (From Schürmann M. et al. (1997) *NeuroReport* 8: 531–534.)

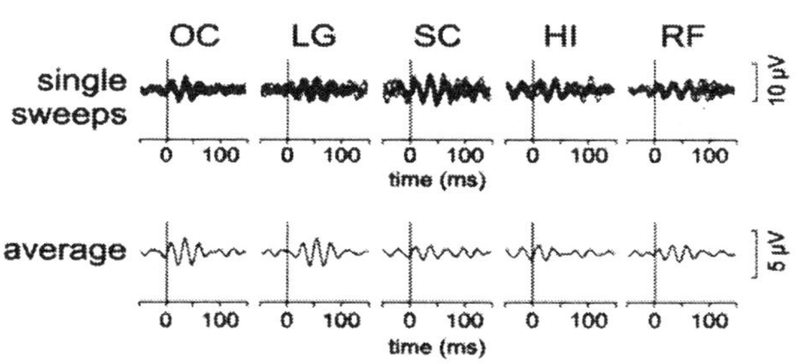

Fig. 3.7. Responses to visual stimuli in different brain structures in a representative cat. *Top row:* superimposed single-trial EPs (gamma-band filtered, 30–48 Hz); *bottom row:* averaged EP (gamma-band filtered). Abbreviations as in Fig. 3.1. (From Schürmann M. et al. (1997) *NeuroReport* 8: 531–534.)

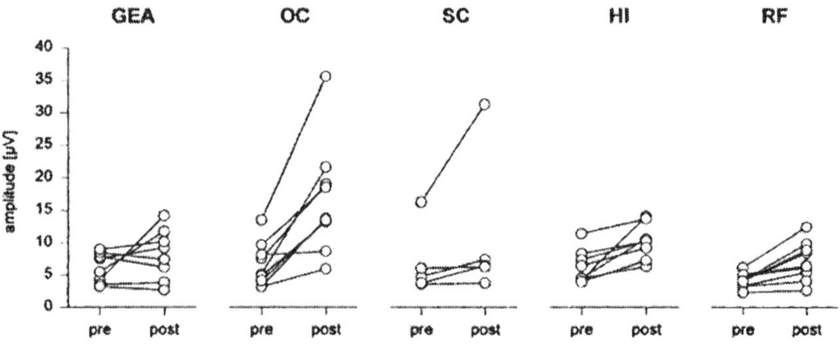

Fig. 3.8. Maximal amplitudes of gamma-band filtered averaged prestimulus EEG ("pre": −1000 to 0 ms) and averaged EP ("post": 0–200 ms) in nine cats and in five brain structures. (From Schürmann M. et al. (1997) *NeuroReport* 8: 531–534.)

differences were largest for the visual cortex (no test possible for LG due to small sample size).

3.3 Gamma Responses – Multiple Functional Correlates

Are gamma oscillations in the EEG functionally related to certain brain processes? Such a relation has been suggested for the processing of complex stimuli (of cognitive relevance). It is an open question whether this relation is a specific one. In contrast to previous studies, we observed gamma rhythms even in response to very "simple" stimuli, measured in cats with electrodes implanted into several parts of the brain; i.e., gamma rhythms occur even when the demand for cognitive processing is minimal. We conclude that gamma rhythms are related to the processing of simple, as well as complex, stimuli. This supports the view that gamma rhythms are elementary signals of the brain, functionally related to diverse (not only cognitive) brain processes.

For an extended discussion of these results the reader is referred to Chap. 26, in which the multiple functional features of gamma will be discussed.

4. Highest Frequency Range in Reticular Formation and Inferior Colliculus (100–1000 Hz)

4.1 Introduction

In Chap. 11 in Vol. I we already have shown examples of strong resonance phenomena in the reticular formation and in the inferior colliculus of the cat upon auditory stimulation. Frequency-locking phenomena were demonstrated in the frequency range up to 1000 Hz for the reticular formation and the inferior colliculus.

In recent years, the analysis of the highest frequency window of evoked oscillation has gained increasing importance. This is due to the fact that new MEG measurements performed by Curio et al. (1996) and by Lopez and Sannita (1997) have demonstrated frequency responses in the frequency range 100–1000 Hz. We also mention earlier publications by Karakaş and Başar (1982) and Başar (1980).

Although in all brain structures that we have studied in this chapter, time-locking and resonance can be shown in the stated highest frequency channel, for frequencies above 100 Hz, the reticular formation and the inferior colliculus have the most marked selectivities. Therefore, we will emphasize here the two brainstem structures, in order to study the brain responsiveness in the highest frequency region between 100 Hz and 1000 Hz.

Important results of this highest frequency window came from chaos experiments. Röschke and Başar (1989) demonstrated that in 25 percent of recordings in cerebellum and reticular formation the spontaneous activity showed a chaotic behavior in the cat brain (see Chap. 10 in Vol. I).

Experimental findings in this section were obtained from the cat mesencephalic reticular formation and inferior colliculus by application of a high-pass analog filter with a cutoff frequency of 80 Hz (pass-band of the overall system between approximately 80 Hz to 2000 Hz at 3 dB points). Background masking of low-frequency, high-amplitude EEG and EP components usually created difficulty in observing the relatively low-magnitude, high-frequency electrical activities of the brain. High-pass filtering provided attenuated transmission below 80 Hz and allowed for high-frequency analysis in the electrical activities of the reticular formation and inferior colliculus. The typical examples of high-frequency responses (recorded in reticular formation and inferior colliculus) were chosen from similar observations during experiments with 8 cats. The stimulation consisted of auditory step functions in the form of a tone burst of 5000 Hz at 90 dB (SPL).

4.2 Selectively Averaged Transient Evoked Potentials

Figure 4.1 shows two typical examples of averaged EPs of the reticular formation, and Fig. 4.2 shows two typical averaged EPs of the inferior colliculus during the waking stage. The EPs of both reticular formation and inferior colliculus depicted marked short latency waves with steep slopes or large slope changes. A positive short latency wave around 3 msec following a marked negative wave of 4.5–5.0 msec latency were the most prominent waves of the averaged EPs of the reticular formation (see Fig. 4–1).

The averaged EPs of the inferior colliculus (Fig. 4.2) depicted a negative short latency wave between 5.0 and 5.5 msec. A further detailed analysis of the latencies and waveforms of the transient evoked potentials is not necessary. As stated earlier, "slope and slope changes and the entire time course of evoked potentials carry the whole information concerning different activities and frequency selectivities of brain structures rather than the number and latencies of the waves" (Başar and Özesmi 1972; Başar and Ungan 1973).

4.3 Amplitude–Frequency Characteristics

Figure 4.3 shows a typical amplitude–frequency characteristic obtained from the reticular formation during the waking stage. The amplitude–frequency characteristics were computed using the averaged EPs presented in Fig. 4.1b

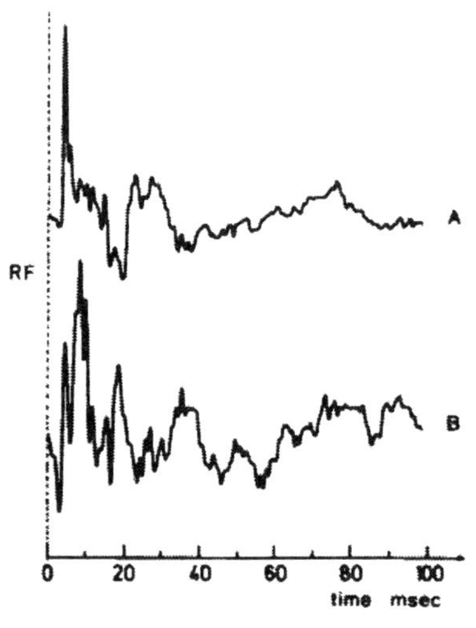

Fig. 4.1. Two typical examples of the selectively averaged EPs of the RF obtained from two awake cats. The number of averaged responses are (**a**) 5, (**b**) 6. The negativety is *upward*. The stimulation consists of auditory step functions in the form of tone bursts of 5000 Hz and 90 dB (SPL). (From Gönder A. (1977) thesis: Hacettepe University, Ankara.)

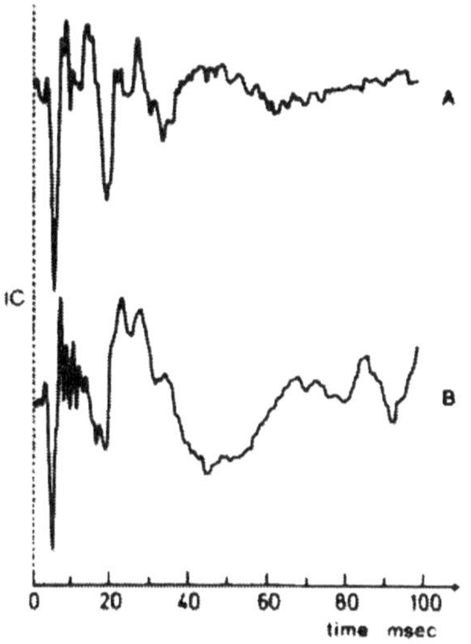

Fig. 4.2. Two typical examples of the selectively averaged EPs of the IC obtained from two awake cats. The number of averaged responses are (**a**) 8, (**b**) 10. the negativity is *upward*. The stimulation consists of auditory step functions in the form of tone bursts of 5000 Hz and 90 dB (SPL). (From Gönder A. (1977) thesis: Hacettepe University, Ankara.)

with the help of the Fourier transform (see Chap. 4 in Vol. I). Along the abscissa is the input frequency in logarithmic scale, and along the ordinate is the potential amplitude, in relative units and decibels. The curves are normalized in such a way that the amplitude at 0 Hz is equal to 1 (= 0 dB on logarithmic scale). Various numbers of distinct amplitude maxima, which

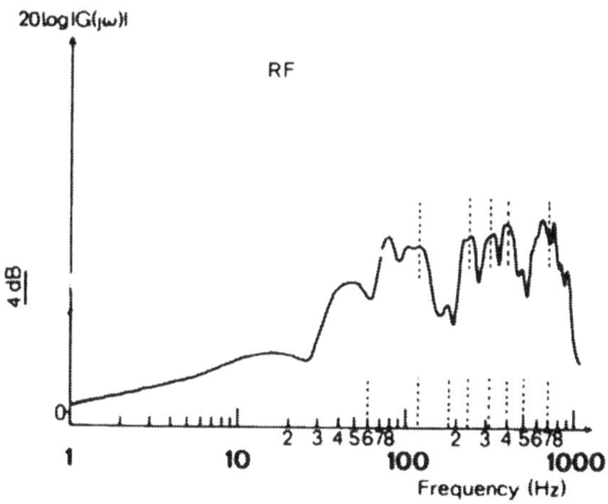

Fig. 4.3. A typical amplitude–frequency characteristic of the RF of the cat determined by the TRFC method and using the selectively averaged EP of Fig. 4.1. *Along the abscissa* is the input frequency in logarithmic scale; *along the ordinate* the potential amplitude, $|G(j\omega)|$, in decibels. The curves are normalized in such a way that the amplitude at 0 Hz is equal to 1 (or $20 \log 1 = 0$). (From Gönder A. (1977) thesis: Hacettepe University, Ankara.)

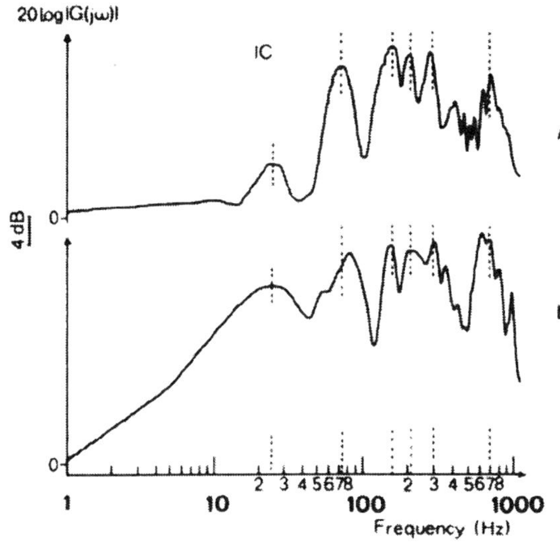

Fig. 4.4. Two amplitude–frequency characteristics of the IC of the cat determined by the TRFC method and using the selectively averaged EPs of Figs. 4.2 and 4.2, respectively. *Along the abscissa* is the input frequency in logarithmic scale, *along the ordinate* the potential amplitude, $|G(j\omega)|$, in decibels. The curves are normalized in such a way that the amplitude at 0 Hz is equal to 1 (or $20 \log 1 = 0$). (From Gönder A. (1977) thesis: Hacettepe University, Ankara.)

were centered at 50 Hz, 120 Hz , 240 Hz, 310 Hz, 410 Hz, and 700–800 Hz, were seen in the frequency characteristics of the reticular formation.

Similar to the characteristics of the reticular formation, the amplitude characteristics for the inferior colliculus shown in Fig. 4.4 also exhibited distinct maxima in higher frequency ranges. Amplitude maxima around 75 Hz, 170 Hz, 220 Hz, 300 Hz, and 700 Hz were the most prominent ones seen in Fig. 4.4. A small amplitude maximum around 20–25 Hz (beta range) was highly attenuated due to high-pass filtering. Minor peaks around 400–500 Hz were also identified. Amplitude maxima around 70 Hz and 180–200 Hz were previously defined by Başar et al. (1975b) as the consistent response of the inferior colliculus. These selectivities were also observed in the results of this section as the most prominent maxima in the amplitude characteristics of the inferior colliculus.

4.4 Consistent Selectivities in the Highest Frequency?

The mean-value amplitude characteristics of the reticular formation and inferior colliculus of all the experiments were computed. The mean-value curves were obtained by averaging the amplitude–frequency characteristics of six experiments on six cats. Figure 4.5 shows the mean-value amplitude–frequency

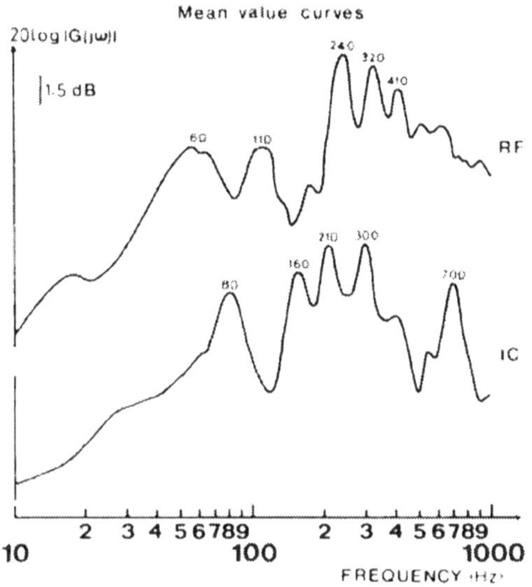

Fig. 4.5. Mean-value curves of amplitude characteristics of the RF obtained from 6 experiments on 6 cats and the IC obtained from 6 experiments on 6 cats with auditory stimulation during the waking stage. *Along the abscissa* is the input frequency in logarithmic scale, *along the ordinate* the mean-value potential amplitude, $|G(j\omega)|$, in decibels (From Gönder A. (1977) thesis: Hacettepe University, Ankara.)

characteristics of the reticular formation and inferior colliculus. Various distinct maxima are seen: 60 Hz, 110 Hz, 240 Hz, 320 Hz, and 410 Hz in the RF; 80 Hz, 160 Hz, 210 Hz, 300 Hz, and 700 Hz in the IC. These are consistent maxima or consistent selectivities. During the waking stage, the probability for maximal signal transfer through the stated consistent selectivity channels is always existent in all experimental conditions.

The higher-frequency selectivities of the reticular formation and inferior colliculus presented above are extensions of previous results obtained by Başar et al. (1975b).

4.5 Very High Frequency Responses in the Human Brain

Başar (1980) has published results on human auditory responses up to a frequency of 900 Hz. Later, Karakaş and Başar (1983) published results with frequency characteristics and filtered EPs upon visual stimuli. Consistent responses up to frequencies of 1200 Hz have been pointed out in these results. Recently, Curio et al. (1994, 1996) used MEG methodology to study frequency responses higher than 100 Hz. These authors found a frequency response of approx. 600 Hz upon somatosensory stimulation. Lopez and Sannita (1997) recorded magnetic oscillatory responses to luminance stimulation in humans in the range of 120 Hz.

All these studies and also our findings in Chap. 10 (in Vol. I) emphasize that the frequency window higher than 100 Hz could gain importance in the study of brain dynamics.

5. The Brain of the Sleeping Cat: Dynamics of Electrical Signals

5.1 Some Sleep Stages of the Cat

There are several possible classifications to describe the various stages of sleep in the cat, which spends about two-thirds of its day asleep. The process of falling asleep for the cat follows a characteristic course in time and is recognizable by easily observable external cues. Typically, the animal curls up into a ball with its neck in a bent position. The flexing of the neck is a clear sign that the neck muscles retain some tonus; i.e., they are not completely relaxed. In this position, the cat lapses into a light sleep from which it is easily awakened (Jouvet 1964). Jouvet states that the first stage of sleep, or light sleep, is characterized by a slackening of electrical activity in the cortex and subcortical structures, by the occurrences of *"spindles,"* or groups of sharp jumps in the brain waves, and by the retention of muscular tension. Ursin (1971) makes a classification in which the so-called slow wave sleep (SWS) stage has two aspects: light slow wave sleep characterized by 12–14 Hz high-voltage sleep spindles and some 1–4 Hz high-voltage slow waves on a low-voltage background. Deep slow wave sleep consists of at least 10 sec of slow waves and spindles covering half of the recording period.

5.1.1 Spindle Sleep Stage

In our studies on the dynamics of potentials during stages of sleep, we definitely differentiated the periods with spindles from the delta-wave period: A stage with cortical spindles is observed in most of the cats before and during the onset of sleep preceding the stage of hypersynchrony with high-voltage delta waves.

During the spindle sleep stage (SS-stage), spontaneous 10–14 Hz high-voltage spindles appear in the cortical EEG recording as illustrated in Fig. 5.1. Usually, spindles were recorded in the cortex; however, thalamic or hippocampal spindles were also observed in some experiments. For selective averaging of evoked potentials during the SS-stage, Başar et al. (1975c) selected only those patterns where spindles were observed in the cortex. Interspindle intervals were not considered for the selective averaging. In other words, for the evaluation of selectively averaged EPs during the SS-stage, only epochs

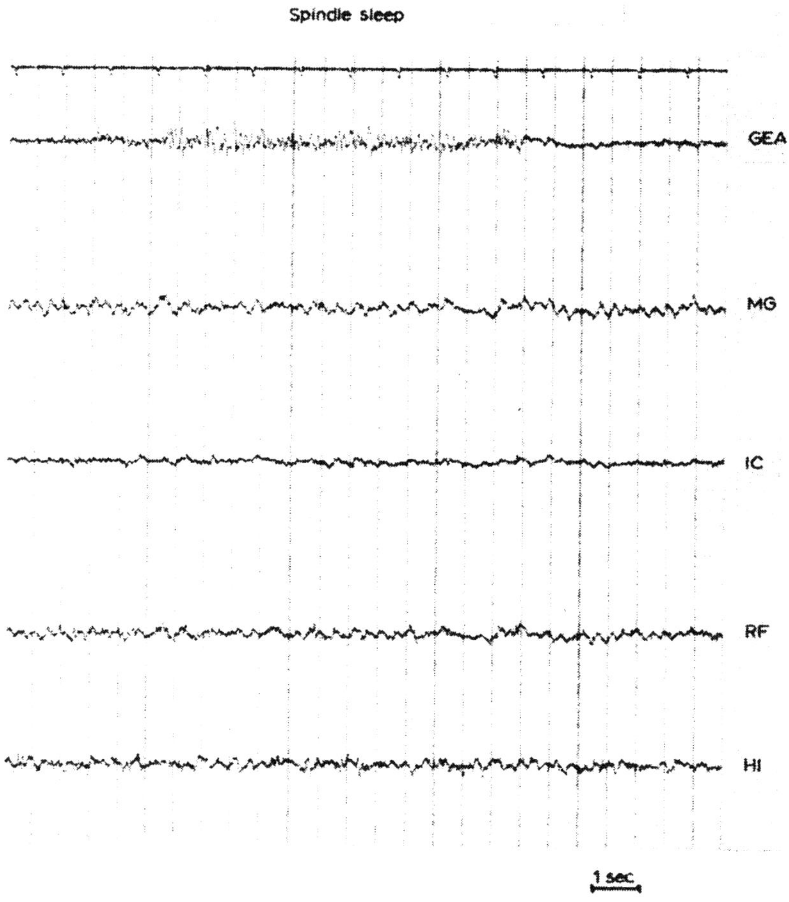

Fig. 5.1. Simultaneously recorded spontaneous activity of different brain nuclei during the spindle sleep stage (From Başar E. et al. (1975) *Biological Cybernetics* 20: 161–169.)

with cortex spindles were averaged. The interspindle periods were eliminated for the evaluation of spindle sleep stage EPs and for all the nuclei under study. We will see in Sect. 8.3 that it is very important to differentiate between the measurements during spindles and slow waves. The systems characteristic of the brain during the SS-stage are completely different from those of the SWS-stage.

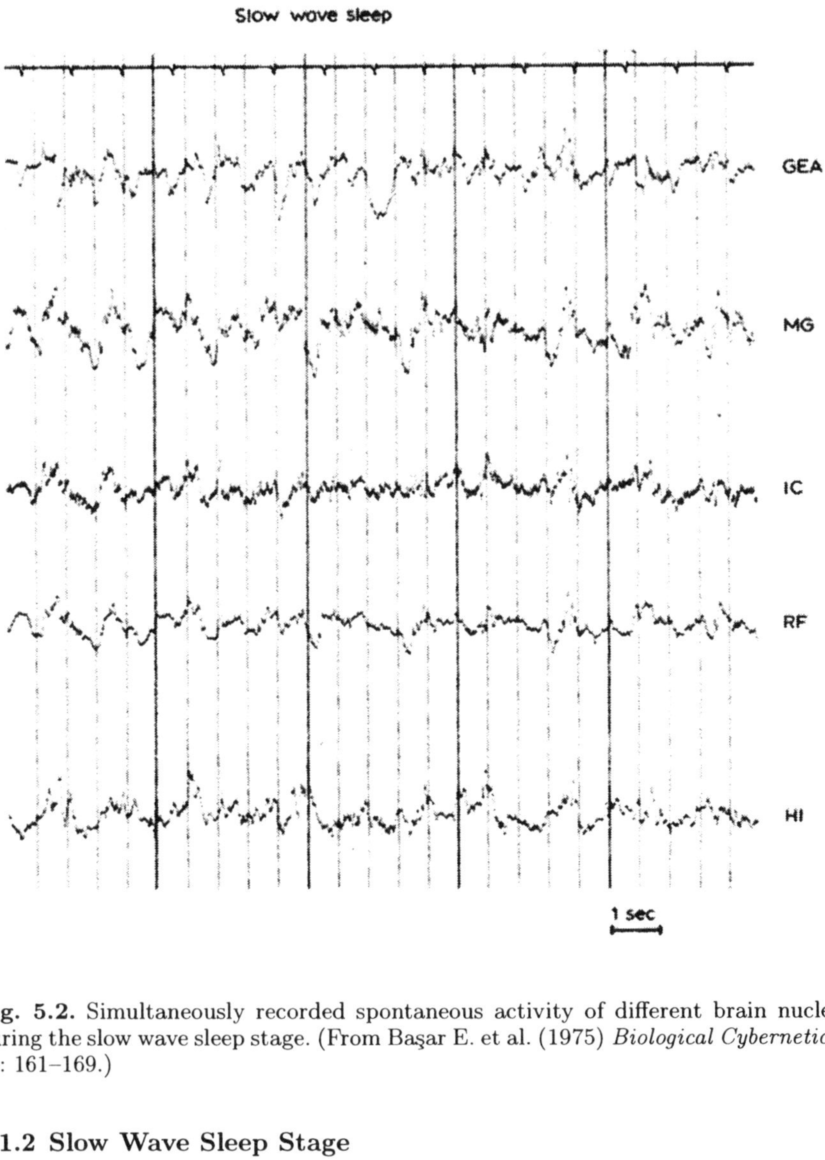

Fig. 5.2. Simultaneously recorded spontaneous activity of different brain nuclei during the slow wave sleep stage. (From Başar E. et al. (1975) *Biological Cybernetics* 20: 161–169.)

5.1.2 Slow Wave Sleep Stage

The invasion of spontaneous high voltage slow waves (1–3 Hz) in the auditory cortex was accompanied by the occurrence of the same waves in all the nuclei under study (Fig. 5.2). We used this kind of period (or epoch) as a sample pattern in selecting the epochs for the evaluation of evoked potentials during the SWS-stage. We selected the EP epochs for the SWS-stage only from a recording period where the state of hypersynchrony shown in Fig. 5.2 was observed in all the nuclei during a sufficiently long sleep period.

5.2 Evoked Potentials During Sleep Stages

Several authors have demonstrated the great variability of potentials evoked in the brain by sensory stimulation at different stages of wakefulness and sleep. In this chapter, we will analyze the change in the dynamics of potentials during sleep stages of the cat auditory pathway, reticular formation, and hippocampus. Moreover, we will compare the data from the cat brains with the data from sleeping human brains and use the same tools and concepts that were applied during the waking stages studies (Chaps. 1 and 2). The extension of the results on the dynamics during sleep is necessary in order to understand the generalized dynamics of the brain.

Figures 5.3 and 5.4 illustrate typical examples of selectively averaged EPs in all the nuclei under study during the defined sleep stages. The stimulation consisted of auditory step functions in the form of 2000 Hz, 80 dB tone bursts (for surgical procedures, refer to Chap. 1).

During SWS, the short latency components of averaged EPs of all these nuclei (especially of the reticular formation and inferior colliculus) have lower magnitudes. On the other hand, 80–100 msec waves (long latency waves) depict the largest magnitudes in all these nuclei (Figs. 5.3, 5.4).

The similarity of simultaneously recorded evoked potentials, especially for waves of middle and long latencies, is striking. The selectively averaged EPs of SWS particularly resemble one another.

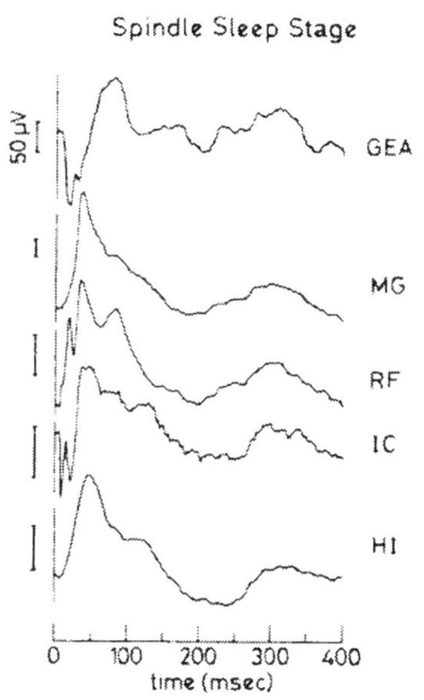

Fig. 5.3. A typical set of simultaneously recorded and selectively averaged evoked potentials in different brain nuclei of chronically implanted cats, elicited during the spindle sleep stage by an auditory stimulation in the form of step function. Direct computer plottings. Negativity *upwards*. (From Başar E. et al. (1975) *Biological Cybernetics* 20: 161–169.)

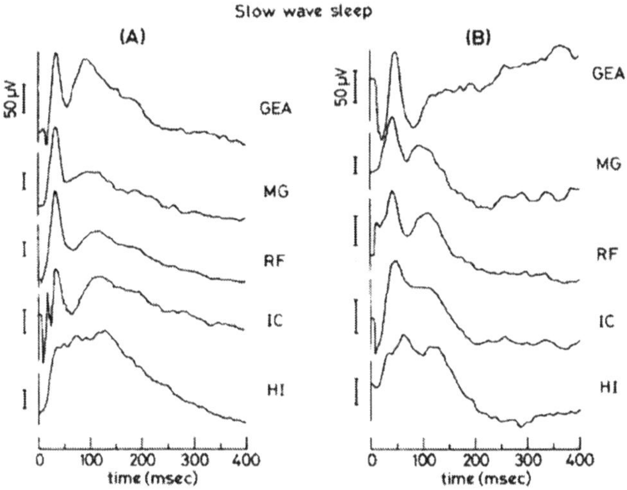

Fig. 5.4. Two typical sets of simultaneously recorded and selectively averaged evoked potentials in different brain nuclei of chronically implanted cats, elicited during the slow wave sleep stage by an auditory stimulation in the form of step function. Direct computer plottings. Negativity *upwards*. (From Başar E. et al. (1975) *Biological Cybernetics* 20: 161–169.)

5.3 Amplitude–Frequency Characteristics During Sleep Stages

5.3.1 Spindle Sleep (SS) Stage

The most common feature of the amplitude characteristics of all the nuclei was the maxima around 4–5 Hz (theta frequency range). This theta maximum was the dominant maximum in the characteristics of all the studied nuclei except the auditory cortex (GEA, see Fig. 5.5). The 50 Hz maximum in the reticular formation (RF) and the 70 Hz maximum of the inferior colliculus (IC), which were usually found to be the dominant maxima during the waking stage, remained unchanged during SS-stages. The dominant theta maximum in subcortical structures and the dominant maximum around 14 Hz in the acoustical cortex can, in any event, be described as constant and important findings at the SS-stage. Figure 5.6 presents the mean-value curves of amplitude characteristics from six experiments performed during SS-stages. As mentioned in the experiments on waking stage, the averaged curves from different cats and different experiments can only give information to a decreased degree. However, the existence of dominant maxima in the theta frequency range, the dominant alpha maximum in the cortex, and the reticular 50 Hz maximum are marked in the averaged curves, thus indicating that consistent maxima (consistent selectivities) are found in these frequency ranges during the SS-stage.

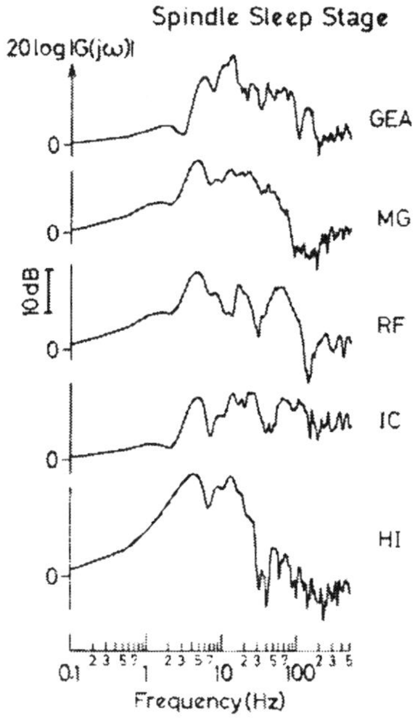

Fig. 5.5. A set of simultaneously obtained amplitude frequency characteristics of different brain nuclei of the cat determined by the TRFC method and using the selectively averaged transient responses of Fig. 5.3. Direct computer plottings. *Along the abscissa* is the input frequency in logarithmic scale, *along the ordinate* the potential amplitude, $|G(j\omega)|$, in decibels. The curves are normalized in such a way that the amplitude at 0 Hz is equal to 1 (or $20 \log 1 = 0$). (From Başar E. et al. (1975) *Biological Cybernetics* 20: 161–169.)

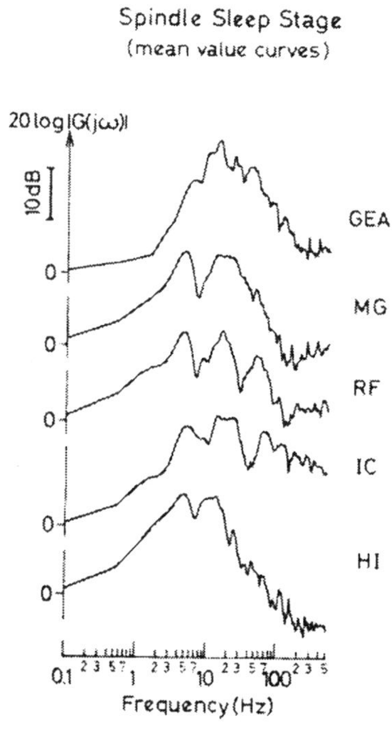

Fig. 5.6. Mean-value curves of simultaneous amplitude frequency characteristics of different brain nuclei of the cat obtained from six experiments with auditory stimulation during the spindle sleep stage. Direct computer plottings. *Along the abscissa* is the input frequency in logarithmic scale, *along the ordinate* the potential amplitude, $|G(j\omega)|$, in decibels. The curves are normalized in such a way that the amplitude at 0 Hz is equal to 1 (or $20 \log 1 = 0$). (From Başar E. et al. (1975) *Biological Cybernetics* 20: 161–169.)

5.3.2 Slow Wave Sleep (SWS) Stage

During the SWS stage, the amplitude characteristics of all the studied nuclei showed two important common deviations from the characteristics of the waking stage: (1) Prominent amplitude maxima in the low-frequency range between 1 and 3 Hz became apparent in all the curves, and (2) maxima in frequency ranges higher than 20 Hz either became highly attenuated or completely disappeared. This was seen even in the nuclei of the inferior colliculus and mesencephalic reticular formation, where high frequency mechanisms were dominant during the waking stage.

Figure 5.7 shows the amplitude characteristics of the nuclei studied. The amplitude characteristics were obtained from simultaneously recorded and selectively averaged evoked potentials using the TRFC method (see Chap. 4 in Vol. I):

1. **Auditory cortex (GEA)**. Maxima in the frequency range of 1–3 Hz and maxima in a broad frequency range covering alpha and beta ranges (7–

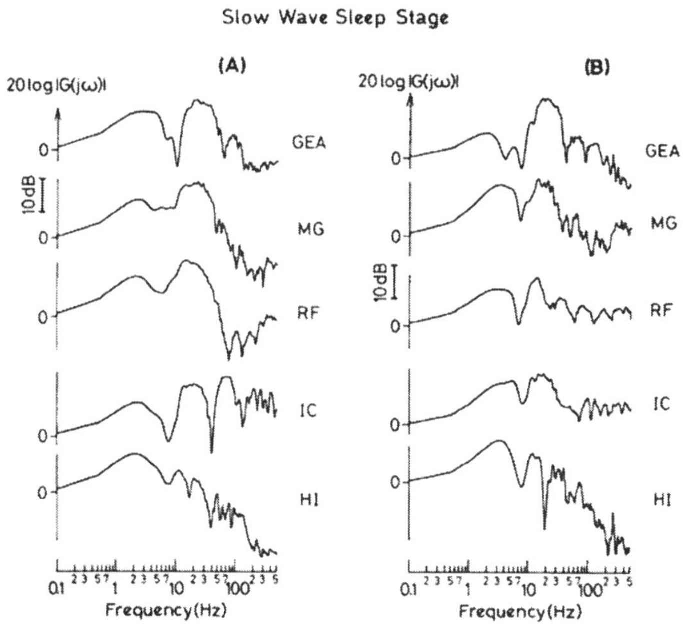

Fig. 5.7. Two sets of simultaneously obtained amplitude–frequency characteristics of different brain nuclei of the cat determined by the TRFC method and using the selectively averaged transient responses of Fig. 5.4. Direct computer plottings. *Along the abscissa* is the input frequency in logarithmic scale, *along the ordinate* the potential amplitude, $|G(j\omega)|$, in decibels. The curves are normalized in such a way that the amplitude at 0 Hz is equal to 1 (or $20 \log 1 = 0$). (From Başar E. et al. (1975) *Biological Cybernetics* 20: 161–169.)

40 Hz) were present in all the amplitude characteristics of GEA ("alpha–beta maximum"). The maximum around 40 Hz, which was present during the waking stage (see Chap. 1), was absent, and the maximum between 50 and 80 Hz was either highly attenuated or nonexistent. The maximum between 8 Hz and 40 Hz with peaks at 12–15 Hz was usually the dominant maximum, the delta maximum being less prominent.

2. **Medial geniculate nucleus (MG).** The amplitude–frequency characteristics of the medial geniculate nucleus (MG) were similar to those of the acoustical cortex during slow wave sleep: Maxima at 1–3 Hz and 7–40 Hz were observed. In this nucleus, too, the high-frequency mechanisms were either highly attenuated or missing. In some of the curves, the maximum in the delta frequency range was the dominant one.
3. **Mesencephalic reticular formation (RF).** A maximum in the 1–3 Hz delta frequency range existed in all the curves. As in the case of the GEA and MG, an alpha–beta maximum in the frequency range of 8–40 Hz was always present. However, this alpha–beta filter in the RF showed narrow filtering properties when compared to the MG and the GEA. The 30 Hz maximum, consistent during the waking stage (Başar et al. 1975b), disappeared. The relevant 50 Hz maximum of the waking stage was unstable during slow wave sleep and sometimes entirely missing.
4. **Inferior colliculus (IC).** Amplitude characteristics were similar to the characteristics of the cortex (see GEA, above) between 1 Hz and 40 Hz. The high-frequency maximum in the range of 70–80 Hz observed during the waking stage was sometimes not existent during SWS. Although higher–frequency maxima were also attenuated in the amplitude characteristics of the IC, this attenuation was not as pronounced as in other studied nuclei.
5. **Hippocampus.** The delta maximum (1–2 Hz) was usually the most prominent maximum in the hippocampus (HI). The maximum around the alpha–beta frequency range had narrower filtering properties in comparison to other studied nuclei.
6. **Consistent selectivities.** We have already described the consistent selectivities during the spindle sleep stage (Figs. 5.5, 5.6). During SWS the following consistent maxima can be indicated (Fig. 5.8). In all the nuclei, a maximum between 1 and 3 Hz is observed. The most important 1–3 Hz maxima are seen in the medial geniculate nucleus and in the hippocampus. In these nuclei, this delta maximum is often the dominant one; while in the inferior colliculus, this maximum is less relevant. The maxima around 80–100 Hz in the cortex and 70 Hz in the inferior colliculus, which are also seen during the waking stage, remain almost unchanged during SWS. The 180 Hz maximum of the inferior colliculus is also a consistent maximum during SWS. The higher-frequency maxima above 40 Hz are absent in the reticular formation. The alpha–beta maximum described in the previous section is among the consistent selectivities. This alpha–beta filter has a narrow band in the reticular formation.

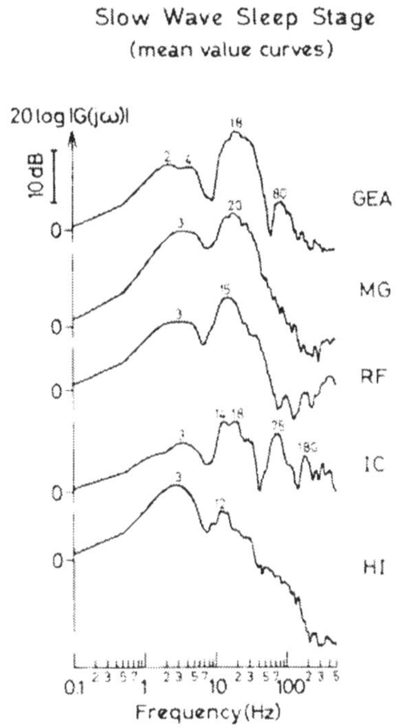

Fig. 5.8. Mean-value curves of simultaneous amplitude frequency characteristics of different brain nuclei of the cat obtained from six experiments with auditory stimulation during the slow wave sleep stage. Direct computer plottings. *Along the abscissa* is the input frequency in logarithmic scale, *along the ordinate* the potential amplitude, $|G(j\omega)|$, in decibels. The curves are normalized in such a way that the amplitude at 0 Hz is equal to 1 (or $20 \log 1 = 0$). (From Başar E. et al. (1975) *Biological Cybernetics* 20: 161–169.)

5.4 Application of Combined Analysis Procedure to the Spontaneous and Evoked Activities

5.4.1 Simultaneously Recorded and Filtered EEG-EP Epochs (1–45 Hz)

Figure 5.9 shows the frequency components of five EEG-EP epochs that were simultaneously recorded from the five brain structures studied. The EEG-EP epochs were pass-band filtered (without phase shift) within the three different band limits of above 1–8 Hz, 8–20 Hz, and 25–45 Hz. The band limits of these filters were chosen according to the band limits of the selectivities depicted in the amplitude–frequency characteristics of the same cat.

The filtered EEG-EP epochs illustrated in Fig. 5.9 present strong resonance phenomena in the 1–8 Hz, 8–20 Hz, and 25–45 Hz frequency ranges: Most of the enhancement factors have values varying between 1.5 and 5.8. The time locking is perfect in most of the nuclei in the presented frequency channels (additional information demonstrating perfect time locking). The congruency in the responses of the various nuclei in the delta–theta and

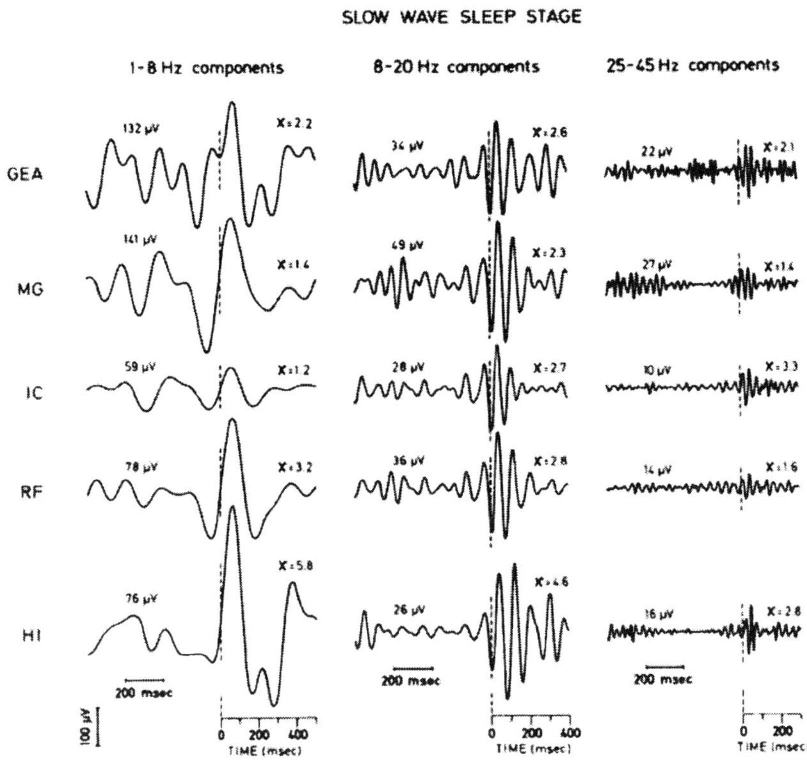

Fig. 5.9. Three components of a typical set of single EEG-EP epochs, which were simultaneously recorded from various brain structures of the cat during the slow wave sleep stage and were filtered without phase shift. The pass-bands of the theoretical filters applied were chosen according to the shared band limits of the selectivities in the respective amplitude–frequency characteristics (not shown). The exact frequency limits of the pass-bands, which are approximately quoted at the top, are as follows for different nuclei: GEA (1–8 Hz, 9–25 Hz, 30–50 Hz); MG (1–7 Hz, 8–20 Hz, 25–40 Hz); IC (1–7 Hz, 8–20 Hz, 25–50 Hz); RF (1–7 Hz, 8–20 Hz, 23–40 Hz); HI (1–9 Hz, 10–22 Hz, 23–47 Hz). The rms values of the spontaneous parts and the enhancement factors are also given above each filtered EEG-EP epoch. (From Başar E. et al. (1975) *Biological Cybernetics* 20: 161–169.)

alpha–beta frequency channels are, apart from the absolute values, almost perfect.

As was the case in studies of the waking stage, the enhancement factors showed a large variability during the same experimental session within a homogeneous SWS-stage. Figure 5.10 illustrates this variability for a given nucleus in the hippocampus for the activity in the 1–8 Hz frequency range. Six filtered EEG-EP epochs in the frequency range of 1–8 Hz are shown. The enhancement factors vary between 1 and 6.4 for the six illustrated examples.

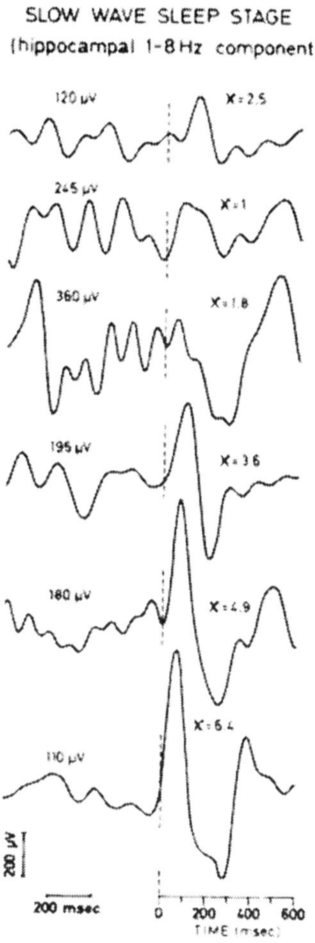

Fig. 5.10. Six examples for the 1–8 Hz frequency component of the EEG-EP epochs recorded from the cat hippocampus during the slow wave sleep stage. All examples are from a single experimental session with the same cat. The rms values of the spontaneous parts and the enhancement factors are given *above* each filtered EEG-EP epoch. (From Başar E. et al. (1979) *Biological Cybernetics* 34: 21–30.)

A time locking exists after stimulation. We should further note that large responses were usually observed in cases with low spontaneous activity prior to stimulus (the last three EEG-EP epochs in Fig. 5.10). In cases of synchronization of the EEG with high-voltage delta–theta activity, the enhancement factors were usually low (see Chap. 1 for results during the waking stage). We must also mention that the time-locked hippocampal responses can reach magnitudes up to 900 µV in this low-frequency range.

5.4.2 The Coherence Functions Between All Possible Pairings of Recording Electrodes

In Fig. 5.11 the coherence functions that were computed from the spontaneous and evoked activities of all possible electrode pairings in the five studied brain

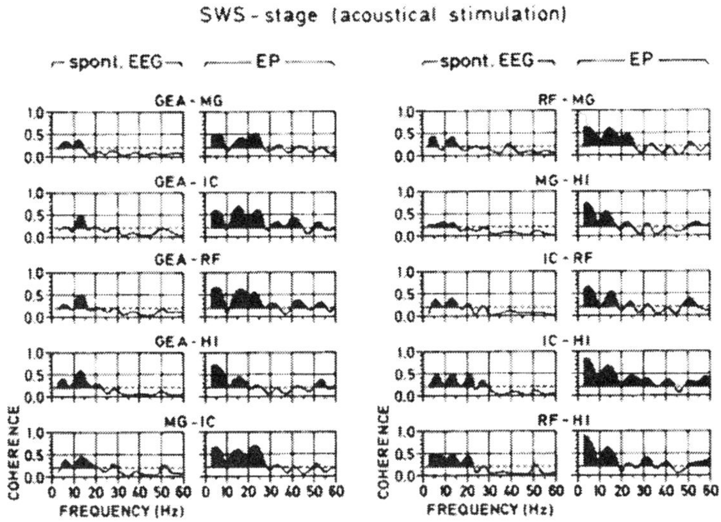

Fig. 5.11. A typical set of coherence functions computed from the spontaneous and evoked potentials of all possible pairings of the studied brain structures during the slow wave sleep stage. The scale is indicated *at the bottom*. *Along the abscissa* is the frequency from 0 to 60 Hz; *along the ordinate* is the coherency between 0 and 1. The *horizontal broken lines* indicate the significance level, which is 0.2 for all the plots. The *area under the coherence function* is darkened only if the curve surpasses this level. In order to facilitate a comparison between the coherence values computed from spontaneous and evoked parts of the EEG-EP epochs, the respective coherence functions are presented *adjacently* as couples for all the pairings of recording electrodes. (From Başar E. et al. (1975) *Biological Cybernetics* 20: 161–169.)

structures are comparatively illustrated for frequencies up to 60 Hz as the results of a typical experiment. For all the coherence functions, a significance level of 0.2 was attained due to the averaging and smoothing processes in their computation. Thus, the area under the coherence function is darkened only if the curve surpasses this value, in order to give emphasis to those parts of the curves above the significance level.

Table 5.1 presents the averaged coherencies for frequencies below 8 Hz (delta–theta) for the frequency band of 8–18 Hz (alpha) and 20–27 Hz (beta). To obtain averaged coherency, we defined the mean value of the corresponding coherencies computed from the results of 11 experimental sessions with 11 cats. Başar et al. (1979b) have also evaluated a mean value of averaged coherencies for all possible pairings of the recording electrodes. The value is called the overall coherency of all the studied nuclei.

During the SWS-stage, the alpha–beta range of 8 Hz to 25 Hz is not the only frequency band where the coherencies between evoked responses can reach significantly high values. For frequencies below 8 Hz (delta–theta) range, where the coherencies between evoked potentials are quite low during

Table 5.1. Averaged magnitudes of coherences in pairings of spontaneous EEG and evoked potentials recorded from the respective brain structures of 11 cats during slow wave sleep. The mean values of the averaged coherences in all possible pairings of nuclei are also given, in the delta–theta, alpha, and beta frequency rnages, as the overall coherency, which may represent the coherence state of the brain as a whole. Note that the maximal increase in overall coherency is in the delta–theta frequency channel during SWS. (From Başar et al.: *Biol Cybern* 34 (1979): 21–30)

Parings of Brain Nuclei	$f < 8$ Hz Delta-theta channel		8–18 Hz Alpha channel		20–27 Hz Beta channel	
	EEG	EP	EEG	EP	EEG	EP
GEA-MG	0.15	0.38	0.18	0.40	0.06	0.23
GEA-IC	0.12	0.36	0.28	0.33	0.11	0.28
GEA-RF	0.21	0.46	0.28	0.53	0.14	0.31
GEA-HI	0.21	0.54	0.31	0.27	0.14	0.11
MG-IC	0.14	0.41	0.21	0.26	0.17	0.28
MG-RF	0.39	0.60	0.28	0.59	0.17	0.32
MG-HI	0.14	0.55	0.13	0.26	0.06	0.12
IC-RF	0.19	0.52	0.34	0.32	0.13	0.29
IC-HI	0.26	0.65	0.29	0.29	0.15	0.15
RF-HI	0.32	0.69	0.22	0.30	0.15	0.13
Overall coherency	0.21	0.52	0.25	0.36	0.13	0.22
Factor of overall coherency increase		2.5		1.44		1.69

the waking stage (see Chap. 1), the coherence functions depict prominent increases during the SWS-stage. This observation indicates that the maximal coherence between the responses of brain nuclei, which takes place in the alpha–beta frequency band during the waking stage, is shifted to the delta–theta frequency range with slow wave sleep. In this sleep stage also, the spontaneous activities of the studied nuclei can show, unlike the waking stage, significant coherence in the delta–theta and alpha–beta frequencies. The coherency values of evoked responses in the alpha–beta band, however, are not as high as those encountered during the waking stage. Therefore, upon stimulation, the overall coherency can increase only by a factor of 1.44 in the alpha range and by a factor of 1.69 in the beta range during the SWS-stage, whereas these factors have been determined as 6.75 and 4.38, respectively, for the waking stage.

5.5 Further Comments on the Component Analysis and the Real Responses in Evoked Potentials

The analysis during the SWS-stage emphasizes once more that the frequency components in frequency ranges higher than 30 Hz are completely separate and independent from the lower-frequency components, and not arbitrarily chosen Fourier components or components that might result from the use of mathematical tools. As the present results reveal, the 1–8 Hz and 10–25 Hz components keep their frequency positions during SWS and show increases in magnitude when compared with the waking stage, whereas the higher-frequency components (40 Hz in HI and GEA, 70 Hz in IC, and 50 Hz in RF and MG measured during waking) do not increase in magnitude and suffer from slight frequency shifts.

5.6 Interpretation of Results on Stereodynamics in the Auditory Pathway During the Slow Wave Sleep Stage

5.6.1 Synchronization and Coupling of Resonances in the Responses of Various Brain Centers in Alpha and Beta Frequency Ranges

The immediate observation of Fig. 5.10 shows that the studied brain nuclei are similar to coupled oscillators without phase lag in the 8–20 Hz frequency range. With respect to amplitude–frequency characteristics, we know that there are usually two distinct selectivities centered on alpha and beta frequencies and that the averaging procedure causes a fusion of these frequency responses. During the SWS-stage, similar effects of synchrony, as in higher frequencies between 8–20 Hz, are observed in the delta and theta frequency ranges (Fig. 5.9). Based on experimental findings, we have already postulated the existence of a central command mechanism that regulates the responsiveness state of the brain during the waking stage (see Başar 1980, Chap. 1). The findings presented in the present Chap. support the existence of similar mechanisms during SWS: There are periods in which the brain responses depict large enhancement factors in all the nuclei. There are, however, other periods in which the enhancement factors of all the nuclei are relatively small (see Başar 1980). In other words, the state of responsiveness in various nuclei are synchronized during SWS also. We must also emphasize that the magnitude of the brain response changes during SWS, as does the absolute magnitude of the spontaneous activity.

The synchronization of responses recorded simultaneously from several brain centers (in alpha and beta frequencies) is visible in Fig. 5.10. (During SWS, responses in the delta and theta frequency ranges are also synchronized.) There is no phase lag between responses of all the studied nuclei, although no significant phase congruency could be observed for the spontaneous activities of these nuclei in the time domain.

The increased coherence induced by stimulation is very significant for the delta–theta components during SWS. The overall coherency increases from 0.21 to 0.52 upon stimulation (see Table 5.1). This fact shows that during SWS, the maximal common power of the synchronized responses is accumulated in the delta and theta frequency ranges, and not in the alpha frequency range. This finding is interesting, since the absolute magnitudes of all the delta responses reach their largest voltages during SWS. This fact leads to the conclusion that the responsiveness levels of all the brain centers affect each other in a significant manner during SWS in this low-frequency channel. The high overall coherency of the responses in low frequencies supports this view.

5.7 Human Frequency Responses During SWS Sleep

Experiments on human auditory evoked potentials during various sleep stages have been carefully and intensively analyzed by Röschke et al. (1995). In this chapter we only mention some of the relevant findings without going into details. Figure 5.12 illustrates amplitude–frequency characteristics of a subject during SWS sleep upon auditory stimuli (Röschke et al. used the same programs for computing the amplitude frequency characteristics as we did for cat data). The human amplitude frequency characteristics also depict dominant delta responses during the SWS-stage, thus confirming the experiments performed on sleeping cats. It can therefore be hypothesized that the dynamics of potentials in the human and cat brains obey general transfer functions in which the information processing in the delta frequency range plays a major role.

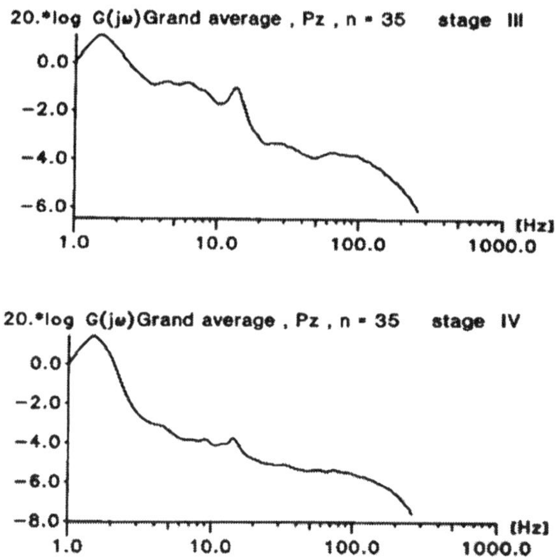

Fig. 5.12. Amplitude–frequency characteristics computed from auditory EPs recorded in human subjects during slow wave sleep (courtesy of Dr. J. Röschke, Johannes-Gutenberg-University Mainz; cf. Röschke J. et al. (1995) *Electroencephalography and Clinical Neurophysiology* 96: 390–397.)

6. Dynamics of Potentials from Invertebrate Brains

Atsuko Schütt, Theodore H. Bullock, and Erol Başar

6.1 Introduction

Since Adrian (1931, 1937) and Adrian and Matthews (1934), among others, drew attention to some evidence of similarity and, mainly, differences between invertebrates (insects) and mammals, little progress has been made. Bullock (1945, 1974, 1983, 1984a, 1984b) and Bullock and Horridge (1965) pointed out evidence for both widespread similarity among vertebrates and general differences between them and invertebrates.

Many invertebrate ganglia are simple-structured nerve cell populations consisting of a relatively small number of cells. Lower-level ganglia, analogous to the spinal cord, plus a few brains of the less advanced invertebrates, consist of up to a few thousand neurons. The brains of most insects, lobsters, and cephalopods have from fairly complex up to quite complex structures and tens to hundreds of thousands of cells. *Helix* is intermediate, probably $> 10^4$. The cat or human brain – a complex, three-dimensional network – comprises an astronomical number of neurons (on the order of up to 10^{12}).

The significance of studying an invertebrate CNS as a brain model may be supported by the following observation made by Bullock (1988): "When ongoing compound field potentials in the brains of invertebrates and vertebrates are compared, the recorded differences between vertebrates and most invertebrates are not due to brain size, cell size, number or density, lamination, or single cell power spectrum, but are primarily due to assembly properties, i.e., to cooperativity." We have therefore addressed the following questions: (i) Is it possible to record, in the extremely small *Helix* ganglion (< 2 mm in diameter), ongoing and stimulus-evoked field potential activities that are reproducible? (ii) What properties do these activities show in frequency and amplitude characteristics? How are they in comparison to those of other species, such as Aplysia, fish, cat, and human? (iii) Are there some commonalities among the characteristics of evoked potentials of the vertebrate and the invertebrate representing different stages of evolution? The rationale of initiating experiments on the gastropod ganglia using systems-theoretical analyses is based on the descriptive efficiency of these methods outlined in Chap. 4 of Vol. I.

6.2 Anatomy and Physiology of the Invertebrate (Gastropods) Nervous System

The nerve cells of gastropods, particularly *Helix*, have been extraordinarily favored for studies in neurocytology and electrophysiology. The sea slug *Aplysia californica*, which has long been favored as well, and the land snail *Helix pomatia* both belong to the phylum of molluscs and the class of gastropods.

The ganglia of the central nervous system (CNS) are organized such that together with the commissures and connectives, they form a circumesophageal ring around the intestine and the esophagus. Figure 6.1 illustrates the anatomy of *Helix* as a prototype. The complement of the ganglia comprises the following series of nodelike clusters of cells and neuropile with connectives and commissures: (i) a pair of cerebral ganglia; (ii) a pair of buccal ganglia; (iii) a pair of pleural ganglia; (iv) a pair of pedal ganglia; (v) a pair of parietal ganglia; (vi) the visceral ganglion. On the upper side of the esophagus lie the cerebral ganglia, from which strands of nerves (connectives) extend from the right and left of the esophagus to the subesophageal complexes, which are easily distinguishable as two masses: the abdominal

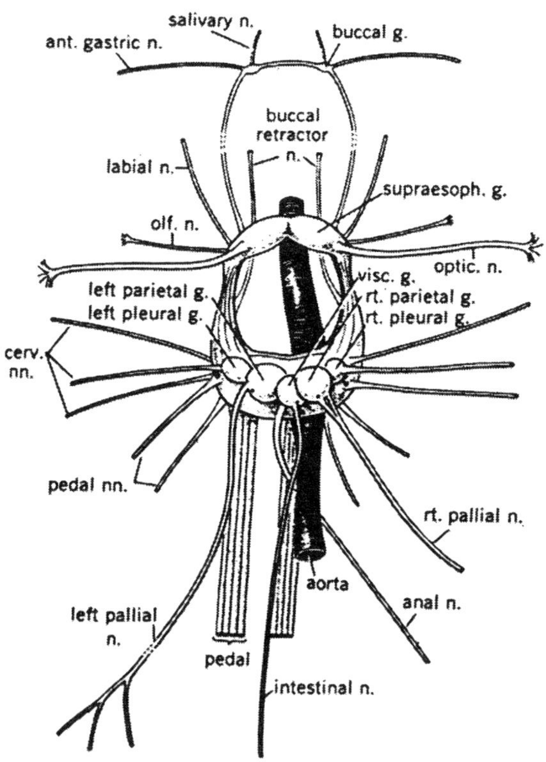

Fig. 6.1. The central nervous system and major nerves of a young *Helix aspersa*. (Courtesy G. Kerkut.)

ganglia complex comprising the pleural, parietal, and visceral ganglia at the dorsal side and the pedal ganglia at the ventral side. The buccal ganglia, the smallest of all ganglia, are located under the pharynx, where the intestines emerge, and are connected with the cerebral ganglia.

The most highly developed gastropod brain known is that of *Helix*. It has a distinct procerebrum with globuli and dense neuropile and postcerebrum with pleural, pedal, and commissural lobes and mesocerebrum (Bullock and Horridge 1965). From the cerebral ganglia nine nerves (eight paired and one nonpaired) emerge: N. olfactorius; N. opticus; N. peritentacularis externus; N. peritentacularis internus; N. acusticus; N. labialis externus; N. labialis medialis; N. labialis internus; N. penis (nonpaired). The cerebral ganglia, from which connectives to the buccals, pleurals, and pedals also spring, innervate eyes; statocysts; head tentacles; skin; some muscles of lips, head, and neck; and the penis region. The procerebrum of the Mollusca has sometimes been designated as the *"intelligence area"* (Smidt 1900; Haller 1913). The postcerebrum, the largest part of the cerebral ganglion, may be considered as the core of the total nervous system with regard to its function: Not only are all nerves of the cerebral ganglia rooted in the postcerebrum, but all centers are connected with this part of the brain as well. Besides the function of higher sensory association center, a cardioregulatory function and regulation of muscle tonus are also involved.

6.2.1 The Abdominal Ganglia Complex

The abdominal ganglia complex and the pedal ganglia are both covered by one sheath, so that the whole mass may look like one unit having roughly the form of an equilateral triangle (2 mm in length and ca. 0.5 mm in thickness; Schmalz 1914). Five ganglionic subunits, bordered by tissue septa, form the abdominal ganglia complex: pleurals, right and left; parietals, right and left; visceral, unpaired).

The commissures within the abdominal ganglia are organized in such a way that the neighboring ganglia are all connected with one another. The pleural ganglia have a very special function, building a kind of relay station between the cerebral ganglia and the abdominal ganglia complex as well as between the cerebral ganglia and the pedal ganglia. From each of the ganglia spring two nerves: N. musculi collumellaris and N. musculi retractoris pharyngealis. The functions of the parietal ganglia lie in the innervation of the mantles and the lung organs. In addition to a minor nerve innervating the aorta, N. aortae, two strong nerves innervating the right mantle leave the right ganglion: N. pallialis dexter externus and N. pallialis dexter internus. From the left ganglion, close to the border of the pleural ganglion, emerges a strong nerve, N. pallialis sinister, which innervates the left mantle and body wall.

The visceral ganglion is characterized by physiological multifunctions: not only innervation of the skin and body wall, but regulation of the reproductive organs, the digestive tracts, kidney, liver, and heart as well. The unpaired visceral ganglion is in connection with the adjacent parietal ganglia. From this spring three nerves: N. cutaneus pallialis, N. intestinalis, and N. analis.

6.2.2 The Pedal and Buccal Ganglia

The pedal ganglia form the ventro-oral part of the subesophageal ganglia mass. The nerves springing from the pedal ganglia can be divided into two groups: (i) skin nerves and (ii) foot nerves. The skin nerves innervate the skin and the reproductive organs, whereas the foot nerves innervate the foot muscle as well as the foot vessels. The buccal ganglia supply an extensive area: Besides pharyngeal walls, muscles, and salivary glands, their nerves are distributed to a plexus on the esophagus and stomach and perhaps other viscera.

6.2.3 Microscopic Anatomy

The most important cellular building blocks of the gastropod central nervous system are the unipolar ganglion cells. The ganglia of gastropods, such as *Aplysia* and *Helix*, are built from only a small number of neurons (several thousand). For the details of the microscopic anatomy and the cytology of *Helix pomatia*, the reader is referred to the pioneering work of Kunze (1917) and, of gastropods in general, to the book by Bullock and Horridge (1965).

Cell size. In the CNS of *Helix* are nerve cells of different sizes (6–400 μm) varying from the smallest cells (6–7 μm; de Nabias 1894) in the globuli cell mass of the procerebrum to the largest (so-called giant cells), ranging from 260 μm (Kunze 1921) to 400 μm (Boehmig 1883) in the parietal. Giant cells are found in all the central ganglia in small numbers: fewer in the buccal (1), pleural (1 to 2), and cerebral ganglia; most in the visceral (22), parietal (9 to 10), and pedal (10; Bullock and Horridge 1965; Kunze 1917).

Synapses. In the rind of the ganglia where cell somata are present, there are usually no nerve endings or synapses. The main synaptic field is, in number and volume, the neuropile of the core. Most synapses are on the collateral processes of the unipolar cells, often called dendrites although not cytologically equivalent to vertebrate dendrites.

6.3 Materials and Methods

Helix pomatia was chosen as the experimental animal due to its accessibility and the well-developed ganglion cell mass. Electrode sites are selected to avoid loci dominated by large unit spikes as well as loci with little activity. In one study, electrical stimulation of submaximal intensity was applied to different nerves (intestinal, anal, right and left pallial), and various recording

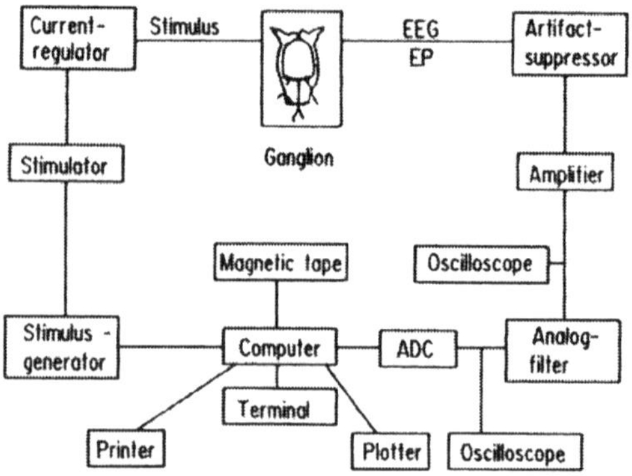

Fig. 6.2. The experimental setup. Recording, electrical stimulation, and off-line data processing can all be handled by this facility.

loci (cerebral, pleural, parietal, and visceral ganglia) were chosen (Sturbeck 1988). In the other study, however, the recording was carried out solely on the visceral, ganglion applying electrical stimulation of two-times threshold (Schütt and Başar 1992).

Figure 6.2 illustrates the experimental setup employed for *Helix pomatia*. The methodological details have been presented elsewhere Schütt et al. 1992; Schütt and Başar 1992).

6.4 Results

6.4.1 Ongoing Compound Field Potentials

In our laboratory, ongoing compound field potentials could be recorded from all the *H. pomatia* ganglia attempted (cerebral, pleural, parietal, and visceral ganglia (Röschke 1988; Sturbeck 1988; Schütt et al. 1992), confirming the results of Bullock (1945) on another pulmonate gastropod, *Ariolimax*). During our recent experiments with more than 70 snails, we observed that spontaneous firing of the visceral ganglion is of two types: mostly with a modest amount of spikes and occasionally (15%) with continuously bursting spikes. Power-spectral density of the former type reaches a peak below 50 Hz (less than 5 dB; RMS (root-mean-square) voltage of activity at 0 dB corresponds to ca. 2 μV; Fig. 6.3), whereas that of the latter type attains a first major peak below 30 Hz and a second one between 50 and 150 Hz. In the latter case, the power extends strongly into the range 200–300 Hz, falling at least 10 dB at 300 Hz (Fig. 6.4). The *H. pomatia* ganglia, though totally isolated from the peripheral organs, can be electrically active by their own intrinsic activity for as long as a few to 15 hours. The records of spontaneous firing reveal pass-band filtered, EEG-like low-amplitude slow waves ($< 10\,\mu$V RMS at < 50 Hz), which are usually superimposed on spikes (Fig. 6.5). The bath noise level, measured with one electrode on an inactive preparation and an-

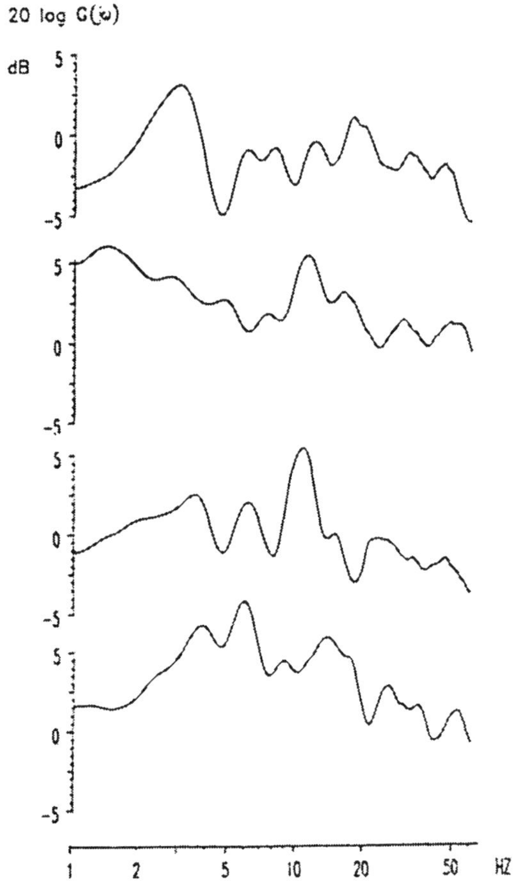

Fig. 6.3. *Helix pomatia.* Power spectrum of a sample of 80 s each of the spontaneous activity in the neuropile of the isolated visceral ganglion. Four different preparations. The signals were wideband filtered from 1–125 Hz. An example of the time signal from the curve second from the bottom is shown in Fig. 6.5. The curves are normalized at the amplitude value of 1 Hz.

other in bath solution, is usually less than 1.5 μV RMS at < 50 Hz (Schütt et al. 1992).

6.4.2 Spikes

Since the records of spontaneous field activity in the *Helix* ganglia are, as in *Aplysia*, dominated by either small or modest-sized spikes, a question was raised: *Are these spikes triggered by field potentials, or are the field potentials results of the spikes (afterpotentials)?* To answer this question, Röschke (1988) examined the correlation between the 5 Hz low-pass filtered time signal of field potential and the envelope of the high-passed part of the signal using the correlation coefficient. When the time signal has some small to modest-sized spikes, the correlation coefficient r is smaller than 0.1, which means that the spike activity and the slow field potential are essentially independent phenomena. However, when the field activity is predominantly of

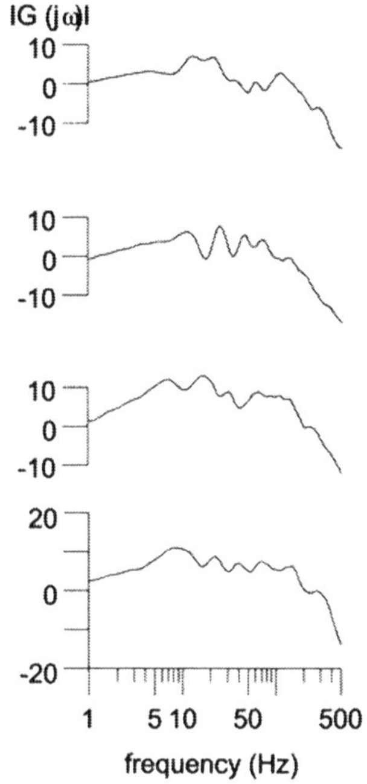

Fig. 6.4. *Helix pomatia.* Power spectrum of a sample of 10 s each of the ongoing activity with a high amount of bursts in the isolated visceral ganglion. Four different preparations. The curves are normalized at the amplitude value of 1 Hz.

large unit spikes, r is 0.5, which means that the slow potential fluctuations are very strongly correlated with the unit spikes observed. This finding agrees with the observation of Bullock and Başar (1988) in *Aplysia* that the low-amplitude slow waves are not attributable to envelopes of spikes, but they may come from the afterpotentials of the larger units.

6.4.3 Relationship Between EEG of Vertebrates and Field Potential Fluctuations of Invertebrates

On comparing the slow rhythmical potential fluctuations of the *Helix* and *Aplysia* ganglia with those of the higher vertebrates, it must be kept in mind that we are dealing with a CNS that is totally disconnected from the peripheral organs. This means that the sensory association fields no longer have exogenous inputs in natural forms. And yet, distinct fluctuations are revealed in the low-amplitude field potentials of the invertebrate ganglia when the signals are arbitrarily pass-band filtered: for *Helix pomatia* in the 1–4 Hz, 4–8 Hz, 8–15 Hz, 15–30 Hz, 30–48 Hz, and 52–125 Hz ranges (Röschke 1988; Schütt et al. 1992) and for *Aplysia* in the 2–5 Hz, 5–10 Hz, 10–20 Hz, and 40–80 Hz ranges (Bullock and Başar 1988).

The different types of pass-band filters chosen for *Aplysia* and *Helix* are simply arbitrary and not due to any consistent, reproducible species differences in fluctuations. These fluctuations are solely of intrinsic cellular activities of small cell populations independent of sensory modality. What

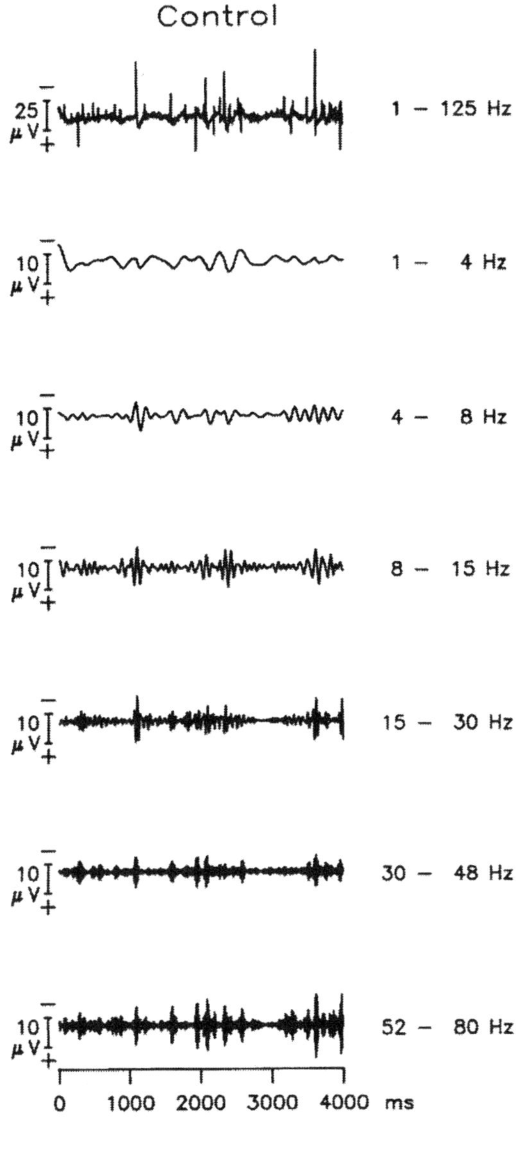

Fig. 6.5a. *Helix pomatia*: A typical time signal (ca. 4 s) of the ongoing activity with a modest number of spikes in the neuropile of the isolated visceral ganglion; wideband (1–125 Hz) and passband components (1–4, 4–8, 8–15, 15–30, 30–48, and 52–80 Hz).

generates the field potential then? According to Bullock (1989), several postsynaptic events, such as a long-lasting hyperpolarization, plateau potentials, pacemaker potentials, subthreshold local potentials, and long-lasting afterpotentials of spikes, as well as potentials caused by the glial cell membranes, largely contribute to the field potential.

The relative weakness of slow waves in most invertebrates might be due to relatively little synchrony among neurons. In contrast to *Aplysia* and presumably most other invertebrates, vertebrates have a significant degree of

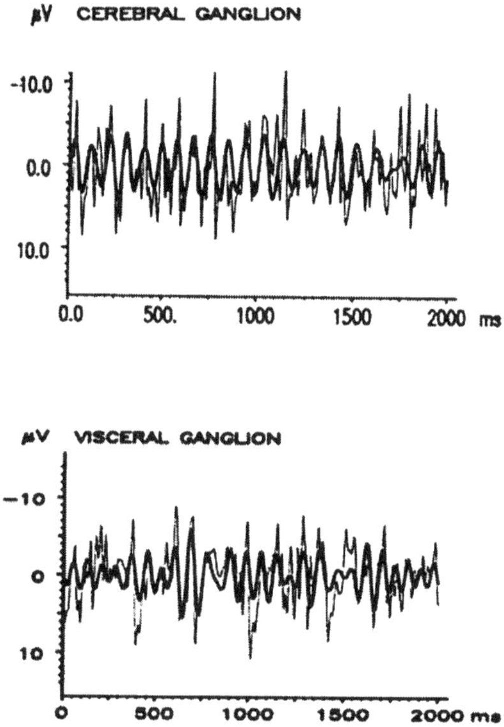

Fig. 6.5b. *Helix pomatia*. Field potential fluctuations in different ganglia; A sample of 2 s each. *Top:* the cerebral ganglion; *Bottom:* the visceral ganglion; *Thin line:* wide-band component (1–50 Hz); *fat line:* narrow-band component (8–15 Hz). Note that the 8–15 Hz fluctuation is not just a filtered white noise, since it is also clearly visible in the wide-band record.

synchronization of low-frequency components (< 25 Hz), judging by the mean coherence between electrodes less than 1 mm apart. Another basic difference is that power spectra of higher centers in arthropods and gastropods are much stronger, from 50–500 Hz, than in the vertebrate higher centers, resembling the cerebellum and spinal cord (Bullock and Başar 1988). Let us now compare the fluctuating activity observed in the invertebrate with the mammalian cortical EEG. At this moment we confine ourselves to the signals from the cerebral and visceral ganglia, which are filtered in the 8–15 Hz range, and compare them with the 8–13 Hz activity in the brains of higher vertebrates. In Fig. 6.5a are illustrated the 8–15 Hz apparent rhythms from both the cerebral and the visceral ganglia of *Helix pomatia* superimposed on the wide-band (1–50 Hz) filtered time signals (Röschke 1988; Schütt et al. 1992). This activity has also been recorded in the *Aplysia* cerebral ganglia (Bullock and Başar 1988). The 8–15 Hz fluctuation is not just a filtered white noise, since it is also clearly visible in the wide-band record. Evidences provided by these findings strongly imply that the 8–15 Hz activity (alpha band), which is known as one of the most important frequency components of mammalian brain activity, is also generated in much smaller and less-developed CNS, such as isolated invertebrate ganglia. It remains to be tested whether there are characteristic valleys on each side of this band.

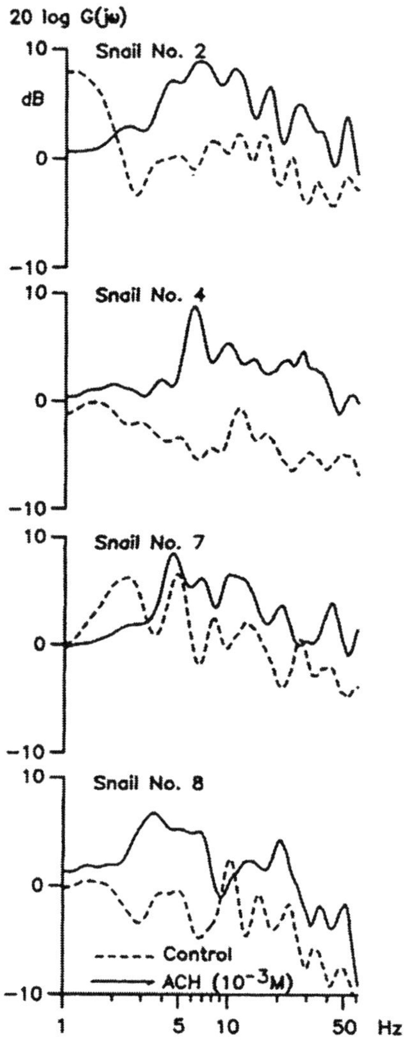

Fig. 6.6. Power spectra of the ACh-induced activity in four preparations. Each curve represents a power spectrum averaged from 20 epochs (81.92 sec).

Our recent experiments with some neurotransmitters (Schütt et al. 1992) have provided additional evidences that the transmitter of the cholinergic system, acetylcholine, which is known to be involved in sensory as well as cognitive processes of the mammalian brain, induces frequency responses in the 2–20 Hz range (Fig. 6.6).

6.5 Potentials Evoked by Means of Electrical Stimulation

6.5.1 *Aplysia*

The potentials evoked in the *Aplysia* cerebral ganglion by electrical stimulation to the connective show characteristic responses that can roughly be grouped into two types. One group ($n = 8$) exhibits clearly stimulus-evoked amplitude enhancement in the range of 2–20 Hz and additional resonance in the range of 40 Hz, as well as 80 and 100 Hz. Another group ($n = 5$) elicits, in contrast, clearly separated resonances in the ranges of 5 Hz, 10 Hz, and 25 Hz and additional local maxima between 60 and 100 Hz. The amplitude characteristics of two typical responses from both groups are presented in Fig. 6.7 (Sturbeck 1988). Sturbeck calls special attention to the 40 Hz resonance, which also exists in this simplest neuronal network.

6.5.2 *Helix Pomatia*

Sturbeck (1988) recorded potentials evoked by electrical stimuli (maximal intensity, 30 mA, of 1 ms duration) from five different ganglia (right and left pleural, right and left parietal, and visceral) stimulating three different nerves

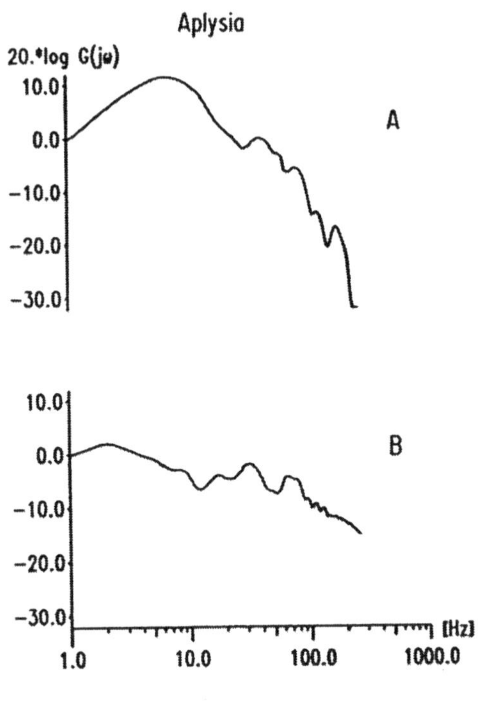

Fig. 6.7. *Aplysia californica*: Amplitude–frequency characteristics of potentials evoked in the cerebral ganglion by electrical stimulation to the connective. Curve A shows a typical experiment of group 1 and B that of group 2.

(N. pallialis dexter, N. analis, and N. pallialis sinister). The right pleural ganglion responds, accompanied by amplitude enhancement, with power increase in the frequency bands of 2–10 Hz and 20–30 Hz. The right parietal ganglion, on the other hand, elicits power increase in the prestimulus spontaneous activity that is evenly distributed over the entire range of 1–100 Hz. The frequency–amplitude characteristics of the poststimulus evoked activity show maximizations in the ranges around 2 Hz, 13–23 Hz, and 40–65 Hz.

The left parietal ganglion stimulated through the left pallial ganglion responds differently: There is amplitude increase in the evoked potentials (in the 2–50 Hz range). However, the frequency amplitude characteristics show no resonance phenomena of discrete frequency components. The left pleural ganglion stimulated through the left pallial nerve responds with amplitude enhancement. The frequency characteristics of the poststimulus evoked activity reveal the amplitude maxima in the ranges of 2 Hz, 8–12 Hz, 15–28 Hz, 35–55 Hz, and 70–85 Hz.

The evidence seems to suggest that the response of the visceral ganglion to the stimulation through the anal nerve appears in the ranges of 2–8 Hz (max. at 3 Hz), 12–22 Hz (max. at 15 Hz), 32–40 Hz (max. at 35 Hz), as well as in the range of 60–80 Hz. Peaks seem to be rather constant (Fig. 6.10, see below; four grand averages of controls) among a number of preparations and over time.

Sturbeck's results on *Helix* show that "unphysiological" electrical stimulation seems to evoke in different, diverse neuronal populations characteristic field potential fluctuations, similar to those in mammals, having peaks at the same frequencies in the bands 1–8, 8–15, 15–30, 30–48, and 52–100 Hz. It is his impression that there are typically ca. five peaks clustering around the centers of these bands, not continuously distributed but with relatively consistent intermediate bands representing valleys. Further tests are obviously in order, especially with more physiological stimuli. Concerning a slow evoked potential caused by an electric shock to a connective in the *Aplysia* cerebral ganglion, Bullock (1988) stated, "In spite of a predominantly spiky ongoing background activity, the neuropile of an invertebrate ganglion can respond to a synchronized nerve shock with a slow evoked potential, presumably due to several factors including conduction velocity, wide soma potentials, long afterpotentials, and imperfectly synchronized postsynaptic potentials."

From our recent study (Schütt and Başar 1992) with more than 70 preparations, we present some of the results in the following. Described are the evoked field potentials recorded from the *Helix* visceral ganglion neuropile, to which stimulus was applied via the anal nerve. The experimental conditions differ slightly from those of Sturbeck's in that electrical stimulation was given not repetitively, but randomly, with intervals of 3 to 5 s and that the intensity of the stimulus was 15 mA (two-times threshold), instead of 30 mA.

A typical evoked potential (EP), averaged from 20 epochs, consists of four major negative peaks at latencies of 25, 100, 200, and 650 ms and three positive peaks at latencies of 50, 100 to 150, and 350 ms (Fig. 6.8a). The Fourier

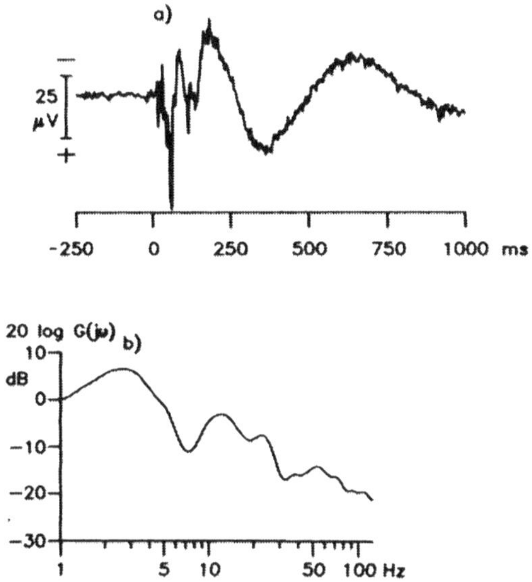

Fig. 6.8a,b. *Helix pomatia.* Potentials evoked in the isolated visceral ganglion by electrical stimulation to the anal nerve. (**a**) averaged evoked potential (AEP) of a typical experiment; (**b**) Frequency–amplitude characteristic of the AEP; (**c**) typical single power spectra of a preparation with an especially high number of bursting spikes. Pre- and post-stimulus activities are separately presented. Note that evoked activity is clearly visible up to 250 Hz.

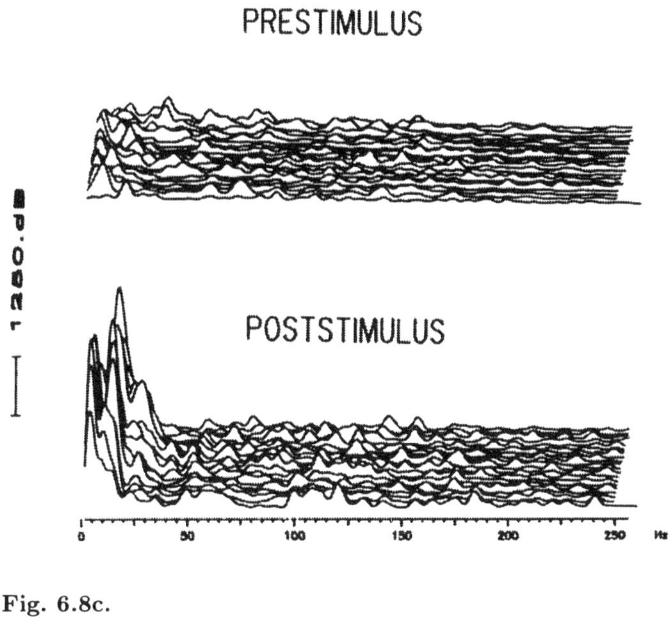

Fig. 6.8c.

transform of the averaged EP (Fig. 6.8b) reveals three major frequency components with maxima at 2.5 Hz, 12 Hz, and 22 Hz, as well as additional minor components in the range 30–80 Hz. These response characteristics are evident also in the Fourier transform of the grand average from six experiments (120 epochs), as illustrated in Fig. 6.10. Figure 6.8c illustrates single power spectra of a preparation with a particularly high amount of bursting spikes. Evoked field activity is, here again, strongest below 50 Hz, but that of the higher-frequency range is clearly visible up to 250 Hz.

6.6 Gamma (30–50 Hz) Activity

The gamma component, more or less pronounced in all the structures examined, should be especially mentioned here, since this band is recurrently discussed in regard to cognitive processes. Figure 6.9 demonstrates 40 Hz burst activity that is evoked by electrical stimulation in the *Helix* visceral ganglion and modulated by dopamine.

The results of the present study lead to a potentially important finding: Prominent activity peaks somewhere in this band appear to be usual in the frequency–amplitude characteristics of evoked potentials from snails, fish, and mammals. On present evidence we cannot assert that this is less true of neighboring bands. There are two opinions about the meaning of this finding, if it is confirmed. Some of us claim that 40 Hz (or any other frequency) is homologous or analogous or equivalent in some sense besides frequency

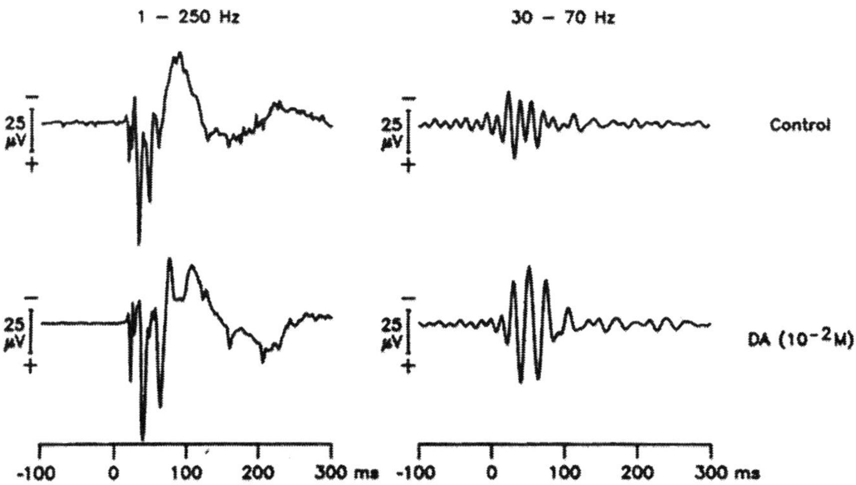

Fig. 6.9. The averaged evoked potentials from an experiment in which 40 Hz activity increases after dopamine: Control, *top*; and dopamine (10^{-2} M), *bottom*. Wide-band filtered from 1–250 Hz, *left*; pass-band filtered from 30–70 Hz, *right*.

in these diverse cases. Others regard the question as completely open until specific commonalities in meaning or mechanism are indicated.

The gamma-band activity seems to occur regardless of the temperature difference, the form of stimulation (direct shock vs. "adequate" or "physiological"), and the place in the pathway (first and second synapse, as in gastropod). This activity is induced or evoked at various brain centers (gastropod ganglia; fish telencephalon and medulla; cat inferior colliculus, medial geniculate, primary cortex, secondary auditory cortex; human cortex, etc.) and in a variety of conditions (well rested, meaningless click, cognitive or ethologically significant event, etc.).

This may indicate a widely ranging function of this frequency component among the different classes of vertebrates and invertebrates. It is obvious that the 40 Hz activity manifests a certain global characteristic in information processing over a wide range of neuronal networks from the simplest of the snail to the most complicated of human. We do not intend to imply that other bands are less important.

The significance of the 40 Hz activity in the brains of different mammals has been hypothesized by several authors (Başar et al. 1987; Başar-Eroğlu and Başar 1991; Eckhorn et al. 1988; Freeman 1975; Gray and Singer 1987; Sheer 1984) as an important coding channel in processing sensory and cognitive information in neuronal networks, and what we report here is an important agenda for future research.

6.7 Neurochemical Modulation

In the regulatory mechanisms of rhythmic phenomena, a significant role is attributed to the chemical substances of the nervous system. These are produced in the neural elements, and after being released from time to time, they can take part in the regulation of rhythms both centrally and at the periphery. For instance, the excitability of the nerve cells is modulated at the synaptic level by release of neurotransmitters such as acetylcholine (ACh), dopamine (DA), noradrenaline (NA), and serotonin (5-HT). These neuromodulators, and others, have long been known to be involved in motor and cognitive processes in mammals.

Concerning invertebrates, chemical control of *"spontaneous"* activity of isolated ganglia in insects was demonstrated by Prosser as early as 1938. In insects, for instance, ACh is involved as a transmitter in the synaptic process of the information processing system (Gerschenfeld et al. 1967). A number of accounts in the literature report the involvement of these neurohormones in endogenous signal processing in the brains of invertebrates such as gastropods and insects (i.e., Gelperin 1989; Hiripi and Salanki 1973; Kerkut et al. 1975; Twarog and Roeder 1957).

Our recent findings on the modulating effects of these neurotransmitters on stimulus-evoked field potentials of the *Helix pomatia* isolated ganglion

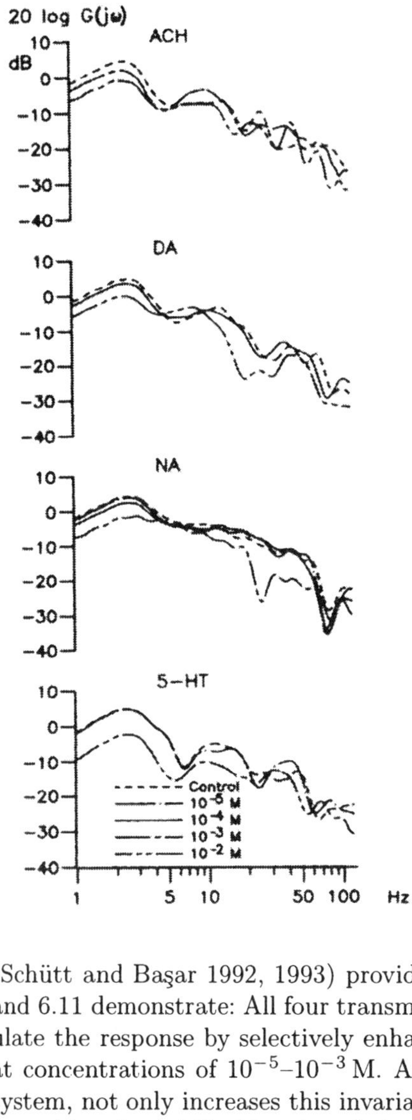

Fig. 6.10. *Helix pomatia*: Amplitude–frequency characteristics of potentials evoked electrically in the isolated visceral ganglion under the influence of neurotransmitters. Grand averages of the experiments shown in Fig. 6.11. ACH: acetylcholine; DA: dopamine; NA: noradrenaline; 5-HT: serotonin.

(Schütt and Başar 1992, 1993) provide the following evidence, as Figs. 6.10 and 6.11 demonstrate: All four transmitters (ACh, DA, NA, and 5-HT) modulate the response by selectively enhancing the 40 Hz activity by 20 to 50% at concentrations of 10^{-5}–10^{-3} M. ACh, the transmitter of the cholinergic system, not only increases this invariant component (40 H activity), but also intensifies the 4–8 Hz (theta) and the 8–15 Hz (alpha) activities by 30%. NA has an additional capability of intensifying the 15–30 Hz activity (beta activity) at 10^{-5} M.

The selective increase of the 4–8 Hz, 8–15 Hz, and 30–70 Hz responses observed in *Helix* under the influence of ACh may be possible evidence for what may happen in the neuronal assemblies of the *Helix* brain by exogenous stimulation. ACh may have a similar function in the mammalian brain. Especially, the increase in the 4–8 Hz and and 30–70 Hz components may add an explanation to the cat hippocampal event-related 3.5–8 Hz (theta) and 40 Hz

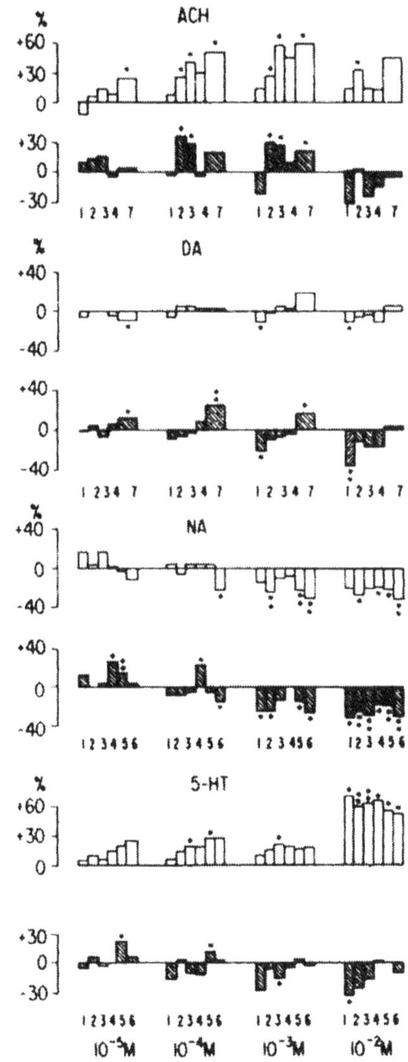

Fig. 6.11. *Helix pomatia*: The isolated visceral ganglion. The effects of ACh, DA, NA, and 5-HT on electrically evoked potentials. Prestimulus ongoing activity (RMS-voltage within 500 ms prior to stimulus) and poststimulus maximum evoked potentials (within 500 ms after stimulus) were separately analyzed. Each value indicates mean percent change from control of prestimulus activity, empty column, and that of poststimulus maximum amplitude, *shadowed column*, averaged from several preparations. Pass-band filters applied are defined as: (1) 1–4 Hz; (2) 4–8 Hz; (3) 8–15 Hz; (4) 15–30 Hz; (5) 30–48 Hz; (6) 52–90 Hz; (7) 30–70 Hz. Differences are based on: ACH, 120 epochs (5 snails); DA, 200 epochs (5 snails); NA, 160 epochs (4 snails); 5-HT, 160 epochs (5 snails). Significance levels: $^{*}P < 0.05$; $^{**}P < 0.01$.

(gamma) activities (Başar-Eroglu 1992), which could be assumed to be of a cholinergic origin.

These basic transmitters tend to excite or to inhibit activity in general. Slow waves (< approx. 50 Hz) are the automatic consequence of activity that is known mostly as firing. The only observation not expected would be differential effects, e.g., enhancing one band and reducing another.

From our observations we may tentatively conclude that all the important invariant components evidenced in the brains of the different species studied here, from mammals to snails, may be modulated by each of the neurotransmitters in a manner similar to what occurs in the *Helix* ganglion.

6.8 Unsolved Problems

Still unexplained is the remarkable ease of recording spikes and the low amplitude of the slow components in most invertebrates compared to the vertebrates.

Still difficult in both invertebrates and vertebrates is to interpret peaks in the power spectrum. The first assumption is usually to think of an oscillation or rhythm, but it is not easy to exclude the null hypothesis that an observed time series is actually unstructured, i.e., "colored noise." Even if the peaks are consistent and reproducible in successive data samples, they could come from a time-domain series of events without rhythms, such as a mixture of stochastic generators of band-limited noise or an irregular repetition of several transient wide-band waveforms, like multiple unit spikes and afterpotentials. Similar spectral peaks can result from radically different causes. In certain places and conditions a few clear rhythms are known in invertebrates, but each candidate rhythm requires its own investigation to justify that designation.

7. Dynamics of Potentials from the Brain of Anamniotes (Vertebrates)

Erol Başar, Atsuko Schütt, and Theodore H. Bullock

7.1 Introduction

Bullock and Başar (1988) report that "when the extracellular ongoing, spontaneous potentials recorded from the surface or the depth of the cerebrum are compared in terms of gross wave form and power spectral density, there are no systematic differences among vertebrate classes (mammals, reptiles, birds, amphibians, bony and cartilaginous fishes) and that the differences in brain electrical activity between fish and mammals are smaller, if any, in spite of the major histological differences, than those in different states and loci within the same individual." For instance, among various bony fishes and elasmobranchs there are no differences in the parameters of ongoing activity. On the other hand, cerebellar cortex, tegmentum, medulla, and cord show quite different forms of ongoing activity (Bullock and Başar 1988).

If we turn our attention to potentials elicited by stimuli or cognitive events – sensory or evoked potentials – the use of descriptors of systems-theory analysis may help in obtaining further insight into the questions of differences and similarities in brain function between the anamniotes (fish) and the mammals. We studied sensory evoked potentials in two representative fishes, ray (*Platyrhinoidis triseriata* from the class of *Elasmobranchia*) and goldfish (*Carassius auratus* from the class of *Teleostei*), with the purpose of comparing them with those of the mammals (cat and human).

7.2 Methods

7.2.1 Ray

Like some other families of the anamniotes, the ray possesses the function of electroreception, i.e., a sensitivity to fractional μV electrical signals of foreign or self origin. The biological significance of electroreception lies in object recognition, orientation, and communication (Bullock and Heiligenberg 1986). Anatomically, the ampullar electroreceptors are part of the lateral-line system. Ampullary organs are highly sensitive to low-frequency (0.1–20 Hz) electrical fields (e.g. weak electrical fields down to $< 0.005\,\mu\text{V/cm}$ of field gradient in the seawater). This ability is useful in passive detection of prey and in orientation to feeble low-frequency fields of inanimate origin (Bullock et al. 1977).

A number of authors have recorded evoked potentials in different structures of the ray brain (Platt et al. 1974; Bullock 1979; Bullock and Corwin 1979; Schweizer 1986; Bleckmann 1989). In the present study, the animal was stimulated with adequate pulses having a gradient field between 20 and 45 μV/cm. The experimental setup to produce this electrical field is described by Schweizer (1986). For the preparation and electrode implantation we refer to Schweizer (1986) and Bullock and Başar (1988).

7.2.2 Goldfish

The fishes were acoustically stimulated by an approximated delta-impulse in the form of clicks at 70 dB through a loudspeaker in the air. In some experiments this stimulus was applied at random intervals of 2 to 3 seconds and in the others at a constant rate. For the preparation techniques the reader is referred to Bullock and Başar (1988).

7.3 Results

7.3.1 Ray

Four experiments carried out with different fish are analyzed in the following. A weak electrical field of 20 to 45 μV/cm was applied as stimulus, and evoked potentials were recorded simultaneously from medulla and mesencephalon. A typical averaged EP is presented in Fig. 7.3c. The frequency characteristics averaged from four experiments are shown in Fig. 7.1.

1. **Midbrain** (Fig. 7.1a). The power maximum lies in the range around 28 Hz and a second, less dominant, maximum in the theta range of 5 Hz.
2. **Medulla** (Fig. 7.1b). Three distinct maxima of similar amplitude at 9 Hz, 16 Hz, and 24 Hz are visible, indicating resonance at these frequencies. Additional maxima at 40 Hz and 50 Hz, lower in amplitude, are also indicated.

From these results it may be stated that evoked potentials with similar resonance characteristics can also be recorded in the ray brain by stimulation of the electroreception. Stable resonance phenomena found in the midbrain were in the theta and beta ranges, those in the medulla in the alpha and beta ranges with an additional indication in the gamma (40 Hz) range.

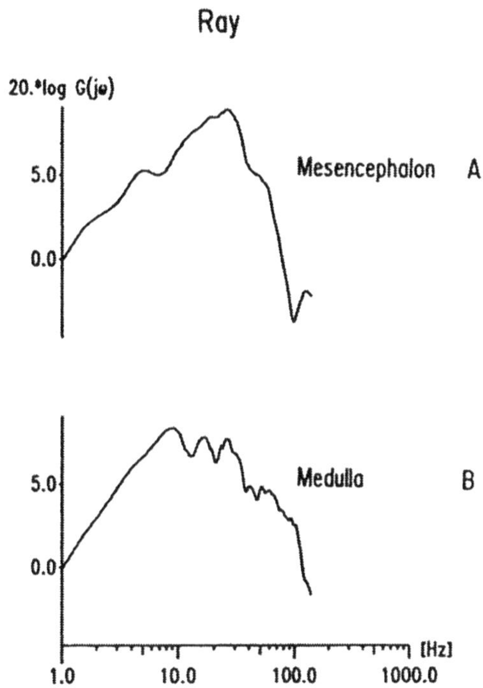

Fig. 7.1. Ray. Amplitude–frequency characteristics of evoked potentials averaged from 4 experiments (470 epochs): (**a**) midbrain; (**b**) medulla.

7.3.2 Goldfish

The experiments were performed with five different animals. After acoustical stimulation (70 dB clicks), evoked potentials could be recorded from three different brain structures: medulla, mesencephalon, and telencephalon. Figures 7.2a–c illustrate the amplitude–frequency characteristics averaged from five experiments each. In each experiment 100 single-evoked potentials were recorded (a typical averaged evoked potential is shown in Fig. 7.3).

In all three brain centers investigated, a double-peaked Fourier component, dominant and stable, appears in the higher-frequency range around 80 to 100 Hz. However, the patterns in the other frequency ranges vary depending on loci. In our particular loci of the telencephalon (a) prominent peaks exist at 3 Hz, 9 Hz, 12 Hz, and 35 Hz. (Echteler (1980) has described AEPs with different dynamics and form in a second region of the telencephalon.) In our loci of the midbrain (b) another stable component is clearly distinguishable as a pronounced peak at 30 Hz. The other components around 4 Hz, 7 Hz, and 13 Hz are also indicated, but only as shoulders in the slope of the curve. In the loci sampled in the medulla, the theta component disappears totally, but alpha as well as beta maxima are recognizable as minor peaks.

From the results shown above a conclusion may be drawn that characteristic frequencies can occur in the ranges of 4 Hz, 10 Hz, 30–35 Hz, and 80–100 Hz in acoustically evoked potentials in different brain structures of

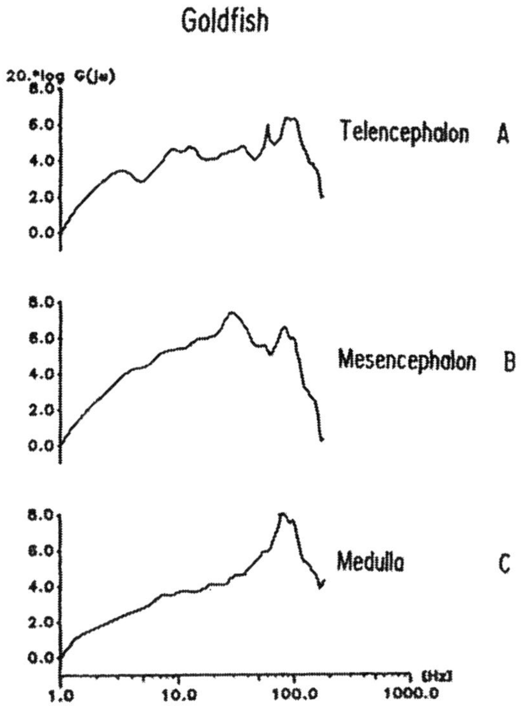

Fig. 7.2. Goldfish. Amplitude–frequency characteristics of evoked potentials averaged from 5 experiments (500 epochs); (**a**) telencephalon; (**b**) mesencephalon; (**c**) medulla.

the bony fish. These resonances, however, elicit different amplitude levels in the transfer function depending on the structures.

A number of studies describe the sensory modality of electroreception in some classes of fishes and amphibians (Fessard and Szabo 1974; Heiligenberg 1977, 1992; Bullock 1982; Bullock and Heiligenberg 1986; Bullock et al. 1983). Lissmann (1958) demonstrated, based on his observation of behavior in fish, that there was likely to be a system of specialized receptors and central processing that mediates the utilization and cognition of weak electrical fields in water by these fishes. In the direct physiological experiment, then, the existence of this special sensory modality was demonstrated in different brain structures such as medulla, mesencephalon, telencephalon, and cerebellum.

The "characteristic" frequency components observed in the samples of different brain structures of the ray and goldfish are presented again in Table 7.1 together with those of the invertebrate samples and the mammals (cat and human). What one notices is that the predominant frequency components of the mammalian cortex are usually found within a range not higher than 40 Hz. However, the reticular formation (cat) shows a dominant frequency component around 85 Hz. It is of interest that a similar high-frequency apparent invariant (80–100 Hz) exists in goldfish medulla and, to a lesser degree

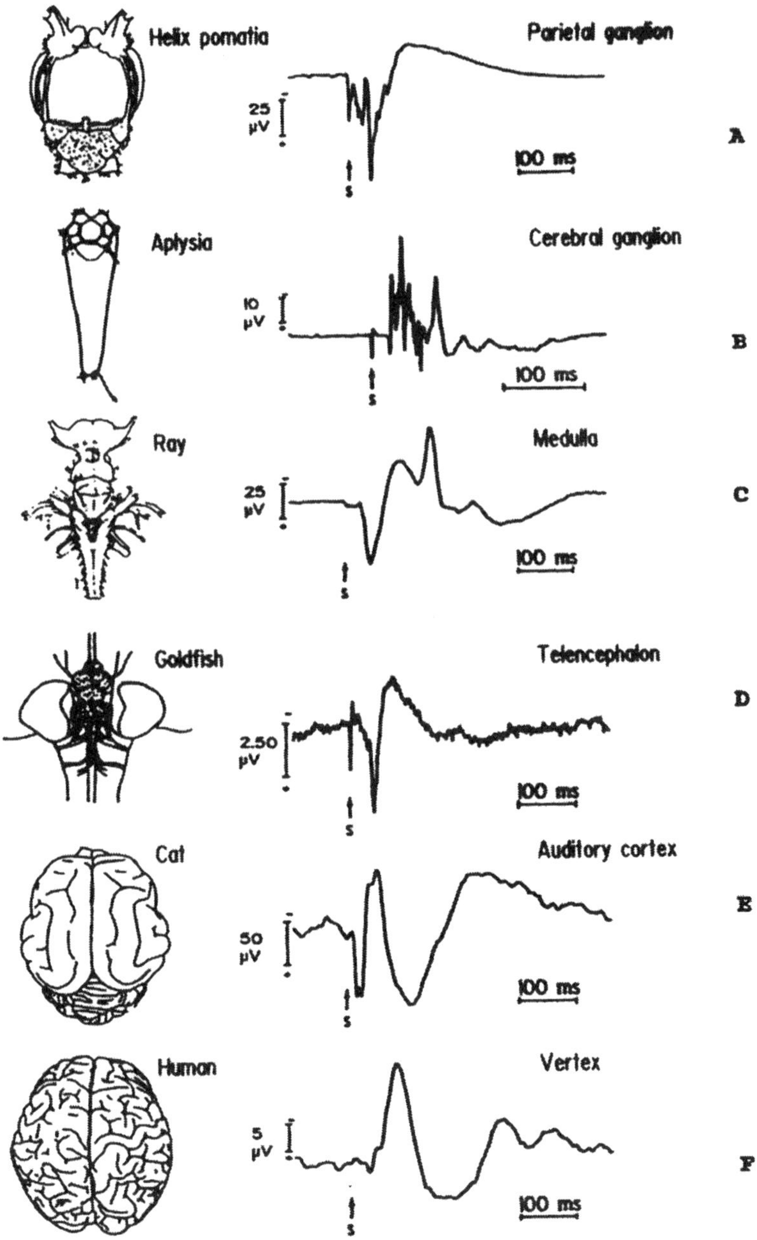

Fig. 7.3. The brains investigated are schematically presented together with the characteristic averaged evoked potentials from different loci. s: stimulus A and B: direct shock to afferent nerve; C: physiological stimulus of electroreceptors; D–F: acoustical stimulation.

Table 7.1. Comparative presentation of maxima in frequency characteristics in different species.

Frequency band (Hz)	1–8	8–15	15–30	30–48	52–100
Species	Maxima in frequency characteristics (Hz)				
Helix					
Pleural r.	5	–	20	–	–
Parietal r.	2	–	16	40	65
Visceral	2	–	15	35	60
Parietal l.	2		to	50	–
Pleural l.	2	9	18	40	75
Aplysia					
Group I	2	to	20	40	100
Group II	3	–	25	–	60
Ray					
Mesencephalon	5	–	28	–	–
Medulla	–	9	16; 24	40	–
Goldfish					
Telencephalon	3	10	–	35	90; 100
Mesencephalon	–	–	–	30	80; 100
Medulla	–	–	–	–	80; 90
Cat					
GEA	5	–	18	40	80
RF	5	11	25	40	85
HI	5	12	–	45	–
Human					
C_z	4	8	25	40	–
O	3.5	10	25	45	75
P	6	10	20	40	70
F	4	8	20	35; 45	–

(only seen as an indication), in ray medulla. The 15–30 Hz invariant also exists in all structures examined.

7.4 The Reasons for Neuroethological Comparison

The idea to perform a comparative analysis of neuronal electrical activity in some selected species from the groups of the vertebrates and invertebrates goes back to Bullock (1945) and (1981), who sees a possibility to clarify questions about the mechanisms of higher brain functions from the neuroethological standpoint. In the present chapter, some *in vivo* experiments were carried out on vertebrates that are partly considered phylogenetically *old.* The reason was to obtain information based on more *physiological* conditions in addition to the *nonphysiological* conditions that exist in the *in vitro* studies of the isolated organs. Our main purpose is to search for *invariants* in brains of various complexities.

7.5 Similarities and Differences

Is there commonality at all stages of evolution, or does the functional mechanism change during development from simple to complicated neuronal networks? Does the development to complexity involve new possibilities of information transfer? We have carried out, based on these questions, comparative analyses of evoked potentials in different invertebrates and vertebrates. The species studied are two invertebrates, sea slug (*Aplysia californica*) and land snail (*Helix pomatia*); two lower vertebrates, ray and goldfish, and two higher vertebrates, cat and human. Except for *Aplysia* and *Helix*, sensory stimulus modalities applied are *physiological* (acoustical or, in the case of the ray, low voltage electrical field). In *Aplysia* and *Helix nonphysiological* electrical stimulation was given to a connective or a nerve. Examples of the averaged evoked potentials of the species mentioned above are schematically shown in Fig. 7.3. It must be pointed out here, however, that the shapes of averaged evoked potentials may vary depending on different conditions. The spectral peaks from different structures of the species are listed in Table 7.1.

As a first approximation, one would like to consider these components *invariants* for the particular recording loci and afferents stimulated, since they are found not only in individual frequency–amplitude characteristics, but also in the average of a number of frequency characteristic records (860 in *Aplysia*). Concerning the highly developed vertebrate brains, not only human cortical field potentials measured by scalp surface electrodes but also cat cortical signals recorded by intracortical semimicroelectrodes show dominant frequency components up to the 40 Hz range. The response signals from our samples of the subcortical structure, reticular formation, in cat exhibit a frequency component around 85 Hz. Interestingly, a similar high-frequency *"invariant"* (80–100 Hz) is especially pronounced in certain loci of the goldfish and ray medulla.

According to Sturbeck's observation in *Helix* (1988), practically all frequency bands seemed to occur except the 10 Hz range. This underrepresentation of the 10 Hz response in *Helix* observed by him may, however, be due to the rather strong stimulus intensity (four times threshold). In our later study with the *Helix* visceral ganglion (Schütt and Başar 1992) we could show that when the stimulus is about two-times threshold (15 mA), this structure always responds to electrical stimulus, under our conditions, with a roughly 10 Hz frequency component accompanied by the other important *invariant* components like 3 Hz, 20 Hz, 40 Hz, and 60 Hz. This additional evidence in *Helix* can perhaps fortify the notion that the 10 Hz component may exist as in the brains of a wide range of species from human and cat to snail. In the *Aplysia* cerebral ganglion in the potential evoked when a certain nerve is electrically stimulated under our conditions, all frequency components, such as 5, 10, 25, 40, 60, and 100 Hz, could be found, implying the corresponding valleys between them.

In summary: Principally, all frequency components exist throughout the species. This observation may be supported by Bullock's statement (1988) that invertebrates (insects, crustaceans, gastropods, cephalopods) also show both fast and slow evoked activity. As our observations have shown, not only the most frequently discussed 8–13 Hz (alpha) and 40 Hz (gamma) components but also other frequency components have equally important weight for the frequency-oriented interpretation of neuronal network characteristics. Not only the 60–100 Hz apparent "invariant" but also the low-frequency components and the 15–30 Hz (beta activity) seem to be maintained in the course of evolution.

7.5.1 Unsolved Questions

In addition to the unsolved questions stated above, in the chapter on invertebrates including the difficulties in interpreting spectral peaks, comparative observations raise questions peculiar to these classes including fish, amphibians, reptiles, birds, and mammals. In brief, why are they so similar and why are they so different?

In spite of the basic diversity in architecture and physiology of the pallium of lower and the cortex of higher vertebrates, the general shape of the power spectra of ongoing electrical activity is remarkably similar – a maximum below 15 Hz and a steep decline at higher frequencies. With the analytical tools used so far, no differences in power spectra have been found between the classes. Are the crucial dynamics that surely distinguish mammalian isocortex from primitive pallium to be found in more subtle measures than crude power spectra? In spite of the similarity of spectral form, a much greater absolute power or RMS voltage characterizes mammals, virtually independent of brain size, cell size, or cell density. Reptiles seem to be intermediate between mammals and fish in EEG amplitude. Could there be an evolution of synchrony? Bullock and McClune (1989) presented preliminary evidence of this, with a new measure of synchrony. It remains to be confirmed with a more adequate sampling of species, brain regions, and states.

8. Frequency Response of the Cat Brain Influenced by Pharmacological Agents

Canan Başar-Eroğlu, and Erol Başar

By application of pharmacological agents the brain was put into a new functional stage. This chapter aims at reaching two categories of results:

1. Can we learn about CNS functions by means of evoked potentials upon application of pharmacological agents?
2. Can we find strong support for the justification of the core concept for the superposition principle described in Chap. 14 in Vol. I?

We will demonstrate that both questions can be answered positively, and for this purpose CCK-8, Haloperidol, and Neostigmine were utilized.

8.1 Effects of Ceruletide in the Brain

The C-terminal octapeptide of cholecystokinin (CCK-8) is widely distributed in the mammalian nervous system and exerts a high number of biochemical and neuropharmacological effects in laboratory animals and man. This peptide is supposed to be a modulator rather than a transmitter of neuronal activity and to be important for the regulation of food intake, pain sensation, motor activity, and psychic states.

Chemically related to CCK-8 is cerulein, a decapeptide detected in the skin of the Australian frog Hyla caerulea (Erspamer and Melchiorri 1983). Caerulein is used in the form of the diethylammonium hydrate (ceruletide), which is more stable and in most tests more potent than CCK-8. Ceruletide has high affinity to the specific CCK receptors and exerts nearly all effects of CCK-8; i.e, it is a *sedative, anticonvulsive, analgesic*, and *antistereotyoic*; it inhibits eating and has inconsistent antipsychotic effect in schizophrenics (reviews: Zetler 1985b; Nair et al. 1986; Albus 1988; Montgomery and Green 1988).

The effect of ceruletide on the human EEG has been rarely studied. In healthy subjects, one study revealed no substantial influence of the peptide (Dolecek et al. 1983), whereas another study found the spectral power density to be increased in the theta band and decreased in both the alpha and beta ranges (Stacher et al. 1982a,b). In similar experiments, CCK (99% pure CCK-33) increased the theta- and beta-range frequencies, while lower doses decreased and higher doses enhanced the amplitude of auditory evoked potentials (Stacher et al. 1979). In schizophrenic patients, ceruletide (0.8 μmg/kg, intramuscularly once weekly for three consecutive weeks) increased the EEG power values in many cortical areas (Mizuki et al. 1988). However, visual ERPs (P300) of schizophrenic patients were not modified by an infusion of ceruletide (Tamminga et al. 1986). In awake rats, CCK-8 (injected i.p.) increased after 20 min the auditory evoked response in the raphe nucleus, lateral hypothalamus, and hippocampus but not in the septum, anterior hypothalamus, amygdala, and caudate nucleus; 30 min later, the response was depressed in the raphe nucleus and hippocampus (Dafny et al. 1975). Furthermore, CCK-8 enhanced the EEG slow-wave activity (Kapas et al. 1988) and induced synchronous or asynchronous wave sleep (Mansbach and Lorenz 1983).

Başar-Eroğlu et al. (1996a) tested the possible influence of ceruletide on the EEG and auditory EPs of selected areas of the cat brain. Reference drugs used were the neuroleptic *haloperidol* and the indirect cholinergic *neostigmine*. Haloperidol, which easily passes the blood–brain barrier, was used because several (but not all) behavioral and biochemical effects of ceruletide are haloperidol-like. Neostigmine, which does not penetrate into the brain (Koelle and Steiner 1956), served as a control for the peripheral effects of ceruletide.

8.2 Methodological Remarks on Experiments with Pharmacological Agents (Haloperidol, Neostigmine, Acetylcholine)

The experiments were performed on nine chronically implanted and freely moving female cats (3–3.5 kg). Stainless steel electrodes of 100 μm diameter were located while the cats were under Nembutal anesthesia (35 mg/kg) in the auditory cortex (gyrus ectosylvianus anterior; GEA), dorsal hippocampus (HI), mesencephalic reticular formation (RF), and cerebellum (acoustical vermis: CE) with coordinates as given in Chap. 1 (Snider and Niemer 1964).

8.2.1 Experimental Procedure and Data Analysis

Auditory EPs were elicited by presenting 100 binaural tones. We used a 2 kHz tone for stimulation with an intensity of 80 dB SPL (sound pressure level) with a duration of 0.8 s and a random stimulus interval of 1.6–3.6 s.

The following drugs were used as solutions in saline and administered subcutaneously: ceruletide, 0.03 mg/kg; haloperidol (Haldol®-Janssen), 0.25 mg/kg; neostigmine (Prostigmin®), 0.01 mg/kg. The time interval between the application of substances in the same cats was at least two months in order to avoid interactions of substances.

The cats were sitting in a cage in a soundproof and echo-free room, which was dimly illuminated, and the EEG was monitored during all the experimental sessions.

The mean-value amplitudes of AEPs were tested for the significance of differences by means of a nonparametric test. For this purpose the Friedman two-way analysis was used (for detailed information on the experiments and data analysis see Başar-Eroğlu et al. (1996a)).

8.3 Auditory EPs (AEPs) upon Application of Cerulein, Haloperidol, and Neostigmine

Figure 8.1 shows the grand average of AEPs from 5 responder cats for auditory cortex (GEA), hippocampus (HI), reticular formation (RF), and cerebellum (Ce), respectively. AEPs under application of cerulein, haloperidol, and neostigmine versus control AEPs are also presented (from left to right) in Fig. 8.1. For the sake of simplicity, in Fig. 8.1 the AEPs 30 min after injection of substances were presented.

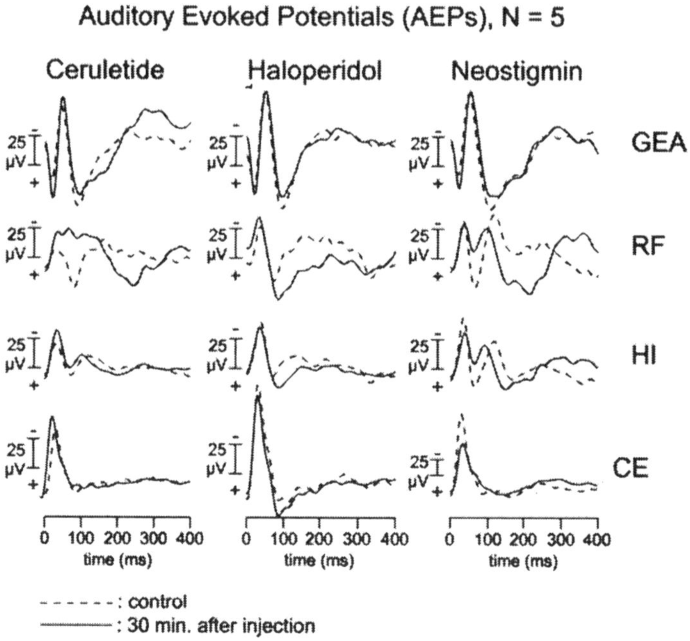

Fig. 8.1. Grand mean averaged AEPs recorded from auditory cortex (GEA), hippocampus (HI), reticular formation (RF), and cerebellum (CE) of 5 cats. *Interruped lines* represent averaged controlAEPs, *solid lines* represent AEPs after applications of ceruletide (*left column*) haloperidol (*middle column*) and neostigmine (*right column*). Stimulus onset at 0 ms.

In regard to all structures, most of the changes by application of substances were observed in the hippocampus. Within 30 min following injection of ceruletide, the auditory EP took the shape of a slow wave with a flat negative peak in the region of 80 ms with some phase reversal (instead of a positive peak at 80 ms in control-EP). Furthermore, the peptide induced an increase in amplitude during the first 100 ms time period with P250 (N1-P2) peak. Such waveform changes were not observed in other structures. Haloperidol had an effect on the amplitude of AEP in hippocampus. The P80 peak of control-EP was shifted to higher latencies, and the N50/P100 amplitude was increased. However, the entire time course remained congruent. Also neostigmine induced some changes in auditory EP.

In addition to hippocampus, other parts of the brain also revealed changes in amplitude values (Fig 8.1). Cerebellum is the structure that showed virtually no change in comparison to all other structures. The effect of cerulein and haloperidol (although small) had the same directions on cerebellum; both of them induced an amplitude increase, whereas neostigmine induced a decrease. The drug-induced changes of AEPs were in reticular formation similar to those in the hippocampus, although not reaching statistical significance.

8.4 Amplitude–Frequency Characteristics

Figure 8.2 shows AFCs computed from the transient responses summarized in Fig. 8.1. The most prominent changes occurred in the hippocampus. Ceruletide induced changes of low-frequency components such as *theta* (3–8 Hz) and *alpha responses* (8–15 Hz) of all the structures, except cerebellum. Haloperidol increased the low-frequency components in hippocampus and reticular formation, but not in auditory cortex and cerebellum. The effects of neostigmine were similar to those of ceruletide.

Since the most pronounced changes were observed in hippocampus, we will present records from this structure of one representative cat named Lucy (Fig. 8.3). In order to compare the effects and their duration of three substances, Fig. 8.3 also presents AEPs of the second and third day following treatment. As can be seen, there were large differences in the amplitudes of N50/P100 deflection on the third day. Ceruletide still induced high amplitude in comparison to control EP, whereas haloperidol slightly reduced it and neostigmine had lost its effect.

We have to note that in all 5 cats the enhancing effect of ceruletide lasted longer than that of the other two drugs. The effect of neostigmine was the shortest one, since already 4 hours after treatment the EPs of 4 cats showed almost the same time course and amplitude as the control EPs.

Figure 8.4 presents a single-sweep analysis of EEG and AEPs filtered in the frequency range between 3 and 8 Hz in hippocampus 24 h after treatment with ceruletide (A) and haloperidol (B) versus control experiments. Each of the quadrants is divided into 3 parts: The uppermost shows the single sweeps

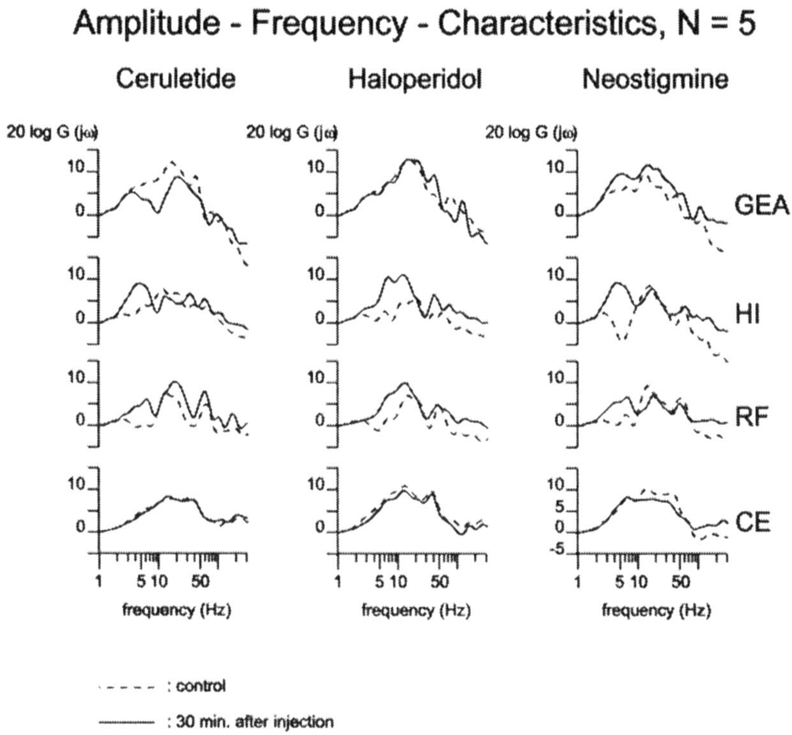

Fig. 8.2. Amplitude–frequency characteristics computed from the AEPs shown in Fig. 8.1. Abscissa: frequency in Hz; *ordinate*: amplitude in relative units and decibels (dB).

(filtered 3–8 Hz), the lower curves present averaged EPs filtered 3–8 Hz, and the lowermost curves show unfiltered EPs. As can be seen, 24 h after application of ceruletide, single EPs show time-locked great enhancements of theta activity, whereas the single sweeps under haloperidol present neither such an enhancement nor time-locked theta response. This observation indicates an important contribution of the theta components to the poststimulus EEG (EP) changes in hippocampus after ceruletide treatment.

8.5 Interpretation of Pharmacologically Induced Changes by Application of Cerulein, Neostigmine, and Haloperidol

Our finding that ceruletide enhanced the auditory-evoked potential in the hippocampus corresponds with the observation of an increase by CCK-8 of the field potential in the hippocampal dentate gyrus evoked by stimulation

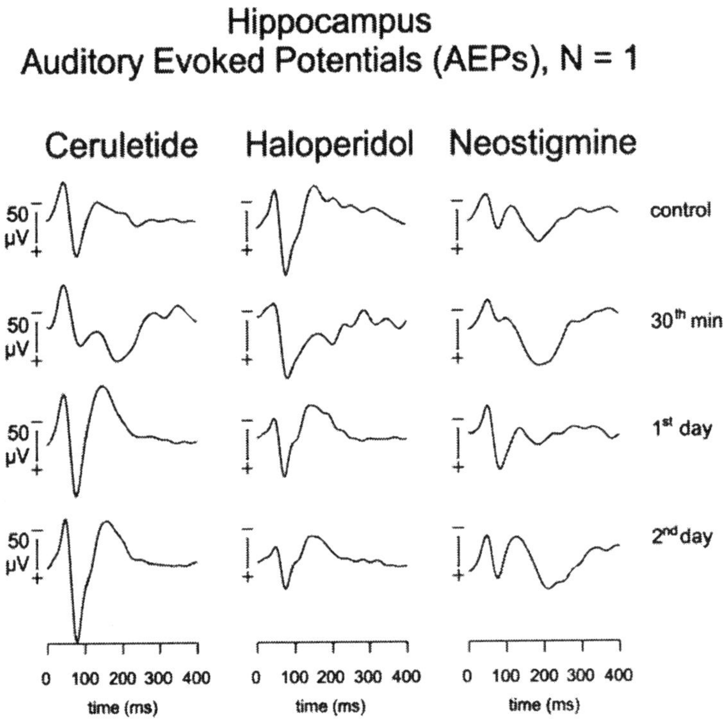

Fig. 8.3. Averaged AEPs from the hippocampus of one representative cat. Records were taken immediately before drug injection (control) and thereafter at the times shown *on the right margin*. Two months elapsed between treatments.

of the medial entorhinal cortex (Dahl 1987, 1988; Sinton 1988). From this statement it was concluded that the peptide modulates hippocampal cell activation via a reduction in the threshold for synaptic cell activation, and therefore plays a role in *gating input* to the hippocampus. Accordingly, we may infer that ceruletide in the present experiments increased the input to those brain areas where it enlarged the EP. On a cellular level, experiments in vivo and in vitro have shown that CCK-8 and related peptides reduce the membrane potential of hippocampal CA1 neurons and increase their tendency to fire spikes (Jaffe et al. 1987; Boden and Hill 1988), and spinal neurons (Jeftinija et al. 1981). Hence, an enhanced neuronal excitability may be the common denominator of the effect of ceruletide on evoked potentials.

Our finding that the cerebellum, in contrast to hippocampus and cortex, virtually did not respond to ceruletide induces the question of whether CCK receptors are present in the cat cerebellum. Unfortunately, this is not known, and marked species differences in CCK receptor distribution have been observed. In contrast to other brain structures such as cortex and hippocampus, the cerebellum of man and guinea pig has very many receptors, but that of

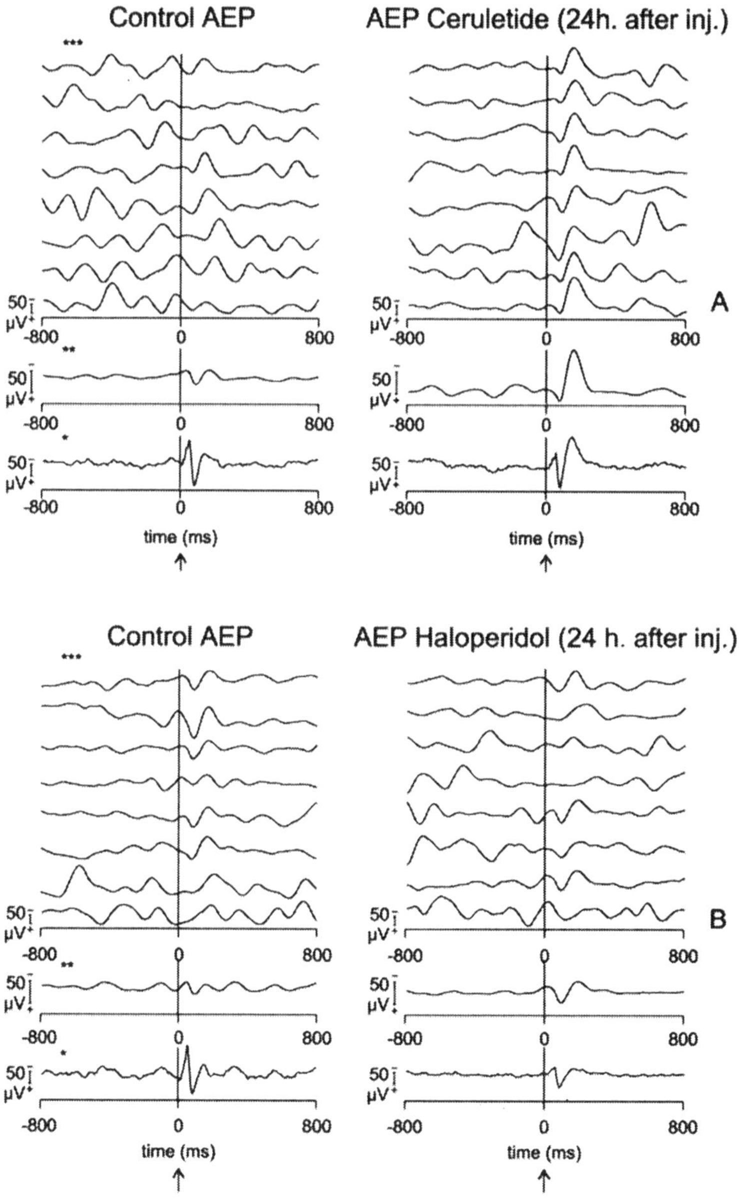

Fig. 8.4. Hippocampal AEPs before (control) and 24 h after the application of ceruletide (**a**) or haloperidol (**b**), recorded in one cat. *Uppermost curves* (***): Single EEG-AEP sweeps (Filtered: 3–8 Hz). *Middle curves* (**): Averaged AEP (Filtered: 3–8 Hz). *Lower curves* (*): Unfiltered AEP, averaged from about 50 artifact-free sweeps.

rat, hamster, and mouse few or none (Zarbin et al. 1983; Van Dijk et al. 1984; Gaudereau et al. 1985; Mantyh and Mantyh 1985; Sekiguchi and Moroji 1986; Williams et al. 1986; Dietl et al. 1987; Niehoff 1989).

Systematically administered tertiary cholinergic drugs, such as physostigmine and caerulein, both of which easily pass the blood–brain barrier, evoked hippocampal theta rhythm (Olpe et al. 1987), whereas the peripherally acting quaternary cholinergic neostigmine did not influence the EEG (Longo 1962). The latter negative finding can be explained by the inability of this drug to reach the brain when applied systemically (Koelle and Steiner 1956). Thus, the neostigmine-induced response of the hippocampus and also that of the reticular formation in the present experiments must be an indirect effect arising from peripheral sites, e.g., in the gastrointestinal tract. We may then consider the neostigmine-induced results as consequences (and indicators) of a nonspecific influx mainly via visceral afferents in the splanchnic or vagal nerves. An extended description of pharmacological effects of cerulein is given by Başar-Eroğlu et al. (1996).

The dose of haloperidol used, although low, can be expected to produce central effects. In cats, the drug in doses of 0.5 to 1 mg/kg induced synchronization of neocortical EEG (slow wave sleep) and hippocampal theta "arousal" wave activity (Tsuchiya et al. 1979) and changes of the frequency distribution in the cortex, hippocampus, and amygdala (Fairchild et al. 1980). Furthermore, via blockade of inhibitory somatodendritic dopamine autoreceptors (Bunney 1988), it increased the activity of dopamine-containing ventral tegmental area neurons (Trulson and Pfreussler 1984) that project primarily to the limbic system (including supra- and entorhinal cortex) and also to the frontal cortex (Domesick 1988).

In the present experiments the haloperidol-induced effects on the AEP were most prominent in hippocampus, though different from those of ceruletide. This supports the view that the several further actions of ceruletide, although neuroleptic-like, are not congruous with those of haloperidol (Zetler 1985b).

8.6 The Utility of Frequency Analysis to Neuropharmacological research

Usually, EPs are described in terms of several arbitrarily defined components such as peak (wave) latencies and wave magnitudes. These arbitrarily defined components generally depend on the location of the recording electrode, behavioral state, or sleep stage of the subject under study, and on the nature of the stimulating signal.

A dual approach in the time and frequency domains has been proposed and described in Chaps. 1 and 2. This approach is based on our working hypothesis that EPs reflect resonance properties of the EEG (see Chap. 4

in Vol. I). According to the results presented, the superposition of evoked oscillations in various EEG frequency channels (4 Hz, 10 Hz, 20 Hz, 40 Hz, etc.) gives rise to compound EP. These frequency channels are related to the main peaks in the amplitude–frequency characteristics (AFC) computed from compound EP. To analyze EPs we have also used digital filters with limits chosen adequately according to AFC peaks.

In our view, the conventional averaged EP, which is widely used, was considered only as a rough estimate of brain's EEG response, and it was claimed that the averaged EP does not take into account dynamic changes in the brain's intrinsic activity. In contrast, we consider single EPs as correlates of brain's quasi-invariant resonant modes containing important brain frequency codes related to central nervous system function.

The present chapter indicates a selective and long-lasting strong theta component of evoked potential in hippocampus after treatment with ceruletide. We define this selectivity as the ability of brain networks to facilitate (or activate) electrical transmission within determined frequency channels, when stimulation signals are applied to the brain. The enhanced theta response of AEPs under ceruletide possibly indicates increased signal transfer to or through hippocampal neurons. This may bear relation to the ceruletide-induced memory enhancement (Itoh and Lal 1990) in regard to the role of the hippocampus for memory processes (Squire 1992).

Acetylcholine also induces low-frequency theta rhythms in the hippocampus. This is easily seen if we compare the amplitude–frequency characteristics of the hippocampus and cerebellum, where important changes are visible in the hippocampus and no changes can be observed in the evoked potentials of the cerebellum. For example, the 10 Hz response of the cerebellum remains unchanged. These results show clearly that the Fourier transform performed does not create a dominant peak in the theta frequency range, but the dominant maxima of amplitude–frequency characteristics are certainly based on physiological properties of the structures under study (see also results in Chaps. 6 and 7).

In Chap. 18 in Vol. I it was already discussed that the dissection of EPs into frequency components is not only a system-theoretical approach. Furthermore, it was assumed that if one does not take into account the minor peaking, the evoked oscillations reflect true biological components.

The effect of haloperidol, cerulein, and scopolamine has completely different oscillatory changes. With regard to cerulein one can see, for example, that hippocampal EPs can be reduced to a homogeneous oscillation in the range 1–5 Hz.

These results show clearly that the Fourier transform itself does not create a dominant peak in the theta frequency range, but the dominant maxima of amplitude–frequency characteristics are certainly based on physiological properties of the structures under study.

It is furthermore interesting to note that various pharmacological agents can break the coupling between frequency responses of various brain structures. In Chap. 18 in Vol. I it was already stated that the application of pharmacological agents is an important indication that the evoked rhythms are of a real physiological nature.

The examples stated in this chapter present only a few cases in the vast domain of pharmaco-EEG. Certainly, it is expected that a more detailed analysis by employing the systematic application of a variety of substances might open new perspectives in this field of functional neurophysiological research.

Part II

The Human Brain: Dynamics of EEG, Evoked Potentials, and MEG

9. Evoked Alpha and Theta Responses in Humans to Auditory and Visual Stimuli

Martin Schürmann and Erol Başar

Like the previous chapters (Part I), the present chapter and the following ones (Part II) are devoted to the analysis and understanding of the transfer function in the brain, which is registered upon sensory stimulation, visual as well as auditory stimulation. We have seen in the chapters concerning experiments with cats (see Part I) that there are differences in the quality of transfer function dependent on the mode of stimulation. Moreover, we found that the frequency coding of the response was also a function of the structure being recorded. We have seen important differences in the transfer functions of the occipital locations, gyrus ectosylvianus anterior, and frontal lobes (Chap. 2). The concept of frequency coding is an important coding mechanism for sensory stimuli and has played an important role in the understanding of the experimental data presented. Since in this book "sensory coding" described in terms of frequencies plays an important role, we will use the same approach for the description of human data as well.

Recordings of evoked responses obtained by means of magnetoencephalography (MEG) will be presented in order to give complementary information (Chap. 13). Furthermore, the joint interpretation of data from experiments, on evoked potentials from the human brain and cat brain, and magnetoencephalography allows to formulate at a new hypothesis concerning information processing of the brain.

In this chapter we are confining our analysis to recordings from various parts of the skull (frontal vertex, parietal, and occipital). Although neither a precise nor a rough mapping of brain potentials can be performed merely by using this small number of recordings, a rough spatial distribution of the frequency coding will be sufficient to permit important functional descriptions of cortical functions from the viewpoint of EEG signals. Due to the major disadvantage of scalp recordings, the spatial "smearing", only global statements regarding brain function are possible by describing EEG and EP measurements.

Nevertheless, combining the results from EEG, EP measurements, and those from MEG/MEP as well as those from field potentials from cat intracranial structures, an important description can be achieved. One can further raise the questions, *What is global?* and *What is precise?* Are recordings from single units in animals under anesthesia more precise? Certainly not. Are single recordings of neuronal response better or more suited to understanding brain function? Certainly not, since single neurons are not representative of the entire CNS. It can be replied that any approach that relies either entirely on single recordings or entirely on EEG recordings offers restricted views and that by combining the views from both approaches a better understanding (but not yet a perfect one) of brain function can be attained.

The experimental foundation of these considerations consists of several studies (overview in Başar 1980). An exemplary study is summarized in the remaining part of this chapter (for details see Schürmann and Başar 1994). For specific aspects of the methods applied, see Chap. 4 in Vol. I.

9.1 Subjects, Methods, Environment

The experiments were carried out on 24 voluntary healthy subjects (20–34 years of age; median: 25 years). The subjects did not have any known neurological deficit and did not take any drugs known to affect the EEG. Ag–AgCl disc electrodes were placed on frontal, vertex, parietal, and occipital recording sites (F_3/F_4, C_z, P_3/P_4, O_1/O_2 in the international 10–20 system). Derivations were against a left earlobe reference in subjects #1–#24. Data were recorded from C_z, P_3/P_4, O_1/O_2 in subjects #1–#11 ("series I" of measurements) and from F_3, C_z, P_3, O_1 in subjects #12–#24 ("series II" of measurements).

EEG Amplification and Digitization. All data were amplified by means of a Schwarzer EEG machine (time constant: 0.5 s) and low-pass filtered with a cutoff frequency of 70 Hz (24 dB/octave). 2 s pre- and 2 s poststimulus EEG were digitized with a sampling rate of 250 Hz and stored on hard disk (subjects #12–#24: 1 s pre- and 1 s poststimulus EEG; 500 Hz sampling rate). Artifacts were rejected as described in Chap. 4 in Vol. I.

Stimuli. Auditory stimuli (tone bursts of 2000 Hz, 80 dB sound pressure level, and a rise time of 0.5 ms) were presented binaurally via loudspeakers. The light stimulator was a 20 W fluorescent bulb that was electrically triggered. The stimulating light source was placed at 2 m distance on a level with the subject's eyes so as to equally expose both eyes to the light. The duration of stimuli was equal to the duration of the poststimulus EEG epoch (2000 ms).

Data were evaluated according to the principles outlined in Chap. 4 in Vol. I.

9.1.1 Evoked Potentials: Auditory and Visual Stimuli

Figure 9.1a shows unfiltered recordings of auditory EPs obtained in the typical subject A.F. (#2): Distinct responses are present in vertex and parietal recordings. The vertex response shows a typical peak configuration with a

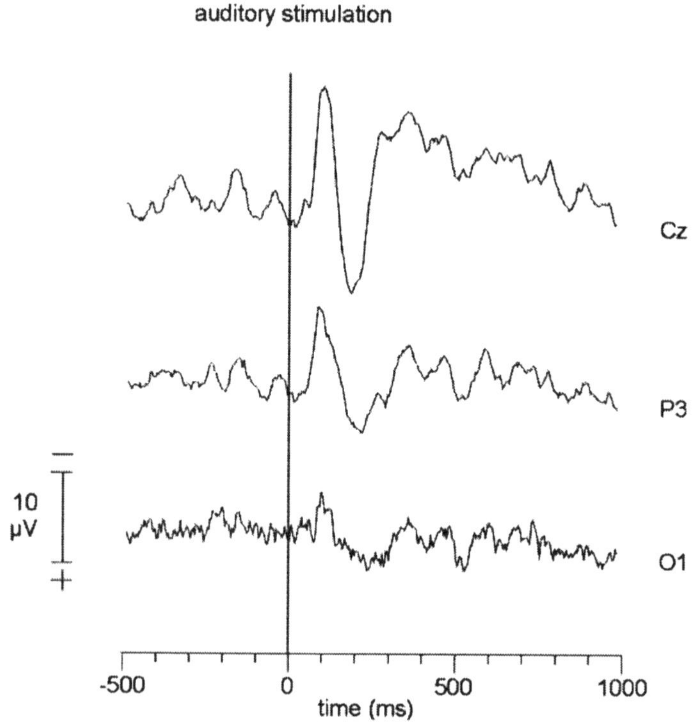

Fig. 9.1a. Averaged auditory EPs in one typical subject (A.F.). Along the abscissa: time in ms; Cz, vertex; P3, left parietal; O1, left occipital. (From Schürmann M. and Başar E. (1994) *Biological Cybernetics* 72: 161–174.)

negative peak around 100 ms and a positive peak around 180 ms. An occipital response is visible, but small in amplitude. Figure 9.1b shows unfiltered visual EPs obtained in the same subject, with responses in vertex and occipital recordings being most distinct.

9.1.2 Frequency-Domain Approach to Evoked Potentials

Amplitude–frequency characteristics (AFCs) computed from the averaged EPs of Figs. 9.1a and 9.1b (subject A.F.) are shown in Fig. 9.2. Solid curves refer to auditory stimulation; dashed curves refer to visual stimulation. Vertex recordings: AFCs computed from the auditory EP were typically characterized by a theta–alpha compound response covering a broad frequency range (peaking frequency 6–7 Hz). Upon visual stimulation, the dominant maximum covered not only the theta and lower alpha ranges, but also the upper alpha and slow beta ranges, the peaking frequency being at 8–9 Hz. Left parietal recordings: Upon auditory stimulation, there was a theta–alpha compound response covering the theta and lower alpha ranges (approx. 3–10 Hz). The

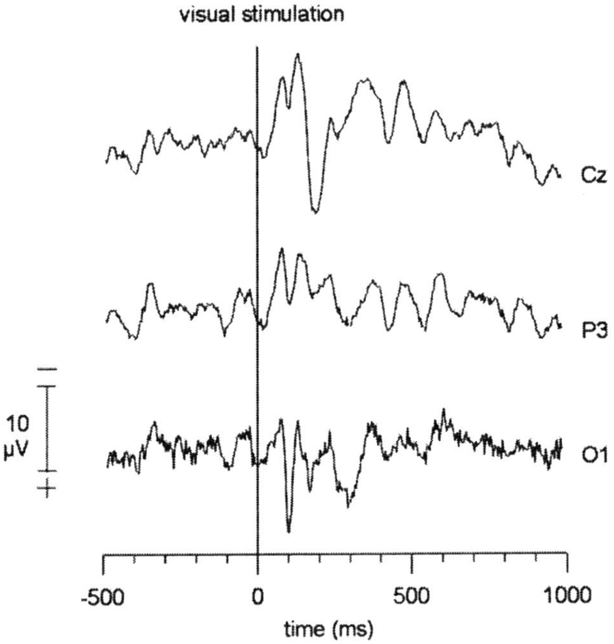

Fig. 9.1b. Averaged visual EPs in one typical subject (A.F.). Along the abscissa: time in ms; Cz, vertex; P3, left parietal; O1, left occipital. (From Schürmann M. and Başar E. (1994) *Biological Cybernetics* 72: 161–174.)

response to visual stimulation was characterized by a maximum covering the upper alpha and slow beta ranges (approx. 7–17 Hz). Left occipital recordings: AFCs computed from auditory EPs were rather flat, the maxima being considerably less in amplitude than upon visual stimulation. AFCs computed from visual EPs show a main maximum in the 10–20 Hz range. The difference between AFCs from auditory stimulation and AFCs from visual stimulation is smallest in the 3–5 Hz range.

AFCs computed from vertex recordings (Fig. 9.3): Auditory stimulation – Fig. 9.3a, left column – elicited dominant maxima in the 5–9 Hz range. Visual stimulation gave rise to dominant maxima in the 5–20 Hz range – Fig. 9.3b, right column. Range and median of the peaking frequencies are given in Table 9.1. Although the interindividual variations in AFCs were large in the visual case, the two columns (auditory vs. visual stimulation) are significantly different with regard to peaking frequencies (Table 9.1).

As shown in the overview of methods (see Chap. 4 in Vol. I), two ways of obtaining group averages for AFCs are available: (i) averaging across subjects in the time domain gives the grand average EP from which the amplitude–frequency characteristics can be computed; (ii) averaging across subjects in

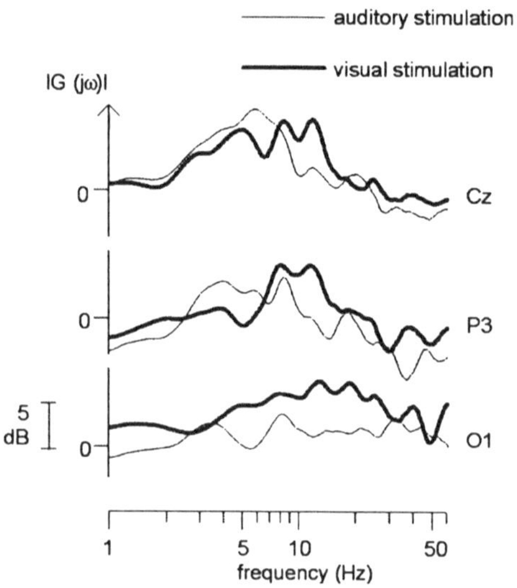

Fig. 9.2. Amplitude–frequency characteristics computed from the averaged EPs shown in Fig. 9.1, (typical subject A.F.) *Along the abscissa*: frequency in Hz; *along the ordinate*: amplitude in relative units. The curves are normalized in such a way that the amplitude at 1 Hz is equal to 1 (or $20 \log 1 = 0$ dB). Two stimulus modalities in each diagram; *thin lines:* auditory stimulation; *thick lines:* visual stimulation. (From Schürmann M. and Başar E. (1994) *Biological Cybernetics* 72: 161–174.)

the frequency domain gives the mean value of AFCs. The combination of both ways of averaging may provide mutual confirmation.

Figure 9.4a shows AFCs computed from grand averages (i.e. with averaging performed in the time domain). Thin lines refer to auditory stimulation, and thick lines refer to visual stimulation.

1. In vertex recordings, auditory stimulation elicited a broad maximum extending from the theta to the lower alpha range (theta–alpha compound response, approx. 5–9 Hz; peaking around 7 Hz). Visual stimulation gave rise to a maximum comprising the theta, lower alpha, upper alpha, and lower beta ranges (approx. 5–18 Hz; peaks around 9 Hz and around 13 Hz). Left parietal recordings: The main maximum of the visual frequency response was again shifted to the higher alpha frequency range. Left occipital recordings: In AFCs computed from visual EPs there was a dominant maximum in both the alpha and slow beta ranges (about 8–23 Hz). In comparison to this maximum, the maxima in the AFCs computed from auditory EPs (3, 9, and 20 Hz) were less ample. The difference between AFCs for auditory stimulation and AFCs for visual stimulation were most marked in the 10–20 Hz frequency range and less marked in the delta and theta ranges.

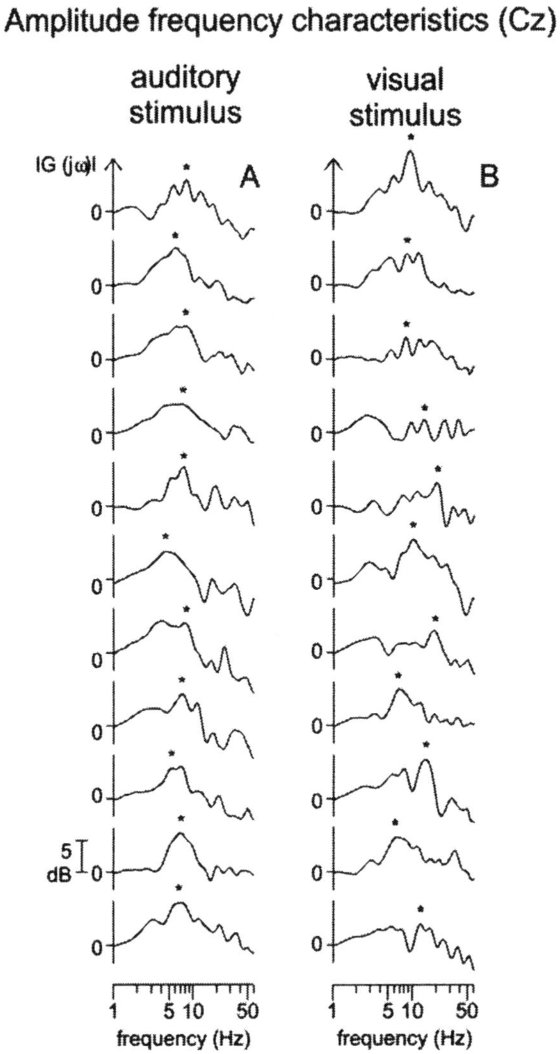

Fig. 9.3. Amplitude–frequency characteristics computed from averaged EPs in single subjects (series I: subjects #1–#11). (**a**) *left column*: responses to auditory stimulation; (**b**) *right column*: responses to visual stimulation. *Along the abscissa*: frequency in Hz; *along the ordinate*: amplitude in relative units. (From Schürmann M. and Başar E. (1994) *Biological Cybernetics* 72: 161–174.)

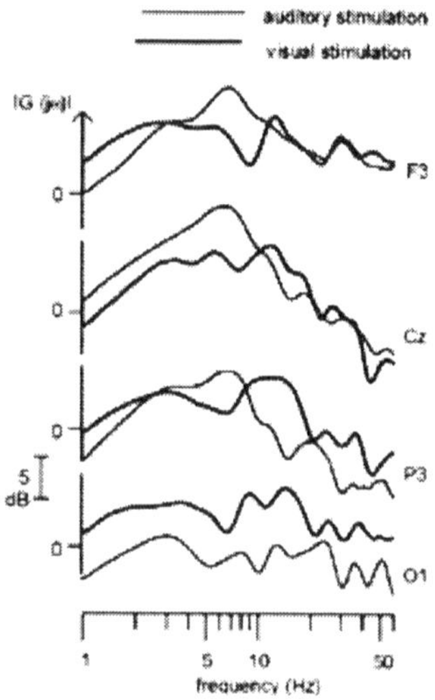

Fig. 9.4a. Amplitude–frequency characteristics computed from grand averages ($N = 24$ subjecs; averaging in time domain). (From Schürmann M. and Başar E. (1994) *Biological Cybernetics* 72: 161–174.)

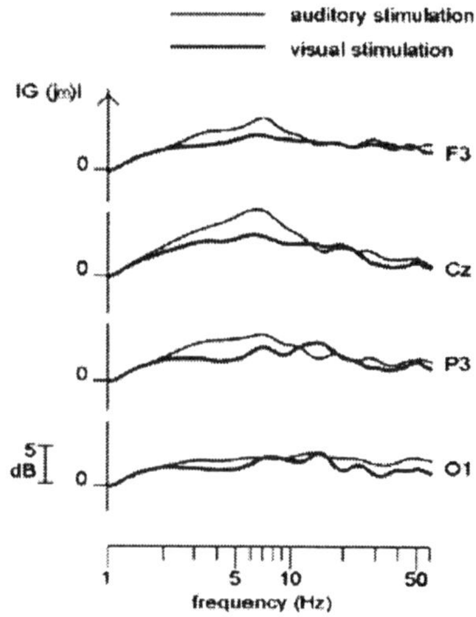

Fig. 9.4b. Mean values of amplitude–frequency characteristics ($N = 24$ subjects; averaging in frequency domain). (From Schürmann M. and Başar E. (1994) *Biological Cybernetics* 72: 161–174.)

Table 9.1. Peaking frequencies of dominant maxima in amplitude–frequency characteristics computed from transient evoked potentials (vertex recordings; subjects #1–#24; $p < 0.01$, Wilcoxon). (From Schürmann M. and Başar E. (1994) *Biological Cybernetics* 72: 161–174)

Stimulus	Median	95% confidence interval
Auditory stimulus	7 Hz	6–8 Hz
Visual stimulus	9 Hz	7–10 Hz

2. Topographic differences within one stimulus modality are also visible, e.g., in Fig. 9.4a: *AFCs computed from auditory Eps – thin lines:* (i) Vertex recordings: a broad maximum extending from the fast theta to the slow alpha range was observed (theta–alpha compound response). The peaking frequency was at about 6–7 Hz. (ii) Left parietal recordings: again, theta–alpha compound responses were observed, the peaking frequency being at about 7–8 Hz. (iii) Left occipital recordings: the maxima were considerably less ample than in the AFCs computed from vertex and parietal EPs. *AFCs computed from visual EPs – thick lines:* (i) vertex recordings: a

rather broad maximum extending from the upper theta range to the slow beta range was dominant. The peaking frequency was observed at 8–9 Hz. (ii) Left parietal recordings: AFCs of dominant maxima extended from the slow alpha to the slow beta range with peaks at approx. 9–10 Hz. (iii) Left occipital recordings: with respect to the dominant maximum, the AFCs were similar to case (ii).

The results mentioned so far were confirmed by computing mean values of AFCs (averaging in the frequency domain, the AFCs of individual EPs are computed prior to averaging) as shown in Fig. 9.4b. As expected, there are only minor differences in comparison with Fig. 9.4a. The shift towards fast alpha–slow beta in the case of visual stimulation can also be seen by comparing thin lines and thick lines. In both measurement series, occipital responses to auditory stimulation were characterized by flat AFCs in contrast to maxima in the alpha range in the case of visual stimulation.

9.1.3 Component Analysis by Means of Digital Band-Pass Filtering

Figure 9.5a shows frequency components computed from averaged auditory EPs (subject A.F., #2). Each column refers to one recording position, each row to a filter component.

Vertex recordings are considered first. A comparison of the frequency ranges 4–7 Hz (theta) and 8–15 Hz (alpha) shows that the theta auditory component (about 15 μV) was distinctly higher than the alpha auditory component (about 5 μV). The theta/alpha ratio in parietal responses (7 μV and

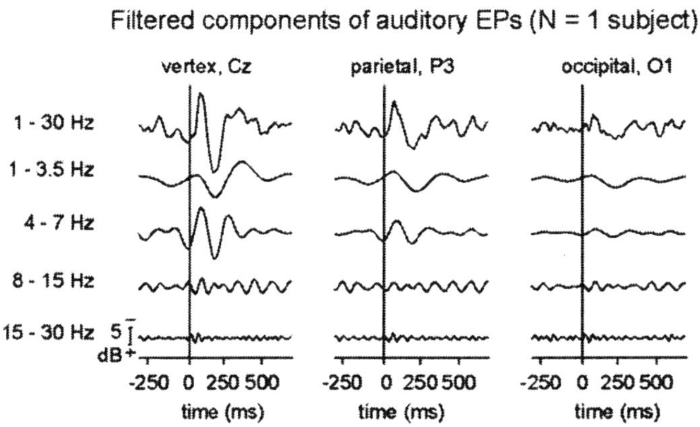

Fig. 9.5a. Filtered averaged EPs in one typical subject (A.F.). Components of vertex, left parietal, and left occipital recordings in various frequency bands were obtained by using band-pass filters with the filter limits given *at the left side* of each respective row. (From Schürmann M. and Başar E. (1994) *Biological Cybernetics* 72: 161–174.)

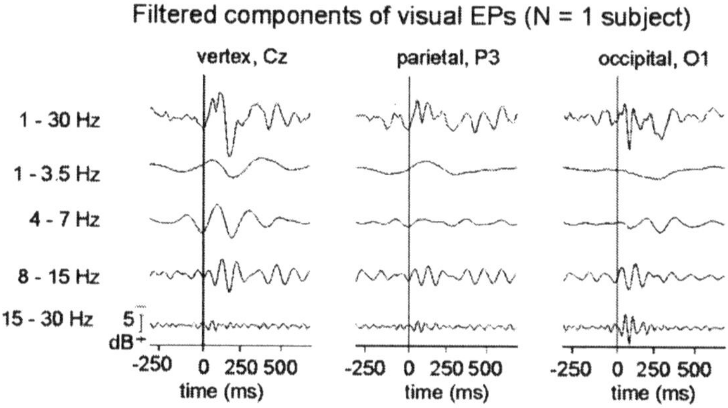

Fig. 9.5b. Filtered averaged EPs in one typical subject (A.F.). Components of vertex, left parietal, and left occipital recordings in various frequency bands were obtained by using band-pass filters with the filter limits given *at the left side* of each respective row. (From Schürmann M. and Başar E. (1994) *Biological Cybernetics* 72: 161–174.)

3 μV, resp.) to auditory stimulation was similar to vertex recordings. In the occipital position the responses – if any – were of low amplitude.

Figure 9.5b shows the respective frequency components of averaged visual EPs (subject A.F.). In the column "vertex," the theta and alpha amplitudes are of similar height (both about 10 μV). Contrary to the auditory case, visual stimulation elicited a large occipital alpha response.

The waveforms of frequency components of grand average auditory and visual EPs (subjects #1–#24; frontal data in subjects #12–#24; Figs. 9.6a,b)

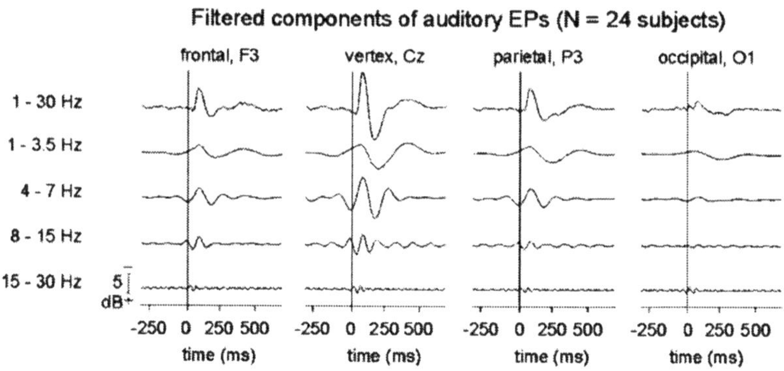

Fig. 9.6a. Filtered auditory EPs (computed from grand averages, $N = 24$ subjects). Components of vertex, left parietal, and left occipital recordings in various frequency bands were obtained by using band-pass filters with the filter limits given *at the left side* of each respective row. (From Schürmann M. and Başar E. (1994) *Biological Cybernetics* 72: 161–174.)

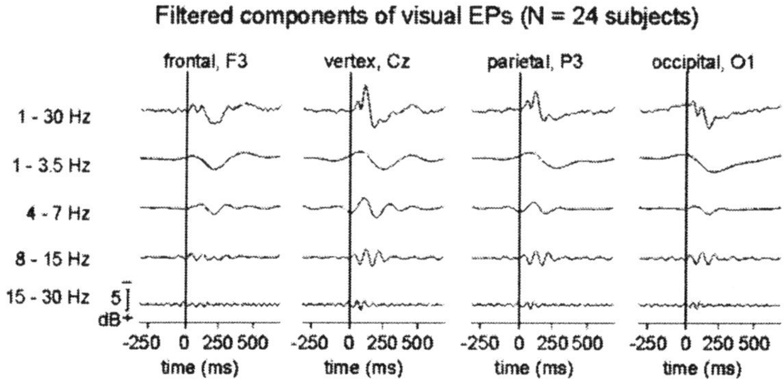

Fig. 9.6b. Filtered visual EPs (computed from grand averages, $N = 24$ subjects). Components of vertex, left parietal, and left occipital recordings in various frequency bands were obtained by using band-pass filters with the filter limits given *at the left side* of each respective row. (From Schürmann M. and Başar E. (1994) *Biological Cybernetics* 72: 161–174.)

were in accordance with the data from subject A.F. provided above. Frequency components of averaged auditory EPs are shown in Fig. 9.6a. Each column refers to one recording position, each row to a filtered component.

Again, in vertex recordings, the theta response to auditory stimuli (approx. 10 μV) was distinctly higher than the alpha response to auditory stimuli (about 6 μV). Parietal responses to auditory stimulation were similar to vertex recordings regarding the theta/alpha ratio (5 μV/2.5 μV). In the occipital position the responses – if any – were of low amplitude. Especially in the alpha range it was nearly impossible to discern the response from the averaged prestimulus EEG activity.

Figure 9.6b shows the respective frequency components of visual EPs. In the vertex case, theta amplitudes (4.5 μV) and alpha amplitudes (5 μV) were of similar magnitude. Left parietal responses had a smaller theta/alpha ratio (2.5 μV/4.5 μV) than vertex recordings. In contrast to auditory stimulation, visual stimulation elicited a large occipital alpha response.

For statistical evaluation the averaged transient EPs were digitally filtered for each subject (#1–#24; frontal data available in subjects #12–#24) within several frequency bands (1–3.5 Hz, 4–7 Hz, 8–15 Hz, 15–30 Hz). Maximal peak-to-peak amplitudes were determined for the obtained oscillatory waveforms. In Fig. 9.7 the medians of these maximal amplitudes are represented by the heights of the respective bars. Open bars refer to auditory stimulation, hatched bars to visual stimulation.

With respect to vertex recordings, the most prominent differences are observed in the theta range. Here the maximal amplitudes of auditory EPs were significantly higher than those of visual evoked potentials. In contrast, there was no significant difference between the amplitudes of alpha components of auditory and visual EPs in the vertex recordings. In the occipital position,

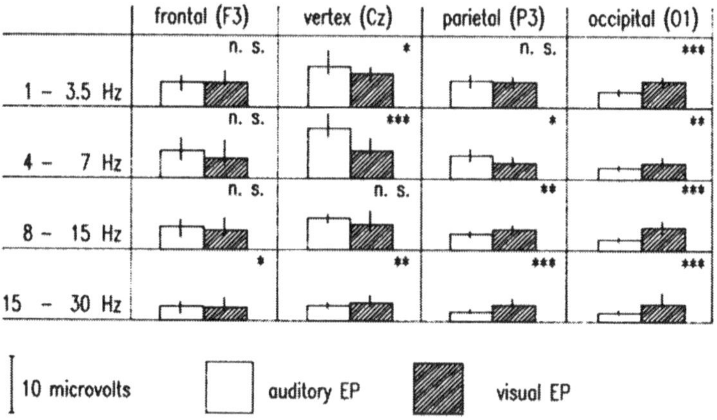

Fig. 9.7. Maximal poststimulus amplitudes in filtered evoked potentials (medians of subjects #1–#24; F3 available in subjects #12–#24). *Open bars*: auditory stimulation; *hatched bars*: visual stimulation. The heights of the bars represent the medians of the maximal amplitudes. The 95% confidence interval for the median is indicated as well. *Columns* refer to left frontal, vertex, left parietal, and left occipital recordings; *rows* refer to different band-pass filters applied. The significance limits for the comparisons auditory vs. visual EP are also given: (*) : $p < 0.05$ (**)$p < 0.01$; (***)$p < 0.001$. (From Schürmann M. and Başar E. (1994) *Biological Cybernetics* 72: 161–174.)

visual stimulation elicited higher amplitudes than did auditory stimulation. This applied to alpha and beta frequency ranges, but not to delta and theta. This finding is dealt with in Chap. 10.

In summary, *theta responses* were elicited both by auditory and visual stimuli, their amplitudes decreasing from vertex to occipital. For vertex and parietal positions, theta responses to auditory stimulation tended to have higher amplitudes than their visual counterparts. In contrast, *alpha responses* were generally higher for visual stimulation. The auditory/visual difference in alpha responses was most marked for occipital recordings. In other words: visual stimulation elicited high amplitude occipital alpha responses. Auditory stimulation did not elicit such responses.

9.2 Brain Resonance Phenomena and Their Manifestation in Evoked Potentials

Before interpreting these results in the framework of resonance phenomena (Başar 1980) and induced rhythmicities in the brain (Başar and Bullock 1992), we briefly summarize our results on topography- and modality-dependent features of EPs:

1. *topography-dependent features of auditory EPs:* theta–alpha compound responses – with dominant theta contribution – were recorded from frontal, vertex, and parietal, but not from occipital, electrode sites;
2. *topography-dependent features of visual EPs:* alpha maxima were typical, the amplitudes of the alpha maxima depending on the recording site;
3. *modality-dependent differences of auditory vs. visual EPs:* upon auditory stimulation, AFC maxima covered lower frequency ranges than upon visual stimulation, the peaking frequencies being higher in the visual case (median values for vertex recordings: auditory stimulation, 7 Hz; visual stimulation: 9 Hz; median values; $p < 0.01$, Wilcoxon). A marked modality-dependent difference was observed in occipital EPs: high-amplitude alpha responses were observed with visual stimulation, but not with auditory stimulation. In contrast, theta responses had comparable amplitudes for auditory and visual stimulation.

The discussion of these results will focus on the implications of these results with respect to different functional roles of theta and alpha EP components. Furthermore, results are to be discussed in relation to resonance phenomena in the brain (see Chap. 7 of Vol. I). As stated previously, such phenomena are not confined to the EEG level but can also be observed at the cellular level and in MEG recordings:

1. *A neuromagnetic study* by Narici et al. (1990) dealt with the spatial structure of brain rhythms enhanced by photic and somatosensory stimulation. In the visual modality, the synchronization was characterized by a potentiation of the subject's alpha. In the somatosensory modality, two different activities were observed: one probably related to the Rolandic mu-rhythm, the second suggested by the presence of two widely separated and time-correlated sources possibly driven by a unique deep clock. These authors conclude that the single frequency resonance elicited by visual stimulation may be located in a region close to the calcarine fissure. Saermark et al. (1992) describe how it is possible to distinguish between induced alpha and theta rhythmicities that can be recorded as more distinct 5 Hz and 10 Hz rhythmicities, depending on the location of the detectors.
2. *With respect to the cellular level,* Llinás (1988) states that if a thalamic neuron is depolarized, it may oscillate at 10 Hz; if it is hyperpolarized, it tends to oscillate at 6 Hz. He concludes that oscillation and resonance allow single elements in the central nervous system to be woven into functional states capable of representing and embedding external reference frames into neural connectivity (see Chap. 9, Vol. I for details). Gray and Singer (1987; Gray et al. 1989) and Eckhorn et al. (1988) recorded oscillatory responses (frequency range: 35–85 Hz) in the striate cortex of the cat brain. These oscillations are synchronous for cells located within a functional column and are tightly correlated with local oscillatory field potentials. It was hypothesized that the synchronization of oscillatory responses in spatially distributed feature-selective cells might establish relations between

features in different parts of the visual field ("binding"). Even neurons in spatially separated columns may synchronize their oscillations, depending on the spatial and orientation preference of the cells.

The measurements here are complemented by cat intracranial recordings of brain resonance phenomena (see Chap. 1). Furthermore, methods from nonlinear dynamics that indicate that the "spontaneous" EEG is not merely "noise" also support the hypothesis outlined above (Babloyantz et al. 1985; Röschke and Başar 1985; for an overview see, e.g., Başar 1990).

9.3 Single EEG-EP Epochs, Averaged EPs, and AFCs for the Study of Brain Resonance Phenomena

Single EEG-EP epochs ("single sweeps") subjected to "response adaptive filtering" (Başar and Ungan 1973) are probably most appropriate and suitable to depict "brain-induced rhythmicities" reflected in EPs. In earlier studies (Başar 1980), however, the existence of maxima in AFCs was demonstrated to be a manifestation of the enhancement (amplitude increase, frequency stabilization) in single responses. Taking this into account, the exemplary investigation presented in this chapter demonstrates that global statements can also be derived from averaged EPs and their AFCs.

In AFCs of mean value curves it is usually not possible to clearly separate theta or alpha peaks. Especially in human vertex responses to auditory stimuli it is difficult to describe a "pure" frequency response. Therefore, the term "compound theta/alpha response" refers to a fusion of resonances in the fast theta and slow alpha ranges (Başar 1980).

9.4 Functional Correlates of Theta and Alpha EP Components in Responses to Inadequate and Adequate Stimuli

As mentioned above, modality-dependent differences in auditory vs. visual EPs were observed, with AFCs for auditory stimulation covering lower frequency ranges than AFCs for visual stimulation and peaking frequencies being higher in the visual case (by approx. 2–3 Hz). Modality-dependent differences in occipital EPs are of special interest because auditory stimuli are "inadequate" and visual stimuli "adequate" for the visual cortex underlying these electrodes. In Chap. 10 these results are discussed in the framework of functional roles of alpha and theta responses.

9.5 Prospective and Future Research

As the results presented in this chapter are confined to frequency ranges below 30 Hz, the reader is referred to other chapters (3 and 26) for investigations on EP components in the 40 Hz range.

Only vertex, left parietal, and left occipital EPs have been presented above. Evaluations of right parietal (P_4) and right occipital (O_2) electrodes showed that laterality differences were small in comparison to modality-dependent differences and to differences of vertex vs. parietal vs. occipital electrodes. Other types of stimuli might elicit laterality-dependent EP waveforms. In our study, only five positions of the ten–twenty system were used. Closer electrode spacing certainly would give additional information. We used a left ear lobe reference, which is popular for EP studies. The effect of further referencing techniques remains to be investigated (Lehmann 1989).

For some remarkable properties of frontal recordings, the reader is referred to Demiralp and Başar 1992 (see Chap. 19): In this study, standard auditory and visual EPs were compared with responses to stimuli preceding an omitted stimulus, with subjects performing a time-prediction task.

In a number of other studies, e.g., by Goff et al. (1977) and Allison et al. (1977), topography- and modality-dependent differences in EP waveforms were investigated (without frequency-domain analysis). The topography of their waveforms for vertex, left parietal, and left occipital EPs are comparable to those presented here. However, only a rough comparison between time series of representative subjects can be made; their further evaluation steps are quite different from ours due to the differences in our approaches. Given these limitations, it is noteworthy that they obtained time series similar to ours.

9.6 Conclusions

The exemplary results summarized in this chapter show topography- and modality-dependent EP differences in the frequency domain:

1. theta–alpha compound responses – with dominant theta contribution – to auditory stimulation were recorded from frontal, vertex, and parietal sites, but not from the occipital location;
2. consistent alpha maxima in the responses to visual stimulation, the amplitudes of the alpha maxima depending on the recording site;
3. a modality-dependent difference in peaking frequencies of dominant AFC maxima: upon visual stimulation, the peaking frequencies were about 2–3 Hz higher than upon auditory stimulation (e.g. in vertex recordings).

Based on a concept of EPs as manifestations of brain resonance phenomena (Başar 1980), we suggest the interpretation that neural structures with different resonance properties may be involved in the processing of the auditory and visual stimuli applied.

10. "Cross-Modality" Experiments in Humans

Erol Başar and Martin Schürmann

As emphasized in previous chapters, the complex dynamics of compound potentials and resonance phenomena of the brain might play one of the most important roles in brain organization. On the basis of earlier research on the dynamics of EEG and ERPs, the present chapter focuses on an experimental approach involving *cross-modality stimulation.* Such experiments help to clarify *functional correlates* of the alpha and theta response, i.e., response components found by applying Fourier analysis to EPs and ERPs. The terms "theta" and "alpha response" refer to relatively short periods of damped oscillations that are not due to filtering techniques or to Fourier artifacts. Most probably they are due to the effects of change of the EEG immediately following stimulation and visible even in single EEG-EP epochs (see Chap. 9). The "alpha response," e.g., is defined as the oscillatory EEG or MEG activity in the first 200–300 ms following sensory stimulation.

Cross modality experiments, e.g., measuring auditory and visual stimuli with electrodes close to the visual cortex, permit one to investigate responses to "inadequate" (here: auditory) and "adequate" (here: visual) stimuli. On the basis of such "cross-modality" experiments it may become possible to describe global interacting areas in the brain or structures that could be defined as a "selectively distributed alpha response system" and a "selectively distributed theta response system." In the present chapter such "cross-modality experiments" with occipital-scalp electrodes in humans beings are described. These data will be compared with corresponding intracranial recordings from the cat brain.

The results of these "cross-modality" experiments demonstrate that the differences between responses to adequate stimulation and responses to inadequate stimulation ("adequate/inadequate differences") are marked in the alpha range and less marked in the theta range. This hints at a possible special role of EP alpha components in the processing of adequate stimuli. In contrast, EP theta components appear to be less dependent on whether the stimulus is adequate or not, which is compatible with a predominantly associative–cognitive role of the theta response (cf. Başar et al. 1991; Demiralp and Başar 1992; Başar-Eroğlu et al. 1992).

10.1 Analysis of Evoked Potentials and Their Frequency Characteristics: Auditory and Visual Stimuli

The experiments were carried out on 11 voluntary healthy subjects who also participated in the measurements presented in Chap. 9. Taking into account the close vicinity of this site to the visual cortex, auditory stimuli will be referred to as "inadequate stimuli" and visual stimuli as "adequate stimuli."

Data were evaluated with the set of methods presented in Chap. 9. Auditory and visual EPs were filtered with band-pass limits of 4–7 Hz (theta) and 8–15 Hz.

Figure 10.1a shows responses to auditory stimulation (left) and visual stimulation (right) recorded in the subject B.S. (averaged EPs; electrode position: left occipital). With visual stimulation, a clear response with a positive peaking around 100 ms poststimulus was seen. The response elicited by auditory stimulation, if any, was of low amplitude.

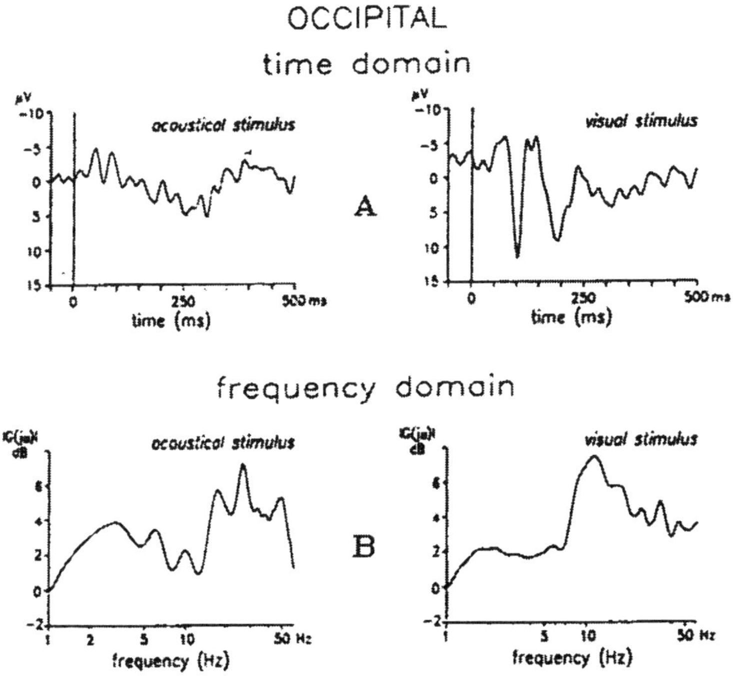

Fig. 10.1. Data for one typical subject; electrode position: left occipital. (**a**): occipital evoked potentials (EPs) in time domain. (**b**): amplitude–frequency characteristics computed from averaged EPs shown in (**a**). *Left*, responses to acoustical stimulation; *right*, responses to visual stimulation. (From Başar E. and Schürmann M. (1994) *Biological Cybernetics* 72: 175–183.)

Figure 10.1b shows amplitude–frequency characteristics computed from the averaged EPs of Fig. 10.1a. In the case of visual stimulation, the most dominant peak was in the 10 Hz (alpha) range. In contrast, there was no such peak for auditory stimulation. On the other hand, the differences for auditory vs. visual stimulation were small in the delta–theta range (3–6 Hz).

Figure 10.2a shows averaged EPs recorded in $N = 11$ subjects (unfiltered, left occipital recordings). Again, the left column refers to auditory stimulation, whereas the right column refers to visual stimulation. In nearly all subjects, distinct responses can be seen following visual stimulation. In

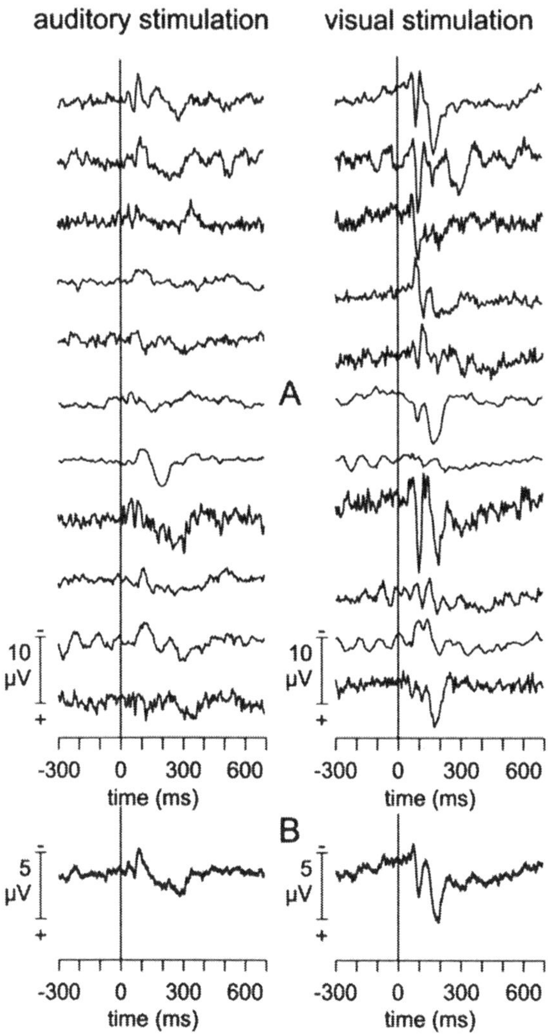

Fig. 10.2. (**a**): averaged EPs for $N = 11$ single subjects; (**b**): grand average ($N = 11$) computed from the curves shown in (**a**). *Left*, responses to auditory stimulation; *right*, responses to visual stimulation. (From Başar E. and Schürmann M. (1994) *Biological Cybernetics* 72: 175–183.)

contrast, the responses elicited by auditory stimulation were generally of low amplitude.

Figure 10.2b shows grand averages ($N = 11$) computed from the curves of Fig. 10.2a. As in single subjects, a distinct response is seen with visual stimulation, whereas the response following auditory stimulation is of small amplitude.

Figure 10.3a shows amplitude–frequency characteristics computed from the averaged EPs of Fig. 10.2a. Again, the left column refers to auditory stimulation; the right column refers to visual stimulation. In the case of visual stimulation, maxima in the alpha range (in the region of 10 Hz) are frequent.

Common features in the amplitude–frequency characteristics of Fig. 10.3a are illustrated in Fig. 10.3b, showing mean values of amplitude–frequency characteristics. These curves were obtained by averaging the curves of Fig. 10.3a in the frequency domain. For visual stimulation, a distinct peak covering the 7–25 Hz range is seen; submaxima in the 7–12 Hz range and in the 12–15 Hz range are present. In contrast, the mean value of amplitude–frequency characteristics for auditory stimulation is a flat curve without distinct maxima.

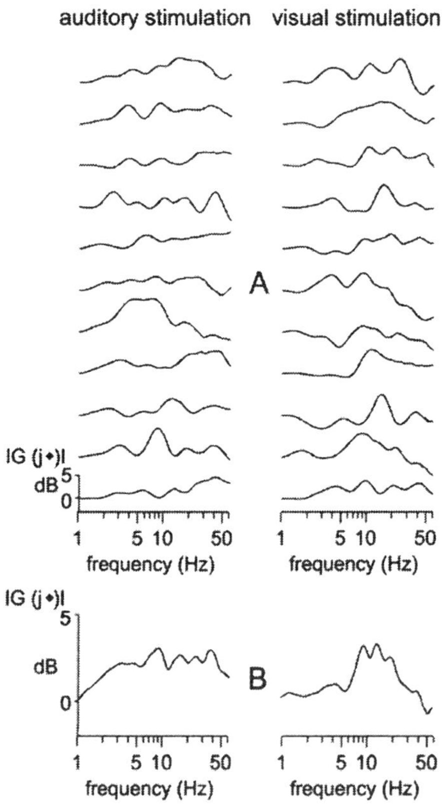

Fig. 10.3. (**a**): Amplitude–frequency characteristics computed from averaged EPs shown in Fig. 10.2; (**b**): mean value ($N = 11$) of amplitude–frequency charateristics shown in (**a**) (averaging performed in frequency domain). *Left*, responses to auditory stimulation; *right*, responses to visual stimulation. (From Başar E. and Schürmann M. (1994) *Biological Cybernetics* 72: 175–183.)

10.2 Filtered Evoked Potentials

The grand average EPs of Fig. 10.2b were filtered with band-pass limits of 4–7 Hz and 8–15 Hz, resp. (for the choice of filter limits, see Chap. 4 in Vol. I). Figure 10.4 is a comparative presentation of theta (4–7 Hz) and alpha (8–15 Hz) responses to auditory and visual stimulation in left occipital recordings. Grand average EPs filtered in the alpha range show a *marked alpha response to visual stimulation* (approx. 3.2 μV). The *alpha response to auditory stimulation*, if any, is of small amplitude (approx. 0.8 μV); it is nearly impossible to distinguish the prestimulus EEG from the EP. In contrast, the difference between the *theta response to visual stimulation* vs. the *theta response to auditory stimulation* was considerably smaller (theta/visual EP: approx. 1.6 μV; theta/auditory EP: approx. 0.8 μV).

In order to facilitate "adequate/inadequate" comparisons, maximal peak-to-peak amplitudes of the alpha components were computed for each subject. The medians and 95% confidence intervals of these values are given in Table 10.1. The medians are also shown in Fig. 10.5, where solid bars refer to acoustical – i.e., "inadequate" – stimulation, and hatched bars refer to visual – i.e., "adequate" – stimulation. As to theta responses, the amplitude difference for auditory vs. visual EPs is small. In contrast, there was a marked amplitude difference for alpha responses – auditory vs. visual EP. In other words, the "adequate/inadequate" difference was significant in the alpha range and insignificant in the theta range.

From the data summarized in Table 10.1, ratios of "maximal amplitude in visual EP/maximal amplitude in auditory EP" were computed. These ratios are referred to as "adequate/inadequate ratios." The "adequate/inadequate ratio" was 1.95 for the EP alpha component and 1.60 for the EP theta component (medians of $N = 11$ subjects; Wilcoxon: $p < 0.05$). This confirms the observation made in the filtered components of the grand averages: marked

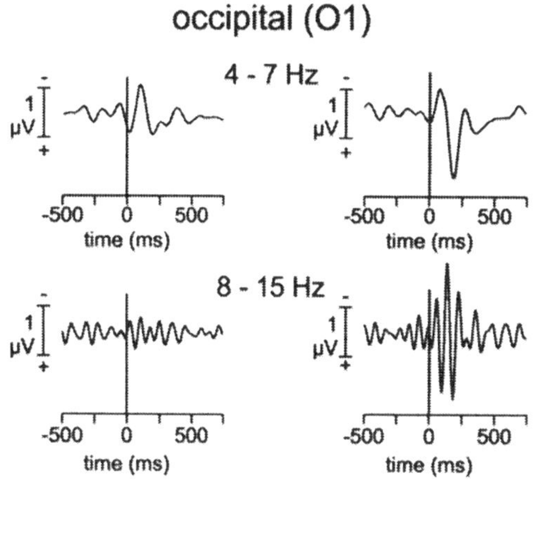

Fig. 10.4. Frequency components of grand averages ($N = 11$). *Top*, filter limits: 4–7 Hz, "theta response"; *bottom*, filter limits: 8–15 Hz, "alpha response." *Left*, responses to auditory (i.e., inadequate) stimulation; *right*, responses to visual (i.e., adequate) stimulation. (From Başar E. and Schürmann M. (1994) *Biological Cybernetics* 72: 175–183.)

Table 10.1. Maximal amplitudes of filtered auditory EPs and visual Eps. (From Başar E. and Schürmann M. (1994) *Biological Cybernetics* 72: 175–183)

Filter range	Median (95% confidence interval) auditory EP	visual EP
theta (4–7 Hz)	2.1 (1.6–2.4)	2.8 (2.2–3.8)
alpha (8–15 Hz)	1.9 (1.5–2.3)	4.0 (3.0–5.2)

"adequate/inadequate" difference in the alpha range; considerably smaller difference in the theta range.

10.3 Cross-Modality Responses Analyzed with Single EEG-EP Sweeps

As a supplement to Figs. 10.4 and 10.5, single-trial EPs (left occipital recordings) obtained with auditory stimulation were compared with single-trial EPs obtained with visual stimulation, as shown for a typical subject (Fig. 10.6, subject K.K.) Single sweeps (a) and the averaged EP (b) have been filtered in the same frequency range, 8–15 Hz (alpha).

In the case of *visual EPs filtered in the alpha range* there were a number of sweeps with a high degree of similarity to the averaged EP with respect

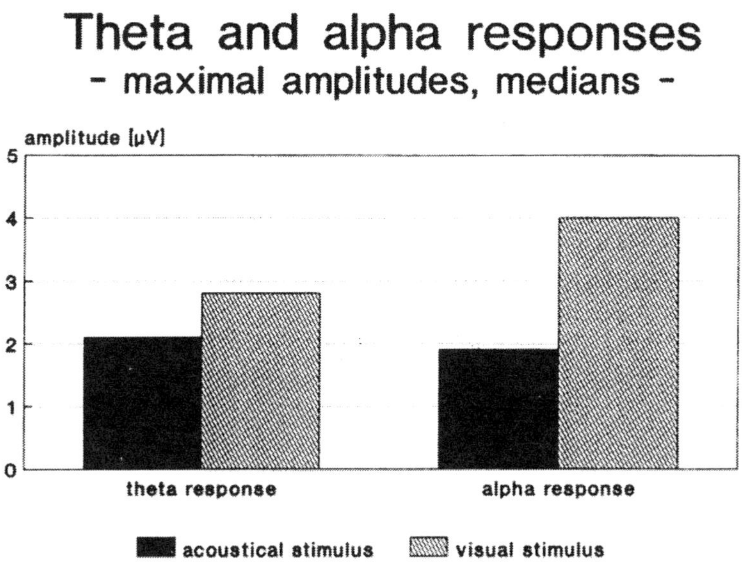

Fig. 10.5. Maximal poststimulus amplitudes of theta and alpha responses (medians of measurements in $N = 11$ subjects). (From Başar E. and Schürmann M. (1994) *Biological Cybernetics* 72: 175–183.)

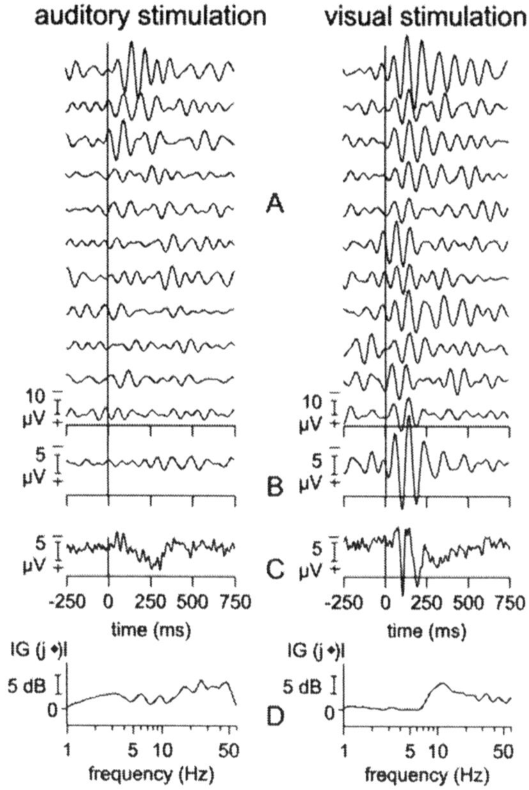

Fig. 10.6. Data for subject K.K. (**a**): filtered single sweeps, 8–15 Hz; (**b**): filtered averaged evoked potentials. *Left column:* auditory stimulation; *right column:* visual stimulation. (From Başar E. and Schürmann M. (1994) *Biological Cybernetics* 72: 175–183.)

to waveform, amplitude enhancement, and time locking. In other words, distinct alpha responses were observed with visual stimulation, but not with auditory stimulation. It is remarkable that in an equal number of sweeps recorded with *auditory stimulation* there were no sweeps similar to the *averaged EP* (Fig. 10.6b), thus confirming the observation made in the filtered grand averages of Fig. 10.4. For a comparison of these data with intracranial recordings in cats and for remarks on the relation between single sweeps and amplitude–frequency characteristics, see below in this chapter.

10.4 Immediate Interpretation of Cross-Modality Experiments

In the scope of resonance phenomena in the brain (Başar and Bullock 1992), it has been hypothesized that topographic differences of EEG and MEG responses might partly reflect the activity of *distinct cortical areas* for which the applied stimulation is either *adequate or inadequate* (Başar et al. 1991, Başar et al. 1992a). The experimental basis of this hypothesis was founded on combined EEG-evoked potential measurements in the cat brain by means of intracranial electrodes (see Chap. 2): Responses to auditory and visual stimulation were recorded from the auditory cortex and from the visual cortex ("cross-modality" experiments, cf. Hartline (1987)).

In the present chapter auditory and visual EPs were recorded in human beings using occipital scalp electrodes. Taking into account the close vicinity of the electrode site to the visual cortex, auditory stimuli were defined as "inadequate stimuli" and visual stimuli as "adequate stimuli." The evaluations showed that the alpha components of responses to adequate stimuli were of a considerably higher amplitude than the alpha components of responses to inadequate stimuli (marked "adequate/inadequate" difference) – both in averaged EPs and in single sweeps. The respective "adequate/inadequate" difference for theta components was smaller. In other words, the amplitudes of alpha responses – but not of theta responses – were strongly dependent on whether or not the stimulus was adequate.

What are the physiological implications of these differences? As to different functional roles of EP frequency components, the hypothesis underlying this approach is as follows: Theta and/or slower responses appear to be involved mainly in associative and cognitive processes, whereas alpha responses are more involved with primary sensory processing.

On the one hand, the present analysis – based on occipital scalp recordings in human beings – extends the scope of previous chapters based on intracranial recordings in cats. On the other hand, cat intracranial recordings – with their good spatial resolution – may compensate for the lack of spatial resolution in human scalp measurements, thus confirming the results obtained in humans (see below in this chapter). The latter goal – to confirm our results by improving the spatial resolution – may also be achieved by MEG recordings in human beings (see also Chap. 13).

10.5 Cat Intracranial Recordings Support the Results from Human Data

The hypothesis on different functional roles of induced oscillations in processing adequate and inadequate stimuli was based mainly on cat intracranial recordings (see Chap. 2). In these measurements, the respective electrodes were placed directly in the primary sensory areas of the visual and auditory cortex. On comparing adequate to inadequate stimulation, similar amplitudes in the delta–theta range and considerable differences in the alpha range were found, thus supporting a functional differentiation of alpha and theta responses. Marked single-trial alpha responses were observed only with adequate stimulation, as demonstrated in Chap. 2 (see Chap. 1 for noncortical data).

10.6 Physiological Implications of "Cross-Modality" Experiments: Possible Functional Roles of Induced Rhythmicities

Possible physiological implications of the results in this chapter and of corresponding results in cat intracranial recordings (Chap. 2) were discussed in an earlier paper (Başar et al. 1991). Briefly, it was hypothesized that "adequate" stimulation of a cortical area (e.g. visual stimulation/visual cortex) is followed by a response with a high-amplitude alpha component and a theta response of small amplitude. "Inadequate" stimulation elicits responses with less difference in the theta and alpha responses; i.e., in the case of inadequate stimulation, the theta response is less reduced than the alpha response. In other words, the alpha response appears to be more dependent on the "adequacy" of the stimulus than the theta response.

Hartline (1987) deals with multisensory representation of space in the central nervous system and states, "The visual cortex is thought to participate in processes subserving the recognition of objects or patterns. Though this part of the brain is considered to be involved exclusively with vision, in cats about one-third of the neurons in area 17 (striate cortex), 18, and 19 are reported to be responsive to sound as well as to visual input."

A more detailed tentative explanation of these alpha–theta differences has been given previously: "Adequate stimulation of the brain elicits maximal alpha responses in appropriate or adequate primary sensory areas. In the case of inadequate stimulation the alpha response is highly reduced. However, this reduction is less striking for the delta/theta responses. ... The alpha responses appear to be more dependent on the adequacy of the stimulus. ... A large number of neurophysiological studies showed that primary sensory stimuli elicit impulses, or volleys converge over thalamic centers to primary sensory areas. On the other hand, the "sensory stimulation of second order" usually reaches the cortex over association areas (see for example Shepherd [1988]). Due to these findings it is conceivable that the responses in the lower frequency ranges (theta, delta) might reflect the responsiveness of various brain areas dealing with association processes involved in global associative cognitive performance. ... The first tentative interpretation of our results led us to formulate that the alpha response component mostly contains the primary sensory processing of signals, whereas the theta response and/or slower responses are mostly involved in procedures of association and cognition" (Başar et al. 1991).

This interpretation is also supported by results obtained with a different paradigm: Especially slow frequencies contribute to differences between EPs obtained in an omitted stimulus paradigm and EPs recorded in a session without cognitive load. In a time-prediction task, selective averaging of responses to the last stimulus before omission showed increased delta–theta amplitudes (Demiralp and Başar 1992). This is noteworthy because the increase was most prominent in frontal and parietal electrodes that are closely related to association areas of the brain (see Chaps. 19, 25).

10.7 Responses to Adequate and Inadequate Stimuli in MEG Recordings in Human Subjects

In a series of MEG measurements (Başar et al. 1992; Schürmann et al. 1992, Chap. 13) auditory stimuli were applied to measure magnetic responses above the temporoparietal region and above the occipital region of the head. These measurement positions were related to "adequate" stimulation (auditory stimuli/temporoparietal position; auditory cortex) and to "inadequate" stimulation (auditory stimulus/occipital position; visual cortex). Our results (see Chap. 13) are in accordance with the results in the present chapter: Alpha responses were present over the temporoparietal cortex and absent – or of low amplitude – over the occipital cortex. Delta–theta (2–6 Hz) components of MEG responses showed less dependence on measurement site.

So far, we have only evaluated data from one scalp position (referring to visual cortex) in human subjects, with the small number of subjects prohibiting final conclusions. We still lack complementary data concerning auditory and visual stimulation of the auditory cortex. It is conceivable that such measurements can better be performed by using MEG recordings; these have already been begun by our group (see Chap. 13).

10.8 Further Thoughts Concerning Functional Correlates of Theta and Alpha Responses

The evaluations summarized in this chapter led to one main result: In recordings from an occipital electrode, marked differences in alpha responses were observed, with low-amplitude alpha responses to auditory stimuli and high-amplitude alpha responses to visual stimuli. Theta responses showed smaller amplitude differences; i.e., they were less affected by the stimulus modality.

With the occipital electrode being close to the visual cortex, this finding may be put as follows: There were low-amplitude alpha responses to *inadequate* stimuli and high-amplitude alpha responses to *adequate* stimuli, whereas the theta responses were less different for inadequate vs. adequate stimuli.

In connection with similar data obtained in cat intracranial recordings (Başar et al. 1991; Başar-Eroğlu et al. 1991a,b) and in human MEG recordings (Başar et al. 1992; Schürmann et al. 1992), these data may indicate *differentiated functional roles of alpha and theta EP components:* Theta and/or slower responses appear to be involved mainly in associative and cognitive processes, whereas alpha responses seem to be more related to primary sensory processing. This supports a view of a close connection of sensory and cognitive mechanisms. The distributed alpha and the distributed theta systems are described extensively in the chapter dedicated to integrative neurophysiology (Chap. 28).

11. The Bisensory Evoked Theta Response – A Correlate of Supramodal Association?

Oliver W. Sakowitz and Erol Başar

The previous chapters have dealt mainly with results in human subjects upon simple stimulation (i.e. event-related potentials). The former analysis was centered on

i. steps of sensory processing (for instance, auditory or visually evoked potentials, i.e., unisensory stimulation) and
ii. the higher cognitive functions involved.

Our real-world experiences do not consist solely of elementary input signals. Accordingly, we ask how the brain processes sensory input of more than one modality, and how, as we will discuss in the following, the process of multisensory convergence could be assessed by electrophysiological means.

In terms of methodology we used a recording setup that included three consecutive measurements of event-related potentials. Stimulus conditions were tone bursts (AEP – auditory evoked potentials) of 2000 Hz with 1 s duration and randomized interstimulus intervals of 1.0–3.0 s, whereas the visual stimulus condition (VEP – visually evoked potentials) consisted of light signals of the same duration (see Chap. 9). The third part of the experiment combined the former two settings in the sense of simultaneous stimulus presentation (BEP – bimodally evoked potentials).

Previous results (Sakowitz and Başar 1996) showed the occurrence of theta-range resonances in amplitude–frequency characteristics predominantly over the association areas of the frontal midline and the temporoparietal locations. By calculating enhancement factors in the 4–7.5 Hz range we were able to demonstrate the topographical dependence of the theta response observed under bisensory stimulation. Figure 11.1 shows an excerpt of single epochs of a representative healthy subject at the Cz recording site. A digital pass-band filter (4–7.5 Hz, i.e., theta) has been applied. The lowest two curves both show the averaged results (computed from $n = 30$ single epochs). Two digital pass-band filters with 4–7.5 Hz and 0.1–70 Hz have been applied. Ten out of thirty single trials are plotted to visualize pre- and poststimulus EEG. The stimulus onset is marked with a guideline. As a measure for resonant behavior, enhancement factors have been calculated (given at the right side of corresponding sweeps).

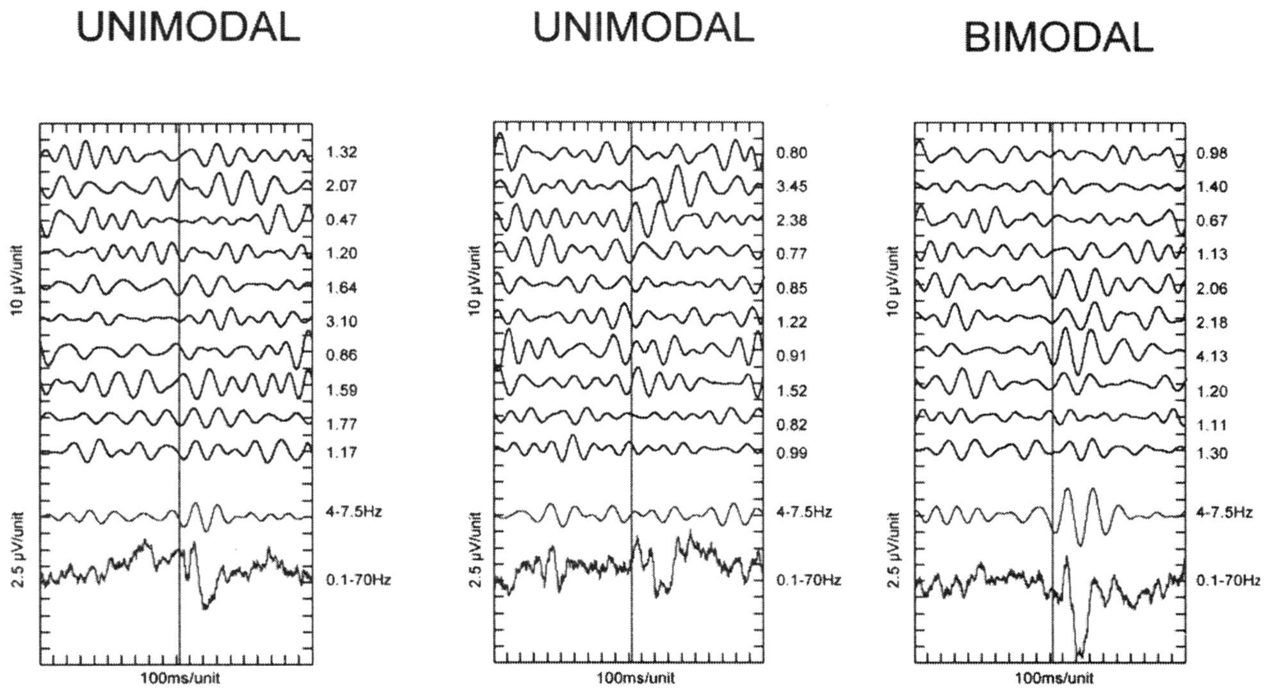

Fig. 11.1. Single-trial and averaged EPs recorded with unimodal and bimodal stimulation in a typical subject. In each column, the lowest two curves show the average of $n = 30$ trials (filtered 0.1–70 Hz and 4–7.5 Hz, respectively). *The upper curves* show ten out thirty single trials (filtered 4–7.5 Hz). Enhancement factors for single trials are given *at the right margin*. The stimulus onset is marked by *a vertical line*.

The enhancement of theta-band components was most stable in fronto-central and was lowest in occipital recording sites. Statistical analysis shows that this "frontality" of audio-visually evoked potentials based on (1) the increment of amplitude per se and (2) the prolongation of this oscillation. It has to be emphasized that the following findings were specific to the theta–delta frequency ranges. Table 11.1 summarizes the results of 15 healthy subjects who were exposed to the respective stimulation categories. To exclude the feature of "novelty," only the last 30 artifact-free trials were chosen out of 120 sweeps per category. Three time windows have been chosen arbitrarily, but based on the average phase pattern of the evoked responses. We calculated the means of the maximal peak-to-peak values and their standard deviation as they were measured over the specified recording sites using an

Table 11.1. Grand average ($n = 15$) of the mean maximal peak-to-peak values and their standard deviations as they were measured from specified recording sites. Single epochs were digitally filtered to extract their theta-range (4–7 Hz) components beforehand. Results were compared between stimulus–conditions according to the Wilcoxon–Wilcox matched pairs rank test (*: $p < 0.05$, **:$p < 0.01$; AEP – auditory evoked potential, VEP – visually evoked potential, BEP – bimodally evoked potential).

Early segment (50–200 ms)							Wilcoxon–Wilcox matched pairs rank tests		
	AEP		VEP		BEP		AEP-VEP	AEP-BEP	VEP-BEP
F3	9.03	(1.73)	7.62	(1.60)	10.74	(2.76)		*	**
F4	9.42	(1.59)	8.28	(1.11)	10.73	(2.26)			**
Cz	14.45	(3.46)	11.03	(3.05)	19.00	(6.96)	*		**
C3	11.14	(3.17)	8.81	(3.21)	14.40	(5.75)		*	**
C4	10.59	(2.01)	8.57	(1.62)	13.34	(4.16)	*	*	**
T3	5.26	(1.60)	4.68	(1.36)	6.16	(1.73)		*	**
T4	5.58	(1.10)	4.97	(0.79)	6.14	(1.36)		*	**
P3	7.81	(2.00)	6.92	(2.42)	9.59	(3.66)		*	**
P4	8.04	(2.18)	7.22	(1.64)	9.38	(3.66)			**
O1	5.85	(1.30)	5.96	(1.09)	6.92	(1.84)			*
O2	6.32	(1.29)	6.38	(1.06)	7.16	(1.77)			
@Middle segment (200–350 ms)							Wilcoxon–Wilcox matched pairs rank		
	AEP		VEP		BEP		AEP-VEP	AEP-BEP	VEP-BEP
F3	8.46	(1.79)	7.90	(1.95)	10.37	(2.51)		**	**
F4	8.88	(1.77)	8.40	(1.53)	10.36	(2.26)		**	**
Cz	13.63	(3.44)	11.47	(3.19)	17.85	(6.88)	*	*	**
C3	10.10	(3.25)	8.98	(3.19)	13.40	(5.83)		**	**
C4	9.81	(2.19)	8.84	(2.13)	12.39	(4.33)		*	**
T3	5.24	(1.77)	4.72	(1.35)	5.82	(1.68)		*	**
T4	5.21	(1.02)	4.92	(0.87)	5.70	(1.28)			
P3	7.28	(2.15)	6.92	(2.25)	9.14	(3.84)		*	*
P4	7.77	(2.39)	7.63	(2.46)	8.99	(4.00)			
O1	5.62	(1.08)	6.15	(1.52)	6.81	(2.18)		**	
O2	6.20	(1.04)	6.48	(1.73)	7.08	(1.96)			
Late segment (350–500 ms)							Wilcoxon–Wilcox matched pairs rank		
	AEP		VEP		BEP		AEP-VEP	AEP-BEP	VEP-BEP
F3	7.09	(1.72)	7.05	(1.80)	7.73	(1.64)		**	**
F4	7.40	(1.59)	7.44	(1.87)	7.78	(1.78)			
Cz	9.82	(2.57)	9.59	(2.52)	11.30	(3.32)			*
C3	7.54	(2.19)	7.36	(2.03)	8.46	(2.60)		*	**
C4	7.40	(1.82)	7.59	(2.02)	8.07	(2.36)			
T3	4.65	(1.76)	4.49	(1.52)	4.67	(1.27)			
T4	4.41	(0.73)	4.71	(1.02)	4.53	(1.07)			
P3	6.06	(1.28)	6.13	(1.54)	6.58	(1.63)			
P4	6.53	(1.75)	6.75	(2.07)	6.78	(2.22)			
O1	5.51	(0.95)	5.58	(1.56)	5.93	(1.55)			
O2	5.92	(0.78)	5.93	(1.68)	6.07	(1.25)			

"early" time window from 50 to 200 ms poststimulus. Before calculation, each single epoch was filtered by means of a digital band-pass (4–7 Hz). As can be easily seen, the means of maximal amplitudes are enhanced in the bisensory stimulation paradigm, with only a slight tendency to fronto-central sites.

These results are significant for all recording sites except occipital. With respect to maximal amplitudes, bisensory stimulation exceeds both acoustical and visual stimulation. Also, the unimodal event-related potentials show differences: The maximal AEP amplitudes exceed the VEP amplitudes significantly in central leads. Although it seems to be trivial that maximal amplitude changes are recorded over primary sensory cortices, i.e., areas that are reactive to "adequate" stimulation, the results in the "mid" time window from 200 to 350 ms poststimulus show that this is only partially true. While unimodal stimulation yields no significant differences except in Cz; the bimodal EPs show a clear superiority that was significant for fronto-central locations and left-hemispheric temporoparietal leads. The results of the "late" time window from 350 to 500 ms show that these differences diminish with further progression along the time axis. Only the bimodally stimulated potentials measured in F3 and C3 remain significantly enhanced in their relation to the unimodal sensory paradigms.

In particular, two issues are of importance:

(1a) For bimodal stimulation, the theta response is not manifested as a simple addition but as a high nonlinear increase.

(1b) This "theta increase" is observed in frontal cortex.

(2) The highly nonlinear response increase takes place in the theta band and not in the alpha band.

With respect to the first issue, Fig. 11.2 demonstrates the frontality of the theta response as it was observed in a typical subject. Median enhancement factors of the given channels were used to plot gray-scale-coded surface

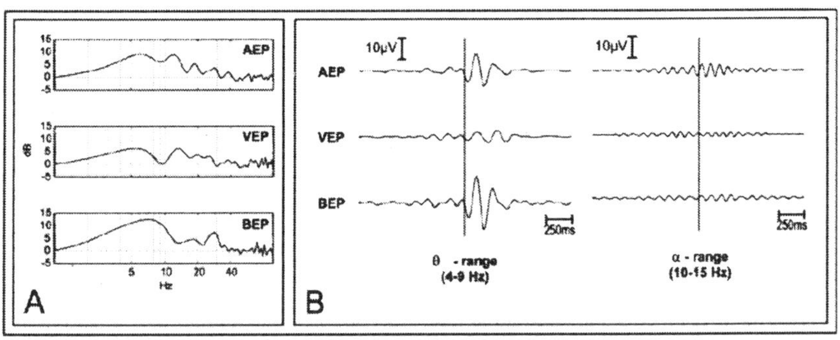

Fig. 11.2. Topography of median enhancement factors ($n = 56$) calculated for subject SW. *Rows* refer to stimulus modality (AEP, auditory; VEP, visual; BEP, bisensory), whereas *columns* take account of the frequency ranges chosen for computation (θ range: 4 to 7.5 Hz, α range: 8 to 15 Hz).

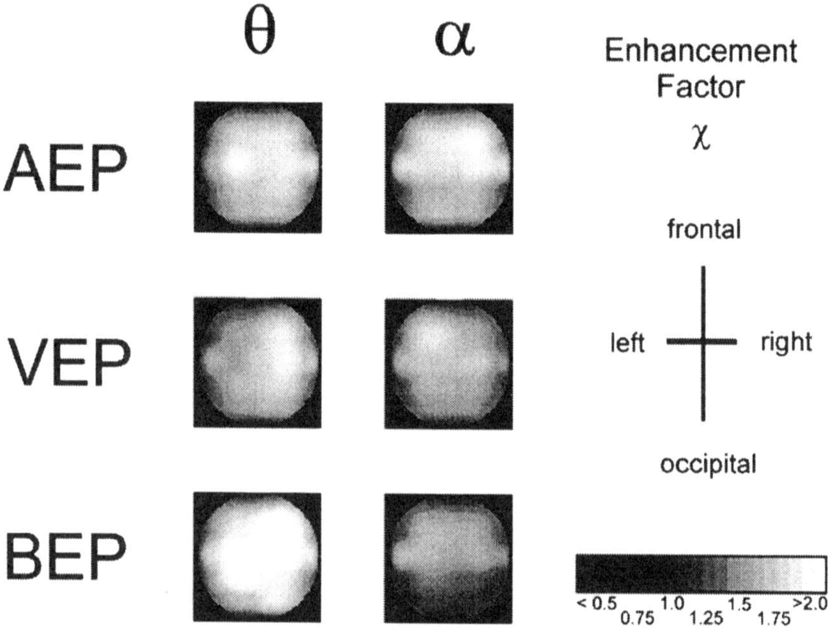

Fig. 11.3. (**a**). Amplitude–frequency characteristics of subject GH recorded from Cz with AEP (acoustical), VEP (visual), BEP (bimodal) stimulation paradigms applied succeedingly ($n = 30$). *y-axis* in dB ($20 \log |G(j\omega)|$), normalized to 1 Hz component ($20 \log 1 = 0$). *Abscissa* in logarithmic scale. (**b**). Averaged responses in Cz after applying digital band-pass filters to the single-trial recordings (with filter limits according to the results obtained in (**a**), θ-range filter with center frequency of 6.5 Hz, α-range filter with center frequency of 12.5 Hz).

maps for the sake of topological orientation. Both frontality and demarcation against alpha responsiveness are clearly visible.

Another example that stresses the issue of frequency-band specifity is shown in Fig. 11.3. A combined analysis approach was used, first computing amplitude–frequency characteristics (a) of the system under stimulation in order to define digital bandpass filters matching the system's resonance properties. As the means of filtered sweeps (b) show, the theta response upon bisensory stimulation is definitely superior to the alpha response in its effective contribution to the recorded surface potentials.

As Başar and Schürmann (1994) have pointed out, the EEG activity over primary sensory cortices was affected by either adequate or inadequate stimulation (cf. Schürmann and Başar 1994 and Chaps. 2 and 10). By applying ERP frequency analysis, a clear distinction could be made, which furthermore led to proposals about functional correlates of evoked oscillatory responses in the alpha range.

An extended view of this emphasizes that the functional correlates of responsiveness in the alpha range might be primary sensory functions, whereas oscillatory responses in lower frequencies such as theta and delta are more correlated with associative functions (Başar and Schürmann 1994, 1996) and, furthermore, topologically frontalized (Demiralp and Başar 1992).

It has to be mentioned that the predominance of theta rhythms in the frontal midline is not a new finding: Arellano and Schwab (1950) reported that theta rhythm appeared in the midline area just anterior to the vertex during problem solving. The investigations of Ishihara and Yoshii (1972) clarified some characteristics of the rhythm they called frontal midline theta (Fm Theta). This distinct theta activity (center frequency: 6.5 Hz) of about 30–60 μV was reported to appear during mental activity (Ishihara and Yoshii 1972; Mizuki et al. 1980) such as arithmetic tasks and reading (see Chap. 25).

While the occurrence of frontal midline theta may be of completely different origin, reviews on rhythmogenesis (for a comprehensive discussion on this see Miller (1991) point toward the involvement of corticohippocampal circuits, which in the case of bisensory stimulation could make sense with respect to the coactivation of different modalities and the intrinsic mechanisms of memory functions. At least on the animal level these phenomena have been extensively studied, and it seems beyond all doubt that not only subcortical structures (e.g., superior colliculus) (Meredith and Stein 1983, 1985, Hartline 1987) but also higher areas of both archi- and neocortical origin (Stein et al. 1993; Wallace et al. 1992; Wallace and Stein 1994; Demiralp et al. 1994) are involved.

Demiralp et al. (1994) found in nonanaesthetized cats that upon omission of a stimulus in a train of tones with fixed intervals oscillatory responses occurred with enhanced amplitudes in the theta range. By use of deep electrodes these could be identified in hippocampus and premotor cortex, which served as an equivalent to the human frontal lobe cortices. In the light of this and other studies (Başar-Eroğlu et al. 1992, Demiralp and Başar 1992) the authors postulated a "distributed theta-response system" in the brain that is involved in the cognitive states of focused attention and expectancy (see Chap. 12 in Vol. I).

Accordingly, we state that the frontal theta response is involved in complex sensory responses. With respect to electrophysiology, the frontal cortex is a dominant area in theta processing. It is also to be noted that the frontal lobe plays a major role in gamma responses to ambiguous pattern reversals (Chap. 26; furthermore, see Chaps. 18 and 19).

In combination with the findings of the bisensory evoked theta responses, this could mean that, as a crude generalization, the "selectively distributed theta-response system" manifests itself whenever the internal concepts of the brain interact with the representations of external sources, namely the streams of primary sensation.

The findings on the bisensory evoked theta response indicate that this processing (both from a physiologist's and a psychologist's point of view) is more than the integration of sensory input or just a "summing up" of information.

12. Evoked Delta Oscillations on the Hearing Threshold

Ralph Parnefjord and Erol Başar

In this chapter two categories of concepts and two types of experiments will be treated. We will report on experiments with human subjects by application of tone bursts of 2000 Hz in various sound intensity levels ranging from the subjective hearing threshold to 80 dB (Parnefjord and Başar 1995; Parnefjord 1996). These experiments are not only useful for the determination of the hearing threshold, but also relevant with regard to cognitive electrophysiology. At the threshold level *decision making* and *increased attention* are processes interwoven with the sensory processes. On the one hand, not every sensory stimulation attracts our attention (this holds in particular for stimuli often encountered in our daily experience). On the other hand, a very low intensity sensory stimulation (which is difficult to detect) puts the subjects into a state of focused attention and readiness to achieve "decision-making." These considerations are related to the following question: Do low-intensity stimuli induce response oscillations of almost homogeneous frequency (which may be difficult to observe by high-intensity stimulation, e.g. 80 dB tones)?

12.1 Slow Wave Oscillations at Hearing Level: An Individual Experiment

One of the subjects in the study of measurements at the threshold level, which will be described in the next sections, showed a seldom encountered behavior. This subject is a staff member and paid great attention in order to detect the signals at the threshold level. The most important behavioral statement was that she had great ambition to deliver a correct report in cases of signals at the threshold level. Figure 12.1 illustrates the time histories of the experiment with this subject. At the bottom of the illustration the EP at the threshold level is given. The averaged response already shows high-amplitude oscillation in the range of 20 μV peak-to-peak. The filtered curve in the range between 1 and 3 Hz is a slow wave oscillation that we might denote as an induced oscillations at the threshold level. In other words, the entire time course of the EP is dominated by this slow wave oscillation showing almost a homogeneous frequency (see Fig. 12.1 for selected single sweeps showing this typical behavior).

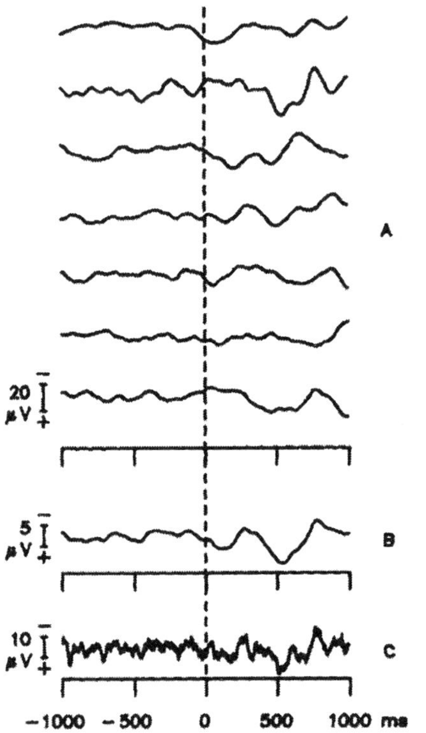

Fig. 12.1. Responses to tones at the auditory threshold that were indicated as heard: (**a**) filtered single EEG-EP epochs (1–3 Hz); (**b**) filtered averaged EP (1–3 Hz); (**c**) wide-band filtered averaged EP. (From Parnefjord R. (1996) thesis: Medical University Lübeck.)

What is the meaning of the results in this particular experiment? The subject tries to perform a signal-detection task and emits slow waves of a regular oscillatory form independent of the intensity of the physical stimulation. This means that the emitted response from the brain is in this case a pure signal due to signal detection or perception at the hearing threshold.

12.2 AEP Investigations at the Threshold Level

The investigations on the late-latency EPs have mainly focused on the relationship between stimulus intensity and EP amplitudes and latencies. In a number of studies the influence of stimulus intensity on the N1–P2 amplitudes have been assessed (Butler et al. 1969; Davis and Zerlin 1966; Picton 1970; Gille and Ullsperger 1987; Adler and Adler 1989). Unfortunately, it proves to be difficult to extrapolate these relationships for stimulus intensities above 70 dB and below 40 dB, where a change in the relation between stimulus intensity and EP amplitudes appears (Adler and Adler 1989; Beagley and Kellog 1970). In summary, the complex relation between stimulus intensity and the cortical response makes the EP amplitude a poor correlate for suprathreshold stimulation (Picton 1986).

With respect to the frequency domain, Sayers et al. (1974a,b, 1979) concluded that auditory EPs are due to a reordering of the background EEG phases following stimulation. At higher stimulus intensities, Beagley et al. (1979) found indications that additive energy could be observed on the phase reordering.

The concept of evoked and induced oscillations within different frequency bands after sensory stimulation or a cognitive task execution has been stated in Chap. 11 in Vol. I. We stated that sensory EPs should be considered as superposition of wave packets with varying degrees of enhancement.

In the coming section we will analyze different evoked frequency components, from the gamma frequency range (30–70 Hz) at higher stimulus intensities down to the theta and delta responses at lower stimulus intensities. Further, we will describe a phenomenon that we call *frequency shift* observed when analyzing the relationship between frequency components and stimulus intensity. It can be shown that high-amplitude auditory stimulation elicits responses in all frequency bands, i.e., from delta to gamma frequency range. This broad frequency response has a smaller number of peaks when the stimulus intensity decreases: First of all, the higher frequency responses disappear, until only the delta response remains as the last one near or at the threshold level.

12.3 Experimental Procedure

Subjects: The experiments were carried out with ten 20–30-year-old, healthy volunteers, all audiometrically tested and having revealed no hearing damage.

Environment: The subjects sat comfortably in a soundproof, echo-free, and dimly illuminated room. With their eyes open they were monitored with a closed circuit TV system. The stimuli were given binaurally over earphone sets with circumaural cushions. No white-noise background was needed. After placement of the electrodes the subjects were given a few minutes of rest to become familiar with the environment. Concerning the principle of data acquisition the reader is referred to Chap. 9.

Stimulus Paradigm: The computer triggered 2000 Hz tones with practically no rise and fall time. Stimulus duration was 1000 ms, and the interstimulus interval was held randomly between 1 and 5 secs. The applied intensities were:

(a) Spontaneous EEG recording (denoted as "SPO")
(b) 80 dB
(c) 60 dB
(d) 40 dB
(e) 20 dB
(f) The threshold experiment

12.3.1 The Threshold Experiment:

> *"The absolute sensory threshold is the lowest stimulus intensity a subject can detect. Threshold is determined statistically. If a subject is given several series of stimuli of different intensities, the intensity at which the stimulus is first detected in each series (i.e,, threshold) will differ slightly. Moreover, if a subject receives a series of stimuli the intensity of which is close to the average threshold, the subject typically fails to detect a certain proportion of stimuli. It is therefore convenient to define threshold as that stimulus intensity detected in 50% of the trials"*
>
> *(Kandel 1976).*

After approximately 50% of the tones were heard during a sufficient length of time to assume stability of the threshold, the actual recording was started. The subjects were not able to predict the onset of tones, and the exact number of tones presented was not revealed. The EP epochs were afterwards divided into two groups: *tone heard* (denoted by TH) and *tone not heard* (TNH).

12.4 Brain Response to Auditory Stimuli with Different Intensities

12.4.1 Time Domain Averages

Figure 12.2a shows the averaged vertex evoked potentials of a typical subject exposed to decreasing intensities of stimulation, filtered in the 0.5–3 Hz band. The last three plots show EPs recorded with *tones heard* and *tones not heard* at the auditory threshold, together with a recording of the spontaneous EEG experiments (TH, TNH, SPO, respectively). Figure 12.2b shows the same auditory EPs without filtering.

The major poststimulus features at 80 dB are a negative potential at 100 msec (N100) and a positive potential at approx. 200 msec (P200). This N100-P200 complex is also well visible in recordings with 60 dB and 40 dB stimulation. With 20 dB stimulation, it loses its strict phase locking. It can be seen that the peak-to-peak amplitudes generally decrease with decreasing stimulation and that the latencies increase. A relative stability in amplitude and latency of the N100 peak in contrast to the P200 peak can be observed. Similar results have been reported earlier (Arlinger 1975; Botte et al. 1975; Picton et al. 1978).

At 20 dB and at the auditory threshold, the auditory EP is not clearly visible. This is possibly due to increasing variability of the latencies (Butler et al. 1969), making the time-domain average weaker in amplitude. The still visible averaged EP at the threshold is a product of the potentials produced by two out of ten subjects and cannot be clearly reproduced with statistical significance for the whole group of subjects.

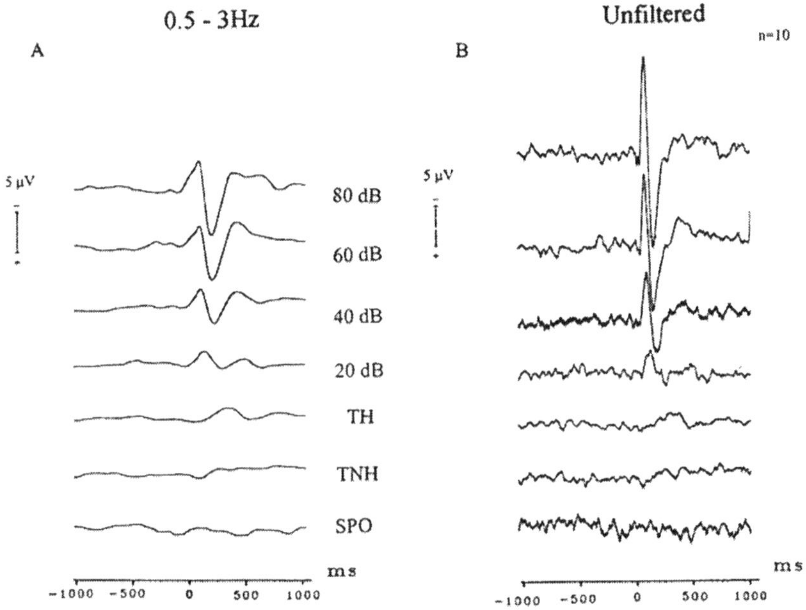

Fig. 12.2a,b. Grand average auditory evoked potential recordings of ten subjects (vertex). *Left:* 0.5–3 Hz digitally filtered AEPs to different stimulation intensities. TH (tone heard) and TNH (tone not heard) are subgroups of the threshold experiments. SPO are the spontaneous activity experiments. *Right:* the same potentials are reproduced unfiltered. Negativity is *upwards*. (From Parnefjord R. (1996) thesis: Medical University Lübeck.)

12.4.2 Digitally Filtered AEPs

Through a digital filtering process of the individual mean auditory EPs of the different stimulus intensities, we have found that the number of evoked frequency bands of the evoked potentials is increased with increasing intensities of stimulation.

In Fig. 12.3, the grand average maximum amplitudes of the digitally filtered peak-to-peak responses are presented (standard deviations are given in the table included in this figure). As can be seen, the 80 dB stimulation leads to evoked responses in all filter bands. As the intensity decreases, the higher-frequency responses lessen and cease. At minimal stimulation, the 1–5 Hz frequency band is left as the last evoked frequency activity. A Wilcoxon/Wilcox test of the 30–50 Hz response at 80 dB stimulus intensity, compared to all other stimulus intensities, shows significant differences ($p < 0.01$).

The evoked frequency range is conditioned by individual types of reaction to stimuli. Thus, the higher-frequency responses do not always include discernible 30–50 Hz activities at 80 dB stimulation in every single individual, but the highest number of frequency bands are always evoked at the strongest

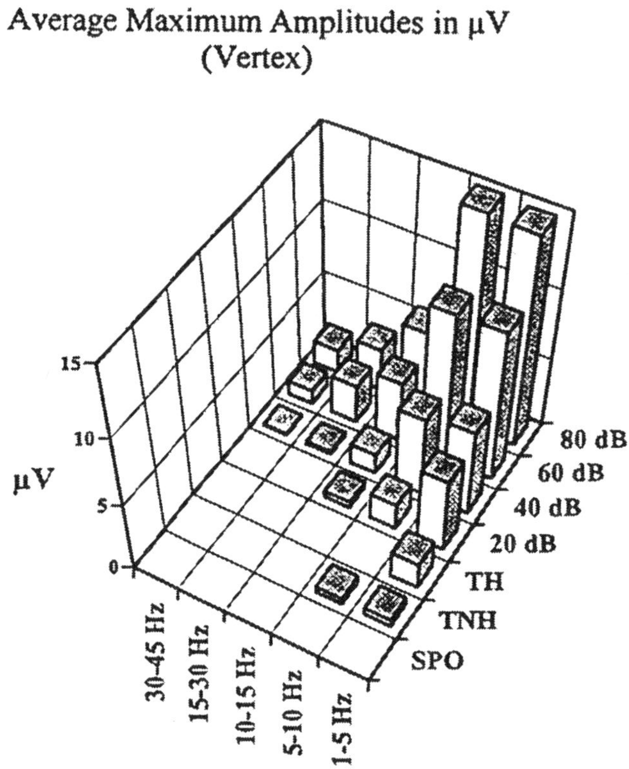

	80 dB	60 db	40 dB	20 dB	TH	TNH	SPO
1-5 Hz	14.84 σ = 5.74	10.87 σ = 3.03	6.31 σ = 3.17	5.38 σ = 2.72	1.88 σ = 1.65	0.50 σ = 1.50	-
5-10 Hz	15.37 σ = 5.52	10.75 σ = 4.82	5.78 σ = 4.25	1.86 σ = 2.12	-	0.44 σ = 1.31	-
10-15 Hz	5.75 σ = 3.75	4.40 σ = 2.20	1.50 σ = 1.53	0.44 σ = 0.92	-	-	-
15-30 Hz	3.12 σ = 2.15	2.87 σ = 1.49	0.19 σ = 0.58	-	-	-	-
30-45 Hz	1.72 σ = 0.95	1.00 σ = 0.98	0.12 σ = 0.36	-	-	-	-

Fig. 12.3. Grand average peak-to-peak AEP amplitudes of ten subjects (vertex recording). *Below* are the absolute values, together with standard deviations, above a histogram representation. TH = tone heard; TNH = tone not heard: Both subgroups of threshold stimulation. SPO = spontaneous activity experiments. The evoked potentials to the different stimulation intensities have been digitally filtered in different frequency ranges. The resulting peak-to-peak amplitudes of the digitally filtered activities are presented. (From Parnefjord R. (1996) thesis: Medical University Lübeck.)

stimulation intensities. Furthermore, the highest evoked frequencies retreat at first when the intensity of stimulation lessens.

What we have, then, is a broad evoked activity or broad forced oscillation at higher levels of stimulation including all bands up to 30–50 Hz. This broadband response narrows with decreasing intensity of stimulation until only the delta and theta bands remain as the last evoked oscillation – these results being founded on peak-to-peak maximum amplitudes only.

12.4.3 Grand Average Amplitude–Frequency Analysis

The grand average amplitude–frequency characteristics are shown in Fig. 12.4 (see below for single subjects). The dominant response activity in the 80 dB experiments is focused in the 8–9 Hz domain. This dominant amplitude–frequency characteristic peak is maintained when the stimulus intensity is decreased to 60 dB, but with lesser relative amplitude. At 40 dB stimulation the dominant frequency activity moves towards 6–7 Hz, dissolves into a 3–6 Hz plateau at 20 dB, and re-forms into a clear amplitude peak at 2–3 Hz, where TH stands for "tone heard" and TNH for "tone not heard" as subgroupings of the sweeps recorded during threshold experiments (see above in this chapter).

Through this analysis of the dominant frequency components, the dynamics of the frequency responses can be followed. Due to the computational limitations of the amplitude–frequency characteristics, this representation has the disadvantage of a weak peak identification of activities with frequencies higher than 30 Hz. This method is, therefore, more suited for the analysis of the dynamics of the dominant frequency components, and not for the evoked responses in higher frequency bands – for which digital filtering has been used.

It can clearly be seen that the dominant frequency components are strongly influenced by the intensity of sensory stimulation. The poststimulus frequency components at 80 dB stimulation showed a diffuse distribution between 2 and 15 Hz, with a dominance in the lower alpha range. At 60 dB stimulation, this diffuse distribution was changed into a "double activity alignment" in the theta and lower alpha ranges. This predominant alpha and theta activity at higher stimulation was altered at 40 dB stimulation intensity. A new distribution seems to take place, presenting a picture of a diffuse theta activity without any clear alignments.

At 20 dB stimulation, a clear theta alignment can be seen. The poststimulus frequency response is centered on 2–5 Hz. This alignment is also present at the threshold (TH), with the difference that the frequency range is even lower (1–3 Hz): In the threshold experiments where the tones were not heard (TNH) and in the spontaneous activity experiment, no frequency alignments can be seen, and the frequency distribution is broad and unspecified, ranging from 1–50 Hz.

When summing up the results of the digital filtering and the amplitude–frequency characteristics, we conclude that there seems to be a frequency

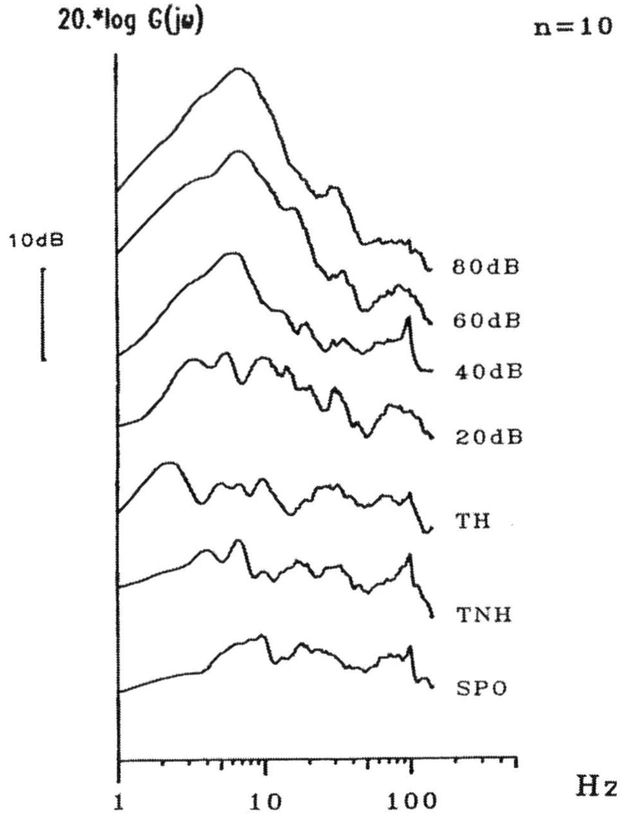

Fig. 12.4. Grand average amplitude–frequency characteristics (vertex recording). TH = tone heard; TNH = tone not heard: Both subgroups of threshold stimulation. SPO = spontaneous activity experiments. See text for comments. (From Parnefjord R. (1996) thesis: Medical University Lübeck.)

shift of the dominant frequency activity as well as the number of evoked frequency bands.

By the term *frequency shift* we understand firstly the reduction of the number of evoked frequency bands, as an effect of reduced stimulus intensity, until only a delta response is seen at threshold stimulation.

Secondly, we observe a change in the dominant frequency components in relation to stimulus intensity. A dominant frequency component in the alpha range is seen at high stimulation intensities. At 40–20 dB stimulation the dominant frequency component is in the theta range, thus presenting a change in the frequency component setup of the evoked potential. A further change is to be seen upon threshold stimulation, where the dominant frequency component is in the delta range.

12.4.4 Selectively Filtered Auditory EPs

We have stated earlier that ERPs should be considered as superposition of event-related oscillations with varying degrees of enhancement within different frequency bands (see Chap. 15 in Vol. I). We will add the following new comparison using the superposition concept.

Figure 12.5 shows digital filtering of a typical 80 dB experiment and threshold experiment (TH) of one subject. The unfiltered AEP is shown at the top, followed by the resulting digital filtering of the same response. It can be seen that the EP at 80 dB is mostly composed of 1–3 Hz and 3–8 Hz activities with a maximum peak-to-peak amplitude at 3–8 Hz. Furthermore, all filter frequencies show signal enhancement with wave packets of frequency-related duration, i.e., approx. 750 ms oscillations within the 1–3 Hz band, down to 100 ms oscillations in the 30–50 Hz band.

On the other hand, at the threshold level, no oscillatory response can be seen in the higher filter bands. The evoked potential contains 1–3 Hz and 3–8 Hz responses only. The maximum peak-to-peak amplitude is in the 1–3 Hz

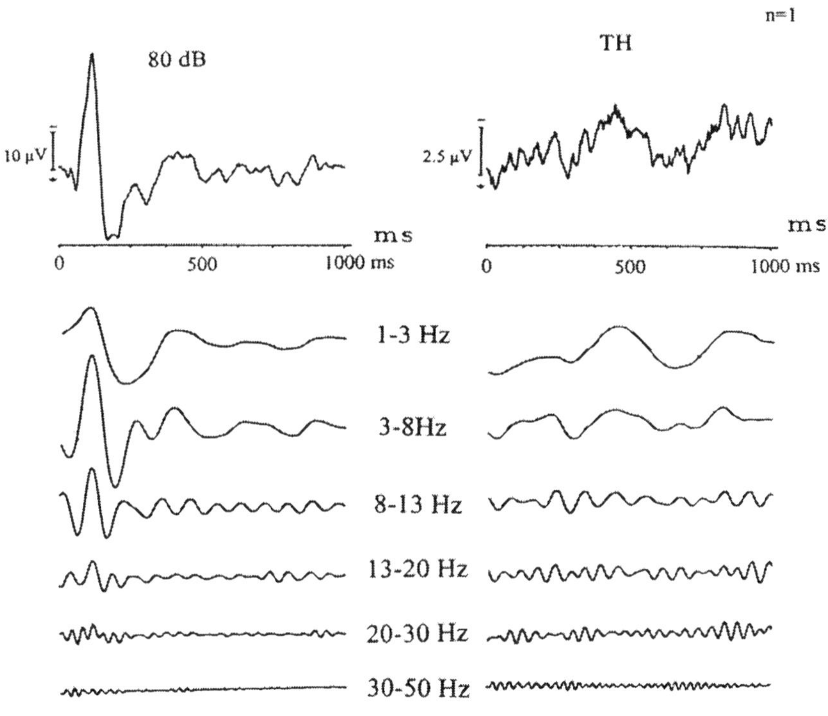

Fig. 12.5. Filtered auditory EPs demonstrating the *superposition principle:* Responses to 80 dB stimulation and "tone heard" responses in a typical subject. *Top*, unfiltered EPs. *Below*, EPs digitally filtered in several frequency ranges. (From Parnefjord R. (1996) thesis: Medical University Lübeck.)

range. This last finding suggests that the dominant frequency components were completely altered. Not only the amplitudes were altered, but also the dominant frequency activity.

12.4.5 Frequency Distribution in Single Subjects

These results were confirmed in at least 20 subjects, and the relationship between stimulus intensity and main frequency components are described. An analysis in the frequency range by means of a fast Fourier transform was also performed.

In order to evaluate single-sweep epochs of all subjects, a graphic representation of the amplitude–frequency characteristics was established. Fig-

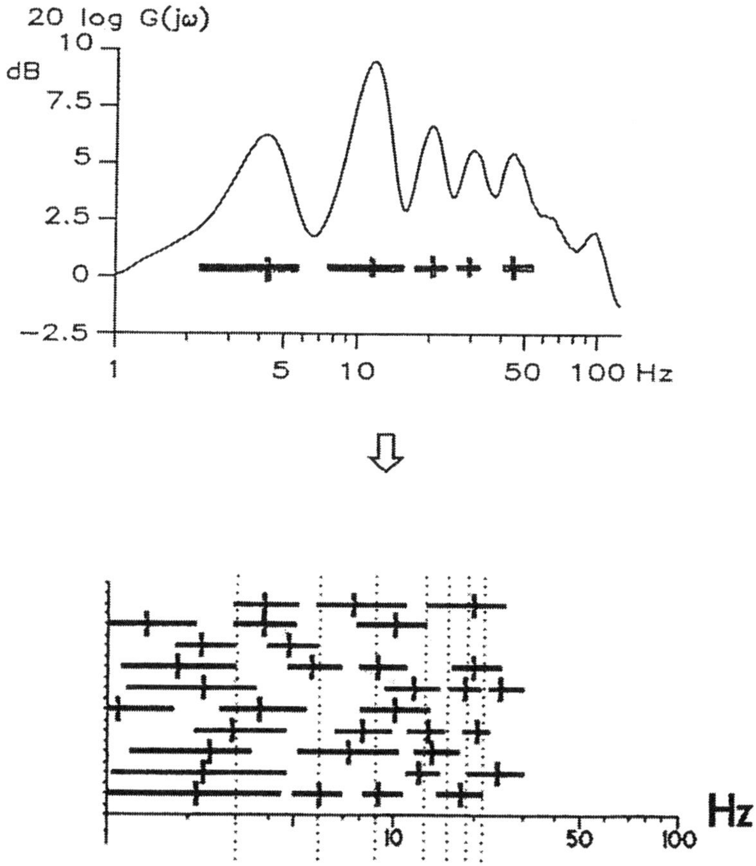

Fig. 12.6. Transform of the amplitude–frequency characteristic's peak into bar-code representation and subsequent numerical presentation. See text for further description. (From Parnefjord R. (1996) thesis: Medical University Lübeck.)

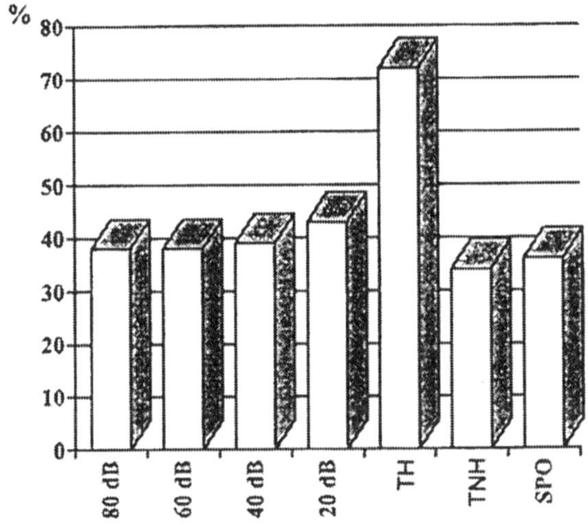

Fig. 12.7. Delta peak occurrence histograms at different stimulation intensities. The increase of delta peaks at the threshold (TH) is clearly seen (increase of up to 45% compared with other experiments). The occurences have been computed according to the earlier mentioned transform of the amplitude–frequency characteristics. (From Parnefjord R. (1996) thesis: Medical University Lübeck.)

ure 12.6 shows the transform of the amplitud–frequency characteristics peak into bar codes and further numerical evaluations. The amplitude maxima are routinely determined by plotting them on a separate paper. Each amplitude maximum with more than 2 dB activity difference to its immediate surrounding is represented by an appropriate bandwidth (horizontal line segment) and a peak frequency (vertical line segment). These bar codes are plotted together, and the number of peak frequencies falling into a set of slots from 1–3 Hz to 18–21 Hz are determined and presented as histograms. In this way, ten random sweeps in each experiment and for each subject, with a total of 100 amplitude–frequency-peak characteristics, have been evaluated. For this method see Chap. 11 in Vol. I.

Figure 12.7 shows the resulting histograms for the delta response at the threshold, in comparison to the other stimulus intensities. A Wilcoxon/Wilcox test gave significant differences ($p < 0.05$) of delta response peaks for threshold experiments vs. all other experiments except 20 dB.

The threshold response pattern is characterized by a strong 1–3 Hz activity. It is to be noted how this delta response dissolves at the threshold stimulation where the tones were not heard. In some subjects, the delta peak was seen at 40–20 dB intensity, but the delta response was always the "last" detectable evoked frequency response. Although these subjects clearly indicated correctly heard tones upon threshold stimulation, the electrical response was no longer more discernible in the form of auditory EPs or evoked oscillations.

12.5 Has the Frequency Shift a Sensory–Cognitive Interpretation?

The frequency shift could be associated with appropriate reaction forms to different sensory perceptions. Sensory perceptions at high intensities evoked not only a cognitive, but probably also an *orienting response*. Donchin et al. (1978) suggested that a surprising stimulus calls for a revision of the active cognitive processes, leading to an "update" of the internal model.

It is to be noted that the time-domain latencies do not change to a greater extent at higher stimulation intensities. What we see is a change in the frequency-component content, where additional evoked frequency responses are added. These higher-frequency responses do not last throughout the auditory EP, but only the first few milliseconds (3–5 complete wavelengths of 30–50 Hz equal about 20 ms).

What we see as a response to higher stimulation intensities is not only an additive power in the already present amplitude–frequency-characteristic peaks, but also a change in the evoked dominant frequency components. It is appropriate to refer to the findings of Beagley et al. (1979), who used a phase-spectral analysis as a method of applying the AEP in objective audiometry. A broad range of components was found to be affected at high levels of stimulation, but at reduced intensities the synchronization became less effective, engaging fewer spectral components. By use of the amplitude–frequency characteristics, we can show that apart from the phase aggregation, an enhancement of existing prestimulus frequency components (Başar 1980) is present. Also, a frequency recruitment and change in dominant frequency components, in the form of the frequency shift, takes place at higher stimulation intensities.

The fact that the attempts to define a relationship between the auditory EP amplitudes and the stimulus intensity presents substantial difficulties in the interpolation of the results to intensities above or below the 40–70 dB range could be related to the dominant frequency component shift. The dominant frequency component at stimulation between 40 dB and 70 dB is in the alpha or theta range. Within this range, different linear and logarithmic relations have been described (Butler et al. 1969; Arlinger 1975).

At stimulation below 40 dB, the EP amplitudes likewise deviate from the expected relation as computed for the 40–70 dB responses. Again, we see a change in the poststimulus response in the frequency components, where the delta range acquires the dominant frequency activity. The fact that the dominant frequency activity is the same in stimulation between 40 and 70 dB could be the explanation why relationships between stimulation intensities and EP amplitudes are found that do not apply for higher or lower stimulation.

The frequency dynamics of the human auditory EPs have been investigated through digital filtering and Fourier transforms in their relation to different stimulus intensities. It could be shown that stimulation at high intensities evokes responses in all filter bands from the delta to the 50 Hz range.

At weaker stimulation intensities, the higher-frequency-band responses disappear, and the overall frequency range of the response narrows down successively until only delta activities are to be seen in the auditory EP at threshold stimulation.

In the frequency domain, it can further be shown that the stimulus intensity is one of the factors determining which frequency component of the auditory EP is dominant:

- At high stimulation intensities, the dominant frequency component is in the alpha to theta range.
- At 40–20 dB stimulation, the dominant frequency component is in the theta range.
- Only at threshold stimulation are the frequency characteristics are dominated by a clear delta activity.

The sensitivity of nonsensory or cognitive factors of the late-latency auditory EP was largely responsible for the discontinuation of using cortical EP recordings in objective audiometry and its replacement with the auditory brainstem response (Ohman and Lader 1977; Hillyard et al. 1973). However, this dependence of cognitive factors has recently turned to advantage in cognitive (Hillyard 1981; Donald 1983; Näätänen 1982) and clinical studies (Näätänen and Mitchie 1979; Duffy et al. 1980; Michalewski et al. 1986; Porjesz 1980). It is becoming clear that cortical auditory EPs are reemerging as a possible tool of investigating more complex and central auditory functions (Regan 1989).

After the relatively discouraging attempts to use late-latency evoked potentials in objective audiometry, most "threshold" investigations have used paradigms that define selective attention as the hearing of given tones against a white-noise background (Donchin and Cohen 1967), against nontarget tones of differing frequencies (Davis 1964; Satterfield 1967) and through acoustical stimulation during attention tasks (Spong et al. 1965). The amount of information concerning the psychophysiological reactions of pure-tone threshold stimulation, without background noise or selectivity tasks, is still unsatisfactory.

Some interesting information on the dynamics of the frequency and phase rearrangement of the AEPs was made by Beagley et al. (1979). Whereas the cognitive aspect was of less interest to these investigators, the presence of phase constraints could be shown as an effect of decreasing auditory stimulation. Başar et al. (1975; Başar 1980) have further shown that frequency enhancements and stabilization play an important role in the central processing and associative management of sensory inputs.

12.5.1 Possible Origin of the Delta Response

Many investigators have focused their attention on the possible *cortical* or *subcortical origin* of theta components associated with the P300:

- Cortical sources have been found by Courchesne (1978), Desmedt and Debecker (1979), Goff et al. (1978), and Pritchard (1981).
- Subcortical origins have been documented by Woods et al. (1982), Halgren et al. (1986), Başar-Eroğlu et al. (1991), and Buchwald and Squires (1982).

So far, no definite statement can be made, and Başar-Eroğlu et al. (1992) have suggested a "selectively distributed theta system" to account for the different reports on subcortical and cortical sources as well as their different reported interplay and projections.

In the theta range, Lang et al. (1987) and Westphal et al. (1990) found slow frequencies increasing in frontal cortex during motor or verbal tasks. Aleksanov et al. (1986) showed that the coherence between hippocampus and prefrontal cortex increased in the delta and theta ranges during the course of alimentary conditioning and at the time of presentation of the conditioned stimuli (see also Chap. 25).

The delta waves have been found to reflect sequences of excitatory and inhibitory processes of cortical neurons (Ball et al. 1977), but the underlying fundamental mechanisms leading to slow wave activity remain unknown (see also Chaps. 18 and 25).

12.5.2 The Decision-Memory System

Mast and Watson (1968) observed that vertex potentials evoked near the threshold auditory stimuli were markedly enlarged and prolonged when the subjects actively listened to the signals. In Chap. 20 a survey of the difference between theta and delta responses will be given. The results led to tentatively assuming that the delta response is mostly involved with the *signal matching* and *decision-making*, whereas the theta response is more involved with complex functions such as focused attention, signal detection, anticipation, and expectation.

In the present chapter, several cognitive components or functions play an important part in the signal processing and decision-making preceding the motor task. Some of the more prominent are to be mentioned in the following:

(a) Selective attention
(b) Threshold perception
(c) Decision-making

In the light of the above-mentioned results we feel confident in the interpretation of the strong delta response as a decision-related cognitive component. For this reason the signal-detection theory at the neural level must be taken into consideration. Experiments are described in Chap. 18, stating that the hippocampus shows a special response for detected signals. We have seen that the signal-detection theory implies that two kinds of processes occur when an organism detects a threshold-level signal. Firstly, the signal must be detected in noise at the level of sensory reception and pathway, and then secondly, a decision to respond must engage the neural populations of the learned response.

Consequently, neural circuits that are differentially activated on detection and nondetection trials are neural circuits for the "decision-memory system." With this argumentation the primary system cannot be considered as a part of the "decision-memory system." The learning-induced hippocampal unit responses were measured by Thompson (1982; see Chap. 18).

We have to state that a comparison of responses from the skull of the human brain and responses from intracranial structures of the cat brain is not possible without restrictions. However, it is highly important to point out that we are confronted with delta responses in signal-detecting experiments such as the Oddball paradigm, hearing at the threshold level, and also with the substructures of the cat brain. All these results form the conclusion that the delta response of different behavioral processes influences the decision-memory very strongly.

We will come to a joint discussion on these results in Chap. 23: Differently induced oscillatory components may have similar functional correlates.

13. Evoked Oscillations in Magnetoencephalography

13.1 Technical Remarks and Advantages of MEG

Magnetic recordings will be used in this chapter for the extension of applications of concepts on dynamics of human potentials. In fact, the dynamics of potentials analyzed by means of evoked magnetic fields confirm and extend the results on EPs described in Chap. 9. Moreover, the evoked alpha oscillations can be elegantly and distinctly shown by this new type of measurement.

MEG devices enabling recordings on the whole cortex have only been recently developed. Forty years after the invention of the human EEG, the recording of weak magnetic fields outside the head by means of MEG was pioneered by Cohen in 1968. This first instrument was an induction coil magnetometer used to detect the magnetic "alpha rhythm" by means of signal averaging. The development to build SQUID technologies and arrays with multisensors has been possible only during the last few years. Several review articles and books can be recommended for further reading (Hari and Kaukoranta 1985; Weinberg et al. 1985; Hari and Lounasmaa 1989; Williamson et al. 1989; Grandori et al. 1990; Sato 1990; Hari 1991; Weinberg 1991; Hari and Salmelin 1997).

As Hari (1993) points out, the main advantages of MEG are the following:

1. MEG has a comparably good spatial resolution in locating cortical events because of its selectivity to the activity of the fissural cortex. Even when the source areas are only approximately known, functional studies are feasible. The scalp EEG is only indirectly related to neuronal generators, since it arises from the volume currents that flow throughout the brain and its coverings. The poor conductivity of the skull and inhomogeneous impedance of the brain attenuate and distort the spatial pattern of the EEG, so it is difficult to locate the source.
2. The advantage of MEG over EEG in source localization results mainly from the fact that the skull and other extracerebral tissues are practically transparent to the magnetic field; however, the current flow (EEG) from the skull is altered substantially. Accordingly, the magnetic pattern outside the head is less distorted than the electric potential on the scalp. In Fig. 13.1 a plot for a single sweep as simultaneously recorded by seven SQUID channels is shown (see Mikkelsen et al. (1989) for details). One

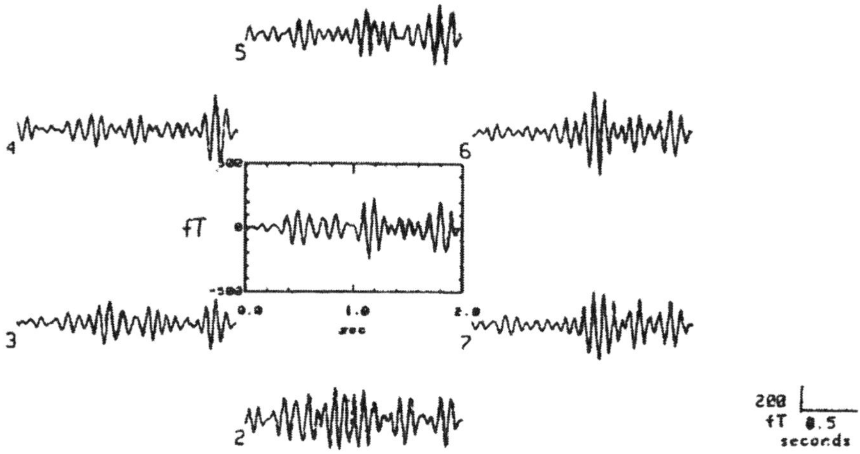

Fig. 13.1. A multiplot of a single sweep, recorded by seven SQUID channels (tone-burst, 1 kHz, random ISI), band-pass filtered in the frequency range 8 Hz to 13 Hz. The SQUID is located at T3. Stimulus occurs at time 1.03 sec. (Courtesy of Dr. K.B. Mikkelsen, Technical University of Denmark, Copenhagen.)

notes that the various channels may show a different behavior, although the interchannel distance is only 2 cm. In fact, in a time window immediately following the stimulus there may be a tendency for some channels to show a "blocking" effect, while others do not.

3. The magnetic recording is reference-free, whereas the electric brain maps depend on the location of the reference electrode.
4. MEG accurately detects superficial tangential currents, whereas EEG is sensitive to both radial and tangential current sources. This means that EEG also reflects activity of the deepest parts of the brain. EEG and MEG are *complementary methods* of investigation: Sources that are radically oriented do not alter the magnetic field (but are the main source of the EEG), whereas MEG is preferentially sensitive to generators lying in the "sulci," which include much of the primary sensory projection areas. Combined EEG and MEG studies, therefore, offer a good possibility of localizing cerebral activity.

Figure 13.2 shows typical auditory evoked magnetic fields in a healthy subject. The main peak, N100m, occurs approx. 100 ms after the sound onset. The location of the "equivalent current dipole" (ECD), found by a least squares fit to the data, agrees with activation of the supratemporal auditory cortex (see Hari (1990) for a review). The success of dipole models is due to the fact that the details of the pattern cannot be discerned at a typical measurement distance of at least 3 cm from the source (Hari 1993).

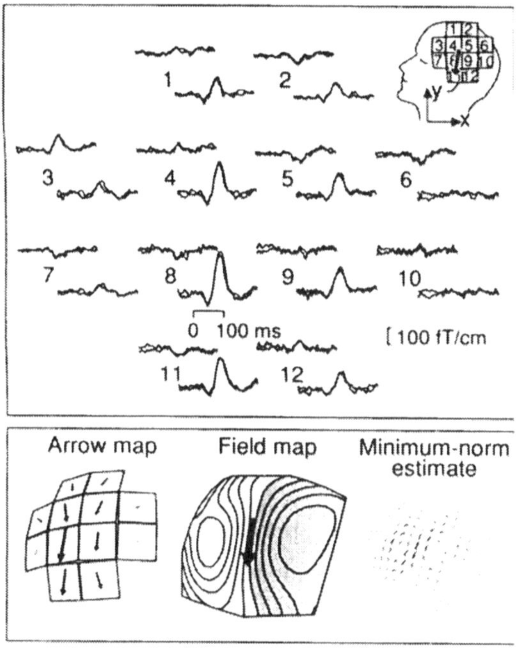

Fig. 13.2. Top, auditory evoked magnetic fields to 50 msec 1 kHz square-wave tones repeated once every 1.2 seconds; 66 single responses were averaged and two averages are superimposed. The passband is 0.05–90 Hz. The recording locations are indicated on the schematic head. The upper traces of each response pair illustrate the $\mathrm{d}B/\mathrm{d}y$ and the lower traces the $\mathrm{d}B/\mathrm{d}x$ derivative. *Bottom*, three different presentations of the signal distribution during the main peak of the response (N100m at 85 msec.). The arrow map shows estimates of the orientations and strengths of currents under each sensor unit; the *arrow* is orthogonal to the resultant gradient detected by the unit. The field map of the radial component is the conventional way to represent the data; the *arrow* shows the location and orientation of the ECD. The minimum-norm estimate on the *right* gives a current distribution responsible for the MEG signal without any assumptions about the source configuration. (From Hari R. (1993) in: Niedermeyer E., Lopes da Silva F.H. *Electroencephalography. Basic Principles, Clinical Applications, and Related Fields*. Baltimore: Williams and Wilkins, pp. 1035–1063.)

13.2 Neural Currents Underlying the ECD. View of Hari

To understand the cellular events underlying the ECD, it is reasonable to divide the currents associated with postsynaptic potentials (PSPs) to transmembrane currents at the active synaptic area, intracellular currents within the neuron, and extracellular "volume" currents. Transmembrane currents are the driving force for both intra- and extracellular current flow. ECD calculated from the measured MEG signals reflects the orientation of the net intracellular current flow. A PSP at the end of one dendrite produces an "elementary" dipole moment of $Q = I \cdot \lambda \approx 10^{-14}$ nA$\cdot$ m, where I is the intracellular current and λ is the length constant of the membrane (Vvedensky et al. 1985). Since it is not known how many PSPs occur synchronously in each pyramidal cell and how much cancellation takes place within the cortex, it is not possible to derive accurate estimates for the source area of a typical evoked response on the basis of cell density and the size of the elementary dipole.

Some estimates can be obtained from measured intracortical current densities. For example, a dipole moment of 10 nA $\cdot$ m during an evoked response probably corresponds to activation of 200–500 mm^2 of cortex (Hari 1990).

13.3 The Electric and Magnetic Alpha: A Comparative Study of Auditory and Visual Evoked Fields

The question of whether several brain areas have their own alpha rhythms has also been raised (Tiihonen et al. 1991). Williamson and Kaufman (1989) discussed in an extended manner possible functional correlates of the 10 Hz magnetic activity. A neuromagnetic study on the spatial structure of brain rhythms enhanced by photic and somatosensory stimulation stated that in the visual modality the synchronization was characterized by a potentiation of the subject's alpha (Narici et al. 1990). Resonance phenomena and synergetic processes in the cortex can be better interpreted by the additional use of phase relationship and concept of phase coherence (Hoke et al. 1989).

In our long-standing studies describing the relation of spontaneous EEG activity to evoked potentials (EPs), we have used the expression "alpha resonance" for states of brain activity during which parts of the cerebral cortex are excited by external sensory stimulations and/or internal excitement resulting from hidden cognitive and integrative CNS afferences. The alpha resonance responses having a duration of approx. 250–300 ms are a consequence of sensory inputs and can be decoded from EPs by means of Fourier analysis and response adaptive digital filtering (for details, see Chap. 9).

According to our earlier results most of the structures of the brain including brainstem, hippocampus, and cerebellum of the cat brain are able to produce this short-duration alpha response whenever exposed to stimulation. Obtaining alpha responses using techniques of digital filtering and spectral analysis is not due to artifacts of such methods, but it can be shown to be responses reflecting sensory and cognitive processes of neural populations (Başar et al. 1991, 1992; see also Chaps. 1, 10, and 18 in Vol. I).

Since extracerebral tissues, such as skull and scalp, are practically transparent to the magnetic field, the field measured outside the head can give accurate information about the sites of locally active cortical area. Therefore, we have undertaken magnetoencephalographic (MEG) experiments to evaluate our working hypothesis on the alpha responsiveness of the brain, which was developed based on results from the cat cortex.

13.4 Evoked Fields to Sensory Stimulation: Alpha Response

13.4.1 Methods

A typical array of responses in frontal–temporal location to acoustical stimuli from 19 MEG channels (Philips Forschungslaboratorien, Hamburg, Germany) is presented in Fig. 13.3 (see inset for measurement position). Evoked responses with greatest amplitudes in this typical example are recorded in channels 5, 6, 7, and 8. Here, we analyze only the responses recorded in the frontal–temporal position, since the alpha response of these areas has not yet been studied in detail (Başar et al. 1992b).

Figure 13.3b illustrates amplitude–frequency characteristics computed from the data in Fig. 13.3a by means of Fourier transform. These curves are characterized by high-amplitude, most dominant frequency responses in the 10 Hz frequency channel. The dominance of the so-called 10 Hz response (or alpha response) is usually reduced in channels near the frontal location. Some of the curves also contain responses in the theta and in the 40 Hz frequency ranges. In this chapter a perfect statistical account of all measurements performed has not been given, since the different sizes of the skulls of subjects do not allow a final statistical evaluation to be performed. However, this type of behavior with dominant 10 Hz responses was found in all subjects.

13.5 Human MEG Responses – Temporoparietal Versus Occipital Alpha and Delta–Theta Responses

We repeat that the "alpha response" is defined as the oscillatory EEG or MEG activity during the first 200–300 ms upon application of a sensory stimulation.

(1) Averaged responses: Figure 13.4a shows MEG responses in a typical subject. The responses recorded in 19 channels are given in a topography-related order corresponding to the coil positions in the sc squid array (see inset for measurement position). Note that channels 8 and 9 are temporoparietal positions, whereas channels 12 to 15 are occipital positions. In temporoparietal channels, the responses had amplitudes of about 600 fT. In occipital positions, amplitudes were around 100 fT.

Figure 13.4b shows amplitude–frequency characteristics computed – by means of Fourier transform – from the evoked responses shown in Fig. 13.4a. There are prominent peaks (about 10 dB) in the alpha, i.e., 8–15 Hz, range in temporoparietal positions, but no relevant alpha peaks in occipital positions.

Figure 13.4c shows filtered curves (range 2–6 Hz). There are marked responses in channels 8 and 9; their amplitudes are in the 200–400 fT range.

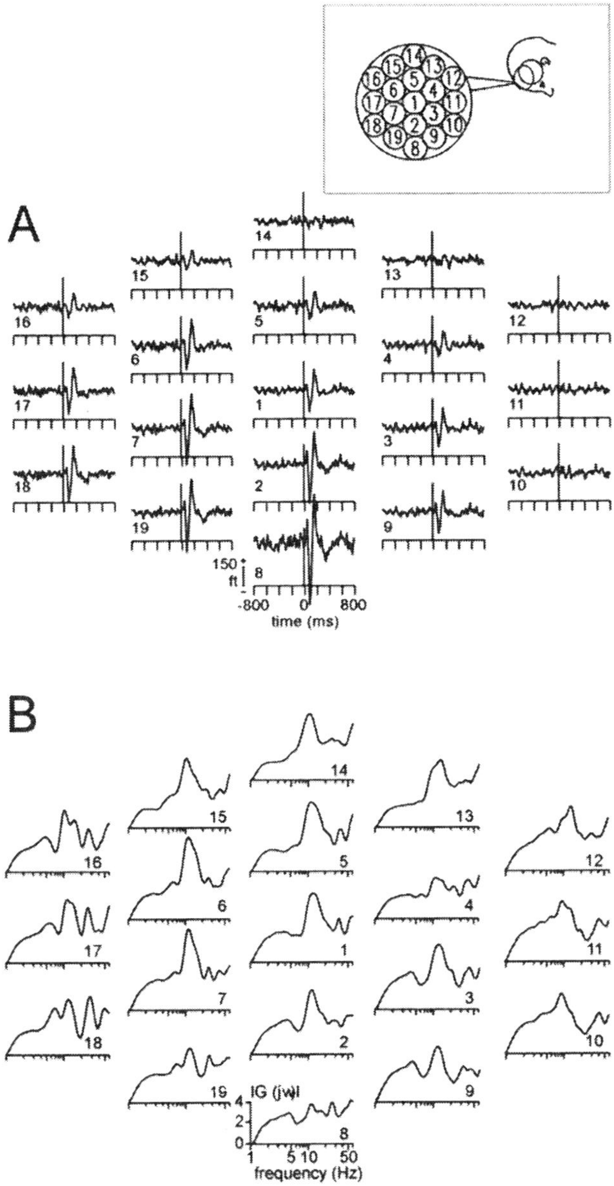

Fig. 13.3a,b. *Inset*: Measurement positions projected onto a sketch of a headshape: arrangement of the 19 channels of the Philips SQUID gradiometer system. (**a**): Averaged MEG responses to 1 kHz tone bursts of 80 dB SPL recorded through 19 magnetic sensors from the left temporofrontal region of subject G.W. (filter: 1–45 Hz, $n = 67$). (**b**): Amplitude–frequency characteristics computed from data shown in (**a**). (Modified from Schürmann M. et al. (1997) *International Journal of Psychophysiology* 26: 149–170.)

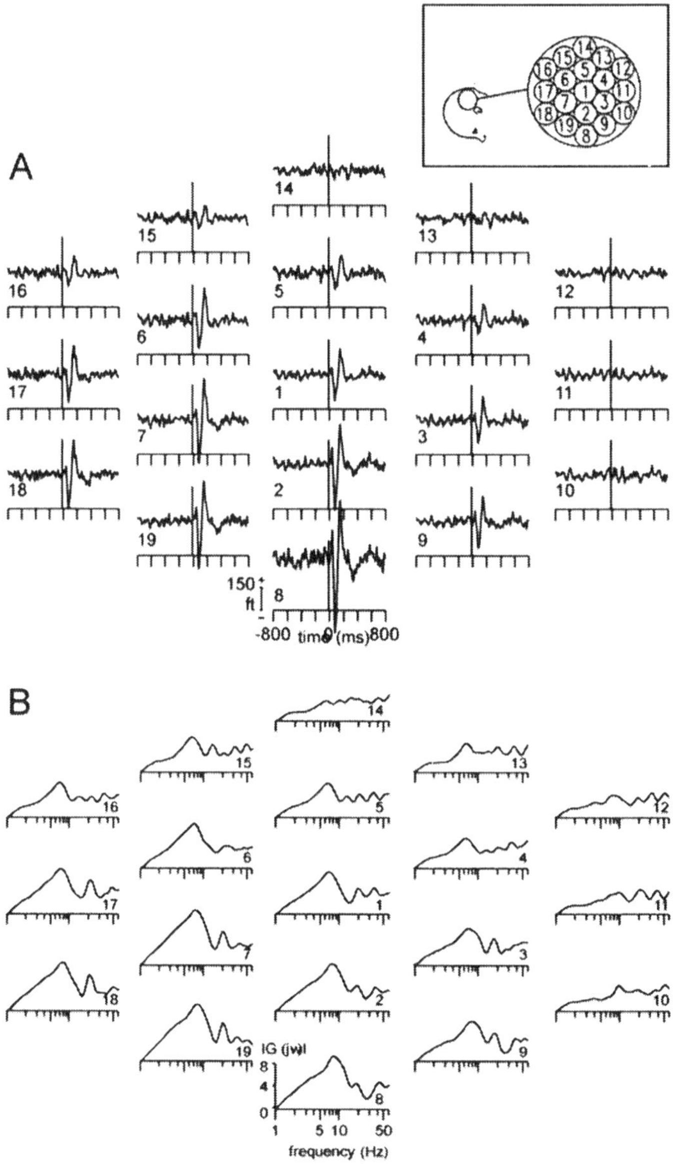

Fig. 13.4a,b. *Inset*: Measurement positions projected onto a sketch of a head-shape: arrangement of the 19 channels of the Philips SQUID gradiometer system. (**a**): Averaged MEG responses to 1 kHz tone bursts of 80 dB SPL recorded through 19 magnetic sensors from the left temporooccipital region of a subject (filter: 1–45 Hz). (**b**): Amplitude–frequency characteristics computed from data shown in (**a**). (Modified from Schürmann M. et al. (1997) *International Journal of Psychophysiology* 26: 149–170.)

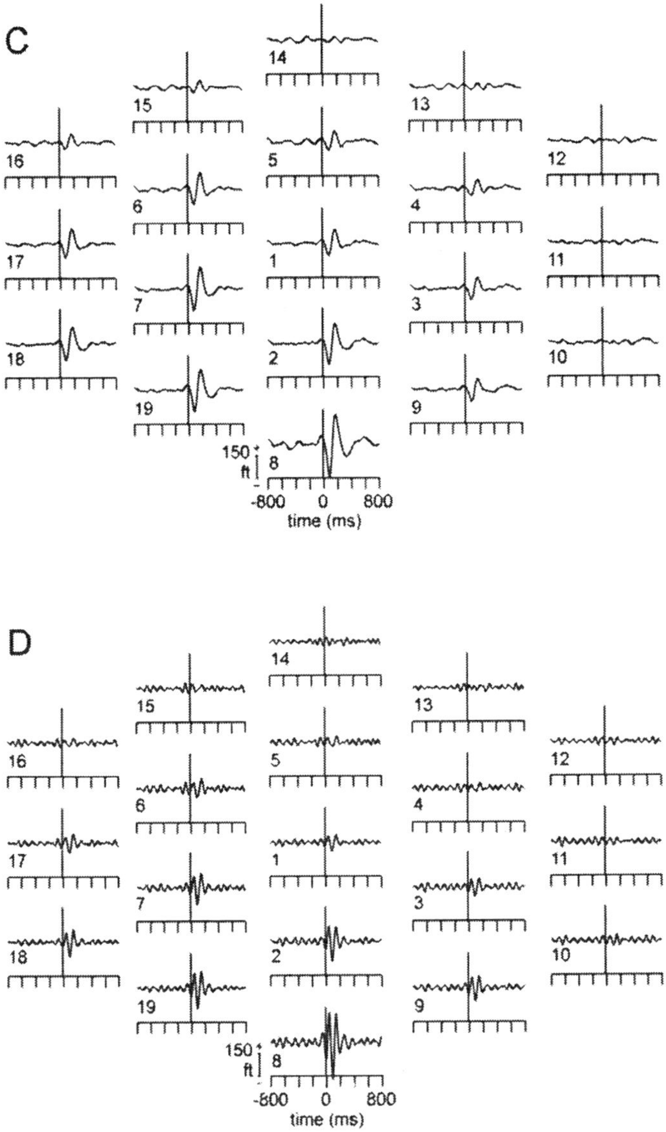

Fig. 13.4c,d. (**c**): Data shown in Fig. 13.4a filtered in the delta (2–6 Hz) range. (**d**): Same data filtered in the alpha (8–15 Hz) range. (Modified from Schürmann M. et al. (1997) *International Journal of Psychophysiology* 26: 149–170.)

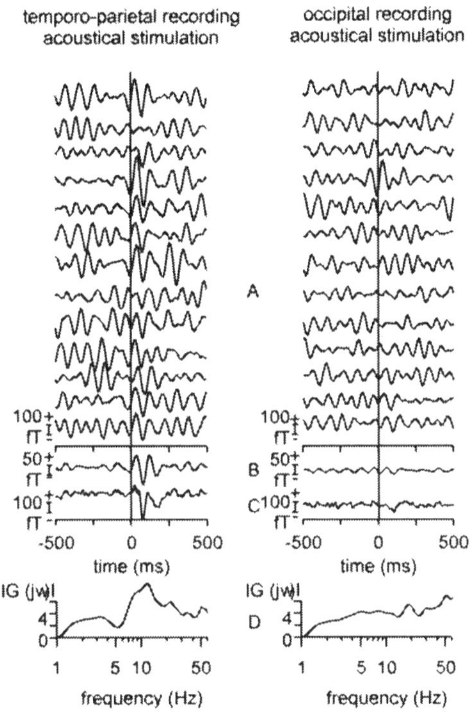

Fig. 13.5. Human MEG responses to auditory stimulation recorded in a temporoparietal position (*left*) and in an occipital position (*right*; subject #1). (**a**). Single MEG responses, filtered 8–15 Hz. (**b**). Averaged MEG response, filtered 8–15 Hz. (**c**). Averaged MEG response, filtered 1–45 Hz ("unfiltered"). (**d**). Amplitude–frequency characteristics computed from averaged MEG response. (Modified from Schürmann M. et al. (1997) *International Journal of Psychophysiology* 26: 149–170.)

The responses in channels 12–15 have considerably smaller amplitudes (about 100 fT); they are, however, present.

Figure 13.4d shows filtered curves (range 8–15 Hz). Marked responses are present in channels 8 and 9 (about 200–500 fT). In the occipital locations, however, responses are barely distinguishable from the averaged MEG; there is no alpha enhancement.

(2) Single MEG Epochs from a Typical Subject: Figure 13.5a shows single MEG epochs (prestimulus MEG segment and auditory evoked response; filtered in the 8–15 Hz frequency range). Epochs from channel 8 (temporoparietal; Fig. 13 5a, left column) showed great enhancements of 10 Hz activity; i.e., the alpha activity is time-locked and enhanced in the first 300 ms following stimulation.

The filtered averaged response (Fig. 13.5b, left column) and even the averaged auditory EP without filtering (Fig. 13.5c, left column) also shows a type of 10 Hz oscillatory waveform. This is in accordance with a peak within the alpha (8–15 Hz) range in the amplitude–frequency characteristics (Fig. 13.5d, left column).

The right column shows single auditory MEG epochs also filtered between 8 and 15 Hz recorded from channel 14 (occipital; Fig. 13.5a, right column). Such measurements are described as "cross-modality" experiments (see Chaps. 2 and 10). Auditory stimulation does not elicit any significant 10 Hz enhancement in occipital recordings. That is, the inadequate stimulation in "cross-modality" experiments evokes mostly a theta response or rhythmic theta activities in the cortex. Figure 13.5d illustrates the amplitude–frequency characteristics computed from the averaged EPs shown in Fig. 13.5c. The amplitude–frequency characteristics computed from the temporoparietal recording (left column) has a prominent maximum at around 10 Hz (approx. 5 dB). Such a maximum is not present in the comparably flat amplitude–frequency characteristics computed from the occipital recording. Both responses are, however, similar in the 1–5 Hz frequency range; i.e., in the 10 Hz frequency range (filter limits 8–15 Hz) the recorded large enhancements of single MEG epochs in temporoparietal recordings are reflected in the amplitude–frequency characteristics – in the shape of a dominant 10 Hz peak.

A comparison of Figs. 13.5a (left column) and 13.5a (right column) clearly shows that acoustical stimulation cannot elicit typical 10 Hz enhancements if recorded on the occipital side. The amplitude–frequency characteristics in the parietal recordings (Fig. 13.5d, left column) also show a sharp 10 Hz peak, which cannot be recorded in evoked potentials (as measured with EEG technology) in the same cortical area. On the other hand, the acoustical stimulation gives rise to evoked potentials (Fig. 13.5c, right column), but these evoked potentials have their main power in the theta–beta frequency ranges. This behavior is an important case, allowing us to pronounce the advantages of MEG versus EEG activity.

13.6 Evidence of 10 Hz and 5 Hz Evoked Magnetic Rhythm

In the following we shall describe the results of only one experiment considered to be representative for most of the subjects under study. We have recorded for each subject the magnetic evoked fields at a number of cryostat positions. The stimulus was a 1 kHz tone burst at 60 dB hearing level. For each channel position the FFT amplitude–frequency characteristic for the averaged auditory evoked fields was computed (for details of the method see Chap. 4 in Vol. I). Depending on the recording position, the amplitude–frequency characteristic showed a distinct maximum at 5 to 7 Hz or a sharp peak at 9 to 10 Hz in temporal locations, and sometimes a compound 5 to 15 Hz structure.

The advantage of using magnetic recordings (instead of electrical recordings) in the computation of amplitude–frequency characteristics is clearly seen from Fig. 13.6. The very pronounced change in the frequency distributions – as a consequence of a spatial displacement of only a few centimeters – cannot be seen on the basis of electrical recordings from the scalp. How does

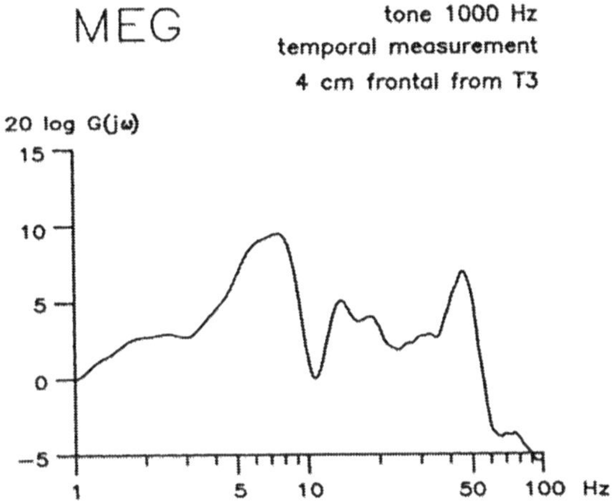

(A)

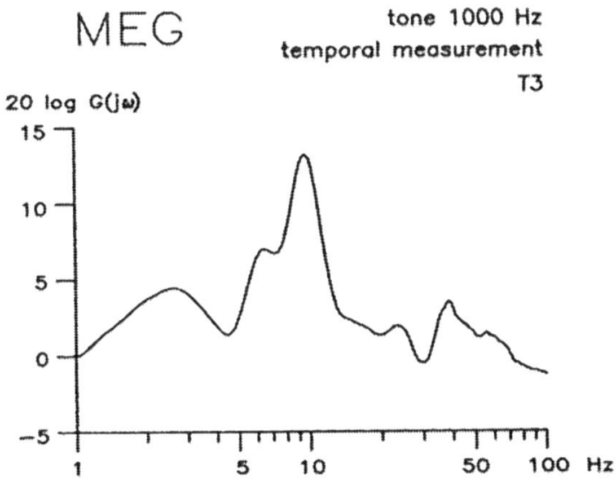

(B)

Fig. 13.6. (**a**): Amplitude–frequency characteristic computed by means of FFT from transient evoked magnetic field for a normal subject. Stimulation: 1 kHz tone burst, 60 dB HL. Location: 4 cm frontal from T_3. (**b**): As in (**a**), but for the temporal lobe location: T_3. (From Saermark K. et al. (1992) in: Başar E. and Bullock T.H., eds. *Induced Rhythms in the Brain.* Boston: Birkhäuser, pp. 129–147.)

the evoked response appear in the time domain for two such different frequency characteristics? Figure 13.7a shows the transient evoked response for the measurement giving rise to the distinct theta maximum in Fig. 13.6a, and Fig. 13.7b illustrates the evoked response (for the same subject) for the measurement giving rise to the high, dominant peak in the alpha frequency range in Fig. 13.6b. Figure 13.7 also shows the recorded data, band-pass filtered in the frequency range of 4 to 8 Hz and in the frequency range of 8 to 13 Hz. A comparison of the two pairs of curves in Fig. 13.7 clearly reveals a rhyth-

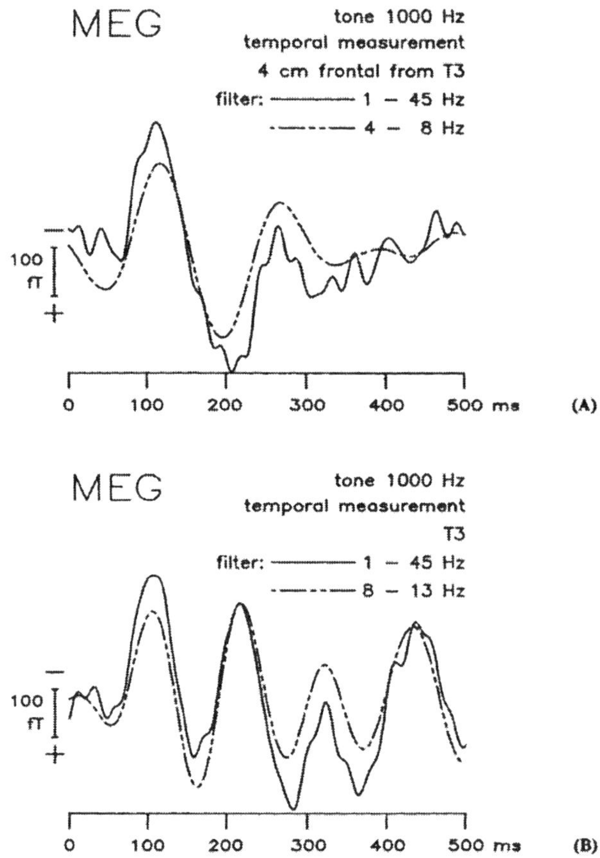

Fig. 13.7. (**a**): Transient evoked magnetic field measured at a position 4 cm frontal from T_3. (**b**): Transient evoked magnetic field measured at the temporal position T_3. Same subject as in (**a**). *Dashed curves*: data band-pass filtered in the band 1–45 Hz. *Solid curves*: data band-pass filtered in the bands 4–8 Hz, respectively 8–13 Hz. (From Saermark K. et al. (1992) in: Başar E. and Bullock T.H., eds. *Induced Rhythms in the Brain*. Boston: Birkhäuser, pp. 129–147.)

micity in the 10 Hz frequency range corresponding to the sharp alpha peak in the frequency distribution of Fig. 13.6b and a damped theta rhythmicity corresponding to the theta peak in the frequency distribution of Fig. 13.6a. It should also be emphasized that the averaged magnetic response in these cases shows almost congruent time courses without any filtering.

As pointed out in Chap. 18 in Vol. I, it is difficult and almost impossible to record *homogeneous* oscillatory responses in a given frequency range by means of electrical recordings. The method of MEG is, so far, an important improvement, since we have now detected such homogeneous oscillatory responses.

The analysis of EPs from temporal and frontal areas of the human brain indicates a dominant peak in the alpha and theta frequency bands, which is officially called the "alpha–theta bound." On the other hand, our measurements show that in recording areas with a distance of only 2 cm, sharp alpha or theta response peaks can be distinguished by means of MEG.

Primary sensory areas are usually surrounded by secondary sensory areas and associated structures. Accordingly, the space resolution with the MEG response will probably allow fine-structured functional brain mapping. At the present time, such an improved sensory–cognitive mapping using alpha, theta, and gamma responses seems to be more feasible than using electrical data.

Part III

Cognitive Processes

14. Selective Attention and Memory: Neurophysiology and Cognitive Psychology

Part III of this volume is devoted to the analysis of EEG and EPs during cognitive processing in the brain. The higher functions of the central nervous system related to *decision-making, memory, learning, surprise,* and *focused attention* have belonged to the most important goals of EEG and EP analysis during the last 20 years. We will try to learn about these functions by analyzing EEG-EP dynamics with the methods and concepts introduced in Chap. 4 in Vol. I and already applied in Parts I and II.

Firstly, the conventional experiments with single-unit recordings will be described in this chapter to bridge over from neural correlates of these cognitive processes to field potentials, as for instance EEG and EPs. At the beginning of this part we will also briefly describe the psychophysiological definitions of cognitive behavior. The rationale of this part consists not only in the description of EEG and EPs but also in finding the links to physiologically fundamental processes.

14.1 Background and Perspective (View of Karakaş)

According to Karakaş (1997) the approches to selective attention and memory can be stated as follows:

> Survival depends on veridical perception of the environment and appropriate response to it. How do high-level organisms perceive, compare, and evaluate stimuli and how do they select and execute responses? How does information processing proceed? How does the mind work?
>
> The functioning of the mind and the manner in which information is processed are not topics only of contemporary concern. Starting with Plato and Aristotles, many philosophers have tried to explain this process, an endeavor that continues today in the field of cognitive psychology. By employing appropriate experimental designs, psychologists attempt to make inferences about the processes that intervene between stimulus and response.
>
> However, more recently, the interdisciplinary field of cognitive psychophysiology has opened a window on these intervening pro-

cesses in the time-domain and the frequency-domain by employing domain-specific electrophysiological methods. In time-domain studies, where the transient responses of an organism to an input stimulus are studied, the studied variables are basically the latency and amplitude of the peaks that usually represent averaged responses. In frequency-domain studies, the emphasis is basically the amplitude and phase of oscillations or frequencies obtained in response to an input stimulus. Through analysis of transient response parameters or frequency and phase characteristics that are inherent in the transient responses, the emerging field of cognitive psychophysiology has provided a reductionistic and a relatively more concrete approach to studying the *black box*, an intervening event studied functionally in cognitive psychology by intricate experimental design. Thus, at present, information processing is described and explained through three separate approaches: cognitive-psychological, cognitive-psychophysiological with emphasis on time-domain responses, and cognitive-psychophysiological with emphasis on frequency-domain responses. ...

The basis for cognitive experimental research that was employed in this chapter can be found in many textbooks of contemporary psychology, examples of which include Baddeley (1990), Ellis and Hunt (1993), and Klatzky (1980). The experimentally based well-known assertions in these works are used to derive the established data and concepts on human auditory information processing. These data and concepts are utilized to identify stages and pathways of communication taking place between the hypothetical stages and in assessing the standing of each stage with respect to a set of issues that is derived as a part of the present study as being critical to information processing. These issues will be the *common denominator* that will facilitate comparisons between the events of the three approaches. ...

At its establishment by Wundt in 1879, the subject matter of the science of psychology was the structure of the mind. By the 1930s, the classical behaviorist Watson and subsequently Skinner demonstrated that an understanding of behavior could only be determined by describing the functional relationship between the stimulus and response parameters. All other intervening processes between the stimulus and response, i.e., the mind, was considered as unknowable and irrelevant (Schultz 1981).

However, spurred by the advent of computer technology and the need to consider the operational characteristics of man–machine interactions, it became evident that understanding behavior demanded the study of the intervening processes. These forces conspired to create the field of *cognitive psychology,* which sought to specify more precisely the mechanisms taking place in the *black box* that behaviorism so ardently denied. This approach demonstrated that mental phenomena could be studied experimentally and, furthermore, spawned the search for the neural substrates of cognition.

These developments led directly to a structural model of auditory information processing that has become the standard viewpoint of cognitive psychology and that can now be related to the underlying neuroelectric events of the brain's activities. The bulk of the experimental work on the structural model that has been replicable and the related paradigms and concepts on which there is general consensus are discussed and reviewed in Baddeley (1996), Ellis and Hunt (1993), Green (1992), Klatzky (1980, 1984), Stern (1985), Sternberg (1985), and Tiberghien (1989).

14.2 Comparative Studies

Following the first discovery of the family of potentials that are called "P300"-Response by Sutton et al. (1965), brain research dealing with cognitive processes has made important progress. During the last few years this new development has been described in several outstanding books (e.g., Regan 1989; Desmedt 1989; Ritter et al. 1968). There are also a number of important review articles that cover this field (Näätänen 1990; Picton 1992).

There are, unfortunately, not yet solid links between scientists working in the field of ERPs with scientists working in the field of conventional neurophysiological spike-recording techniques. Both groups of research scientists use such common expressions as *memory, attention, selective attention, orienting reflex, short-term memory, iconic memory, perception.* They use the same expressions, but the methods used for arriving at these states differ.

Our strategy in this book consists mainly in a trial to bridge both types of views (see also Chap. 9 in Vol. I). Should we choose the conventional way of arriving at explanations on the basis of single-cell recordings of defined locations, as in the fascinating work performed by Mountcastle and coworkers (1984) on visual information processing in the parietal lobes of the monkey or as in the relevant results derived from studies of Fuster and coworkers (1991) on frontal lobes of the monkey? Should we not recognize the importance of experiments by Galambos and Hillyard (1981) concerning ERPs as well as the experiments and the respective treatises of Näätänen and Mitchie (1979), Hillyard and coworkers (1973), and Donchin and coworkers (1978) on selective attention, expectancy, and surprise? Can we try to find a common way to correlate several of these experimental approaches?

In order to develop the Brain Dynamics Research Program, explained in Chap. 1 in Vol. I, experiments have been performed in our laboratories where recordings of ERPs in the cat brain and in the human brain have been made in addition to our experiments on sensory EPs obtained from intracranial structures of the cat brain. Accordingly, a large ensemble of data on cognitive processes has been accumulated in the last 15 years in our laboratories. The evaluation of the accumulated data gave us the possibility of comparing the information obtained under the different approaches with the ensemble of experiments described in this book.

We do not assume that our data on ERPs and our paradigms reach the comprehensive coverage of a psychology laboratory where other techniques of measurement, like psychometric testing, are utilized. However, since our data were steadily compared with data from the cat brain or data from invertebrates, the lack of perfection in the conventional techniques of psychology should not be considered as a shortcoming. We used a number of techniques such as the *combined analysis procedure*, which includes filtering techniques, and amplitude–frequency characteristics. Then by comparison of all types of components we could gain insight into the brains of various species.

In the present part, our systems-theoretical strategies are further applied in our search for a general brain coding. We have recorded EEG, EPs, and MEG from various parts of the brain. Can common brain codes be detected by evaluation of these experiments? As the results of the following chapters will show, the answer is yes. We will see that the parietal lobe of the human cortex, frontal lobe of the cat cortex, and the hippocampus of the cat brain use similar frequencies in connection to similar types of behavior (see Chaps. 25 and 27).

14.3 Concept of Selective Attention and P300

Selective attention is defined as a central nervous system function that allows perceptual or motor responses to be made selectively to one aspect of a total stimulus configuration in preference to other aspects.

In such a competitive process, "irrelevant" messages are either partially or completely rejected from perceptual experience, and the resulting configuration controls behavior. By virtue of their selective influence upon sensory processing, attentional processes are to be distinguished from general activation, alerting, mobilization excitability, arousal, and other similar processes that influence a broad spectrum of sensory inputs in a nonselective fashion.

James (1890) differentiated between two varieties of attention: *Passive attention* involves the inquisitory orientation reaction to an unexpected novel stimulus. *Active attention* is mediated by a variety of mechanisms. James identified "the accommodation and adjustment of the sensory organs; and the *anticipatory preparation* from within of ideational centers concerned with the object to which the attention is paid."

The P300 has been found to correlate with many variables. These include "task relevance," "meaningfulness," "information delivery," "resolution of uncertainty," and "decision-making." Regan (1989) objected that some of these psychological terms overlap, whereas others are not sharply defined, the situation thus leading to controversy. According to Woods (1990), selective attention refers to the preferential detection, identification, and recognition of selected stimuli in an environment containing multiple sources of information. In P300 experimental designs, the eliciting stimulus is task relevant; that is, attention is paid to it. The contradictory claim that P300 can be recorded in a situation where low-probability stimuli are ignored (Roth 1973) is suspect, because it assumes that subjects are able to follow instructions to ignore irrelevant stimuli while they just sit with their eyes shut (Ritter et al. 1968; Squires et al. 1977).

According to Hillyard and Picton (1979), "The P300 seems to index the operation of adaptive brain systems." A rather different line of thought, derived from autonomic psychophysiology, uses the language of innate stimulus–response systems such as *orienting, startle,* and *defense* reflexes (Donchin et al. 1984).

In general, a P300 will be produced by *task-relevant* (Chapman 1973; Chapman and Bragdon 1964) stimuli that occur somewhat unexpectedly and require a motor response or cognitive decision (Donchin 1981; Ritter et al. 1968).

Woods (1990) claims that studies on the psychological basis of selective attention have little influence on each other. On the one hand, cognitive psychologists have typically studied selective attention in human subjects performing complex tasks and have monitored attention through simple or complex responses. In contrast, cellular neurophysiologists have examined the effects of selective attention on the firing of single neurons in animals responding to simple stimuli that are significant. Demiralp and Başar (1992) tried to bridge the gap between these two different approaches. It was their aim to link both types of experiments, by including the discussion of a neurophysiological signal that is common in both categories of analysis: the changes in theta frequency component (especially of the hippocampus and cortex) during learning, association phenomena, memory retrieval, and attentive behavior. We emphasize that during increased attention (the responses to third stimuli) the significant increases in the theta responses were recorded in frontal and parietal locations. Such an overall theta activation in the brain were surveyed by Başar-Eroğlu et al. (1992). The relation of the theta component to single-neuron firing is established in a number of relevant studies (see Lopes da Silva 1992; Buzsaki 1992).

14.4 Visual Selective Attention

One area of research where strong parallels are emerging between ERP and primate neurophysiological studies is that of *visual selective attention.* As shown in Fig. 14.1, when a person focuses attention on a flashing light in one visual half-field, the ERP to that stimulus shows an enlarged negative component peaking at 150 to 200 msec., which is absent when attention is directed to the opposite field. This attention-related negativity is prominent over parietal scalp areas, largest contralaterally to the attended visual field.

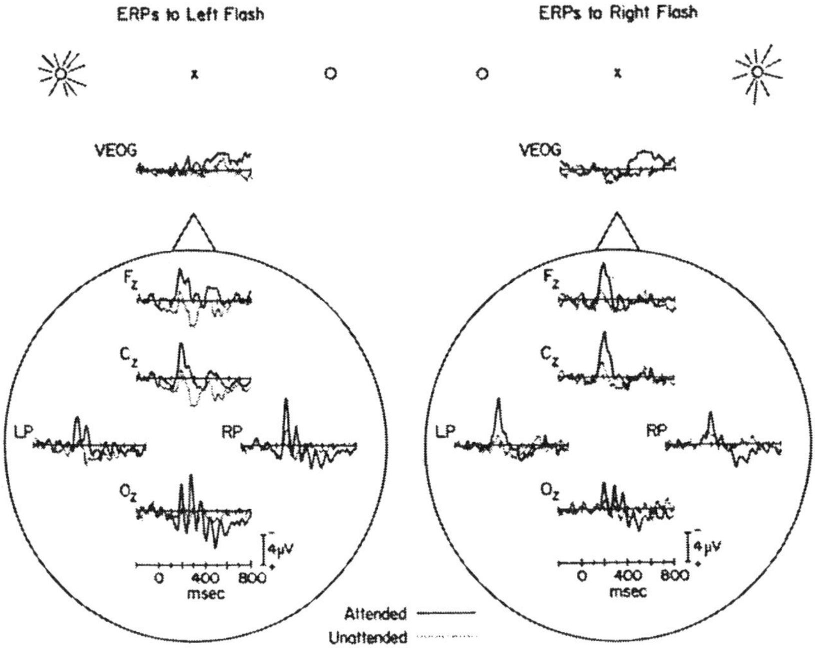

Fig. 14.1. Visual ERPs during a task where the subject attends selectively to sequences of flashes in one visual field at a time. Flashes were presented to the right and left fields (20 degrees from the fixation point) in random order at ISI of 250 to 400 msec. ERPs are shown to flashes in each field that were attended (*solid lines*) and unattended (*dotted lines*), recorded from frontal (F_z), central (C_z), occipital (O_z), and left and right parietal (LP and RP) scalp sites. VEOG = electrooculogram. (From Hillyard S.A. (1981) in: Galambos R. and Hillyard S.A., eds. *Electrophysiolgical Approaches to Human Cognitive Processing.* [Neuroscience Research Program Bulletin, vol. 20].)

Similar studies in rhesus monkeys have shown that an evoked single-unit discharge in the parietal lobe occurs when an animal makes a saccade to a visual target (Mountcastle et al. 1975). Robinson and coworkers (1978) showed that this discharge was enhancement of a baseline-like visual response. In area 17 this enhancement occurred whenever the animal attended to the visual stimulus, regardless of the nature of the response. In the following we will give a detailed account of these experiments. Although the animal experiments differ in certain ways from the human experiments, where the N1 wave shows enhancement, the parallels between them are striking enough to warrant a serious effort at bridging these two lines of investigation.

14.4.1 Selective Attention: Experiments with Monkeys

The parietal association area (A5) in the superior parietal lobule contains a group of neurons that have no apparent sensory input but that discharge vigorously (100/sec for 200–300 msec) in relation to a movement that aims to reach an object that the monkey intensely wants, but not to other types of movements involving the same muscles. In parietal association area 7 of the inferior parietal lobule, more than 50% of the investigated neurons are visual fixation cells that are active at high discharge rates as long as the animal looks at visual targets that are drive-related. Other elements in area 7 are light-sensitive neurons (Yin and Mountcastle 1977; Robinson et al. 1978). The static properties of light-sensitive neurons in area 7 are, however, dissimilar to those of striate cortical neurons.

Mountcastle et al. (1981, 1984) determined how different levels of general arousal and selective attention are correlated with differences in the intensity of responses of light-sensitive neurons in the posterior parietal cortex. In these experiments, the monkey is in one of three behavioral states: a *no-trial state* (state of waking without behavioral task), a *trial state* (a condition of attentive fixation on a target during which the animal is engaged in a detection task) and an *intertrial state*. The increased excitability of light-sensitive neurons of the inferior parietal lobule when the monkey attentively fixates on the target in the trial state (task mode A) is shown in Fig. 14.2 in comparison to periods without attentive fixation (no task mode but during waking).

The proportion of unresponsive parietal visual cells in the no-trial state is virtually equal to that in the intertrial state, leading to the conclusion that the enhanced excitation (responsiveness) does not merely occur with changes in general arousal but is more specifically related to the visual attention directed to the target. Robinson et al. (1978) emphasized that in the process of selective attention "a shift of attention precedes a visually guided movement, but movement is not required for attention."

Figure 14.2 shows a comparison between the attention-related modulations of the scalp-recorded ERP in man and the discharges of a parietal lobe unit in the rhesus monkey. These results at the cellular level are of importance for studies in human visual attention. First, the negative latency wave N100 varying between 40 and 240 msec (mean value of about 100 msec) is comparable to those of the N1 in parietal scalps when human subjects attend to sequences of stimuli in one visual field.

Comparison of effects of visual–spatial attention upon unit activity in the posterior parietal cortex of a rhesus monkey and upon the ERP recorded from the parietal scalp contralateral to the visual field of presentation of a human subject was found to be as follows. In the monkey experiment (Wurtz et al. 1980a), the animal was trained to fixate a central spot while a peripheral flash elicited unit discharge in the contralateral parietal lobe. The two poststimulus time histograms (PSTH) compared conditions where the evoking peripheral

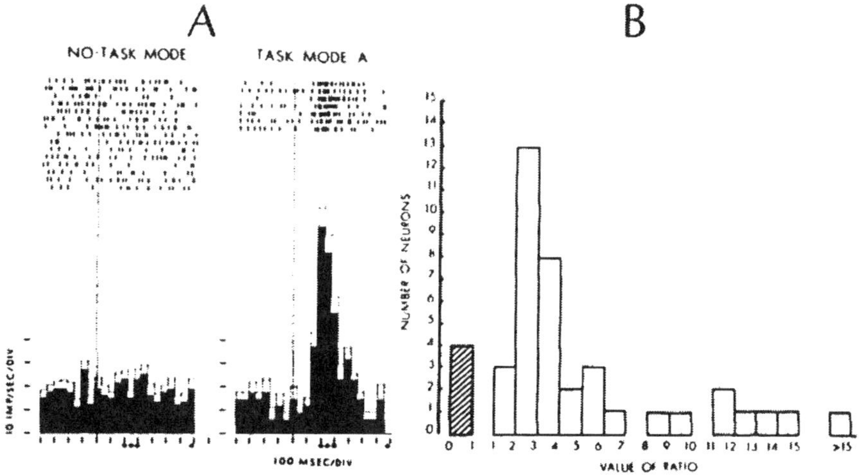

Fig. 14.2. Facilitation of the monkey parietal visual neurons by attentive fixation, (**a**): Comparison of responses of parietal light-sensitive neurons to visual stimuli in no-task and task modes. There is an obvious absence of responses during the no-task mode as compared to the strong responses in the task mode. Summing histograms: Standard error of mean calculated for each bin of the histograms; value shown by *dotted line*. Corresponding bin pairs within the histograms tested for significant differences (Student's t test); bin pairs marked with diamonds differed at the 5% level of significance. The overall response in the no-task and task states differed in the following way: The rate of impulse discharge in the prestimulus period was subtracted from that in the poststimulus period for each trial, and populations of remainders were tested for significant differences ($p < 0.05$ required) and used to form a facilitation ratio for each neuron. (**b**): Ratios for neurons with significant differences. The facilitation ratio is between the net increment in response evoked by a light stimulus while the monkey is in a state of interested fixation over to that evoked by physically identical and retinotopically similar stimuli delivered in the no-task or intertrial state. Fifty-one neurons showed ratios ≥ 1.0, and of these the difference was significant at the 5% level (Student's t test) for 38; values plotted in histrogram. The ratio was fractional for the four neurons indicated to the *left*, whose response was significantly greater in the no-task state than during interested fixation. (Adapted from Mountcastle et al. (1981) *Journal of Neuroscience* 1: 1218–1235.)

flash was not attended vs. when it was attended. Note that the unit discharge at a latency of about 100 msec was greatly enhanced when the evoking flash was made relevant.

The human ERPs shown are recorded from a left-parietal scalp site to right-field flashes under conditions where those flashes were attended and when attention was directed to the opposite field. Note that the modulation of evoked activity occurs with a similar time course and magnitude in the two species. (This comparative view has been emphasized and outlined by Galambos and Hillyard (1981)).

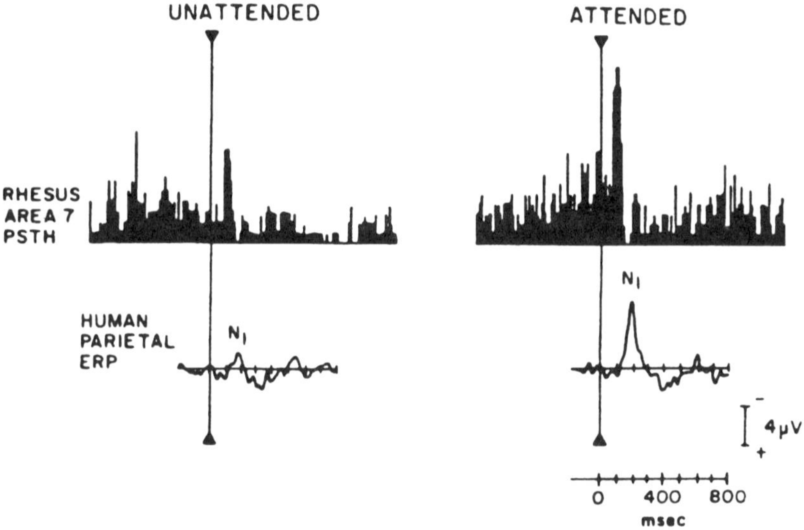

Fig. 14.3. Comparison of effects of visuospatial attention upon unit activity in the posterior parietal cortex of a rhesus monkey and upon the ERP recorded from the parietal scalp contralateral to the visual field of presentation of a human subject. In the monkey experiment the animal was trained to fixate a central spot while a peripheral flash elicited unit discharge in the contralateral parietal lobe. The two poststimulus time histograms compare conditions where the evoking peripheral flash was not attended vs. when it was attended. Note that the unit discharge at a latency of about 100 msec was greatly enhanced when the evoking flash was made relevant. The human ERPs shown are recorded from a left parietal scalp site to right-field flashes under conditions where those flashes were attended and when attention was directed to the opposite field. Note that the modulation of evoked activity occurs with a similar time course and magnitude in the two species. (From Galambos R. and Hillyard S.A., eds (1981) *Electrophysiolgical Approaches to Human Cognitive Processing*. [Neuroscience Research Program Bulletin, vol. 20.2].)

Conventional experiments with single-unit recordings as described in this chapter serve to bridge over from neural correlates of cognitive processes to field potentials, as for instance EEG and EPs. At the beginning of this part we will also briefly describe the psychophysiological definitions of cognitive behavior. The rationale of this part consists not only in the description of EEG and EPs but also in finding the links to physiologically fundamental processes.

With regard to the results from P300 potentials in the human and cat brains not only one paradigm, but several, will be used for the description of the so-called family of P300 potentials. We use two new paradigms, not yet applied extensively by neuroscientists working on cognitive processes, as an extention to the conventional types of ERPs, for instance oddball, P300, and paradigm with omitted stimuli. The first one is a modified oddball paradigm with "alternating targets" during which the subjects are told that every second signal is a target one, thus increasing the probability of the task's occurrence to 100 percent. The second paradigm used is a new version of omitted stimuli whereby the subjects pay attention to repetitively applied third signals prior to omitted ones (Chaps. 19 and 18).

According to analyses of the data obtained from these paradigms, the subjects sometimes behave with "surprise," "decision making," (oddball) "high expectancy," and/or "selective attention." It will be shown that the combined use of all these paradigms will allow us to find functional correlates of the so-called theta, delta, and 40 Hz responses.

14.5 Stages of Memory Processing: Encoding, Storage, and Retrieval

Any system for storing information, whether biological or artificial, will need

1. to be able to *encode, or register, information*;
2. to *store* it, preferably without too much loss or forgetting; and subsequently
3. to access, or *retrieve*, that information.

While these three stages are closely linked, making it difficult to pin down any phenomenon as exclusively occurring at a single stage, nevertheless, this division into processing stages continues to be useful in helping us to understand the working of memory systems.

14.6 Encoding and Sensory Register

The term "encoding" refers to the initial processing of the information that is to be learned or memorized. Immediate memory for arbitrary sequences of verbal material, such as occurs in digit span, typically relies on encoding in terms of the phonological or sound characteristics of the material.

Sensory register processes decode the specific sensory codes into a neural representation of receptor activity, which is considered in behavioral literature as "veridical." Sensory register processes are automatic, preattentive, and occur simultaneously via parallel pathways.

The resulting sensory memory store (e.g., echo) is preconscious and precategorical, with a duration of 150 msec to 2 sec. Maintenance of memory is not controllable. The mechanism of trace displacement is passive decay or erasure by succeeding stimuli in the same modality.

In sum, the functions of the sensory registration process is the storage of the sensory trace and the decoding of the receptor-specific codes to derive information about the physical characteristics of input stimulus. After an interactive process between the short-term memory (STM) and the sensory register, these processes result in pattern recognition (Karakaş 1997).

14.7 Memory

Baddeley and Hitch (1974) have proposed a division into at least three subsystems: Part of the system is an attentional controller, the *central executive,* which forms an interface between long-term memory and two or possibly more slave systems. These systems combine the capacity for the temporary storage of information with an active set of control processes that allow information to be intentionally registered and maintained within the subsystem. One of these, the *visuo-spatial scratchpad*, or *sketchpad*, is specialized for maintaining visuo-spatial information, while verbal information is held using the *phonological*, or *articulatory, loop*. These subsystems correspond to the visual and verbal short-term memory systems.

Baddeley (1986) proposed adopting the concept of a *supervisory attentional system* (SAS) as formulated by Norman and Shallice (1986) as the basis for the central executive. In this model, routine behavior is assumed to be controlled by well-learned habit structures, or *schemata*. When novel action is required, for example in order to cope with an unexpected problem or to override an old habit in an emergency, the SAS overrides the schemata. It is also responsible for planning and coordinating activities, and is assumed by Shallice (1982) to be the system that is disrupted by extensive bilateral frontal lobe damage.

Tulving (1983) originally defined two subsystems, namely *episodic* and *semantic* memory.

- Semantic memory refers to knowledge of the world; knowing the chemical formula for salt or the name of the French capital or how many inches there are in a foot are all examples of semantic memory.
- Episodic memory refers to the system involved in recollecting particular experiences or episodes; remembering what you had for breakfast or setting off for a holiday last year are examples of episodic memory. This is the system that is associated most closely with traditional approaches to learning and memory. It is characteristically accompanied by an awareness of the learning experience and is sensitive to depth of processing. Episodic memory is strongly influenced by degree of attention and organization, which reflects the importance of these processes for setting up memory structures that are accessible to retrieval.

It is the episodic memory system that is readily disrupted by brain damage, and whose impairment forms the basis of the classic amnesiac syndrome. While it is crucially dependent upon the circuit linking the temporal lobes, hippocampus, and frontal lobe, virtually anything that comprises the functioning of the brain will impact on rate of learning and speed of retrieval, suggesting the operation of a massively parallel system capable of graceful degradation when brain tissue is lost, with the hippocampal region playing a crucial role in learning but not itself being the seat of memory.

Working memory refers to a transient type of memory processing, of the order of seconds, in which we can hold "on-line" the relevant stimuli, rules, and mental representations that are needed to execute a particular task (Baddeley 1992; Goldman-Rakic 1987) – representations in a mental workspace long enough so that we can make appropriate responses to stimulus configurations or contingencies in which some, or even all, of the basic ingredients are no longer extant in perceptual space (Fuster 1989; Goldman-Rakic 1987). The concept of working memory has some overlap with short-term memory, as both are considered relatively transient and of limited capacity.

14.7.1 Short-Term Memory

Under focused attention, information is selected from the sensory register and transferred to short-term memory. The short-term memory processes recode the "veridical" representations into basically acoustic/phonetic/verbal codes. Under conscious manipulations that necessarily engage cognitive effort, rules and strategies are employed by retrieving related algorithms from long-term memory (LTM). The information in STM is organized so as to assimilate it into LTM. The immediate experience or working memory is categorical. These processes allow for the attainment of meaningful objects. The duration of STM is on the order of 2–20 sec. However, STM duration can be increased by means of repetitive rehearsal which maintains the stimulus representations or the codes in STM. If such mechanisms are not invoked, information in STM is displaced due to interference or disuse. Another way of keeping the information is to directly transfer the information to LTM by means of elaborative rehearsal.

In sum, the basic function of STM involves the conscious experience and processing of events of the present and the recent past. The related information is recoded through working memory operations; it is maintained in STM through repetitive rehearsal or is transferred to LTM through elaborative rehearsal (Karakaş 1997).

14.7.2 Long-Term Memory

As has been outlined, elaborative rehearsal promotes information transfer for more permanent storage. LTM processes then encode the STM codes into prepositional or analogical representations. LTM has potentially an unlimited capacity. This capacity is used through a process of restructuring which typically involves organization, elaboration, and imagery. LTM experience is encoded long-lasting memory trace of semantic or episodic form. It has been demonstrated that memory traces are maintained by an in-depth analysis via semantic or cognitive processes. Findings support the conclusion that in semantic memory, retrieval from LTM is dependent on the number of pairings of the target stimulus with the cue, the previously stored representations in semantic memory. In the case of episodic memory, it depends on the amount of elaborating analysis of the cue with the episode (target). Memory traces can be displaced by interference or, because of the above mechanisms, by cue-dependent forgetting. LTM processes involve focal attention and conscious processes for new events. On the other hand, for well-practiced events, LTM processes involve passive attention and preconscious processes.

In short, the basic functions of LTM are related to encoding, storage, and retrieval (Karakaş 1997).

14.8 Pattern Recognition

Pattern recognition results from the interaction between the sensory register and LTM. First, context of the sensory trace is analyzed to produce an expectancy effect with respect to the stimulus, based on predispositions, task, and environmental demands. This process provides for an initial normalization of the sensory event. The normalized trace is than transferred to LTM for information extraction in which the trace or the stimulus representation is compared with LTM codes and a decision is reached respecting the issue of best match. The resulting output forms the basis of object perception or of pattern recognition (Karakaş 1997).

15. Memory Templates in Event-Related Oscillations, P300, MMN

By "memory" I will mean any set of events that makes available to an organism something of a situation after that situation no longer obtains. "Novel" I will define as any aspects of a situation which differ sufficiently from prior situations to produce recordable physiological changes in the organism. By the term "thought" I will refer to the active uncertainty produced when an ordered set of memories mis-matches the current novelties of the situation. And "choice" I will use to designate processes of resolution of uncertainty that lead to action.

Pribram (1963, p. 54)

15.1 Remarks on Family of P300 Responses: ERPs

Many studies have reported human brain potentials elicited by unexpected, as well as expected but absent, sensory stimuli. Such "emitted," or "endogenous," potentials are generally regarded as signs of endogenous brain processes such as memory, internal time estimation, and other cognitive mechanisms (Ruchkin and Sutton 1983).

When a task-relevant stimulus occurs unpredictably and infrequently, it is often followed by a late positive potential at a latency of about 300–400 msec, following the point of information delivery. This wave has been variously termed the "late positive component," "P300," or "P3" and constitutes a well-known example of an emitted, or endogenous, potential. From various studies carried out to date, it has been suggested that the P300 change reflects the activation of brain mechanisms selectively concerned with the evaluation of stimulus relevance and probability, and that this change is relatively *independent of stimulus modality, intensity, and presentation rate.* Several authors have shown that the P300 wave can be related to the detection of an infrequent target stimulus that requires some cognitive response, e.g., counting target stimuli occurring randomly and infrequently in a train of nontarget stimuli (Desmedt 1979; Hillyard and Picton 1979).

The studies cited above and indeed most of the studies in this field have been based mainly on latency and amplitude measurements derived from averaged EEG data. The limitations of the averaging technique in the study of sensory EPs and cognitive ERPs have been discussed by many authors. Frequency-domain analysis may hold greater potential for understanding the genesis of these potentials, especially if attempts are made to relate poststimulus changes to prestimulus EEG activity. Such an approach offers greater opportunities for an understanding of brain cognitive dynamics, since it is possible neither to measure response variation in averaged data nor to investigate the relationship between the EP and ERP response and the prestimulus EEG activity.

The P300 response obtained by means of the oddball paradigm has been used successfully for psychological research and clinical diagnostics. Since the earlier applications there have been several excellent reviews concerning methodological, psychological, and clinical aspects of this paradigm (see Hillyard and Picton 1979; Johnson 1988 1990; Näätänen 1988; Regan 1989; Woods 1990).

An important aim in the analysis of physiological correlates of the P300 response consists in searching for sources of generators giving rise to the P300 component of the event-related potential. Accordingly, several investigators have used intracranial electrodes to be able to localize the sources in human recordings and animal models (see Halgren et al. 1986; Başar-Eroğlu et al. 1991a). Several of these authors indicate the existence of multiple generators including sources in hippocampus, parietal, frontal, and several other areas of the association cortex (for a review see Knight et al. 1981; Smith et al. 1990; Paller et al. 1988).

In this chapter an ensemble of paradigms and the combined analysis procedure of EEG and EP are applied to ERPs. The combined evaluation of results that were obtained with unconventional and new techniques is called the *Brain Dynamics Research Program* (see Chap. 4 in Vol. I). This program, which includes the application of a chain of paradigms interwoven with the analysis techniques cited in this chapter, has been applied extensively in Lübeck until now for the analysis of ERPs of both human and cat brains. (The reader should also compare the significant progress performed in this field by Polich and coworkers; their recent publications could not be described in detail in this volume.)

15.2 Experimental Setup and Paradigms

Electroencephalographic activity (EEG) was recorded electrode site in the international 10–20 system, with the reference electrode placed on the left ear. A Schwarzer EEG machine (Encephysioscript) was used to record the EEG activity. The filter band-pass was set to 0.1–70 Hz. The EEG was monitored and recorded continuously during the experiments. The responses of the subjects were monitored through a closed-circuit TV system. Experiments were carried out on 12 healthy, volunteer subjects (4 females, 8 males) whose ages ranged between 25 and 45 years.

Stimulations: Auditory stimuli of 80 dB, 1500 and/or 1550 Hz tones with 0.5 ms rise time and 800 ms duration were presented binaurally. In each recording session, firstly, the spontaneous EEG was registered for a few minutes to determine global characteristics of subjects' spontaneous EEG activity and arousal state at the beginning of the experiments. This period also helped the subjects to become familiar with experimental conditions. Thereafter, on the first group of subjects auditory EPs, omitted stimulus paradigm, and oddball paradigm were applied with short resting periods in between. The auditory EP experiments consisted of the presentation of 1500 Hz tones.

Selective averaging of EPs: The stored raw single EEG-EP or EEG-ERP epochs were selected with specified criteria after the recording session: EEG segments showing movement artifacts, sleep spindles, or slow waves were eliminated.

Recording of EEG-EP epochs: With every stimulus presented, a segment of EEG activity preceding and the EP or ERP following the stimulus were digitized and stored on computer disk memory. This operation was repeated approx. 100–200 times. In recording EEG-ERP epochs, 1.6 sec of EEG activity preceding each stimulus and 1 sec of EEG activity subsequent to the stimulus were digitized, labeled, and stored on computer disk memory of a Hewlett-Packard 1000 F data acquisition system.

15.2.1 Paradigm 1 – Oddball

The oddball paradigm is illustrated in Fig. 15.1. In the oddball paradigm the tones were presented in a pseudorandom sequence, with 1600 Hz tones occurring 20% of the time and 1500 Hz tones occurring 80% of the time. The interval between tones varied randomly from 2.5 to 4 s with a mean value of 3 s as in the auditory EP experiments. The subjects were instructed to keep a mental count of the number of 1600 Hz tones (infrequent target tones).

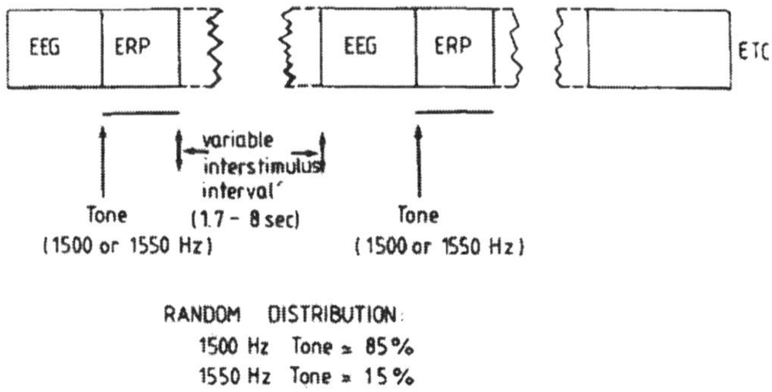

Fig. 15.1. Schematic presentation of the EEG-ERP recordings. The *horizontal thick lines* indicate the duration of the stimulation. (From Stampfer HG. and Başar E. (1985) *International Journal of Neuroscience* 26: 181–196.)

15.2.2 Paradigm 2 – Oddball with Increased Certainty of Alternating Targets

For further analyses the conventional auditory oddball paradigm used was modified to demonstrate the development and variation of prestimulus "preparation changes." The experimental modifications were as follows:

1. Repetitive frequent oddball tones were presented *without informing subjects* that the tones would be presented regularly. 1550 Hz target tones were presented alternately with 1500 Hz nontarget tones. Subjects were asked to count the number of target tones and were promised a cash reward if they counted correctly. From a physiological point of view, the brain was stimulated repetitively at fixed intervals. From a psychological point of view, subjects were uncertain as to when the target tones would arrive, although it was expected that they would soon recognize the regular pattern. However, even when they felt more certain that every second tone would be a target tone, they did not know whether the pattern would change. Figure 15.2 is a schematic representation of the paradigm.
2. Repetitive, frequent oddball tones were presented *where subjects were informed* that every second tone would be a target tone. Stimuli were presented as in (a), and subjects were again promised a cash reward if they counted the correct number of target tones. Subjects were further asked to mentally count the occurrence of target tones. The duration of tones for both (a) and (b) was 800 msec, and the fixed interval between tones was 1.65 sec.

Conventional auditory EPs were obtained with randomly applied step function tones of 1500 Hz at 70 dB above threshold and duration of 800 msec. These sensory EPs were recorded to establish a baseline to which ERPs in task-relevant experiments were compared.

Subjects were seated comfortably in a quiet, soundproof room and were asked to keep their eyes closed. They were instructed to keep a mental count

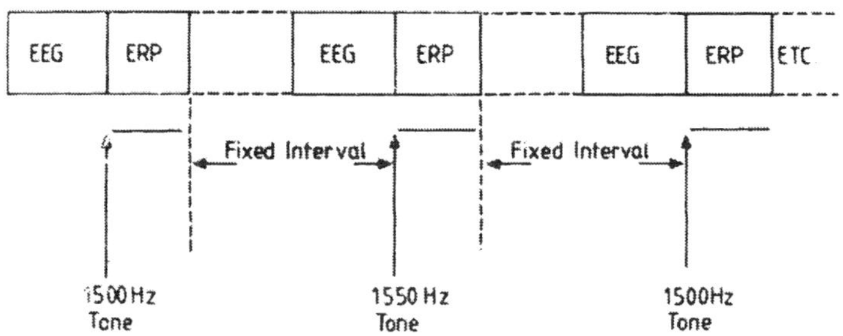

Fig. 15.2. Schematic presentation of the EEG-ERP recordings. The horizontal thick lines indicate the duration of the stimulation. (From Başar E, Stampfer HG (1985) International Journal of Neuroscience 26: 161–180)

of the number of high tones (target tones). A cash reward was promised if subjects counted the correct number of target tones. This was done to enhance experimental cooperation.

15.3 Frequency Analysis of ERPs: Preliminary Results

Comparison of EPs and ERPs: Figure 15.3 shows typical findings from one of our 12 subjects. Figure 15.3a shows the sensory EP (control EP). Figure 15.3b shows the corresponding ERPs in the oddball experiment. Dashed-line curves show the averaged ERP elicited by frequently occurring 1500 Hz tones that subjects were asked to ignore, whereas the solid-line curves show the averaged ERP elicited by infrequently occurring 1550 Hz tones that subjects were asked to count.

Characteristically, the sensory EPs show a negative peak in the region of 100 ms (N100) and a positive peak in the region of 200 ms (P200). The N100 wave is usually preceded by a small, positive wave at around 50 ms. (The reader is also referred to Chap. 14 to compare the N100 type of response from single-unit recordings of the animal brain.)

From an inspection of Fig. 15.3a,b it can be seen that the averaged responses to 1500 Hz tones show a configuration that is similar to their corresponding sensory evoked potential. By contrast, the averaged response to 1550 Hz target tones shows a large positive deflection at around 380 ms, which represents the "late positive component," or P300. All 12 subjects showed a similar type of deflection.

In the preliminary descriptions, only examples of vertex recordings are shown. In Chap. 20 we will give a comparative discussion of the results in which those of vertex, parietal, frontal, and occipital ERPs will be interpreted.

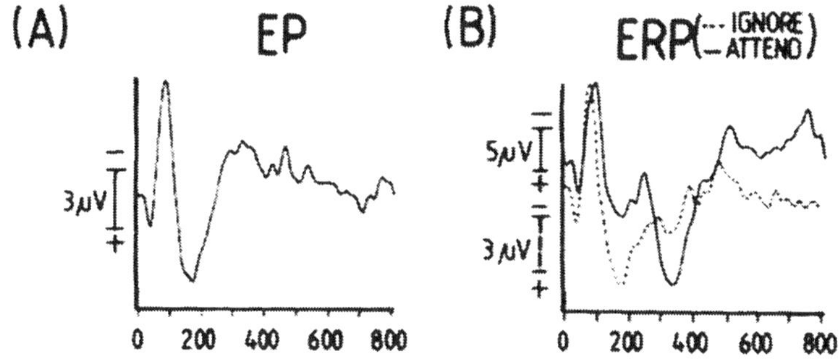

Fig. 15.3. (**a**) Typical averaged auditory EP recorded from vertex. (**b**) ERPs recorded from the same subject. *Solid curves* present the ERPs to rare target tones, whereas *dashed curves* are ERPs to frequent nontarget tones. (From Stampfer HG. and Başar E. (1985) *International Journal of Neuroscience* 26: 181–196.)

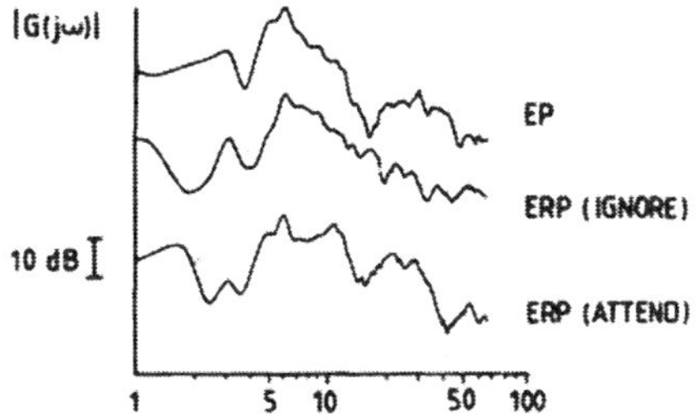

Fig. 15.4. Amplitude–frequency characteristics of the vertex EP and ERPs of one subject. The curves were obtained by applying Fourier transform to averaged evoked potential (EP) and event-related potential (ERP ignore and ERP attend). *Along the abscissa* is the frequency in logarithmic scale; *along the ordinate* is the potential amplitude, or gain, $G(j\omega)$, in decibels. The curves are normalized such that the amplitude at 1 Hz is equal to 1 (or $20 \log 1 = 0$). (From Stampfer H.G. and Başar E. (1985) *International Journal of Neuroscience* 26: 181–196.)

Figure 15.4 shows the amplitude–frequency characteristics computed from the averaged EP and ERPs shown in Fig. 15.3. The top curve was obtained from the sensory EP (control EP), the middle curve from the ERP response to 1500 Hz (ignore), and the bottom curve from the ERP response to 1550 Hz target tones that subjects were asked to count.

Figure 15.5 shows the averaged response and filtered components of the sensory-evoked potential (a), the ERP to 1500 Hz nontarget tones (b), and the ERP to 1550 Hz target tones (c) for three subjects. The top traces are the averaged responses. Below each one are the respective frequency compo-

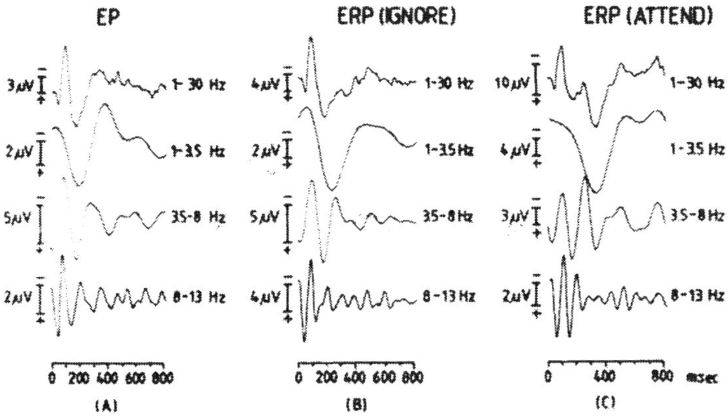

Fig. 15.5. Averaged response and filtered components of sensory evoked potential (EP), ERP to 1500 Hz nontarget tones (ignore) and ERP to 1550 Hz target tones (attend). (From Stampfer H.G, and Başar E. (1985) *International Journal of Neuroscience* 26: 181–196.)

nents, obtained by filtering the averaged response according to the amplitude–frequency characteristics. Figure 15.5 illustrates the different contributions of the three given frequency bands to peaks in the averaged curves. Assuming that these peaks are formed by the superposition of different frequency components, it can be seen, for example, that the N100 peak is formed mainly by 3.5–8 Hz and 8–13 Hz activity; the P200 peak is formed by the superposition of the first positive peak of 1–3.5 Hz activity and the first positive peak of 3.5–8 Hz activity.

15.3.1 Comparative Analysis of Poststimulus Frequency Changes Under Different Experimental Conditions and Their Contribution to Different Latency Peaks

Figure 15.6 shows the filtered frequency components of Fig. 15.5, arranged to provide a comparison within each of the three frequency bands under different experimental conditions.

1–3.5 Hz activity (delta frequencies). It can be seen that the configuration of activity in this frequency band is grossly similar to the EP, nontarget, and target ERPs, except for a progressive latency shift to the right of the major positive deflection, which is maximally delayed in the response to target tones. The delay between the EP and nontarget ERP was in the range of 20–30 msec, whereas the delay between the nontarget and target ERP responses was in the range of 70–90 msec. It was observed that this maximum delay in target responses was associated with a tendency for greater positivity immediately poststimulus; the small negativity that can often be seen in EP and nontarget ERP responses tended to be absent in target ERP responses. It was further noted that the maximum positive deflection to target tones was followed by a negative deflection, peaking at around 600 sec. This peak was

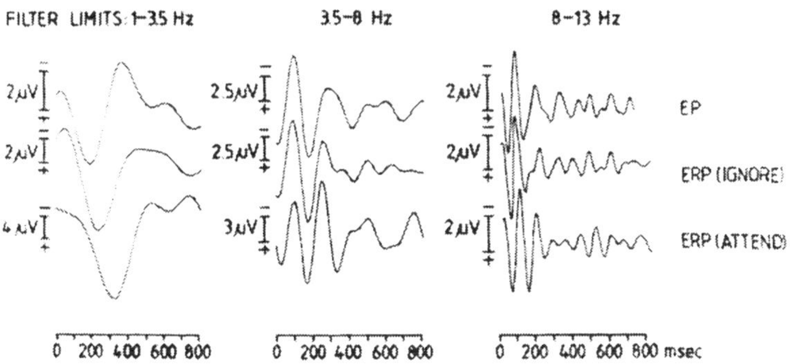

Fig. 15.6. Filtered frequency components of sensory evoked potential (EP) and of ERPs (ignore and attend); curves also shown in Fig. 15.5, but here in different arrangement. (From Stampfer H.G. and Başar E. (1985) *International Journal of Neuroscience* 26: 181–196.)

usually larger for target tones than for EP and nontarget tones. Our analysis suggests that this N600 deflection should be considered as a continuation of the positive deflection at around 380 msec and as part of the mechanism giving rise to changes in the 1–3.5 Hz range. There was a tendency for N600 amplitudes to be larger if the P380 deflection was relatively small.

3.5–8 Hz activity (theta frequencies). Activity in this frequency range showed the greatest changes in the pattern of oscillation and damping. A "reverse" pattern of oscillation during approximately the first 400 msec can be seen in the ERP response to target tones, and it was observed in all 12 subjects. In the EP and nontarget ERP, the maximum response amplitude (enhancement) is usually seen in the first poststimulus wave, whereas in the target ERP response, the maximum amplitude is to be seen in the second (Fig. 15.7a,b) and sometimes third wave. More generally, there is a longer duration of theta oscillation in target ERP responses, such that this frequency band often makes a significant contribution to N200 and P380 deflections.

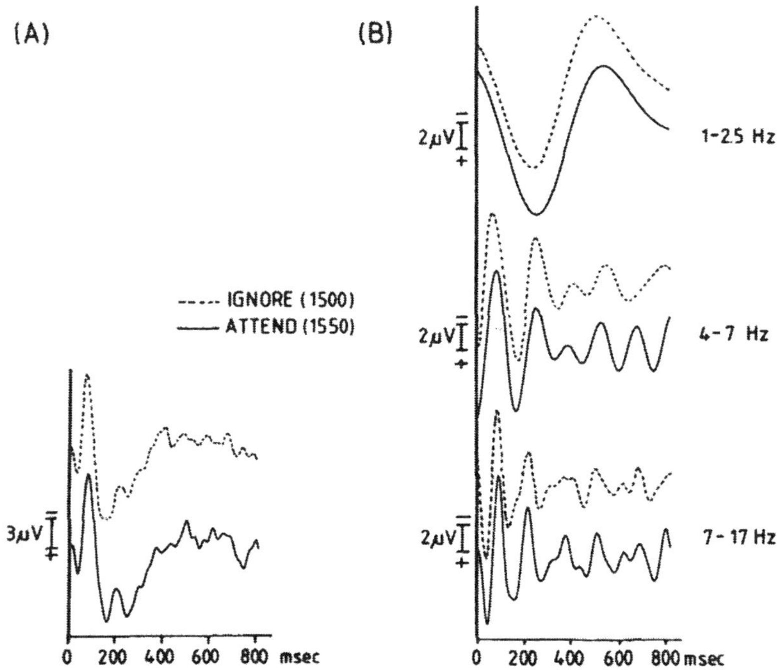

Fig. 15.7a,b. (**a**) Typical averaged ERPs obtained by applying alternating nontarget (*dashed curve*) and target (*solid curve*) tones. Recordings are from vertex. (**b**) Filtered components of the averaged ERPs shown in (**a**) in three different frequency bands. *Solid curves* present the ERP-components to target (attend) stimuli and the *dashed curves* present the responses to nontarget (ignore) stimuli. (From Başar E. and Stampfer H.G. (1985) *International Journal of Neuroscience* 26: 161–180.)

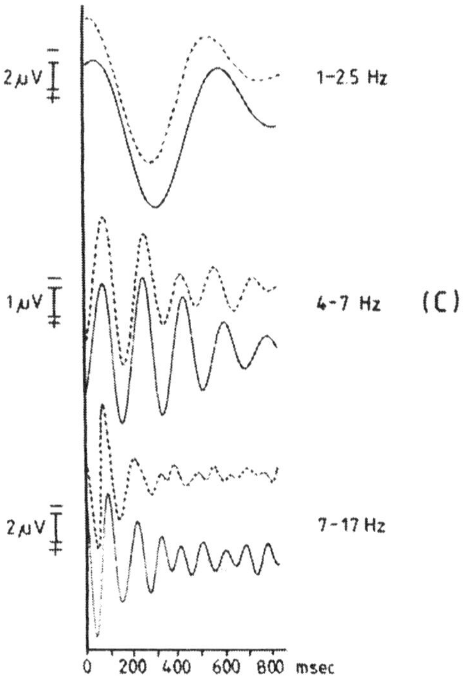

Fig. 15.7c. Comparison between filtered ERP components to randomly applied infrequent target tones (solid line) and averaged ERP response to regular, frequently presented target tones (*dashed line*). (From Başar E. and Stampfer H.G. (1985) *International Journal of Neuroscience* 26: 161–180.)

8–13 Hz activity (alpha frequencies). Event related oscillation in this frequency band does not appear to make any direct, obvious contribution to the positive wave at 380 msec and is usually maximally desynchronized at this latency. However, Fig. 15.6 shows a frequently observed finding, namely that the second major oscillation is greater for target than for EP and nontarget ERP responses. This second oscillation contributes to the N200 wave, which will be discussed below. Our findings suggest that while alpha activity does not appear to contribute to the P300 wave, it does make a contribution to task-relevant ERP changes.

15.3.2 Formation of Peaks

N200. A negative peak around 200 msec was a conspicuous feature of the averaged ERP response to nontarget and, especially, target tones (see Figs. 15.3 and 15.5). Frequency analysis of the averaged response shows that this peak results mainly from the delayed theta enhancement and the more prominent second oscillation in the alpha band, although the delayed positive delta deflection is also an indirect contribution. This peak cannot be observed in the sensory evoked potential, and it is suggested that it should be regarded as part of the complex P300 changes.

P165. In the sensory EP, a positive peak at around 200 msec is considered to be formed by the superposition of 1–3.5 Hz, 3.5–8 Hz, and, more variably, 8–

13 Hz time-locked oscillations (see Figs. 15.3 and 15.5). In the task-relevant ERP, it would appear that a "P200" peak occurs earlier, in the region at 165 msec (P165, Goodin et al. 1978). According to our analysis, this P165 peak is formed by contributions of time-locked 3.5–8 Hz and 8–13 Hz activities, which have become "separated" from the 1–3.5 Hz activity, whose major positive deflection has shifted in latency to the right. Frequency decomposition of averaged data serves to illustrate how the peaks in the averaged data are formed by alternations in the time course of delta, theta, and alpha response oscillations. Such an approach may lead to new perspectives that might facilitate an understanding of the response mechanisms operating under different experimental conditions.

We have performed above a straightforward component analysis that describes how individual peaks in the ERPs are shaped from the superposition of various oscillatory (and elementary) waveforms.

15.3.3 Comparison of ERP Responses to Regular and Random Infrequent Target Stimuli

Alternating tones: In this section we will turn to another type of problem by increasing the probability of target occurrence. We have slightly changed the oddball paradigm. Here we are engaged in the application of a new step: Subjects increased their attention to detect a target stimulus, but not a rare one: The surprise reaction as well as the decision-making are eliminated. How do ERPs react when the probability of target occurrence is increased to 100%?

Figure 15.7a shows typical averaged vertex ERPs obtained under the paradigm of alternating nontarget and target tones where the subjects were not informed that target tones would be presented regularly. The dashed-line curve represents the ERP to nontarget tones ("Ignore"), while the solid-line curve represents the ERP to target tones ("Attend"). The nontarget ERP is characterized by a negative peak at around 100 msec (N100) and a positive one at around 200 msec (P200). The target ERP has a similar shape, except for a marked positive peak at around 300 msec (P300 change).

The amplitude–frequency characteristics obtained from the ERPs shown in Fig. 15.7a are shown in Fig. 15.8. The dashed line represents responses to nontarget tones. The solid line represents responses to target tones. The amplitude–frequency characteristics obtained from both curves show response peaks in the frequency ranges of 1–2.5 Hz, 2.5–4 Hz, 4–7 Hz, and 7–17 Hz.

For the sake of simplicity, we have confined our band-pass filter analysis to the first three major peaks of the amplitude–frequency characteristics shown in Fig. 15.8. Figure 15.7b shows the filtered poststimulus oscillations within these frequency bands. It can be seen that the averaged N100 peaks of both the target and nontarget ERP are formed mainly by 4–7 Hz and 7–17 Hz oscillations. The large positive peak at around 200 msec (P200) is mainly formed by 1–2.5 Hz activity. The smaller positive peak at around 180 msec is

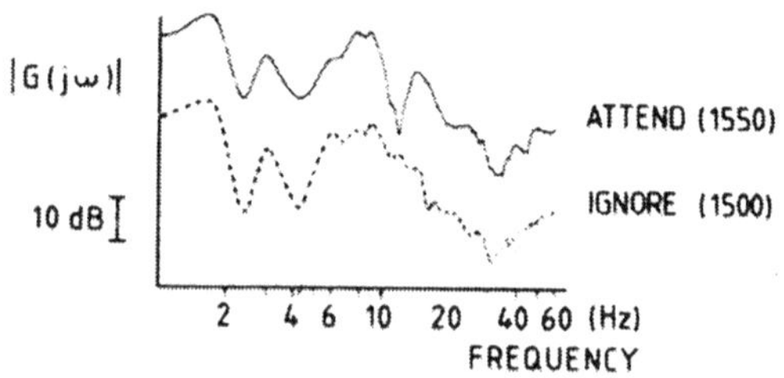

Fig. 15.8. Amplitude–frequency characteristics obtained from vertex ERPs of one subject to target tones (*solid curve*) and nontarget tones (*dashed curve*). The curves were obtained by applying Fourier transform to averaged event-related potentials of Fig. 15.7. *Along the abscissa* is the frequency in logarithmic scale; *along the ordinate* the potential amplitude or gain, $|G(j\omega)|$, in decibels. The curves are normalized in such a way that the amplitude at 1 Hz is equal to 1 (or $20 \log 1 = 0$). (From Başar E. and Stampfer H.G. (1985) *International Journal of Neuroscience* 26: 161–180.)

formed mainly by 4–7 Hz activity, while the small peak at around 280 msec is formed solely by 7–17 Hz activity. The early negative peak at around 50 msec (N50) is formed by the 7–17 Hz activity.

From the above analysis, it can be seen that the averaged ERPs in Fig. 15.7a can each be closely approximated by a superposition of the post-stimulus, time-locked oscillations within the first three frequency bands derived from the amplitude–frequency characteristics.

Figure 15.7c shows a comparison between the filtered ERP components to random, infrequently presented tones (taken from an analysis involving the same subject) and the averaged ERP response to regular, frequently presented target tones, without the subject knowing that the target tones would be presented alternately. While the subject presumably attended closely in both experiments in order to earn the reward, it can be seen that the ERP change is much more pronounced in the response to random, infrequent tones. Solid curves present the responses to random target tones, whereas the dashed curves present responses to regular repetitive tones. It can be seen that 1–2.5 Hz activity to random tones shows a "latency shift to the right" in the ERP response to random, infrequent target tones. Also, the 4–7 Hz activity shows a prolongation of oscillation and a delay in maximum enhancement, or amplification, compared to 4–7 Hz activity of the ERP response to regular, frequent target tones. In the 7–17 Hz response the prolonged oscillation to random target stimuli is greatly reduced by application of regular target stimuli. An interpretation of the above findings is given below.

15.4 Orientation Reaction and Learning During Repetitive Stimulation

Figure 15.9 shows averages of the first and last twenty sweeps for both target and nontarget tones from the experiment in which subjects were not informed that target tones would be presented regularly and alternately with nontarget tones. It can readily be seen that there are minimal differences between the target and nontarget averages of the first twenty sweeps. However, clear differences can be seen between the respective averages of the last twenty sweeps. It would appear, therefore, that despite the physiological regularity of the target stimuli, the cognitive uncertainty and possibly the incentive for gaining the reward resulted in a very similar P300 change to both target and nontarget stimuli during the initial stages of the experiment. The averages of the last twenty sweeps show that the P300 change persisted in the response to target tones but disappeared in the response to nontarget tones. Upon questioning after the experiment, most subjects reported having felt confident about the regular stimulus pattern after 10–20 sweeps. It appears that as subjects became more certain of the stimulus pattern, they were able to identify and ignore the nontarget tones more confidently, and this was reflected in the variation of the P300 change to nontarget tones during the course of the experiment.

Close inspection of Fig. 15.9 reveals that the negative peak at around 200 msec, which is formed by theta and alpha oscillations (see above in this chapter), is more pronounced in the average of the first twenty compared to the last twenty sweeps of target responses. Since this peak depends largely on delayed 4–7 Hz (theta) enhancement, it would appear that this effect becomes less pronounced as subjects become more certain that every alternate tone will be a target tone. Hence, we have tentatively hypothesized that this change reflects a learning process following the initial orienting response (Pavlov 1927). The progressive change can be seen quite clearly in the single sweeps.

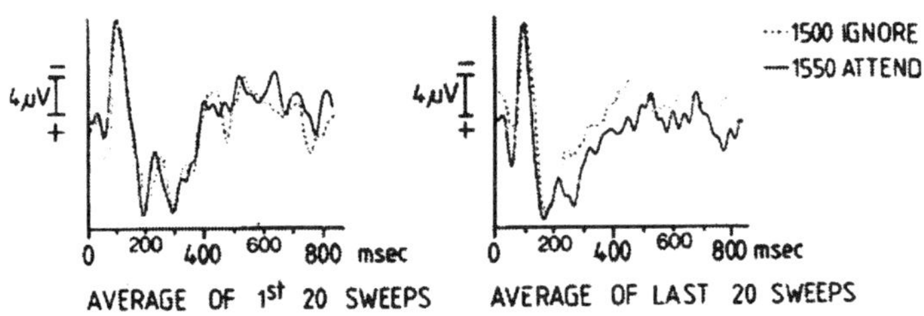

Fig. 15.9. Averages of first and last twenty sweeps for both target and nontarget tones from experiments in which subjects were not informed that target tones would be presented regularly and alternately with nontarget tones. *Solid curves*, ERP to target (attend) tones; *dashed curves*, ERP to nontarget (ignore) tones. (From Başar E. and Stampfer H.G. (1985) *International Journal of Neuroscience* 26: 161–180.)

15.5 Analysis of Pre- and Poststimulus Activity in Single Sweeps: "Preparation Rhythms"

Figure 15.10 shows single-sweep EEG-ERP epochs filtered in three different frequency ranges. The sweeps were recorded during various experimental conditions. Subjects were further asked to mark mentally the occurrence of target tones.

Figure 15.10a shows 0.5–3.5 Hz activity in the responses to target tones (sweeps 43–61). The sweeps were recorded during the experimental paradigm in which subjects were informed that target tones would be presented regularly and alternately with nontarget tones. It can be seen that the general configuration of activity is very regular in both pre- and poststimulus segments. Furthermore, there is evidence of a stable *phase-reordering*, or *preferred phase angle* at the point of stimulation. In the examples shown, there is evidence of a maximum negativity at the time of stimulation.

Efforts were made to exclude methodological errors by recording baseline "EEG epochs" from every subject. In particular, 80 "single sweeps" of EEG were recorded with the same timing and data collection methods but without stimulation. Figure 15.10b shows a sample of baseline sweeps for comparison with Fig. 15.10a. There is no evidence of the pre-and poststimulus regularity seen in Fig. 15.10a.

It should be emphasized that the percentage of single sweeps showing phase reordering at the point of stimulation can vary from 35% to 85%

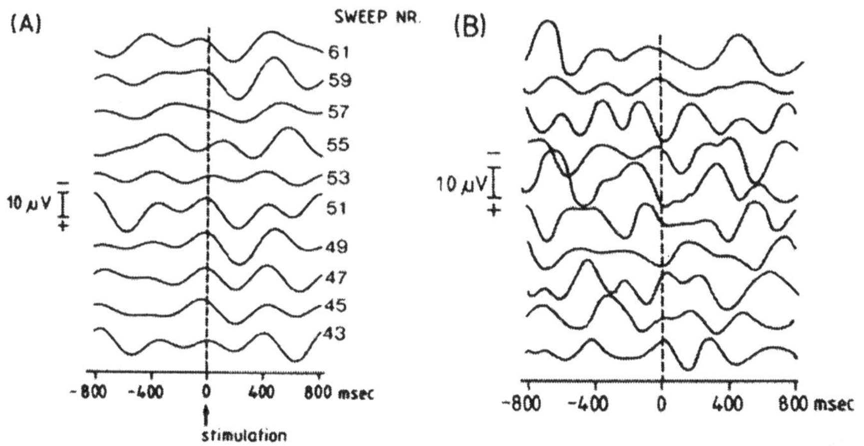

Fig. 15.10a,b. (**a**) Single-sweep EEG-EP epochs obtained in response to repetitive regular target tones are filtered in the 0.5–3.5 Hz frequency band. Sweeps No. 43–61 presented. (**b**) Filtered EEG baseline epochs. (No stimulation has been applied during these recordings.) The curves are pass-band filtered in the 0.5–3.5 frequency band. (From Başar E. and Stampfer H.G. (1985) *International Journal of Neuroscience* 26: 161–180.)

between subjects. The subject's cooperation at the time of the experiment and/or his capacity to focus attention may be important when determining this percentage (see also Chap. 22).

15.6 Event-Related Theta Oscillations

Figure 15.10c shows filtered theta (3.5–8 Hz) activity from single-sweep EEG-ERP epochs recorded during an experiment in which target tones were presented randomly and infrequently (data taken from companion study). Prolonged regular oscillations and delayed enhancement can be seen after sweep No. 49, i.e., the tenth target tone, which also marks a reduction of prestimulus theta amplitude. The onset of these changes varied between and within subjects and also under different experimental conditions. Further studies are required to explain this variation more fully. For present purposes, we wish merely to draw attention to the more obvious changes in pre- and poststimulus activity during the course of various ERP experiments. In this regard, a

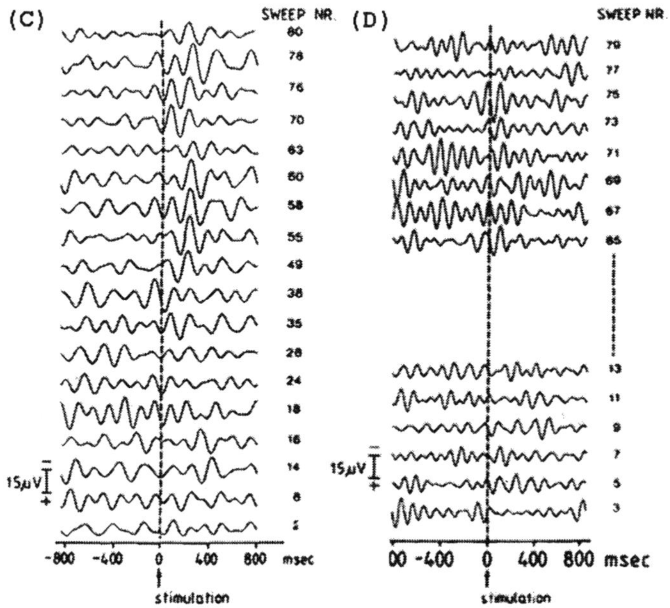

Fig. 15.10c,d. (**c**) Filtered EEG-ERP epochs following randomly applied target tones are filtered in the 3.5–8 Hz frequency band. Sweeps No. 2–80 are shown. The prolonged and enhanced 3.5–8 Hz (theta) deflection is observed after sweep No. 49. (**d**) Filtered EEG-ERP epochs to repetitively applied rate target tones. Filter limits are 8–13 Hz. Group of sweeps (3–13 and 65–79) are separately illustrated to show the relevant changes in EEG and ERP activities. (From Başar E. and Stampfer H.G. (1985) *International Journal of Neuroscience* 26: 161–180.)

general observation was of an increase in theta amplitude in experiments in which subjects were presented with regular, frequent target tones. The regular, frequent pattern is associated with magnitudes of 20 μV, whereas the random, infrequent pattern is usually associated with magnitudes of less than 10 μV. However, as mentioned above, the theta response to regular, frequent stimuli diminishes progressively during the experiment.

The increased prestimulus theta activity associated with regular, frequent target stimulus is unlikely to be an example of the driving phenomenon described by John (1967). In our experiment, the stimulation intervals were around 10–12 times longer than the intrinsic theta frequency. Hence, it seems very unlikely that our findings can be explained by a "driving" phenomenon.

15.7 Event-Related 10 Hz Oscillations

Figure 15.10d shows filtered 8–13 Hz (alpha) activity from single-sweep EEG-ERP epochs, recorded during an experiment in which the subject was not informed that target tones would be presented regularly and alternately with nontarget tones. It can be seen that the magnitude of the prestimulus alpha activity increased progressively during the experiment and is associated with increasing phase reordering at the point of stimulation. These changes were evident in most subjects by approximately the 20th target tone, although there was again variation in this regard, both within and between subjects.

At this point it should be pointed out that while there are similarities in the changes within different frequency bands during any one experiment, the timing or onset of these changes can vary between the different frequency bands. This suggests a degree of functional independence between the different frequencies, and this will be an issue investigated in future studies.

Interim summary: The general conclusion drawn from results presented so far is that when subjects perform tasks in relation to regular, frequently presented target stimuli, the regular pattern of stimulation induces more stable, predictable pre- and poststimulus activity. The onset of these changes can show variation within and between subjects as well as variation between the different frequencies in any one experiment. Informing subjects of the stimulus pattern appears to augment the regularity and synchronization of their EEG, as though the knowledge is somehow able to influence a subject's EEG rhythmicity (see also Chap. 22).

Figures 10a–d provide evidence for our hypothesis of an active functional relationship between prestimulus EEG activity and the event-related potential changes that are manifested as N100, N200, P200, and P300 peaks in the averaged curves; they illustrate the progressive regularity of pre- and poststimulus activity as well as the increasing phase reordering at the point of stimulation.

According to Ruchkin and Sutton (1983), the P300 ERP change is determined more by the subject's expectancy concerning an event and the information provided by it than by the physical characteristics of the stimulus that signals the event. The results in this chapter support this conclusion. However, we will give more information about the mechanisms underlying pre- and poststimulus response variation under different experimental conditions in order to gain a better understanding of the brain dynamics involved (see Chap. 20).

15.8 The Modulation of P300 Activity by Preparation Rhythms

Stampfer and Başar (1985, 1988) found evidence of specific changes in the theta response to unexpected target stimuli, namely, a delay of poststimulus enhancement and a prolongation of oscillation (see previous sections). These findings have been supported by results presented in this section. Both paradigms have shown that the above frequency components are also mainly responsible for the development of the P200 waves seen in the averaged response to nontarget stimuli and in sensory evoked potentials. It would appear therefore that the same sensory mechanism is involved in both the P200 and P300 peaks, except that the reaction time is delayed and prolonged in the case of unexpected target stimuli. Unexpected target stimuli do not elicit a "new" response but appear to trigger additional information processing in the brain, which is reflected in a delayed and prolonged sensory response. If target stimuli are expected, the reaction time is not delayed and is less prolonged. This was demonstrated by experiments in which subjects were informed that target tones would be presented regularly and alternately.

In these experiments with repetitive stimuli there was evidence of *preparation rhythms* in the prestimulus EEG activity, namely, regularity of activity within different frequency bands and alignment of phase angles at the time of stimulation. It would seem that the expectancy and experience of regular target stimuli leads to a physiological prestimulus synchronization, which in turn reduces the poststimulus delay and prolongations that form the P300 waves.

In previous chapters (11 in Vol. I, 9, and 10) it was demonstrated that sensory-evoked potentials were generated by enhancement and frequency stabilization of EEG activity. The findings presented here extend this concept; rare, unexpected stimuli also give rise to enhancement and frequency stabilization, but the enhancement is delayed and prolonged. This suggests that the rare target stimuli have exposed further functional dimensions of the enhanced EEG oscillations. These dimensions are the delay and duration of the enhanced EEG synchronization within delta, theta, alpha, and higher frequency bands (for a summary see also Chap. 30).

15.9 P300, Prestimulus EEG Activity and Their Relation to Short-Term Memory: Memory Templates

It has become increasingly difficult to give a clear definition of the P300 wave, or change. The original P300 component described by Sutton et al. (1965) has been succeeded by a variety of late positive components (Donald 1983). It has not yet been established that all the late components possess the same properties. For example, it has been suggested that responses elicited by certain on-line decision tasks (Kutas et al. 1977) and responses elicited by highly novel stimuli (Squires et al. 1975) are outputs that originate from different mechanisms. In a review of several relevant studies, Donald (1983) concluded that in order to explain how stimulus rarity could affect the P300 amplitude, it would be necessary to place the P300 wave at the output stage of a memory-based system that contains a record of prior stimulation.

One possible version of such a memory record, the echoic trace, would register a template of the most recent stimulus features. This template would then feed back into a comparator that also received each new input. A mismatch would be registered whenever a low-probability stimulus occurred, since an echoic match to that stimulus would be unlikely to exist in storage. One of the results of a mismatch would be the P300 wave. This model provides a theoretical explanation for the P300 wave elicited both by presented and omitted "oddball" stimuli. Donald (1983) suggested that the choice mismatch model was compatible with observations that the P300 wave is very large following the first stimulus in a new series (Ritter et al. 1968; Vaughan and Ritter 1970) because there would be no memory template of the first novel stimulus that would elicit a mismatch response. On the other hand, we have presented evidence that the P300 change is affected by subjects learning the pattern of stimulation. Findings of Donchin et al. (1978) are similar in this regard. They presented subjects with different sequences of target stimuli possessing the same probability of occurrence but varied the degree of regularity (Donald 1983). In the studies cited above, "learnable sequences," in which stimuli alternated in some predictable order, produced smaller P300 responses than irregular sequences that were unfamiliar and unpredictable. The findings of our studies and those of Donchin et al. (1973) suggest that feed-forward from memory can influence the P300 amplitude. If memory correctly predicts the input, the P300 response is reduced; if not, a mismatch is registerd, and a large P300 wave develops. For memory-related oscillations see also Chap. 22.

Considering the findings of this chapter, we can ask whether various EEG frequencies (delta, theta, alpha, and higher) might function as templates of such memory records. The development of prestimulus regularity and phase reordering might reflect short-term memory templates that could feed into a comparator. When the target signal is predictable, the preparation waves are aligned and have a predictable phase reordering at the time of stimulation. In such cases the P300 wave is *moved back* to the P200 position seen in nontarget responses and sensory-evoked potentials. It would appear that the abolition of the P300 responses in single sweeps and the reduction of the P300 amplitude in the averaged response is the result of the regularization of the prestimulus EEG. Furthermore, if the EEG components prior to stimulation show large amplitudes or if the input does not coincide with the "preferred" phase angle, the P300 wave and other ERP components are diminished. Taking into consideration the range of EEG frequencies and the various ERP changes, it is possible that several "templates" of different frequencies are involved in the evaluation of input.

15.10 Theta and Alpha Oscillations in Klimesch's Memory Model

As the writing of this book was almost finished, a new, tentative, model on memory was described by Klimesch (1996). This new memory model is described on three levels: cognitive, anatomical, and physiological. According to the trends in the present book it is important to outline briefly this interesting model (see also Chap. 14 for descriptions of different memory types).

Klimesch (1996) discussed experimental findings that support the following hyphothesis: *Short-term (episodic) memory* demands lead to synchronization in the theta band manifested in an increase in band power. However, *long-term (semantic) memory* demands lead to a task-specific synchronization in the upper alpha band (decrease of power).

Further, it is suggested that short-term memory processes are reflected by oscillations in an anterior limbic system, whereas long-term memory processes are reflected by oscillations in a posterior thalamic system. Oscillations in these frequency bands possibly provide the basis for encoding, accessing, and retrieving cortical codes that are stored in form of widely distributed but intensively interconnected cell assemblies (see Klimesch (1996), p. 85, for an overview).

Two remarks seem to be appropriate to extend this model: (1) The hippocampal 10 Hz oscillations as described in Chaps. 2 and 27 must be also considered in a memory model based on oscillations. (2) In such a distributed memory model the role of cerebellar oscillatory activity might be also taken into account (see new trends of research on cerebellar cognitive performance in Chap. 27). The reader is also referred to memory-related oscillations in Chap. 22.

15.11 Habituation

In several chapters of this book we have been able to demonstrate that if the prestimulus activity in various frequency bands shows high-amplitude regular rhythmic behavior, then the response amplitude bears an inverse relationship to the prestimulus amplitude (see also Başar 1980, Başar et al. 1976ab, 1979ab, 1983b, and especially Chap. 12 in Vol. I).

In terms of our experiments, habituation implies a temporary storage of specific information about previous stimulation, which results in a decline of the response to repeat stimulation (Donald 1983). Temporary storage may be a function of or be reflected in the regularized or synchronized prestimulus EEG activity. Our results do not allow us to draw a definite conclusion. However, we can definitely conclude that the disappearance of the P300 changes are causally related to the development of regular, synchronized prestimulus EEG activity in various frequency bands, and that the mechanism of habituation is somehow connected to this development.

We will come back to this discussion in Chap. 22.

15.12 Appendix: Frequency Analysis of MMN

This appendix gives unpublished results of Karakaş, Polich, and Başar.

> **Early mismatch negativity (MMN),** also specified as the supratemporal component, occurs with a latency of 150–200 msec in response to a change in stimulus properties and is largest over the Sylvian fissure. It is generated from loci in the primary auditory cortices that are specific to the physical attributes of the stimulus. The coherent and extensive studies of Näätänen and coworkers have attempted directly to develop a model/theory (Näätänen 1992) for cognitive peaks/potentials/components that are related directly to information-processing events.
>
> Early MMN is related to fine-grained basic sensory-specific analysis of nonsemantic features, whereby decremental or incremental "mismatch" is detected in physical features of input. Early MMN reflects the change-detection function of the brain, for which it stores a model of the repetitive stimulus. This is the sensory trace (the echo for the auditory version). Then, an incoming input is compared to the trace of the repetitive input, and an MMN is produced in case a change is detected. This process is automatic and reflects a preattentive process driven by the stimulus properties. The passive neural model of the repetitive input is preconscious and denotes a physical level of analysis. The related system has a large capacity that can engage in parallel processing. Trace maintenance is uncontrollable; the sensory trace that develops in about 150 msec decays in about 2–10 sec.
>
> Late MMN, also named the frontal component, occurs with a latency of 200–300 msec. It is recorded predominantly from the Fz location. It is larger over the right hemisphere. Late MMN is triggered by the early MMN and is related to the initiation of the orienting response. The process itself is automatic and preattentive. The related processing system has a large capacity and possibly provides

for parallel input. Since early MMN originates from auditory primary cortex, involves physical input analysis of nonsemantic features, is attention-independent and preconscious, and involves automatic and parallel processing, it engages the sensory register. If late MMN is triggered by the change-detector system, sensory memory enters the limited-capacity system and makes contact with information in LTM, the event sometimes known as the "breakthrough of the unattended" (Karakaş 1997).

In sum, MMN contributes to the initiation of an organismic state involving orientation response to change in physical features of input. Hence, this component represents focusing of attention to change. When the late MMN is triggered, a bottom-up process provides the related input into STM where, under focused attention, it is subjected to conscious discrimination. During the last stages of writing the present book, Karakaş and Başar began frequency analysis of MMN. The results are as yet unpublished. Accordingly, we will append a brief summary to this chapter.

15.12.1 MMN Formation of Peaks

The theory related to the formation of peaks in EPs based on filtered frequency components was published in detail in earlier publications. Also, in Chaps. 18 and 11 in Vol. I these descriptions are given in sufficient detail.

We will describe in the following the modified MMN paradigm that is used by Karakaş and coworkers. Subjects were young, healthy students 20–24 years of age.

MMN paradigm. This paradigm is named after the mismatch negativity components of Näätänen et al. (1992). In this paradigm, the subject is asked to perform an irrelevant task while 1000 Hz standard stimuli and 2000 Hz irrelevant stimuli are presented via headphones. The irrelevant task was to count missing numbers that were printed on an A4 page. Every now and then, a digit would be skipped. The subject's task was to count these interruptions and to report the number at the end of the task condition. In all other paradigms the subject was required to mentally count the target (deviant) stimuli and to report their number after each task condition was completed. Subjects were encouraged to perform the counting task accurately and were given feedback about their accuracy.

Two types of signals were applied: 1000 Hz, 65 dB tones with a duration of 50 ms as *nontarget* signals, and 2000 Hz, 65 dB tones with the same duration as *target* tones. The target tones were randomly applied with a probability of 20%.

In this short description of MMN data in terms of their frequency properties we will only present grand averages ($N = 20$ young, healthy subjects; frontal location). For the sake of simplicity we compare only amplitude–frequency characteristics and filtered results for the oddball and MMN paradigms.

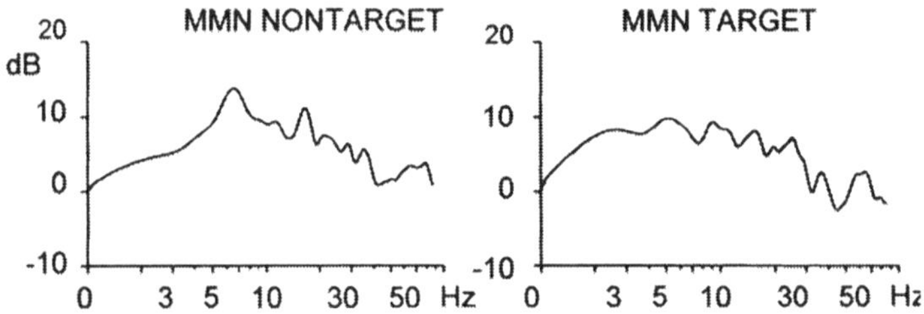

Fig. 15.11. Mean-value amplitude–frequency characteristics of target and nontarget responses.

Karakaş et al. (1997) performed experiments with 6 different paradigms on 60 healthy young subjects. Their measurements included also the application of the MMN paradigm described above. The results that we include here are roughly analyzed preliminary results obtained from 15 of the subjects.

The mean-value amplitude characteristics of target and nontarget responses (deviant and standard) are presented in Fig. 15.11. Dominant peaks are at the theta (5–8 Hz) and low beta (16–20 Hz) frequency ranges for target tones.

Figures 15.12a and 15.12b present the filter analysis of MMN grand average ($N = 15$) responses with delta (0.5–3.5 Hz), theta (4–7 Hz) and wide band (0.5–46 Hz) filters.

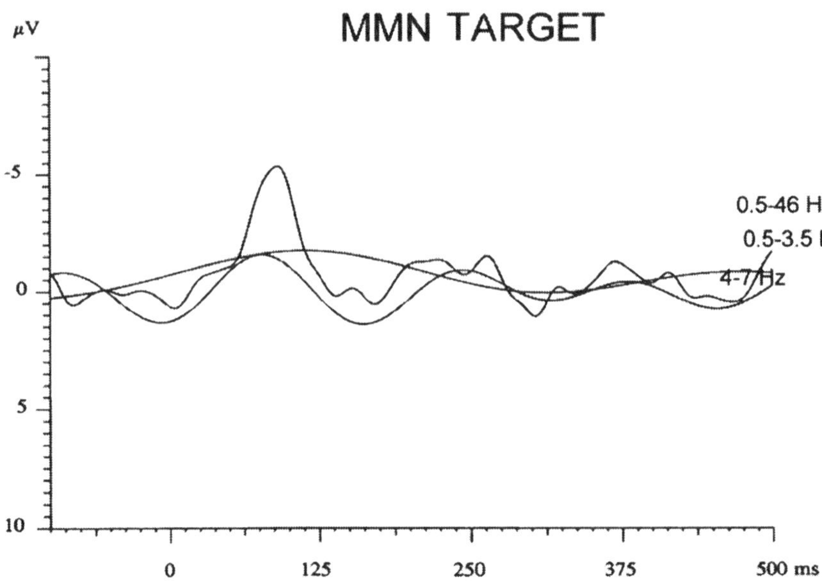

Fig. 15.12a. Grand average ($N = 15$) MMN responses for target stimuli, filtered in several frequency bands.

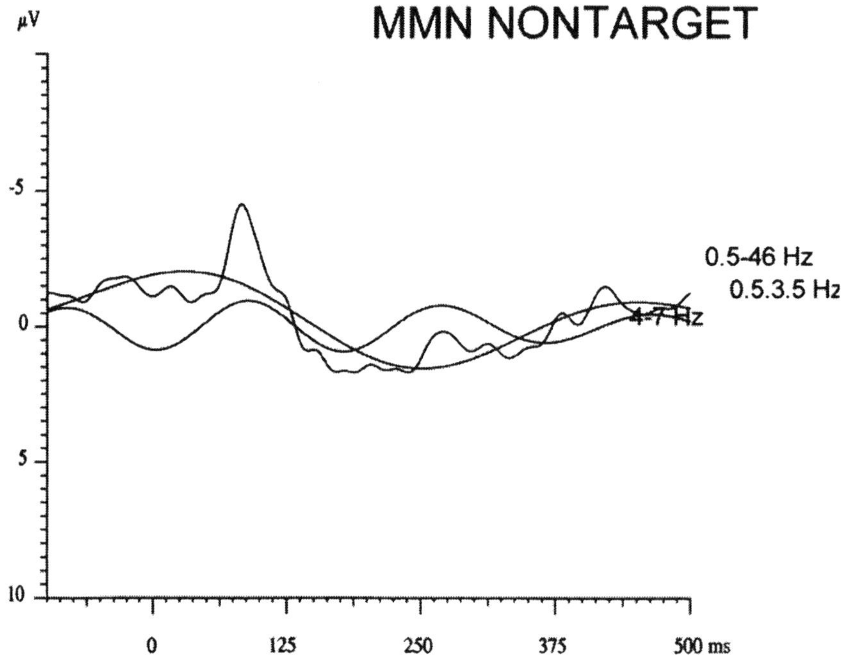

Fig. 15.12b. Same as Fig. 15.12a, but for nontarget stimuli.

It is important to note that the N200 peak of the target response at approx. 180–250 ms latency is shaped mainly by the second positive-going oscillation of the theta response. In this time interval the delta oscillation shows a slightly negative-going trend. On the other hand, in the nontarget (standard) response the delta response has a negativity at approx. 250 ms where the theta response has a maximum with opposite phase (positivity). Clearly, in this case delta and theta responses cancel each other, and a negative MMN peak does not appear in the wide-band response. In other words, the N200 response in deviant responses does not exist, since the theta response that constitutes the N200 is completely obscured (or canceled) by the delta response. It is also important to note that the delta response delay of approx. 100 ms is similar to results in the target response of the oddball ERP (see above in this chapter).

16. Dynamics of Compound Potentials (P300) in Freely Moving Cats

Canan Başar-Eroğlu and Erol Başar

16.1 Introduction

In the preceding chapters we have stated results of the Brain Dynamics Research Program concerning data from the human brain. In the present chapter we will demonstrate some tentative models of comparative data obtained from experiments with the cat brain. Results on ERP recordings in the auditory cortex (GEA), reticular formation (RF), and hippocampus (HI) of 9 freely moving cats will be described. The interpretation of these results will be based on analysis of ERPs in the frequency domain. For an overview of measurements of P300-like potentials in animals, see Paller (1988).

16.2 Methods and Paradigms Utilized for Obtaining P300 from Freely Moving Cats

The experiments described in this chapter were performed on 9 chronically implanted and freely moving female cats. Stainless steel electrodes of 100 μm diameter were located under Nembutal anesthesia (35 mg/kg) in the auditory cortex (anterior gyrus ectosylvian: GEA), dorsal hippocampus (HI), mesencephalic reticular formation (RF) with the coordinates (Snider and Niemer 1964) as shown in Table 16.1. The derivations were against a common reference that consisted of three stainless steel screws in different regions of the skull. A David Kopf 1760 instrument was used for stereotaxic surgery.

Figure 16.1 shows the omitted stimulus paradigm utilized in eliciting ERPs. Every experimental session consists of three experiments in the fol-

Table 16.1. The coordinates (in mm) of implanted eelctrodes. GEA, gyrus ectosylvianus anterior; HI, hippocampus; RF, formatio reticularis. (From Başar-Eroğlu C. et al. (1991) *International Journal of Neuroscience* 60: 215–226.)

	GEA	HI	RF
Frontal	15.5	3.5	3.0
Lateral	12.5	6.5	4.0
Horizontal	10.0	8.8	−1.0

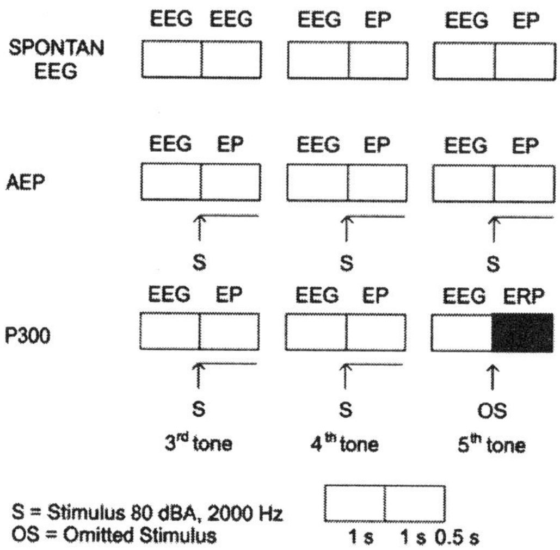

Fig. 16.1. Omitted stimulus paradigm. Sequence of experiments used in triggering event-related potentials (ERPs) in cats. Every experimental session consists of 3 experiments: (**a**) control-EEG (**b**) acoustical EPs (**c**) ERPs with omitted stimuli. Every 5th stimulus is omitted. (Modified from Başar-Eroğlu C. et al. (1991) *International Journal of Neuroscience* 60: 215–226.)

lowing order: (1) *Recording of EEGs* for control and comparison with ERPs obtained in response to the omitted stimulus; (2) *Recording of auditory evoked potentials.* The tones were presented repetitively so as to produce an anticipation in the cats of the time of occurrence of the omitted stimulus in the respective ERP experiment. For stimulation a 2 KHz tone with an intensity of 80 dB SPL, a duration of 1 s and a stimulus interval of 2.5 s was used.

The cats were kept sitting in a cage placed in a soundproof and echo-free room that was dimly illuminated, and the EEG was monitored throughout the experimental sessions. Epochs containing movement artifacts and stages in which the EEG of cortex recordings showed sleep spindles or slow waves in the cortex were eliminated off-line after the recording sessions. The cats were naive; they had not been previously exposed to any conditioning or training trials. Long experimental sessions were avoided to eliminate the effects of adaptation and fatigue that might develop in the cats.

In order to analyze the ERPs in the time domain, again a "combined analysis procedure" of EEG and ERPs was used (see Chap. 4 in Vol. I),

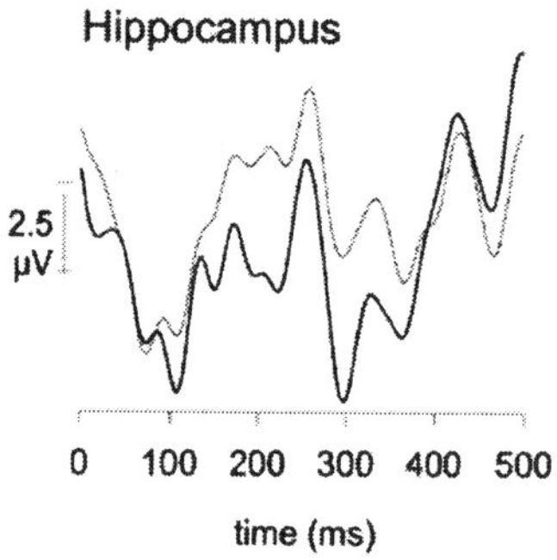

Fig. 16.2. The P300 to omitted stimuli in the hippocampus from nine cats. The ERP obtained by selective averaging (*thick line*) is shown along with that obtained without the use of selective averaging (*thin line*). The ERPs were pass-band filtered in the 1–30 Hz frequency band. (Modified from Başar-Eroğlu C. et al. (1991) *International Journal of Neuroscience* 60: 215–226.)

which was basically the same except for minor modifications. Application of selective averaging is necessary, as explained in detail by Başar-Eroğlu et al. (1991a,b); without this procedure it is not possible to detect proper and clean ERP while the cats are moving freely (Fig. 16.2). In order to elicit ERPs, "omitted stimuli" (paradigm as explained above) were used as targets (Başar-Eroğlu and Başar 1987).

16.3 Systematic Analysis of the Effect of Omission Rate on ERPs Recorded from the Cat Hippocampus

In order to analyze the influence of repetition rate on the ERP, we have performed a systematic study in which we varied the omission rate of stimuli. In the first series of experiments with five cats we applied a paradigm where every third, fifth, eighth, or tenth tone was omitted. Since this type of experiment requires very long periods for recording and analysis, only 5 cats were used as subjects.

The hippocampal ERPs showed that the procedure with either every fifth or every eighth tone gave the most satisfactory results, as can be seen in Fig. 16.3 (Başar-Eroğlu et al. 1991a,b). In order to be certain that this typical ERP cannot be obtained by averaging of EEG segments, we have plotted the average of control EEGs obtained in nine experiments from nine cats (Fig. 16.4). The thin curve in the illustration shows the grand average of the averaged EEG. We have compared the amplitudes of the positive waves between 250 ms and 400 ms. These amplitudes were significantly greater in ERP recordings.

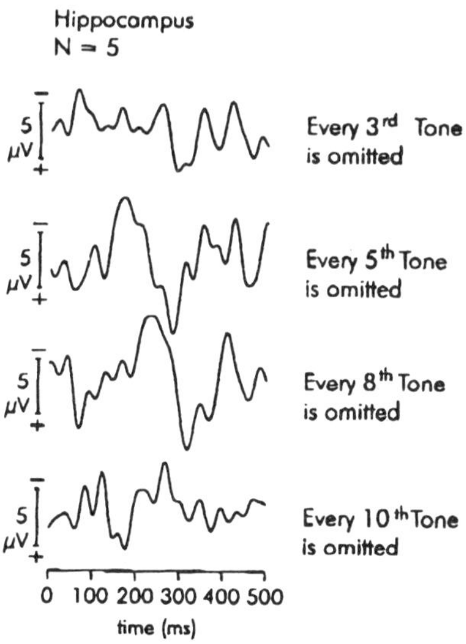

Fig. 16.3. Effect of the rate of omissions on ERPs recorded from the hippocampus. Grand average of 5 cats. Filter range: 1–30 Hz. (Modified from Başar-Eroğlu C. et al. (1991) *International Journal of Neuroscience* 60: 215–226.)

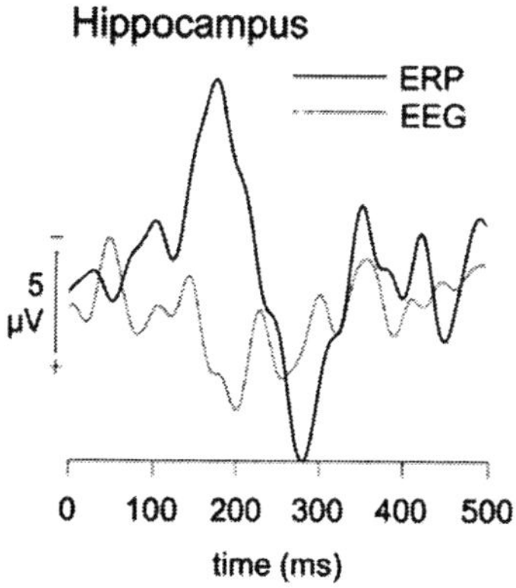

Fig. 16.4. Hippocampal ERPs (grand average of nine cats). *Thick line* represents responses to omitted stimuli, *thin line* average of the control EEG (Modified from Başar-Eroğlu C. et al. (1991) *International Journal of Neuroscience* 60: 215–226.)

Figure 16.5 shows the amplitude–frequency characteristics of the hippocampus computed from data represented in Fig. 16.4. The marked change observed in the amplitude–frequency characteristics is the existence of a maximum at around 5 Hz (theta frequency range). A less distinct peak is recorded between 10 Hz and 20 Hz. Another prominent peak is detected at 40 Hz. We will come back to the analysis of the 40 Hz response in Chap. 17.

Figure 16.6 presents results of experiments on the auditory cortex, hippocampus, and reticular formation of nine cats. At the left of the illustrations, the grand averages obtained from the transient ERPs recorded from the cortex, hippocampus, and reticular formation of nine cats are given. At the right side of the illustration standard deviations of these grand averages are presented. The most important information in this illustration is the characteristic frequency response coding that was revealed in the amplitude–frequency characteristics (in the middle column). The marked, prominent frequency response of the hippocampus is at 5 Hz (theta response).

The cortical response and the response of the reticular formation, however, had the most prominent response in the 10 Hz frequency range. In other words, although in the transient ERPs a P300 response was – as a rule – measured in the auditory cortex, reticular formation, and hippocampus, fre-

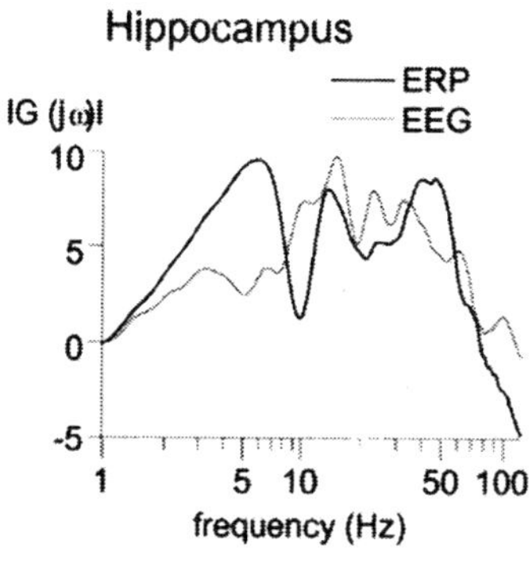

Fig. 16.5. Amplitude–frequency characteristics of the hippocampus computed from the curves in Fig. 16.4. *Thick line* represents ERP, *thin line* control EEG. *Abscissa*: frequency in Hz. *Ordinate*: amplitude in relative units and decibels (dB). (Modified from Başar-Eroğlu C. et al. (1991) *International Journal of Neuroscience* 60: 215–226.)

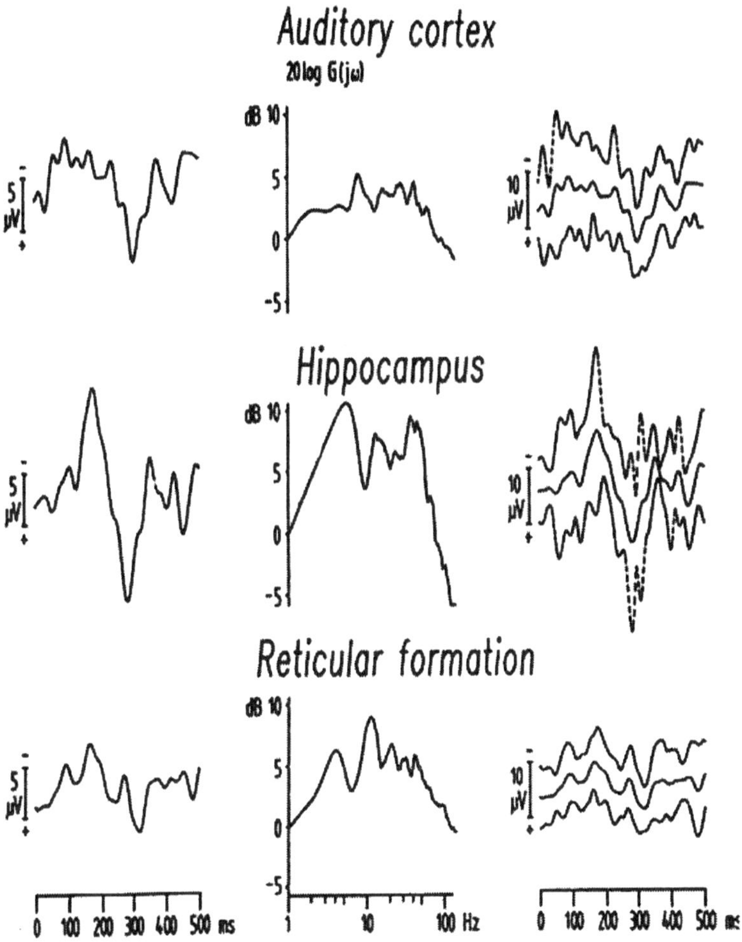

Fig. 16.6. Averaged ERPs recorded from auditory cortex, hippocampus, and reticular formation of 9 cats (*left column*), amplitude–frequency characteristics computed from ERPs shown in left column (*middle column*), and corresponding standard deviations (*right column*). (From Başar-Eroğlu C. et al. (1991) *International Journal of Neuroscience* 60: 215–226.)

quency contents were not the same: While hippocampus showed resonant properties at the theta frequency range, reticular formation and auditory cortex responded at the 10 Hz frequency range.

Figures 16.7a and 16-7b show latency and amplitude histograms of ERPs obtained in three experimental sessions from the three brain structures of eight cats. At the left side is a latency histogram, the vertical axis representing the number of experiments, each containing three experimental sessions. In the auditory cortex of the cats we observed ERPs with a flat distribution

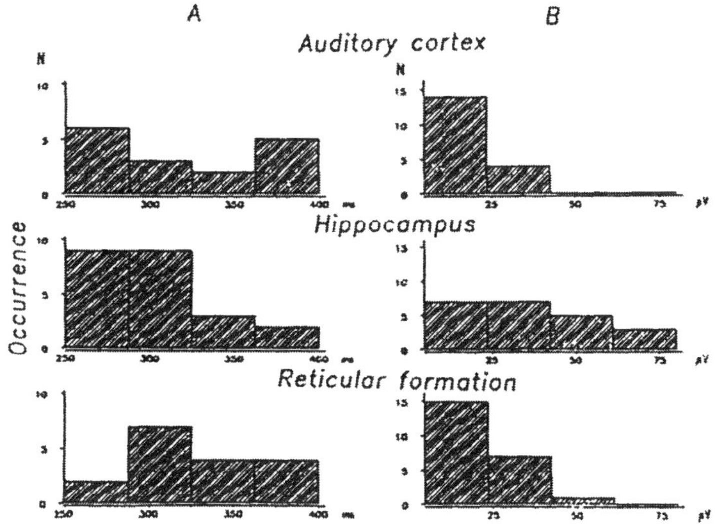

Fig. 16.7a,b. Histograms of ERPs of auditory cortex, hippocampus, and reticular formation. (**a**) Latency histogram. The *vertical axis* represents the number of experiments (3 experimental sessions with each of the 8 cats), the *horizontal axis* time (ms). (**b**) Amplitude histogram: The *vertical axis* again shows the same number of experiments, but the *horizontal axis* shows amplitude values (μV). (From Başar-Eroğlu C. et al. (1991) *International Journal of Neuroscience* 60: 215–226.)

of the latencies between 250 and 400 ms, whereas in the reticular formation most of the latencies were centered at around 300 ms. According to these results, hippocampus is the most distinct among three brain structures we investigated because of the short latency, high amplitudes, and excellent reproducibility of the P300 response. Although these results did not demonstrate that neural sources were in the hippocampus, they showed that the P300 response in cats might be strongly moderated in the hippocampus.

16.4 Utility of Analysis in the Frequency Domain

In Chap. 15 we suggested the importance of the use of the combined analysis procedure including the computation of amplitude–frequency characteristics and digital filtering. Using this procedure in the present analysis, we could clearly show that various brain structures do not differ only with respect to latencies and amplitudes of the peaks in the ERPs but also with respect to their frequency components. These components are preferred frequency channels of brain structures. Depending on the brain structures under study, different frequency components such as theta (3–8 Hz) and alpha (8–15 Hz) contributed to the P300 response (the 40 Hz response has been dealt with in Chap. 17).

Stampfer and Başar (1985) and Başar et al. (1984) demonstrated in human responses (using both the oddball and omitted-stimuli experiments) that the P300 response is characterized by a large increase in the theta response component as revealed by the amplitude–frequency characteristics. Accordingly, the frequency analysis of the P300 response presented in this chapter provides a promising model for the comparison of human and animal ERPs. The relationship of hippocampal spontaneous theta activity with cognitive processes has been emphasized by several investigators (Adey 1966; O'Keefe 1976; Creutzfeldt 1983; Vanderwolf and Leung 1983; Buzsáki 1985). The hippocampal theta enhancement presented in this study may be discussed as the strongest component of the cat P300 response.

16.5 Multiple Electrodes in the Hippocampus

Event-related potentials (ERPs) of the cat hippocampus were investigated using multielectrodes implanted in various layers of the hippocampus. The experimental and mathematical procedures were the same as those of Sect. 1.3 of the present chapter. It is shown that the most ample N200-P300 complex was recorded close to the CA3 region of the hippocampus.

In the last decade, hippocampus and/or limbic system were postulated as a neuronal generator of ERPs of cats (Wilder et al. 1981; O'Connor and Starr 1985) because of a polarity reversal of the P300 response in hippocampus.

The purpose of the description in this section is to determine whether the P300 response varies in latency, amplitude, or polarity according to hippocampal layers. A multielecrode array with four tips was placed in the right hippocampus. The study was carried out on 8 cats. The diameter of the electrodes was 25 μm and the distance between the tips of electrodes was 0.7 mm. Figure 16.8 shows a schematic presentation of the position of the array in the right hippocampus of the cat. Of the electrodes that were labeled HI1, HI2, HI3, and HI4, the first electrode was located in the upper pyramidal layer of the hippocampus (CA1), the second between the upper pyramidal layer and dentate gyrus, the third and fourth on or in, respectively, the lower pyramidal layer (CA3 and CA4). For details about the accuracy of the electrode positions see Başar-Eroğlu (1990).

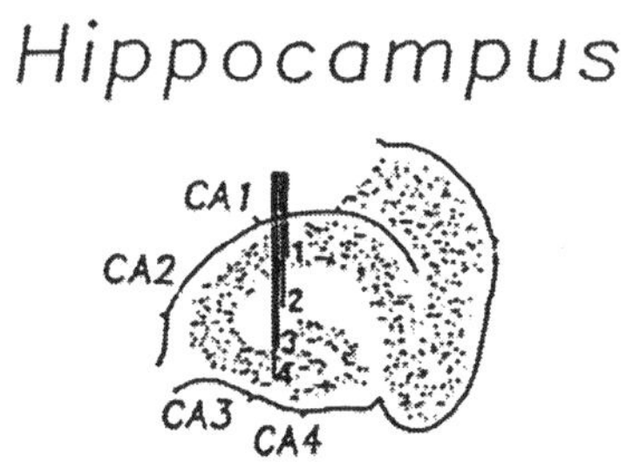

Fig. 16.8. A schematic cross section of hippocampus and the localization of multielectrodes. CA1: upper pyramidal layer; CA2: lower pyramidal layer. (From Başar-Eroğlu C. et al. (1991) *International Journal of Neuroscience* 60: 239–248.)

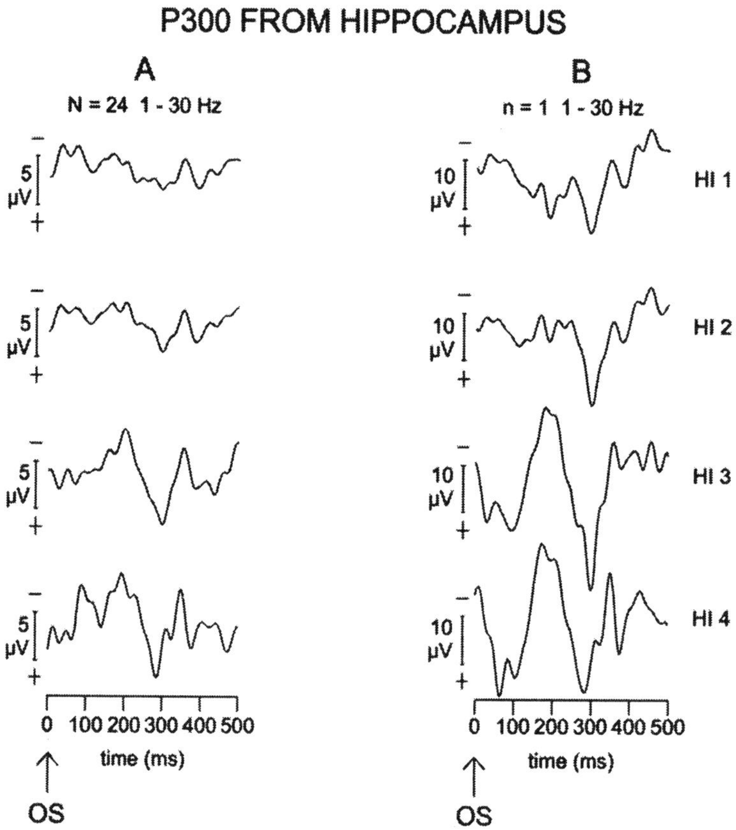

Fig. 16.9. Event-related potentials of hippocampus between CA1 and CA3 (labeled as HI1–HI4, respectively). (**a**): grand average ERPs (eight cats, three experiments each). (**b**): typical averaged ERPs of a single cat. (Modified from Başar-Eroğlu C. et al. (1991) *International Journal of Neuroscience* 60: 239–248.)

Figure 16.9 illustrates the ERPs in various layers of the cat hippocampus with the configuration of electrodes as described in Fig. 16.8. Figure 16.9a illustrates the grand averages obtained from 8 cats. The grand average is the mean value of ERPs of all 8 cats. Each cat was used in the experiments three times (therefore $N = 3 \times 8 = 24$). The ERPs of Fig. 16.9 were filtered with a digital filter between 1 and 30 Hz. In the second part of the illustration (Fig. 16.9b) typical results from one cat are presented. In all the hippocampal positions, ERPs show waves around N200 and P300. However, the HI3-HI4 responses, which are in the vicinity of CA3 of the hippocampus, are most marked. The grand average curves ($n = 24$) show that there is no marked N200 response in upper layers of the hippocampus and only a flat P300. The clearest responses are again in the HI3 and HI4 positions.

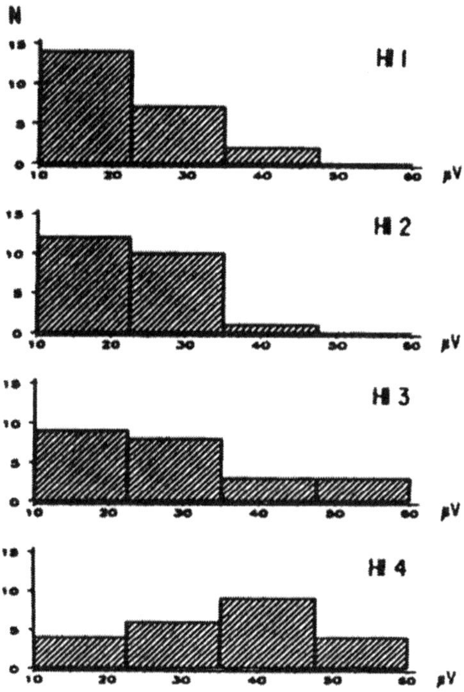

Fig. 16.10. Histograms of amplitude values (μV, peak-to-peak) of P300 response in the four hippocampal layers. (From Başar-Eroğlu C. et al. (1991) *International Journal of Neuroscience* 60: 239–248.)

Figure 16.10 shows the results of Fig. 16.9 in histograms. As can easily be seen, low pyramidal layers of hippocampus (HI3 and HI4 corresponding to CA3) showed the largest amplitudes.

Figure 16.11 shows ERPs from a distant electrode that was placed outside the hippocampus in the lateral thalamus (5–7 mm deeper than the HI4 electrode). Here neither a marked ERP nor a P300 component was observed.

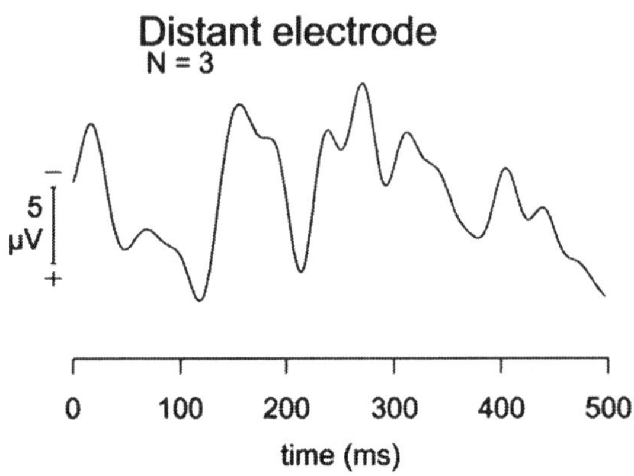

Fig. 16.11. ERP of distant electrode. The electrode is implanted at a level that is 5–7 mm deeper than CA3 of hippocampus ($N = 3$ cats). (Modified from Başar-Eroğlu C. et al. (1991) *International Journal of Neuroscience* 60: 239–248.)

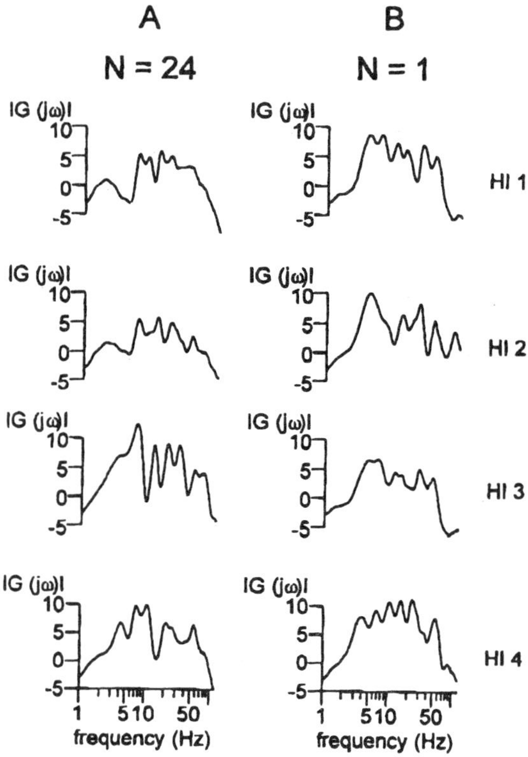

Fig. 16.12. Amplitude–frequency characteristics obtained from the averaged ERPs of the hippocampus. *Abscissa*: frequency in Hz. *Ordinate*: amplitude values in dB. (**a**) 8 cats with 3 experimental sessions each; (**b**) one cat with one experimental session. (Modified from Başar-Eroğlu C. et al. (1991) *International Journal of Neuroscience* 60: 239–248.)

These results emphasize the possibility of the existence of a strong P300 generator in the hippocampus, especially near the CA3 layer.

Figure 16.12 presents the amplitude–frequency characteristics of hippocampus computed from the curves shown in Fig. 16.9 (for computation of the amplitude–frequency characteristics, see the mathematical procedures in Chap. 4 in Vol. I).

We emphasize here that the method using an array of electrodes with a distance of 0.7 mm gives the opportunity of an optimal space resolution, not yet achieved in similar studies. It is important that there are large differences in the amplitudes of the N200/P300 deflection. Figure 16.12 also clearly demonstrates that the magnitudes of the frequency responses of the ERPs in several layers also depict significant differences. In the upper pyramidal layer (CA1) of the hippocampus (HI1 and HI2) the theta frequency response components were not prominent. In the layer CA3 usually the most marked theta event-related responses were observed. Again higher-frequency components (the so-called beta-gamma range; 25–40 Hz frequency range) were mostly higher in the CA3 (HI3 and HI4) positions than in the upper layers. The response in the HI4 position was centered at the 10 Hz frequency range with a shoulder in the theta frequency range. It can be stated here that the

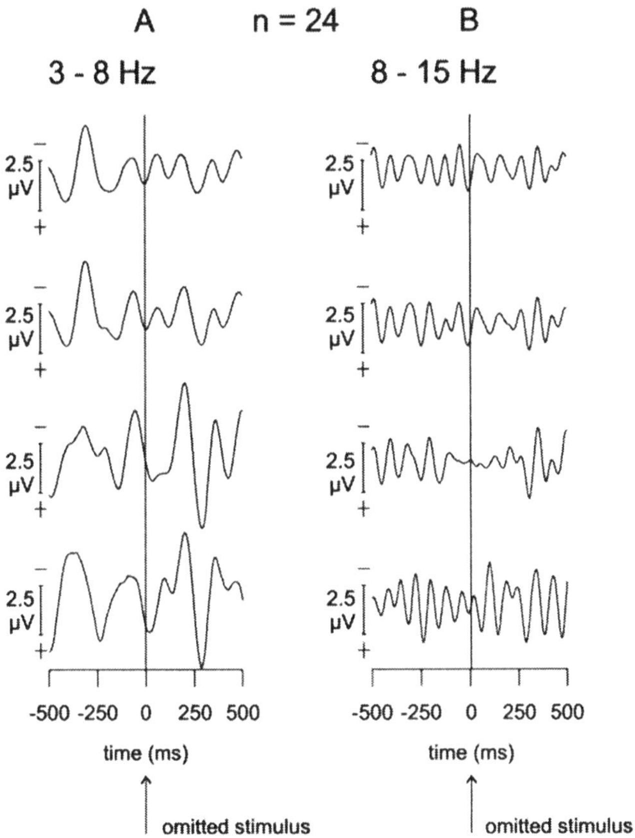

Fig. 16.13. Filtered grand average ERPs of hippocampus computed from data shown in Fig. 16.9. (**a**) filtered in theta range (3–8 Hz); (**b**) filtered in alpha range (8–15 Hz). (Modified from Başar-Eroğlu C. et al. (1991) *International Journal of Neuroscience* 60: 239–248.)

HI4 response is mostly a mixture of theta and alpha components, whereas the HI3 response is an almost homogeneous theta response.

The grand average as an important methodological approach provided valuable parameters that globally described the occurrence of potentials in various brain structures. The grand averages demonstrated differences between various layers. The difference between HI3 and HI4 of the hippocampus is a demonstration that marked differences even within a given structure with interelectrode distances of merely 0.7 mm may be obtained. The results of digital filtering also excellently confirm the conclusions derived from the amplitude–frequency characteristics curves (see Fig. 16.13).

16.6 Hippocampal P300 and Its Cognitive Correlates: The Theta Component in the CA3 Layer

ERPs were elicited by means of omitted stimuli in various hippocampal layers of the cat brain. They showed increasing amplitudes towards deeper layers. In earlier evoked potential studies (Başar 1980) it was shown that the frequency response depended on the electrode location in hippocampus. In the amplitude–frequency characteristics obtained from the upper pyramidal layer, the 4 Hz and 40 Hz maxima were dominant in comparison to other structures (see also Chap. 18 in Vol. I).

The results on dynamics of P300 in the cat brain described in this chapter demonstrate a differentiation of ERPs in time and frequency domain in various hippocampal layers. In the time domain, the amplitudes of P300 waves increased towards lower pyramidal layers in comparison to upper pyramidal layers. The largest P300 was observed in the CA3 region. Neurophysiological theories suggest large amplitude and/or inverted polarity of P300 as criteria for generator processes in certain brain structures. By means of frequency analysis a direct relationship between the amplitude of enhanced theta response and amplitude of P300 could be shown. We further suggest the "theta response" as a major subcomponent that plays an important role in hippocampal P300. Furthermore, evidence that the P300 theta response generators are not homogeneously distributed in various neuronal populations within the hippocampus could be shown. An excellent review on neural generators in CA3 is given by Buszáki (1985).

Cognitive correlates of hippocampal electrical activity have been strongly discussed in the literature: Sokolov (1975), Vinogradova (1975), John (1967), Vinogradova and Dudaeva (1972) have pointed out that the cells in the hippocampus act like comparator cells, and fire only to novelty and soon habituate. However, they come to life again if the expected stimulus is altered. For an interesting report of comparative neurology of expectation we refer to a review of Bullock (1988a,b). We also have to mention the model of Sokolov (1975) for the orienting response. There are expectation cells, which fire according to the expected input; sensory-reporting cells, which fire according to actual stimulus; and comparator cells, which fire whenever there is a discrepancy between the other two.

Smith et al. (1990) gave evidence for a neocortical P300 generator in studies of the intracranial topography of the P300 elicited with the oddball paradigm. Furthermore, these authors mentioned that related activity is also occurring in the hippocampus and probably in the frontal cortex. These authors conclude that activity generated in the hippocampus and frontal lobe may make only minor contributions to scalp recordings and that the scalp-recordable P300 might be conceptualized as only the most readily observable aspect of synchronous activity occurring across a widely distributed yet highly integrated cortical network supporting cognitive activities.

Pineda et al. (1987) discussed a late positive component in the frontal cortex of the monkey brain. The differences observed to date between squirrel monkey and human ERPs in terms of scalp topography and recovery periods are most likely a consequence of differences in brain structure as well as possible differences in function. These authors also concluded that the marked similarities observed in morphology, responses to the specific stimulus parameters, and the presence of analogous subcomponents suggest that a nonhuman primate model of P300 could be useful in the investigation of the anatomical structures and physiological mechanisms underlying human P300 activity.

Harrison et al. (1990) showed that polysensory association cortex and also primary auditory cortex (Harrison et al. 1986) are not essential for the generation of the cat P300 but the limbic septo-hippocampal system is. Although primary results of this group are very important for the P300 analysis in the animal brain, we assume that the expression cat-P300 may not be the most adequate one; most probably, the P300 response has several components and various generators depending on neural populations from which these potentials are recorded.

The results of this chapter indicate the necessity of studying the P300 response differences in a given structure (such as hippocampus) with methods of systems theory and a larger number of simultaneous recording electrodes. Extended discussion about the 40 Hz frequency response of the hippocampal P300 is given in Chap. 17.

The present analysis shows that at least one of the generators might be located in the hippocampus. Knowledge about the dipole location necessitates analyses using current source density (CSD) data. Our preliminary results indicate the possibility for the location of the dipole in CA3.

We emphasize the existence of multiple generators in various brain structures, such as cortex and, indeed, also in the brainstem. The importance of frequency analysis should be strongly taken into account, since possible distributed generators do have – according to our analysis of frequency responses – different weights. Thus latency of ERPs alone does not describe perfectly the entire coding of the response. Without analysis of cross-correlation and coherence functions, the relation between generators cannot be properly established.

17. The Compound P300–40 Hz Response of the Cat Hippocampus

Canan Başar-Eroğlu and Erol Başar

17.1 The P300–40 Hz Compound Potential

The methods used have been explained in Chap. 4 in Vol. I. The cats were stimulated with tones of 2000 Hz and 80 dB. Every fifth stimulation was omitted. The omitted stimulation is the target signal. Mathematical algorithms, experimental procedures, and neurophysiological methods (location of electrodes) and the paradigms for obtaining the P300 responses in the hippocampus of the cat brain were described in Chap. 16.

Experimenters who might be willing to replicate these results must pay attention to the fact that without a selective averaging procedure where movement artifacts are discarded, such P300 potentials cannot be recorded effectively.

Figure 17.1 shows 10 single EEG epochs recorded during a P300 experimental session that were digitally filtered. The filter limits were in the range of 30 to 50 Hz. At the bottom of the illustration, the averaged curve of 50 sweeps (also filtered between 30 and 50 Hz) from the same experimental session is shown together with the wide-band filtered ERP. The wide-band filtered averaged curve shows a marked ERP in the region of N200 and P300. Without filtering it is not possible to recognize the 40 Hz components, which are probably masked by the low-frequency activity. However, the filtered single epochs and the averaged curve show a 40 Hz burst between 270 and 300 ms following the omitted (target) stimulation. The observation of single sweeps shows that in a large number of single sweeps, in fact, a 40 Hz wave packet can be seen. The maximal peak-to-peak amplitudes of 40 Hz wave packets usually do not exceed 50 µV; as a rule they were in the range of 20 µV. The time locking is weak.

At this point, it should be emphasized that at the beginning of the experiments (the first 10–20 sweeps) single sweeps depicted weak phase locking or no phase locking. As a rule, a better and sometimes almost strong time locking was observed toward the end of the recording sessions.

Figure 17.2 illustrates results of experiments from 8 cats together with the curves filtered between 30 and 50 Hz. This illustration shows that in all the cats, a significant 40 Hz packet at around 300 ms was recorded in the

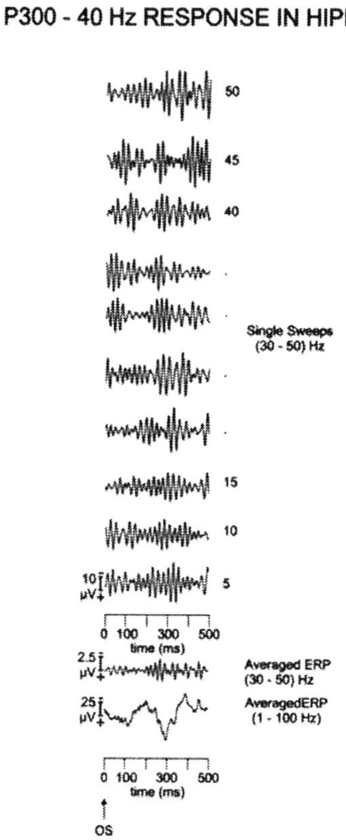

Fig. 17.1. Event-related potentials of lower pyramidal layer (CA3) of hippocampus (one cat). *Top*: Single ERP sweeps (epochs) filtered at 30–50 Hz. *Middle*: Averaged ERP filtered at 30–50 Hz. *Bottom*: Unfiltered ERP, average of 50 artifact-free epochs. (Modified from Başar-Eroğlu C. and Başar E. (1991) *International Journal of Neuroscience* 60: 239–248.)

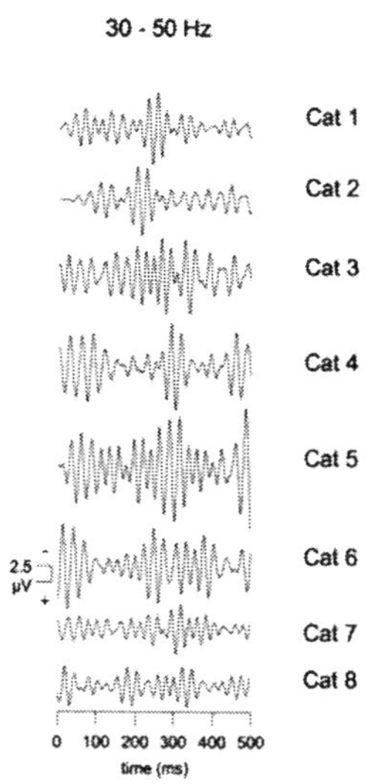

Fig. 17.2. ERPs of 8 cats recorded from lower pyramidal layer (CA3) of hippocampus. Waveforms filtered between 30 and 50 Hz. Each filtered ERP is the average of about 50–60 artifact-free single epochs. (Modified from Başar-Eroğlu C. et al. (1991) *International Journal of Neuroscience* 60: 239–248.)

CA3 region of the hippocampus. The position of the 40 Hz packet showed slow fluctuations around the time axis.

Close inspection of the 40 Hz responses (at P300 latency) recorded from the four electrode locations showed that the responses with the highest amplitude were found in CA3 (HI3 and HI4, see Table 17.1). This is supported by the statistically significant differences found between amplitudes of HI3 and HI4 responses on the one hand and amplitudes of HI1 responses on the other (Wilcoxon–Wilcox test; see Table 17.2).

In Fig. 17.3 grand averages of wide-band filtered ERPs and the 40 Hz responses are shown together. The grand averages were obtained from the ERPs of 8 cats and were digitally filtered at 30–50 Hz. This illustration again demonstrates that the slow wave activity (the so-called N200-P300 response) is mostly marked in the CA3 region, whereas the response in the region of the

Table 17.1. Amplitude values (peak to peak, in μV) of 40 Hz response (250–400 ms after omission of stimulus) in various layers of hippocampus. (From Başar-Eroğlu C. and Başar E. (1991) *International Journal of Neuroscience* 60: 227–237.)

Cat	Number of Experiment	HI_1	HI_2	HI_3	HI_4
Timi	1	3.40	2.50	2.07	3.46
	2	1.32	1.70	2.83	1.40
	3	3.29	2.63	3.44	2.92
Mini	1	0.25	3.94	3.08	6.75
	2	2.58	5.19	5.21	5.28
	3	5.51	4.14	7.16	9.66
Hope	1	6.59	5.07	8.84	6.36
	2	7.45	4.70	7.31	9.78
	3	5.76	6.25	10.40	11.61
Jeny	1	8.26	7.86	4.90	22.83
	2	3.45	6.13	6.27	30.16
	3	3.91	6.38	6.23	16.01
Lina	1	2.78	3.21	2.77	4.60
	2	2.59	2.70	1.28	5.99
	3	4.84	5.58	6.49	5.67
Anna	1	2.42	3.84	4.23	4.84
	2	2.03	3.34	6.91	11.86
	3	1.89	2.11	2.59	5.63
Nina	1	4.13	7.47	7.74	8.32
	2	5.49	4.27	7.27	1.46
	3	5.79	3.71	3.27	8.06
Maya	1	2.60	7.39	12.78	15.75
	2	1.30	1.34	5.65	6.24
	3	2.47	2.83	6.92	4.37

Table 17.2. Statistical analysis of amplitude values given in Table 17.1 using Wilcoxon–Wilcox test (Sachs 1992). (n.s.: not significant, $p < 0.05$). (From Başar-Eroğlu C. and Başar E. (1991) *International Journal of Neuroscience* 60: 227–237.)

	HI_2	HI_3	HI_4
HI_1	n.s.	$p < 0.05$	$p < 0.01$
HI_2		n.s.	$p < 0.01$
HI_3			n.s.

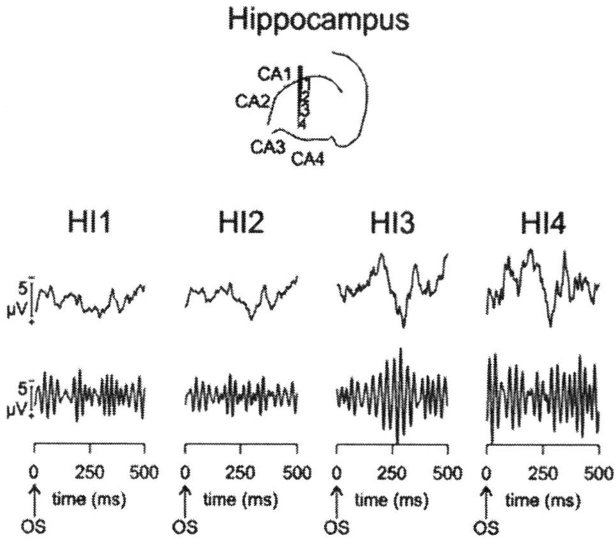

Fig. 17.3. Grand averages (mean values from 8 cats) of ERPs in various layers of hippocampus. *Top*: Location of multielectrode; CA1 and CA3 corresponding to HI1 and HI3/HI4 respectively. *Middle*: Unfiltered ERPs. *Bottom*: Filtered ERPs (30–50 Hz; OS: omitted stimulation). (Modified from Başar-Eroğlu C. et al. (1991) *International Journal of Neuroscience* 60: 239–248.)

first, or upper hippocampal, electrode (HI1) was less significant (cf. Chap. 16 and Başar-Eroğlu et al. 1991b). Moreover, the waves recorded in the first and second hippocampal locations (HI1 and HI2) also did not show a prominent N200 response. The P300 wave and the accompanying 40 Hz oscillation were not as marked as those in the CA3 region.

It is also to be emphasized that in a large number of sweeps, just at the moment of the occurrence of omitted stimulation (the first 100 ms), a small 40 Hz wave packet was also seen, but usually it was not as high as the 40 Hz wave at the P300 location. Earlier results have already shown that the hippocampus depicted significant 40 Hz responses to sensory stimulation that reached response amplitudes around 200 µV in single sweeps (Başar 1980, 1988; see also Chaps. 3 and 26). The question whether the appearance of a slow wave with the N200–P300 complex is usually combined with a wave package of 40 Hz activity cannot yet be answered. We may add to this point, however, that in our earlier reports we could show that the grand average of 15 human subjects had a 40 Hz enhancement at around 300 msec following the omitted stimulation (Başar et al. 1989; see also Chap. 22).

17.2 Gamma Activity in Earlier Studies

The first observation of "40 Hz resonance" to acoustical stimuli was reported in the cat auditory cortex (Başar 1972) and in the cat hippocampus (Başar and Özesmi 1972; Başar and Ungan 1973). The 40 Hz human auditory response was first published by Başar et al. (1976b). The 40 Hz resonance phenomenon (or 40 Hz enhancement) was further analyzed as a distributed property of various brain structures including cerebellar cortex, and it has been indicated as one of the principal components of the middle latency responses of the auditory pathway evoked potentials (Başar 1980). One of our studies has shown the enhancement of the 40 Hz EEG activity elicited with auditory stimuli and demonstrated the important relation between 40 Hz spontaneous activity and the middle-latency peaks of the human EPs (Başar et al. 1987; see also Chaps. 3 and 26).

The P300–40 Hz response is very important from the viewpoint of cognitive processes. This gamma response is not phase-locked, but time-locked, and it appears with a considerable delay similar to the gamma responses observed by Singer and coworkers (Gray and Singer 1987; Gray et al. 1989) and by Eckhorn et al. (1988). For an extended discussion of the results the reader is referred to Chap. 26.

18. Event-Related Potentials During States of High Expectancy: Results on the Cat Hippocampus, Cortex, and Reticular Formation

We recommend reading this chapter in parallel with Chaps. 16, 19, 21, and 25.

Bullock (1993) generally discussed the psychophysiological function *expectancy.* The nervous system shows adaptation by preparedness in a variety of ways. This state can be passive and active; it comes from genetic factors and can be learned both for normal and unimportant input.

Learned expectancy includes short-term and long-term memories. According to Bullock (1993), the term "expectation" means an "inferred state" of the memory system's accounting from any of a variety of behavioral or physiological sciences that are more or less specific. Sensory input has been anticipated as a familiar event. A surprise may be either a familiar unanticipated movement or an unfamiliar stimulus. Expectation is often a form of learning. Associative learning always requires an expectation. Bullock further states that expectations are at least as widespread in the animal kingdom as habituation and associative learning.

In this chapter we will describe experiments with chronically implanted cats and apply a similar paradigm to that explained in Chap. 19 on human ERPs during focused attention states. The cats have been stimulated with repetitive auditory and expectancy stimuli ERPs, and changes of ERPs were recorded in various intracranial structures. Multiple electrodes were placed in various layers of the hippocampus. The results will be further interpreted in Chaps. 21 and 20, which provide an integrative discussion combining results on human ERPs and the results of the oddball experiments included in various chapters.

18.1 Neuronal Activity of the Hippocampus During Learning, Searching, and Decision Making

The aim of the descriptions in this section is to focus the attention of the reader on the changes in neuronal activity of the hippocampus, so that they can be later correlated with the theta components that exist in event-related potentials at the single-cell level. We have chosen examples from experiments that seem to be the most pertinent ones with regard to the studies on event-related potentials described in this book.

18.1.1 Unit Activity and Behavior

Single-unit recording from the hippocampus of intact, behaving animals began with the pioneering studies of Vinogradova (1970), Olds and associates (e.g., Olds et al. 1969; Segal and Olds 1972), O'Keefe and Dostrovsky (1971), and Ranck (1973). All these workers noted that learning appears to exert powerful influences on hippocampal unit activity. Among the most fascinating results to date are the studies of "spatial" units in the hippocampus by O'Keefe and associates (O'Keefe 1976, 1979a,b; O'Keefe and Nadel 1978). These workers described two types of responsive hippocampal units in the freely moving rat: *Displace* units and *place* units. The displace units are Ranck's *theta cells*; during hippocampal theta activity they show an increase in overall firing rate in relation to theta waves. In the rat, hippocampal theta activity is most commonly associated with behavioral movements of an exploratory nature (Vanderwolf 1969; Vanderwolf et al. 1975; Black 1975). Place units, on the other hand, respond when the rat is in particular places in a maze. Both cell types are found in the fields CA1 and CA3 of the hippocampus (Ammon's horn), but place cells, at least, are not found in the dentate gyrus (O'Keefe and Nadel 1978). The second class of units in the CA1 sector of the dorsal hippocampus are those whose firing is dependent on the rat's location in an environment. The part of the environment where the unit fires, or fires maximally, is called the "place field." Some units fire maximally when the rat is in the place field and, in addition, performing a specific behavior, such as "sniffing" or receiving a specific stimulus. Some of these units are almost certainly identical to the units that Ranck (1973) has called "approach–consummate–mismatch" cells.

18.1.2 Multiple Sensory–Behavioral Correlates in Single Neurons – Theta Cells in the Hippocampus: View of Ranck

Best and Ranck (1982) noted that almost all reported sensory–behavioral correlates of hippocampal complex-spike neurons involve some aspect that is dependent on the past experience of the animal. This dependency on past experience is different from the situation with many other neurons in the brain and with the *theta cells in the hippocampus*, which fire with a predictable pattern associated with sensory stimuli, specific motor events, or behavioral state. Ranck (1982) suggested two possible explanations for these observations:

1. One possibility is that the firing of neurons in the hippocampus is related to some more pervasive or fundamental characteristics than anything that has yet been described.
2. The other possibility is that these neurons are involved in learning either as a cause or result in such a way that most complex spike cells fire in *association* with some aspect of every experimental situation.

18.1.3 Training-Induced Increase in Hippocampal Unit Activity: View of Thompson

Thompson and coworkers introduced a training and learning paradigm for conditioning, and in their studies of the hippocampus, they discovered a rather extraordinary learning-induced neuronal response (Berger et al. 1976; Berger and Thompson 1977; Berger and Thompson 1978a,b). Neurons in fields CA3 and CA1 increase their activity very rapidly and form a "temporal model" of the learned behavioral response (see Fig. 18.1). This hippocampal response does not develop in unpaired control animals. Furthermore, under a wide range of conditions that alter, impair, or block the development of the hippocampal unit response, the unit response is an invariable and predictive concomitant of the development of the learned behavioral response.

The growth of the hippocampal unit response is an invariable and strongly predictive concomitant of subsequent behavioral learning. If the hippocampal response does not develop, the animal will not learn. If it develops rapidly, the animal will learn rapidly. If it develops slowly, the animal will learn slowly. Further, the temporal form of the hippocampal response predicts the temporal form of the learned behavioral response.

18.1.4 Signal Detection in the Hippocampus

Signal detection theory implies that two kinds of processes occur when an organism detects a threshold level signal:

1. The signal must be detected in noise at the level of sensory receptors and pathways.
2. A decision to respond must engage the neuronal circuitry of the learned response – the response the animal has been trained to make to indicate that the stimulus has been detected.

Consequently, neuronal circuits that are differentially activated on *detection* and *nondetection* trials are candidate circuits for the *decision–memory* system. By the same token, neuronal circuits that do not respond differentially on detection and nondetection trials are not a part of the decision–memory system (compare also the decision-making system in Chap. 12).

An example of the strongly predictive nature of the hippocampal unit response for learned behavior is provided in current work on signal detection. Rabbits are trained, overtrained, and taken to threshold, using a white-noise conditional stimulus (CS) (Kettner et al. 1980). A staircase procedure is used so that each animal approaches asymptotically at 50% detection a constant-intensity threshold level CS. Interestingly, the behavior detection response is dichotomous at the threshold, being clearly present on detection trials and completely absent on nondetection trials. In this sense it is very similar to the all-or-none response typically used in human threshold studies. (Here the reader is referred to human threshold studies in Chap. 12.)

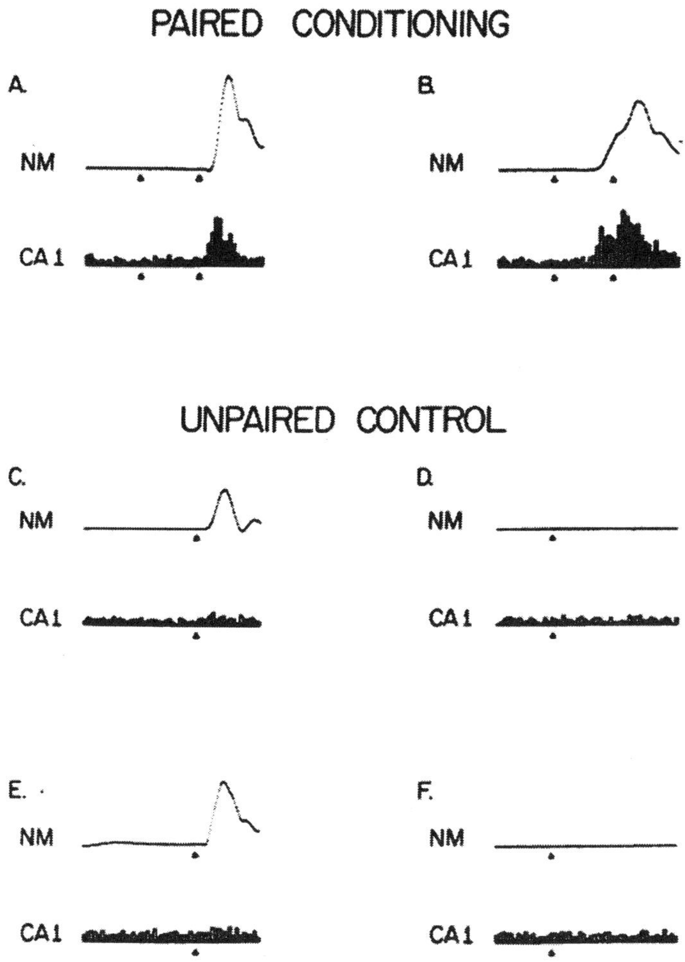

Fig. 18.1. Examples of 8-trial averaged behavioral nictitating membrane (NM) responses and associated multiple unit histograms of hippocampal activity for a conditioning (**a, b**) and a control (**c–f**) animal at the beginning and end of short-delay training. Note the very large increase in hippocampal unit activity that develops in the conditioning animal. CS, conditioned stimulus; UCS, unconditioned stimulus. *Upper trace:* average NM response for 1 block of 8 trials. *Lower trace:* hippocampal unit poststimulus histogram (15 msec time bins) for 1 block of 8 trials. (**a**). First block of 8 paired conditioning trials, Day 1. (**b**). Last block of 8 paired conditioning trials, Day 1, after conditioning has occurred. (**c**). First block of 8 unpaired UCS-alone trials, Day 1. (**d**). Last block of 8 unpaired UCS-alone trials, Day 2. (**e**). First block 8 unpaired CS-alone trials, Day 1. (**f**). Last block of 8 unpaired CS-alone trials, Day 2. (From Berger T.W. and Thompson R.F. (1978) *Proceedings of the National Academy of Science of the USA* 75: 1572–1576.)

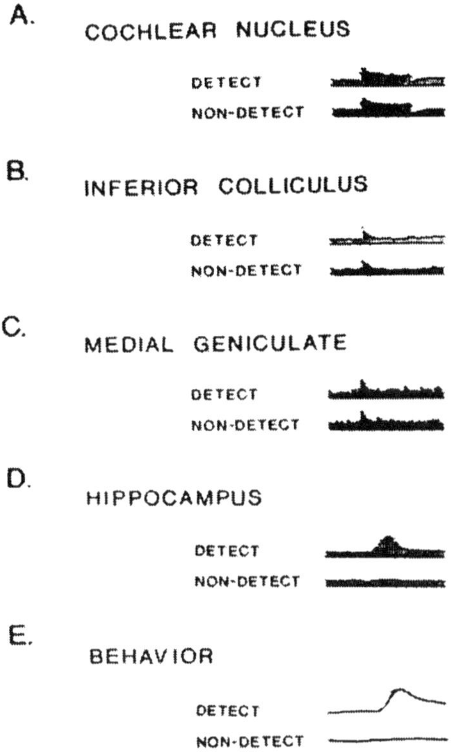

Fig. 18.2. Comparison of multiunit responses from auditory relay nuclei and hippocampus during detect vs. nondetect trials. NM, nictitating membrane; CS, conditioned stimulus. (**a**) through (**d**): Average poststimulus histograms (15 msec time bins) created by averaging from 200 to 300 trials (obtained from several testing sessions) for cochlear nucleus (**a**), inferior colliculus (**b**), medial geniculate (**c**), and hippocampus (**d**) on detect (upper histogram) vs. nondetect (lower histogram) trials. (**e**): Average NM response for detect (upper trace) vs. nondetect (lower trace) trials. Detect trials were defined as trials in which at least 0.5mm of NM extension occurred within 500 msec after CS onset on CS-alone trials. Note the large difference between hippocampal responses during detect vs. nondetect trials in comparison with the slight or nonexistent differences in responses of the auditory relay nuclei. (From Thompson R.F. (1982) in: Swanson L.W., Teyler, T.J., Thompson R.F. *Hippocampal Long-Term Potentiation: Mechanisms and Implications for Memory.* [Neurosciemces Research Program Bulletin, vol. 20.5].)

Unit responses in the primary auditory relay nuclei – the anteroventral *cochlear nucleus*, the *central nucleus* of the *inferior colliculus*, and the ventral division of the *medial geniculate body* – are substantial and identical on both behavioral *detection* and *nondetection* trials (see Fig. 18.2 and Kettner et al. (1980); Thompson et al. (1980). The auditory system reliably detects the stimulus on all trials at behavioral threshold. By this argument the primary auditory system is not a part of the decision–memory system.

In marked contrast, the learning-induced hippocampal unit response is present and well developed on detection trials and completely absent on non-detection trials (Fig. 18.2; see also Kettner and Thompson 1982). It is actually possible to predict the occurrence of behavioral detection and nondetection responses on a trial-by-trial basis from the occurrence of the learning-induced neuronal response in the hippocampus.

In terms of the types of neurons within the hippocampus that generate the learning-induced neuronal model of the learned behavioral response, results to date indicate that it is primarily the pyramidal neurons of fields CA3-CA1 that are responsible (Berger and Thompson 1978b; cf. results on P300 responses in the CA3 layer of the hippocampus in Chap. 16).

18.2 Event-Related Potentials in Cortex and Hippocampus in a P300-like Paradigm

Our approach is now slightly different, although the paradigm used is similar to the one used in Chap. 19 on human subjects. What does it mean when a cat emits a P300 wave following the omitted stimuli? The cat has heard four tones repetitively with constant intervals of 2.5 sec. The cat must, therefore, have performed a type of cognitive process: Probably, the cat is in an expectancy state. It should also be repeated that the fifth tone that should follow the fourth tone was omitted. In other words, the cat has to develop a type of *expectancy* or *selective attention*, and its brain must be able to perform a reaction that is called the P300 wave when the fifth tone does not occur (see also Bullock's definition of expectancy at the beginning of this chapter).

If we accept this view, we can further admit that the evoked potentials of the cat brain should have *various levels of cognitive processing during repetitive stimuli* in comparison to the conventional evoked potentials where stimulation is applied randomly. The question is now precisely whether the cat hippocampus will generate different types of evoked potentials with changing amplitudes in these various levels of cognitive processing with repetitive stimuli.

We again underline our assumption that during such a paradigm the cats are put in a complex state of expectancy, attention, and short-term memory. This means that in such a process all states of the brain that are defined as alertness, attentiveness, arousal, sensory memory, short-term memory, are in play and interaction.

The experiments were carried out on 8 freely moving female cats with chronically implanted electrodes. Stainless steel electrodes of 100 μm diameter were located while the cats were under Nembutal anesthesia (35 mg/kg) in the auditory cortex (gyrus ectosylvianus anterior, GEA) and mesencephalic reticular formation (RF). A multielectrode array having four tips was placed in the right hippocampus. The distance between the tips of electrodes was

0.7 mm (see Chap. 16 for a schematic presentation of the position of the electrode array). The electrodes were labeled HI1, HI2, HI3, and HI4. The first electrode was located in the upper pyramidal layer of the hippocampus (CA1), the second electrode between the upper pyramidal layer and gyrus dentatus, the third and fourth electrodes on or in the lower pyramidal layer (CA3). Detailed analysis of the accuracy of the electrode position is given in references (Başar-Eroğlu 1990; Başar-Eroğlu et al. 1991a,b). In four of the cats an electrode was also present in motor cortex (MC).

The derivations were against a common reference that consisted of three stainless steel screws in different regions of the skull. The cats were sitting in a cage in a soundproof, echo-free room that was dimly illuminated. For methods see Chap. 16 and Demiralp et al. (1994).

The stimulus was a 2000 Hz sound with an intensity of 80 dB and duration of 1 s. The interstimulus interval was 2.5 s. An earlier investigation had shown that the optimal rate for the target signal is the paradigm with every fifth tone omitted. The design for four applications of the stimuli and the data acquisition intervals in the three experiments mentioned above are shown in Chap. 16.

18.3 Frequency Responses During States of "High Expectancy"

Figure 18.3 shows the grand averages of unfiltered time-domain EPs to the four stimuli preceding the omitted stimulus in auditory cortex (GEA), reticular formation (RF), various layers of hippocampus (HI2, HI3, HI4), and motor cortex (MC), respectively.

18.3.1 Time-Domain Analysis of the Responses to the 1st, 2nd, 3rd, and 4th Stimuli Preceding the Omitted Stimulus

The maximal amplitudes of the ERPs recorded from the auditory cortex (GEA) and reticular formation (RF) by 1st, 2nd, 3rd, and 4th stimuli preceding the omitted stimulus (the 5th one) did not show any relevant differences. However, the responses recorded from the HI2, HI3, and HI4 electrodes in the hippocampus showed a prominent increase in the peak-to-peak amplitude following the 4th stimulus (i.e., the predecessor of the omitted stimulus). Similarly, in all four cats with electrodes in the motor cortex (MC) it could be seen that the peak-to-peak amplitude of the response to the 4th stimulation was prominently higher than those to previous stimuli.

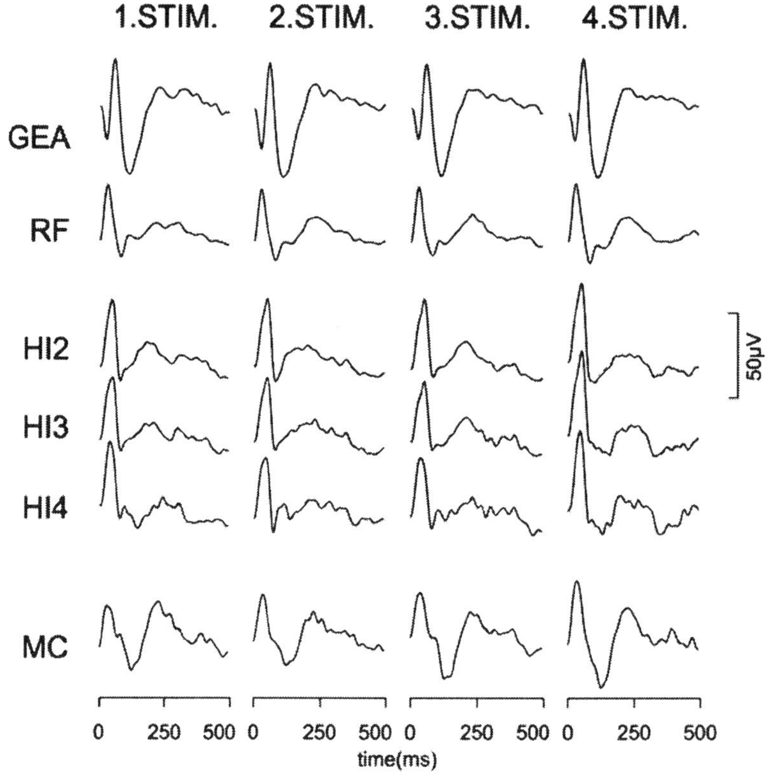

Fig. 18.3. The grand averages of the AEPs elicited by the 1st, 2nd, 3rd, and 4th stimuli in the omitted stimulus paradigm with every 5th tone omitted in GEA, RF, H12, H14, and MC. (Modified from Demiralp, T. et al. (1994) *International Journal of Psychophysiology* 18: 35–48.)

18.3.2 Frequency-Domain Analysis by Means of the AFCs

To obtain the general frequency characteristics of the ERPs in different structures, the AFCs are calculated for the grand averages of the responses of the 1st–4th stimuli preceding the omitted stimulus (Fig. 18.4). The AFCs obtained from GEA, RF, HI2, HI3, HI4, and MC showed different weights of frequency components:

The GEA response showed a broad peak around a center frequency in the alpha range (12 Hz) with small shoulders in theta (7 Hz), delta (3 Hz), and beta (20 Hz) bands. A small separate peak was also observed in the gamma band (40 Hz). In the response of the reticular formation (RF) a clearly dominant peak was observed in alpha band (12 Hz). Secondary peaks are in theta (5 Hz), beta (20 Hz), and gamma (40 Hz) bands. In all hippocampal recordings (HI2, HI3, HI4) the AFCs had similar characteristics: a compound theta–alpha–beta peak (with approximately equal weights of the three frequency

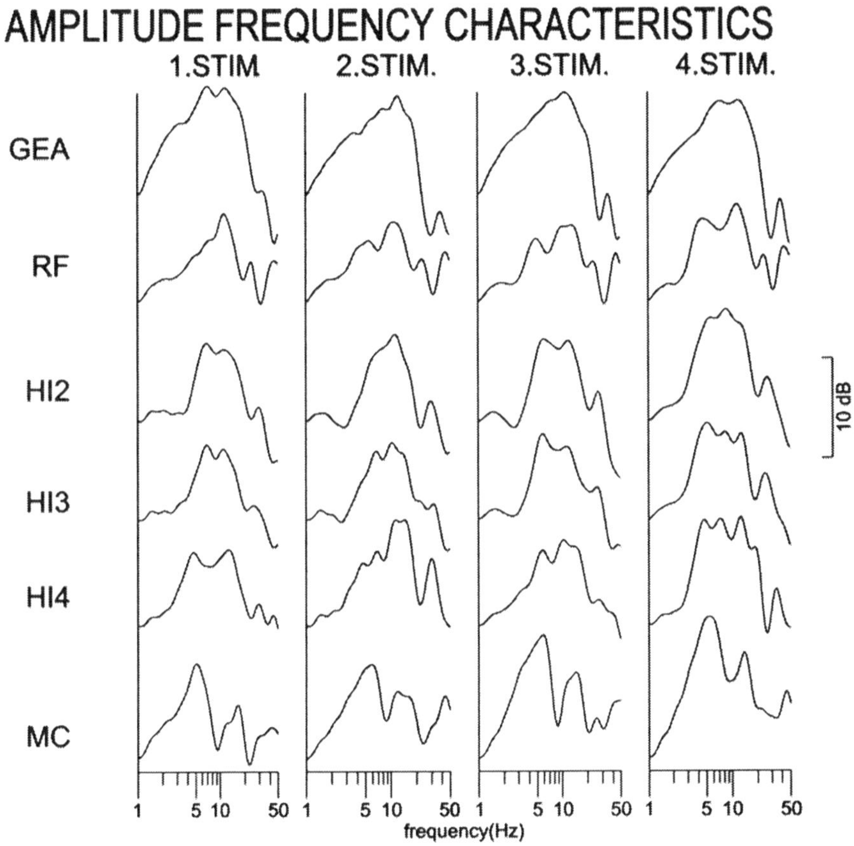

Fig. 18.4. The amplitude-frequency characteristics computed from the responses shown in Fig. 18.3. (Modified from Demiralp T. et al. (1994) *International Journal of Psychophysiology* 18: 35–48.)

components in the regions of 5–7 Hz, 8–15 Hz, and 15–30 Hz) and a separate peak around 40 Hz. In motor cortex (MC) the dominant peak was in the theta band (5 Hz) followed by a secondary alpha–beta compound peak between 8 and 30 Hz and a gamma-band peak around 40 Hz.

18.3.3 The Differences Between AFCs of Responses to 1st, 2nd, and 4th Stimuli Preceding the Omitted Stimulus

The amplitude increases following the 4th stimulus were correlated with an amplitude increase in the theta-frequency band in the AFCs. In HI2, HI3, and HI4 recordings a clear increase of the peak around 5 Hz (3–8 Hz) could be detected. An evident increase in the theta (5 Hz) response amplitude was observed in the MC responses of all of the 4 cats after the 4th stimulation.

In auditory cortex (GEA) there are no differences between the AFCs of the responses of the 1st, 2nd, 3rd, and 4th stimuli.

18.3.4 Adaptive Digital Filtering of the Responses and Statistical Testing of the Results

The ERPs were filtered using digital band-pass filters with no phase shift. The band-pass limits of the digital filters were selected according to the bandwidths in the AFCs of the grand average waveforms to 1st, 2nd, 3rd, and 4th stimuli at all of the peaks. For the sake of simplicity only the theta (3–8 Hz) and alpha (8–15 Hz) band's filtered responses are shown in Figs. 18.5 and 18.6, respectively.

The significance of the differences between the maximum peak-to-peak amplitudes of filtered responses of the 1st, 2nd, 3rd, and 4th stimuli preceding

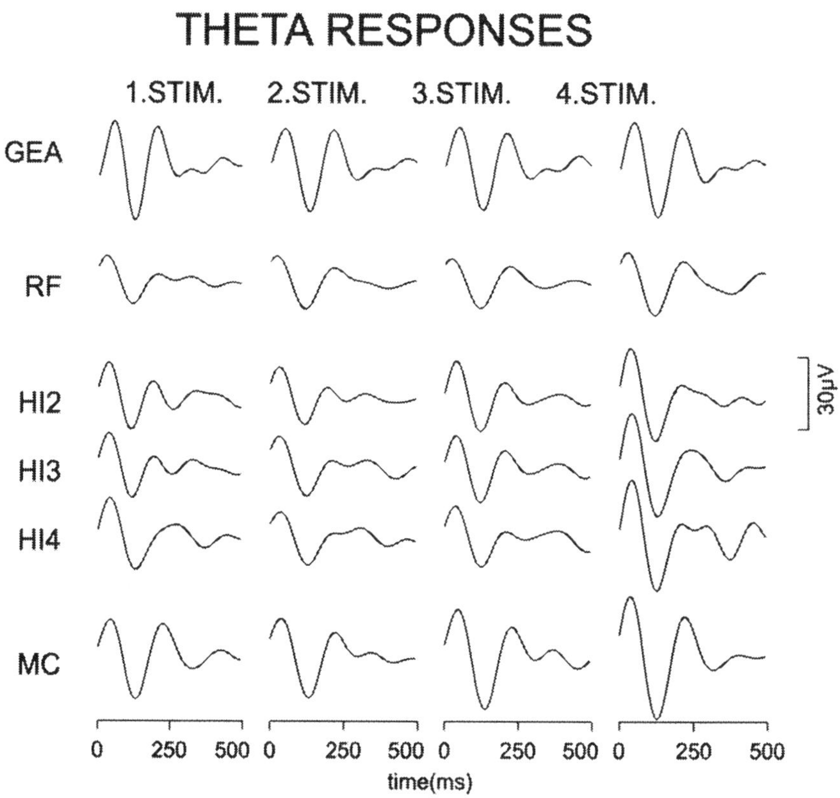

Fig. 18.5. The theta response components (3–8 Hz) of the responses elicited by the 1st to 4th stimuli preceding the omitted 5th stimulus in GEA, RF, various layers of the hippocampus, and MC. (Modified from Demiralp T. et al. (1994) *International Journal of Psychophysiology* 18: 35–48.)

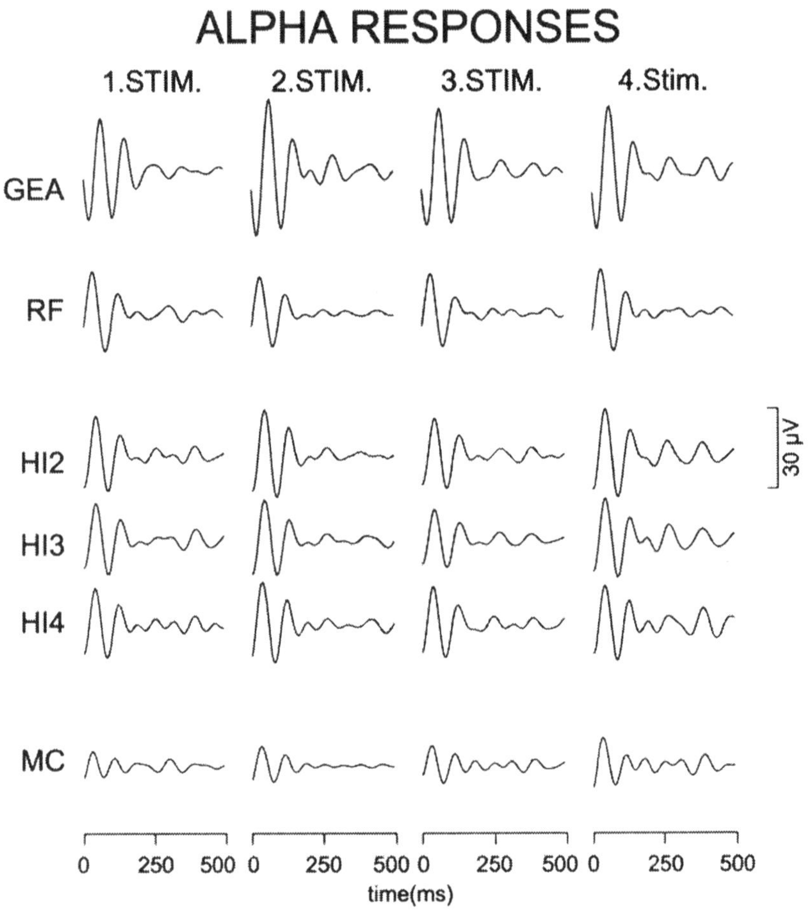

Fig. 18.6. The alpha response components (8–15 Hz) of the responses elicited by the 1st–4th stimuli preceding the omitted 5th stimulus in GEA, RF, and various layers of the hippocampus and MC. It is given as an example for the insensitivity of frequency components other than theta to the stimulus repetition. (Modified from Demiralp T. et al. (1994) *International Journal of Psychophysiology* 18: 35–48.)

the omitted stimulus have been tested by a one-way ANOVA analysis and post hoc tests. The results of the statistics are shown in great detail by Demiralp et al. (1994) and in Fig. 18.7, where the mean values of the maximal peak-to-peak amplitudes of the band-pass filtered responses are shown in histograms.

As can be seen from Figs. 18.5 and 18.6 and from the results of the statistical tests (Demiralp et al. 1994), significant differences were obtained only in the theta components of the hippocampal responses to the 4th stimulus compared with the responses of the 1st, 2nd, and 3rd stimuli. The percent values of significant theta-response increases (as the percent increase of theta

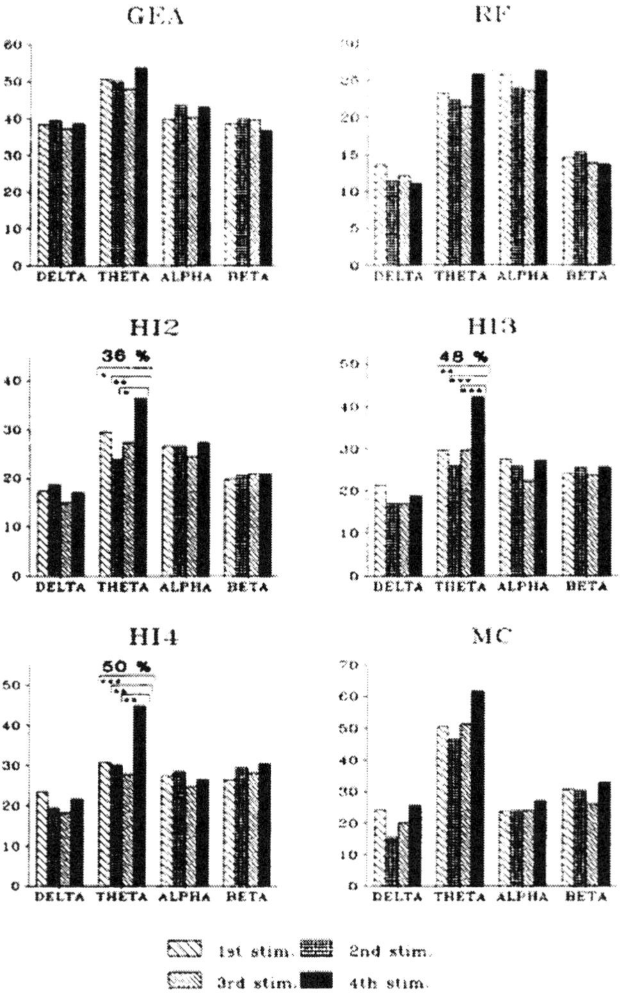

Fig. 18.7. The mean values of the maximal amplitudes of various frequency components of AEPs elicited by 1st, 2nd, 3rd, and 4th stimuli in various brain structures (in μV). The significant differences are indicated. There were only significant differences betweeen hippocampal theta responses following the 4th stimulus and those elicited by each of the 1st, 2nd, and 3rd stimuli in H12 and H14. The percentual values of the significant theta increases following the 4th stimulus are calculated by assuming the mean of the theta responses to 1st, 2nd, and 3rd stimuli as 100%. (From Demiralp T. et al. (1994) *International Journal of Psychophysiology* 18: 35–48.)

response amplitude following the 4th stimulus in comparison to the mean of the theta-response amplitudes after the first three stimuli) are displayed on the histograms. The HI2 recording showed a mean theta increase of 36% (1st vs. 4th, $p < 0.03$; 2nd vs. 4th, $p < 0.01$; 3rd vs. 4th, $p < 0.02$), whereas the HI3 and HI4 recordings from the CA3 layer showed more prominent theta-response increases of approximately 48% and 50%, respectively (for HI3: 1st vs. 4th, $p < 0.002$; 2nd vs. 4th, $p < 0.0002$; 3rd vs. 4th, $p < 0.0007$; for HI4: 1st vs. 4th, $p < 0.0007$; 2nd vs. 4th, $p < 0.01$; 3rd vs. 4th, $p < 0.005$).

The consistency of the theta response increases in HI2, HI3, and HI4 and their dominance in HI3 and HI4 are also shown in Fig. 18.8, where the responses of single cats are superimposed in unfiltered form and band-pass filtered in the theta range.

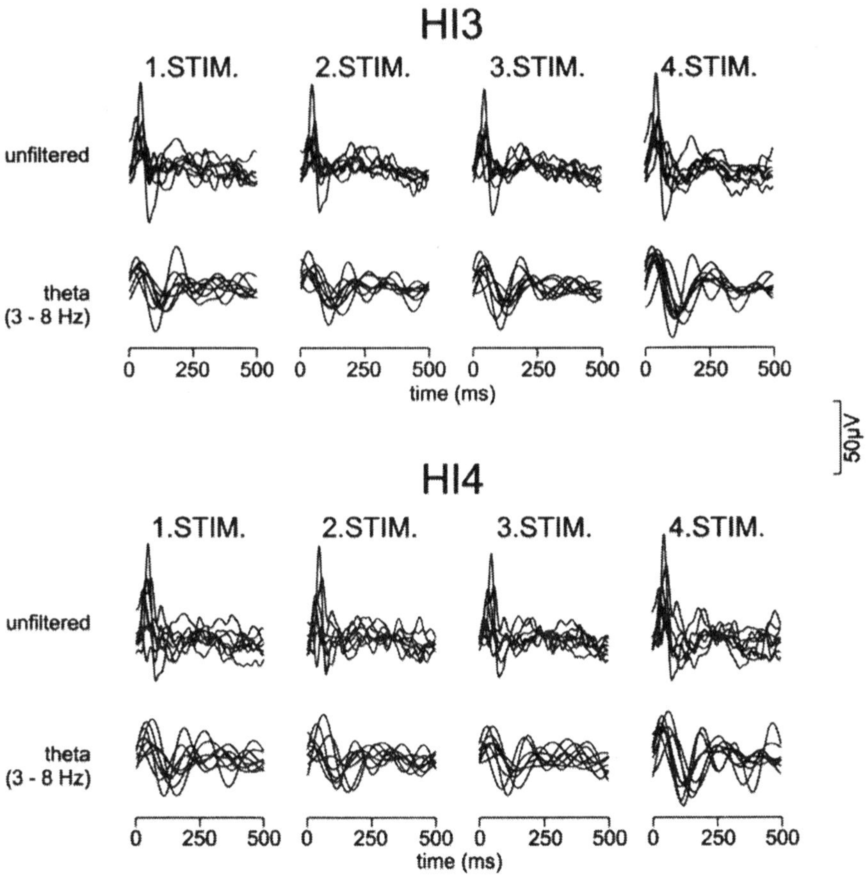

Fig. 18.8. The superimposed unfiltered (*upper rows*) and theta band (3–8 Hz) filtered (*bottom rows*) responses of H13 and H14 (CA3) of 8 cats. (Modified from Demiralp T. et al. (1994) *International Journal of Psychophysiology*18: 35–48.)

18.4 Selectively Distributed Theta System of the Brain: The Limbic, Frontal, and Parietal Areas Are Mainly Involved

The omitted stimulus P300 responses of freely moving cats with chronically implanted electrodes in various brain structures have been studied in Chap. 16. Furthermore, the same paradigm applied to human subjects showed a considerable enhancement of the theta components of the responses to the stimulus preceding of the omitted one (Demiralp and Başar 1992 and Chap. 19). The theta response increases were significant over frontal and parietal areas in the auditory modality. In visual modality significant theta increases were more diffuse, over frontal, parietal, occipital areas and at vertex, but again more pronounced at frontal and parietal sites.

In light of the results mentioned above we are now searching for similar changes in the responses to the stimuli that precede the omitted one in cats with chronically implanted electrodes. Certainly, the cognitive contents of the omitted stimulus experiments carried out on *passively learning cats* and *instructed human subjects* are not exactly comparable. However, the presence of a P300-like response of the cat brain to the omitted stimulus (see Chap. 16) enables us to assume that the cats may also, to a certain extent, develop *expectancy* and *focused attention* to the repetitive stimulations and the regular omission of a stimulus.

18.4.1 Frequency Selectivity of the Amplitude Enhancements in Hippocampus

These results are in accordance with the results obtained from mainly frontal and parietal recording sites of human subjects in Chap. 19. In the human study, a 44% increase in the theta-response amplitudes in the frontal recording site was found. In the present results the theta-response increases in the CA3 layer of the hippocampus are also at a similar level (48–50%).

The amplitude increases in the responses of the hippocampus and motor cortex to the stimulus preceding the omitted one were correlated with a selective amplitude increase in the theta frequency band in the amplitude–frequency characteristics (AFCs). In HI2, HI3, and HI4 recordings the statistical tests revealed that these amplitude increases in the theta band were significant. In the motor cortex this result could not be tested statistically. However, it was consistent in all four cats with the implanted electrode in the motor cortex. In all other recording sites and frequency bands no significant difference occurred among the responses to the 1st–4th stimuli.

The primary auditory cortex (GEA) responses in the cat brain showed no significant changes, which seems partly to contradict the results obtained in human subjects, where a slight theta increase could also be observed in scalp locations reflecting the activities of the primary sensory areas, even though the most prominent theta enhancement was over frontal and parietal areas. However, considering the spatial smearing (blurring) effect of scalp recorded EEG and the refined recording method of implanted electrodes used by the present study, it can be stated that the slight theta increses in human scalp recordings were possibly due to the reflection from frontal and parietal association areas. Accordingly, we assume that the experiments with the cat add a highly significant explanation regarding the location of the theta increase. *Theta rhythms as a common feature of the limbic structures are also involved in the mechanism of selective attention.*

The hippocampal electrical rhythm in the theta frequency range (4–7 Hz) has been one of the most studied rhythmic activities of the mammalian brain (Vanderwolf 1969; Lopes da Silva and Arnolds 1978; Vanderwolf and Robinson 1981). All limbic areas are capable of displaying theta rhythm. The appearence of the hippocampal theta activity correlates in different weights with large body movements, the alert posture in the immobile animal, the specific orienting movements (in the absence of larger body movements), during eye movements, and during fixed staring in various species (for a review see Miller (1991) and Chap. 25 and Chap. 3 in Vol. I).

Important results on the functional significance of the hippocampal theta rhythm are experimental observations showing that the theta activity in hippocampus can facilitate the transmission of information between the hippocampus and target structures, such as the nucleus accumbens (Lopes da Silva et al. 1984), and that stimulation at the theta frequency may induce LTP in hippocampal formation (Larson and Lynch 1988; Rose and Dunwiddie 1986; Greenstein et al. 1988).

The learning and LTP are enhanced if the delay between the pair of stimuli is in the range of the theta period. In our findings the increase in theta response was highly prominent in the CA3 location; it can thus be stated that theta is generated in the CA3 pyramidal layer.

Furthermore, results of Başar-Eroğlu et al. (1991b) and Chap. 16 have shown that the P300-like response of the cat hippocampus to omitted stimuli has a strong theta response component, again in the CA3 layer. Accordingly, the hippocampal theta response, which is especially dominant in the CA3 layer, seems to have a general functional role in processes of focused attention.

18.4.2 Comments on the Anatomical and Physiological Links Between the Hippocampal Formation and the Association Areas of the Neocortex

The specific thalamic projection nucleus to the prefrontal cortex is the nucleus medio-dorsalis. The medial magno-cellular part of the nucleus medio-dorsalis receives its main input from nuclei amygdalae and the temporobasal cortex including the hippocampus. Hence, the main thalamic input to the orbito-frontal cortex stems from the limbic system (Nauta 1971, 1972). Also, lateral parts of the nucleus medio-dorsalis, and in this way the dorsal part of the prefrontal cortex, received input from hippocampus–septum system (Creutzfeldt 1983).

There are direct efferents of the hippocampal formation that reach the isocortex (for a review see Miller 1991). CA1 cells send efferents that reach the perirhinal cortex, retrosplenial cortex, and various parts of frontal and temporal regions (Schwertfeger 1979; Swanson 1981). The subicular components also project back to posterior cingulate cortex, dorsal prefrontal cortex, temporal and prepryform cortex (Irle and Markowitsch 1982a,b; Cavada et al. 1983).

The hippocampal formation is most directly linked with the cingulate cortex and the prefrontal areas, which send efferents to and receive afferents from the rest of the cortex. Efferents from the posterior cingulate cortex have been seen projecting to the inferior and posterior parietal cortex, dorsal and lateral prefrontal cortex, orbito-frontal cortex, parieto-temporal cortex, medial and temporal cortex (Mesulam et al. 1977; Pandya et al. 1981; Baleydier and Maugiere 1980). Each sector of parietal cortex is connected in parallel with a particular sector of the principal sulcus and presumably transposes some of its sensory–limbic specializations to these areas accordingly (Goldman-Rakic 1988). Areas of the temporal neocortex also project topographically to the inferior convexity and orbital prefrontal cortex (Jones and Powell 1970; Pandya and Kuypers 1969; Markowitsch et al. 1985; Moran et al. 1987; compare also Chap. 27).

Aleksanov et al. (1986) showed that during the course of alimentary conditioning, and at the time of presentation of the conditioned stimulus (CS), the coherence between hippocampus and prefrontal cortex increased. This increase in coherence was observed especially in the delta and theta frequency ranges. Some recent evidence that the midline prefrontal region of the cortex can generate theta activity was supported by Mizuki et al. (1980) in certain cognitive states. Lang et al. (1987) used, in the same way as Westphal et al. (1990), spectral analysis of frontal EEG and showed that theta frequencies (3–7 Hz) were increased during motor or verbal learning tasks.

In light of the findings described above we come to the following conclusion: The theta response increases in hippocampus and frontal cortex of the cat brain and theta response increases in human frontal and parietal locations probably reflect the general responsiveness of the hippocampal–frontal–parietal system during focused attention and expectancy.

Recently, the concept of theta resonance has been analyzed extensively by Miller (1991), who proposed that Hebbian processes of synaptic strengthening select patterns of loops passing from hippocampus to cortex and back to the hippocampus. The view of Miller will be explained in detail in Chap. 25. The theta response presented in this chapter supports the interpretation of Miller.

18.4.3 The Integrative Analysis of the Increased Theta Response in the Brain: Diffuse Theta Response System in the Brain

The results of the present chapter demonstrate that in the cat hippocampus a selective strong theta-component increase occurs in response to stimuli preceding the omitted one in an omitted stimulus paradigm. Probably, an expectancy and increased attention state should be developed by the cat to the onset time of the repetitive stimuli similar to theta-response increases in frontal and parietal responses of human subjects (Demiralp and Başar 1992).

We postulated a *selectively distributed theta system* in the brain that is involved in the cognitive states of focused attention and expectancy. The parallelism between the theta response increases in these studies recorded over human neocortical areas and those in the present chapter recorded intracranially from cat hippocampus show that the hippocampus plays an important role in this postulated theta system. The close anatomical relationships between the hippocampus and neocortical association areas, especially the frontal and parietal association areas, suggest that the interaction of the hippocampus with the neocortical association areas in the theta frequency band might be the basis of the theta-response system, which is involved in focused attention and expectancy. (See also Chap. 12 in Vol. I, where we point out that the major operating rhythm of the frontal cortex is theta and that theta controls the visual EP in frontal recordings.)

18.5 Interpretation of Changes in ERPs

What has changed in the hippocampus during these types of experiments? What could we learn in experiments with the recording of evoked potentials and also the evaluation of the amplitude–frequency characteristics?

The hippocampus is a *supramodal structure* involved with motivation, emotion, and short-term memory, in other words, in all types of reactions, where human beings and animals have the task to compare all peripheral and endogenous stimulation arriving at the CNS with events that have happened before, i.e., with events that were seemingly stored in short-term or in long-term memory (see Chap. 3 in Vol. I). The primary cortex, for example the auditory cortex, has to process information immediately and react to a sensation. Hippocampus would not react to a sensation without signal detection and comparison. In this case, the cat as a living being would be supposed to perform judgments: What does the first stimulation evoke, what does the second one evoke? As the EPs recorded in this case are related not only to sensation but also to learning, attention, and reasoning processes, one would expect that the hippocampus should react in a completely different way in comparison to cortex and to reticular formation. Indeed, the results of this type of experiment confirm this ability of hippocampus, which shows a type of plasticity.

In the following we will try to give a brief summary of the experimental situation in human and cat experiments. As we stated elsewhere, the use of the terms *alertness, emotion, attention, short-term memory,* are often expressions that are extensively discussed by psychologists, and physiologists and these are expressions that are descriptive and often subjective. Accordingly, we will try to relate experiments to the observed electrophysiological changes.

What is going on in experiments with human subjects? They were told that they have to observe an omitted stimulation. This is the fourth or fifth one after the first stimulation. They all have reported that in order to focus their attention to the fourth omitted one, they had to increase their expectancy to the third one. In comparison to conventional EPs (recorded with random stimulation), frequency components of EPs for "third tones" showed a significant increase in the theta response in frontal and parietal locations. It is to be strongly emphasized that it is not possible to ask the cats about their expectancy and feeling by presenting them repetitive stimuli and then the omitted fifth one. We have only an objective criterion, namely that the omitted stimulation gave rise to a type of P300 wave (see Chap. 16). In other words, the cats might also have a type of attentive stage that is *somewhat comparable* to the human experiments with the same parameters and design.

Although there is no strong way to compare human and animal experiments, a similar physiological design and a common significance of the electrophysiological permits a comprehensive interpretation. What has changed in the cat brain? What has changed in the human brain during repetitive stimulation? In vertex and occipital recordings of humans there are not huge changes, whereas in the parietal and frontal recordings a significant increase of the theta response was observed. In the cat brain, there are not important or significant changes in the auditory cortex (in the primary sensory area). However, there are important changes in the hippocampus and in the frontal cortex. Both significant changes were observed with an increase of 4 Hz response (40%).

Our immediate interpretation from these experiments is now the following: If human subjects or cats ("seemingly") pay attention to a sensory stimulation, the EP to the attended stimuli shows a significant increase in the theta-frequency response. This is also in good accord with our results where we applied the 3rd signal paradigm to human subjects. Here again we had a significant increase in the theta-frequency range in the attended channel. The significance was again high in frontal and parietal regions.

18.6 Why We Compare EP Results with Conventional Experiments on Hippocampus

In the beginning of the present chapter we described in detail experiments on signal detection of the cat hippocampus on functional firing of displace units and place units. The conventional experiments demonstrated that these place units are *theta cells*. The hippocampal theta activity is in most of the experiments described as the dominant rhythmic one. Training also induces unit activity in the hippocampus, and in the decision–memory system hippocampus plays a more important role compared to collicus inferior and medical geniculate nucleus. As will be assumed in Chaps. 27 and 28, hippocampus and the limbic system are dominant structures of the selectively distributed theta system of the brain.

Why do we describe in the present chapter the conventional neurophysiological experiments demonstrating the induced theta activity in hippocampus? The rationale is based on the following issue: As will also be explained in Chap. 25, the 4 Hz activity is involved with short-term and long-term memory, expectation, and learning. With these new results we reach another new window to explore the brain function with the help of ERPs. The stimulation of the brain induces in hippocampus significant *theta enhancements* as a major response that is highly influenced by cognitive tasks.

Not only movements of cats or decision-making processes induce theta rhythms. A target (cognitive) stimulation also gives rise to high-amplitude theta components manifested as evoked potentials. As we have repeatedly assumed, the EEG is a record stemming from unknown inputs to the central nervous system that can also be considered as internally evoked or internally induced potentials. This view is strongly supported by pioneering results of Adey (1966) that showed *induced theta rhythms in the hippocampus of learning and exploring cats.*

We will return to an integrative discussion of theta oscillations of the brain in Chap. 25.

19. Event-Related Potentials During States of High Expectancy and Attention in Human Subjects

Tamer Demiralp and Erol Başar

19.1 Selective Theta Distribution

In this short chapter experimental results only are presented, which will be interpreted in a combined manner in Chap. 21. A parallel reading of Chap. 18 is recommended for further physiological interpretation of the results.

From the physiology literature it is well known that CA3 layers of the hippocampus, frontal lobes, and parietal lobes have important neural connections. This system is called the hippocampo-fronto-parietal system. Our findings respecting the significant increases of the theta response in the hippocampus, frontal cortex (see Chap. 21), and the parietal cortex support the concept of a selectively distributed theta system of the brain.

Although a late ERP component following the early exogenous potentials with a latency shift of 250–400 ms called the "P300 wave" is the first important example of the endogenous-type potentials, there is still evidence for faster endogenous components that probably cannot be detected in most of the paradigms used, because they overlap with early exogenous components (Picton and Stuss 1980; Desmedt et al. 1983; Näätänen 1988). Many solutions have been proposed to isolate spatiotemporally overlapping components (for a review see Picton and Stuss (1980)). According to the authors, isolation on account of experimental manipulation and factor analysis have been the most commonly used methods that yielded objective results. However, these methods lack the physiological interpretability of the isolated components.

In the following we will present a paradigm that will be an attempt to detect, by means of a method that is sensitive to changes in frequency components of the ERPs, endogenous components in ERPs that may not necessarily differ from the exogenous components in time and space. Başar et al. (1989) carried out a series of ERP studies on human subjects by applying a modified form of the omitted stimulus paradigm of Sutton et al. (1965). The paradigm consisted of auditory or visual stimulations with *regular interstimulus intervals* where some stimuli were *omitted* in a random or regular order with various degrées of probability. The subject's task was to mark mentally the time of the omitted stimulus. With this type of paradigm especially when the stimulus omission occurred in a regular manner (for example every 4th stimulus was omitted), quasi-deterministic, reproducible patterns of EEG signals occurred in anticipation of the omitted stimulus. (see Chap. 22). In such a paradigm, the subjects reported that they paid attention to the rhythm of preceding stimuli to be able to fulfill the task.

In the following analysis we apply the same paradigm to test whether event-related changes occur in the responses of different brain areas to the stimuli that precede or follow the omitted stimulus. It can be shown that whereas the time-domain analysis of the same responses show no prominent differences, the frequency-analysis approach may differentiate the responses to the stimuli that are coupled to a cognitive task from the standard EP, which requires detecting some specific changes in frequency components.

19.2 Experimental Paradigm

As auditory stimuli, 80 dB, 1500 Hz tones (0.5 ms rise time and 800 ms duration) were presented binaurally. The light stimulator was a 20 W fluorescent bulb that was electrically triggered with steps of 800 ms duration. A regular recording session involved three steps, with short resting periods in between, and was carried out in one day:

1. First, the spontaneous EEG was registered for a few minutes, which served to determine the global characteristics of the subject's spontaneous EEG activity and arousal state at the beginning of the experiment. It also allowed the subject to become familiar with the experimental conditions.
2. One group of subjects were exposed to the auditory EP and omitted stimulus paradigm with auditory stimulation.
3. The second group of subjects were exposed to the same paradigm but this time with visual stimulation. Thus, the visual EP and omitted stimulus paradigm with visual stimulation were applied to the second group of subjects.

Two different groups were used so as to avoid any learning effect that might easily have occurred because of the similarity of the cognitive tasks in both sensory modalities. A schematic illustration of both paradigms is shown in Fig. 19.1. In standard EP recordings, the interstimulus interval varied randomly between 2.5 and 4 s with a mean value of 3 s, whereas in the omitted stimulus paradigm, the stimuli were delivered at 3 s regular intervals. In the omitted stimulus paradigm, every 4th stimulus was omitted, and the subject's task was to predict and to mark mentally the time of occurrence of omitted signals. Subjects had not been informed of the task beforehand to avoid their unwanted cooperative efforts during the preceding (EEG, EP) experiments.

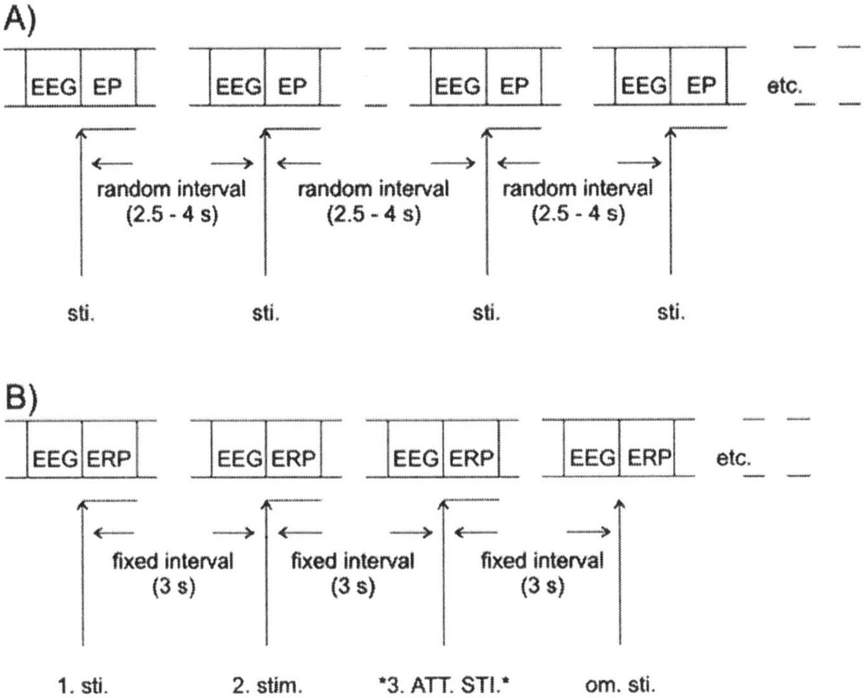

Fig. 19.1. Schematic illustration of the (**a**) evoked potential and (**b**) omitted stimulus paradigms (sti.: stimulus; om. sti.: omitted stimulus). (Modified from Demiralp T. and Başar E. (1992) *International Journal of Psychophysiology* 13: 147–160.)

19.3 ERPs to Repetitive Stimuli

Although no strategy was imposed on the subjects for the fulfillment of the time prediction, at the end of the experiments all subjects reported that their *expectancy* was highest for the third stimulus (the predecessor of the omitted stimulus) in the omitted stimulus paradigm. They said that they could attain the rhythm of the stimulations just after the second stimulus by comparing the onset times of the first and second stimulus, and that they attended to the 3rd stimulus to test their feeling for the rhythm. The first stimulus after the omission served also as a control of their performance in the prediction of the virtual onset of the omitted stimulus.

The responses to the first stimulus after the omission were incomparable with other responses because, the stimuli were followed by an ISI which was two times longer. These responses can be compared in another study with EPs recorded with longer ISIs (for example a pseudorandom series of intervals with a mean value of 6 s). In the preliminary data analysis, the responses to second stimuli showed slight differences compared to the standard EPs. Considering these points and the subject's reports, we focused our attention on the comparison between the responses to third attended stimuli and the standard EPs.

For both sensory modalities those responses that were obtained from four recording sites (F3, Cz, P3, and O1) were analyzed. Due to the limitations of space, data obtained from F3, P3, and O1 in visual modality and from F3, P3, and Cz in auditory modality are shown in the respective figures. Thus data from two recording sites close to association areas (frontal-F3 and parietal-P3) and data for each modality from the location most commonly used to record the responses (occipital-O1 for visual and vertex-Cz for auditory modality) are given.

19.3.1 Averaged Responses

The bottom row of Fig. 19.2 shows the superimposed averaged responses of 10 subjects recorded from F3, P3, and O1 leads with the standard visual EP

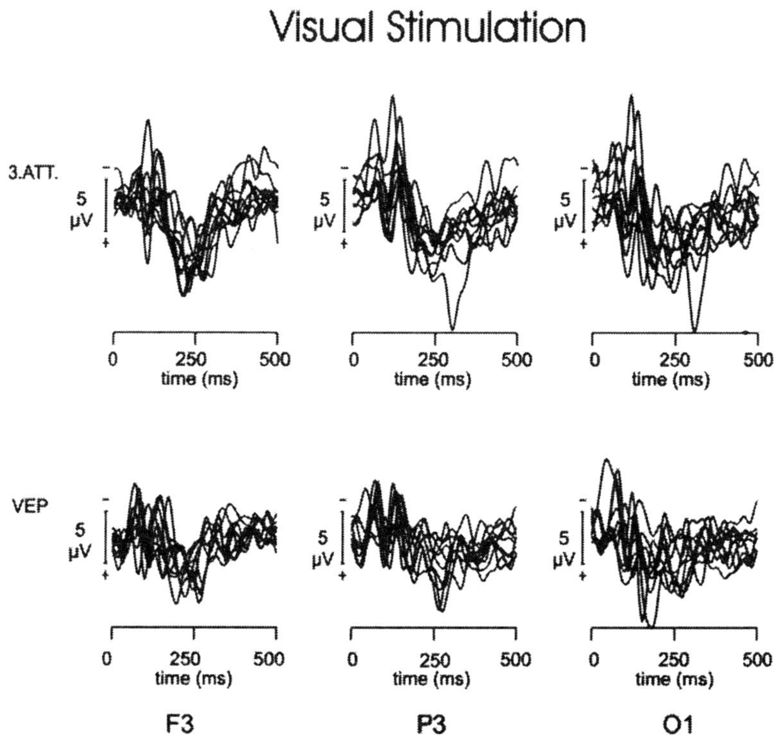

Fig. 19.2. Superimposed standard VEPs (VEP) and responses to 3rd attended light stimuli in the visual omitted stimulus paradigm (3.ATT) of 10 subjects obtained from frontal (F3), parietal (P3), and occipital (O1) regions. (Modified from Demiralp T. and Başar E. (1992) *International Journal of Psychophysiology* 13: 147–160.)

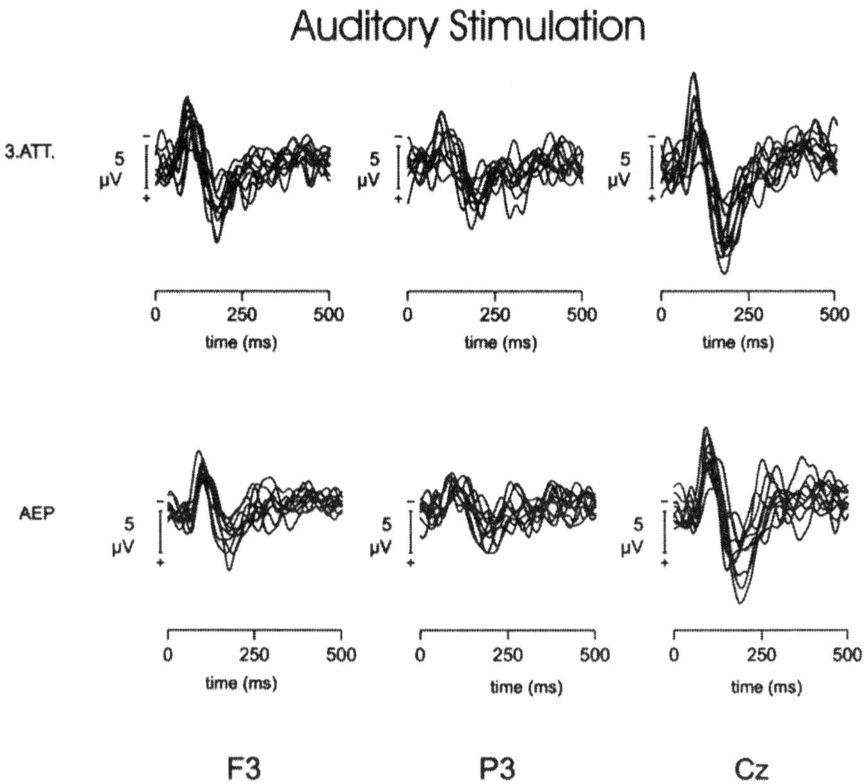

Fig. 19.3. Superimposed standard AEPs (AEP) and responses to 3rd attended tones in the auditory omitted stimulus paradigm (3.ATT) of 10 subjects obtained from frontal (F3), parietal (P3), and vertex (Cz) regions. (Modified from Demiralp T. and Başar E. (1992) *International Journal of Psychophysiology* 13: 147–160.)

paradigm. The upper row of the same figure illustrates the responses of the same subjects to the third attended visual stimulus (3.ATT) in the visual omitted stimulus paradigm.

Figure 19.3 shows analogous results derived from F3, P3, and Cz leads in similar paradigms but with acoustic instead of visual stimulation (auditory EP and 3.ATT). In both sensory modalities amplitudes of the responses increased when the subjects had to perform the time-prediction task.

In Figs. 19.4 and 19.5 the grand averages in the time domain and the AFCs obtained from the corresponding grand averages are shown for several electrode positions.

- In both sensory modalities the AFCs showed differences in the frequency contents of evoked responses obtained from various recording areas. In visual EPs the alpha-band activity was most prominent in occipital and parietal areas. In frontal visual EP the difference between theta-band amplitude and alpha-band amplitude was smaller. On the other hand, the low theta peaks with center frequencies around 3.5 Hz showed further differences in frontal, parietal, and occipital regions. In the parietal area the theta-band activity showed a homogeneous single peak, whereas in occipital recording a shoulder at 4 Hz and in frontal recording a similar side peak at 5 Hz were identified (Fig. 19.4, solid lines).

- The auditory EPs showed a peak at 7 Hz in vertex with a shoulder in the alpha range (10 Hz), whereas in the parietal region 7 Hz as well as 10 Hz activities were at the same level of magnitude with a side peak occurring at 4 Hz. Frontal area showed a concave shape in the 3–6 Hz band (Fig. 19.5, solid lines). Differences in the frequency contents of the EPs obtained from primary sensory areas and nonprimary areas of the brain in these sensory modalities are described in Chap. 1.
- Differences between the grand averages and between the AFCs of evoked and 3rd attended responses indicated increases in the amplitudes of the N100-P200 complex and N140-P200 complex in responses to 3rd attended stimuli compared with the standard EPs in auditory and visual modalities, respectively. However, the wave shapes were quite similar, and any compo-

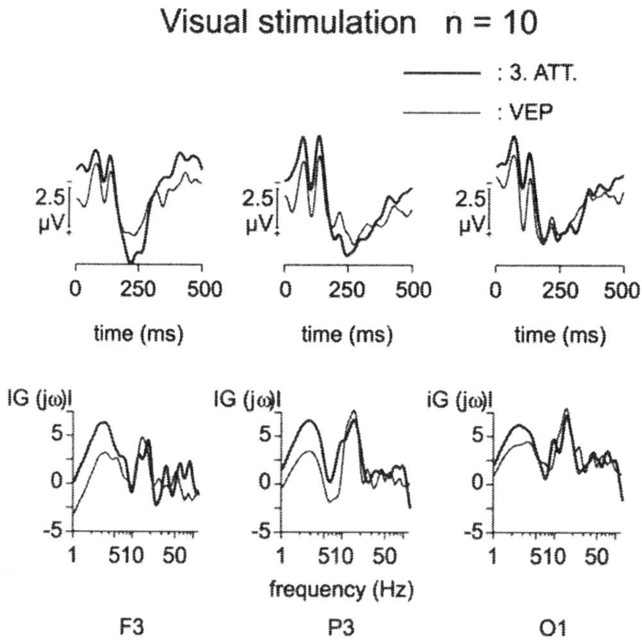

Fig. 19.4. Time-domain grand averages ($n = 10$) of standard VEPs (VEP) and responses to 3rd attended light stimuli in the visual omitted stimulus paradigm (3.ATT) obtained from F3, P3, and O1 leads (*upper part*) and corresponding AFCs (*lower part*). (Modified from Demiralp T. and Başar E. (1992) *International Journal of Psychophysiology* 13: 147–160.)

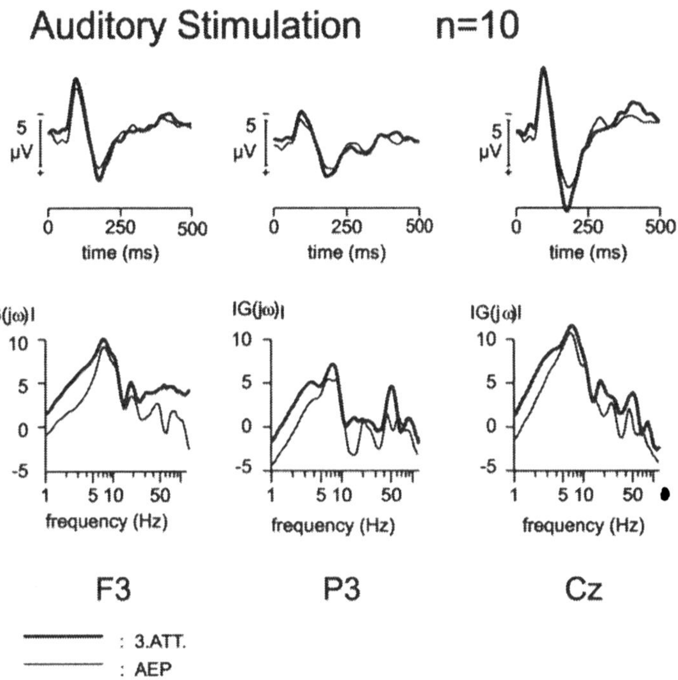

Fig. 19.5. Time-domain grand averages ($n = 10$) of standard AEPs (AEP) and of responses to 3rd attended tones in the auditory omitted stimulus paradigm (3.ATT) obtained from F3, P3, and Cz leads (*upper part*) and corresponding AFCs (*lower part*). (Modified from Demiralp T. and Başar E. (1992) *International Journal of Psychophysiology* 13: 147–160.)

nents that differed temporally from the peaks and valleys of the standard EPs were not detected. In AFCs an increase in the theta frequency band between 3 and 6 Hz was observed. However, alpha response amplitude (6–13 Hz to acoustical stimulation and 8–15 Hz to visual stimulation) kept the same magnitude or decreased in all recording locations.

19.3.2 Adaptive Filtering of Respectively Applied Evoked Responses

The averaged responses were filtered in various frequency bands by means of digital filters with zero phase-shift. The filter band limits were selected according to the maxima in AFCs. Figure 19.6 shows the visual EPs and the responses to the 3rd attended light stimuli (3.ATT) in the omitted stimulus paradigm (superimposed) and the grand averages obtained in both conditions filtered in the theta frequency band (3–6 Hz).

Figure 19.7 shows the analogous results in auditory modality. In the case of the experiments where subjects increased their attention to the third stim-

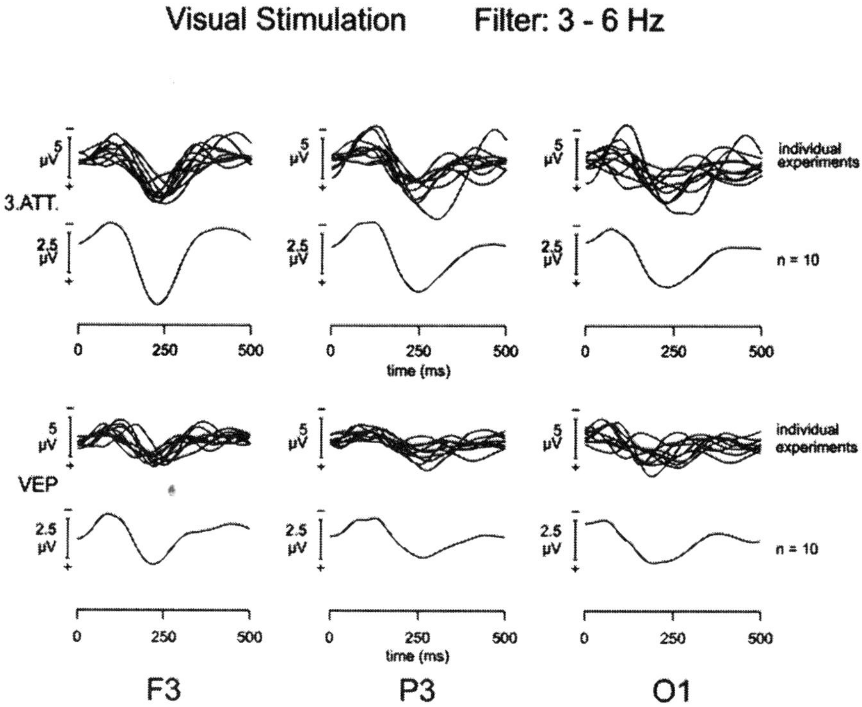

Fig. 19.6. Superimposed standard VEPs (VEP) and responses to the 3rd attended light stimuli in the visual omitted stimulus paradigm (3.ATT) of 10 subjects and their grand averages filtered in theta frequency band (3–6 Hz). (Modified from Demiralp T. and Başar E. (1992) *International Journal of Psychophysiology* 13: 147–160.)

uli, a consistent increase was observed in the amplitude of theta oscillation, especially in the frontal (F3) and parietal (P3) areas. For statistical evaluation of these results, the maximum amplitudes of the oscillations of a certain frequency band in the first 250 ms of the evoked responses were measured, and the differences between the values obtained in standard EPs and responses to the 3rd attended stimuli were tested via the Wilcoxon test for paired comparisons. The median values and 95% confidence intervals of amplitudes of theta, alpha, and beta components of EPs and the responses to 3rd attended stimuli obtained in all 4 recording sites are displayed in histograms in Fig. 19.8 and Fig. 19.9. The medians of the amplitudes of these frequency components, the percentage change between both conditions as a percentage of the standard EP amplitude and corresponding significance levels, are given in Tables 19.1 and 19.2 for visual and auditory modalities, respectively. The percentage values are based on the median values.

The amplitudes of theta oscillations in responses to the 3rd attended stimuli in visual modality showed an increase of 48% with respect to the standard

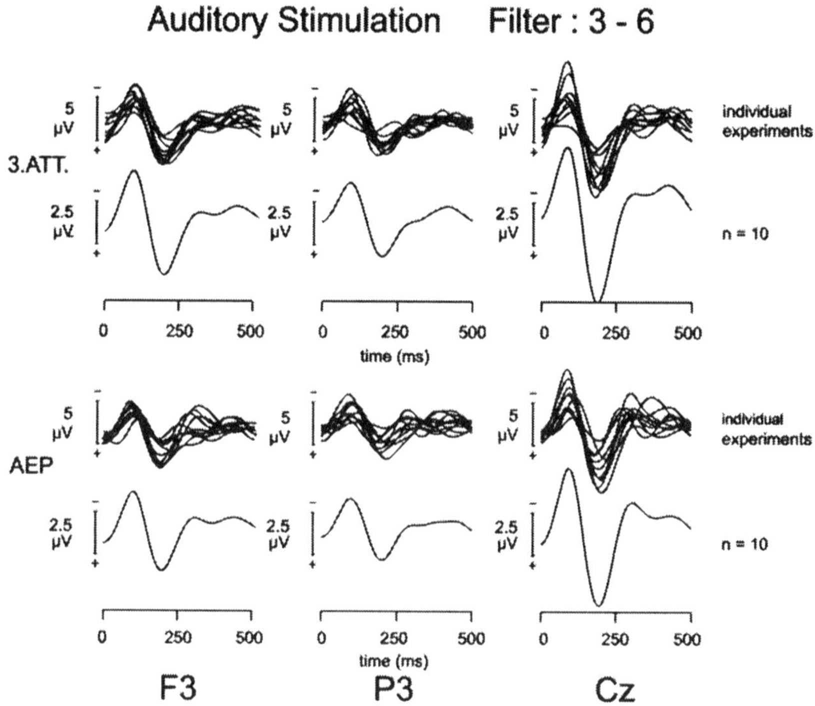

Fig. 19.7. Superimposed standard AEPs (AEP) and responses to the 3rd attended tones in the auditory omitted stimulus paradigm (3.ATT) of 10 subjects and their grand averages filtered in theta frequency band (3–6 Hz). Alpha-and beta-band responses showed no consistent change. Therefore, these waveforms are not illustrated, but these frequency components are included in the following histograms and tables. (Modified from Demiralp T. and Başar E. (1992) *International Journal of Psychophysiology* 13: 147–160.)

VEP in the frontal region ($p < 0.01$) and 45% with that in the parietal region ($p < 0.05$). In vertex and occipital areas there were also significant increases of theta components, but they were less ample in comparison to frontal and parietal responses (Cz: 26%, $p < 0.05$ and O1: 11%, $p < 0.05$). The increase in alpha- and beta-band amplitudes were insignificant for recording sites.

In auditory modality there were also significant increases of theta components in frontal and parietal regions (F3: 44%, $p < 0.01$ and P3: 10%, $p < 0.05$). The parietal theta increase was, however, less in comparison to responses in visual modality. In vertex and occipital regions no significant changes in theta components were observed between AEPs and responses to 3rd attended tones. Likewise, the changes in alpha and beta components were not found to be significant for any of the recording sites.

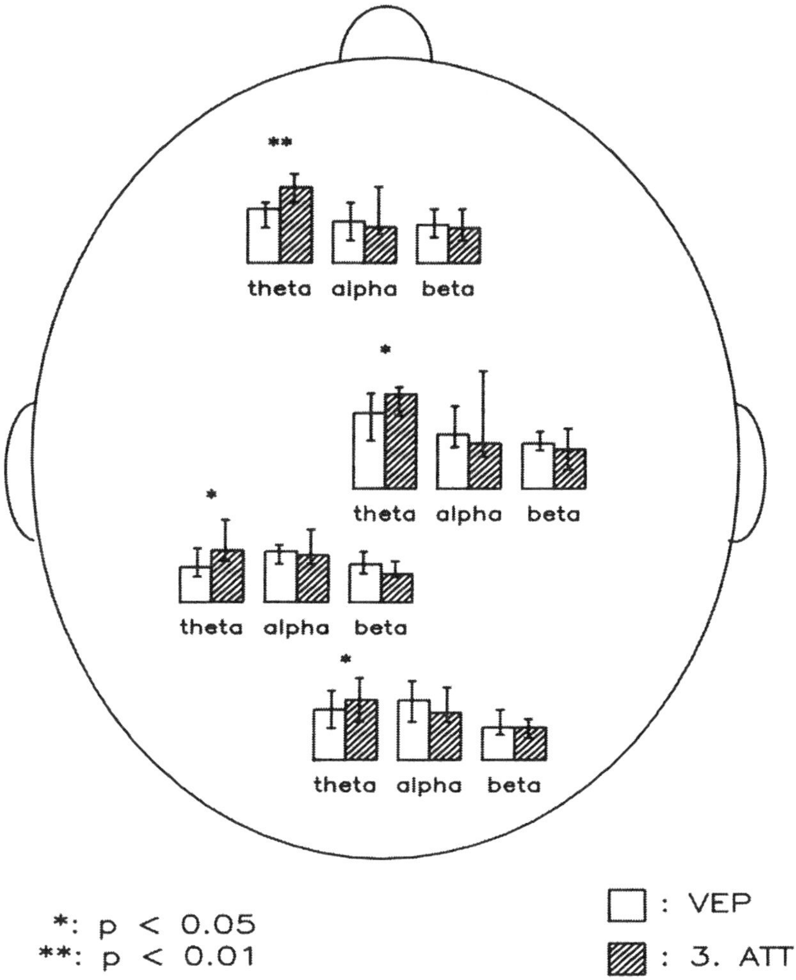

Fig. 19.8. Medians and 95% confidence intervals of the amplitudes of theta (3–6 Hz), alpha (8–15 Hz), and beta (15–30 Hz) components of standard VEPs (VEP) and responses elicited by 3rd attended light stimuli in visual omitted stimulus paradigm (3.ATT). (Modified from Demiralp T. and Başar E. (1992) *International Journal of Psychophysiology* 13: 147–160.)

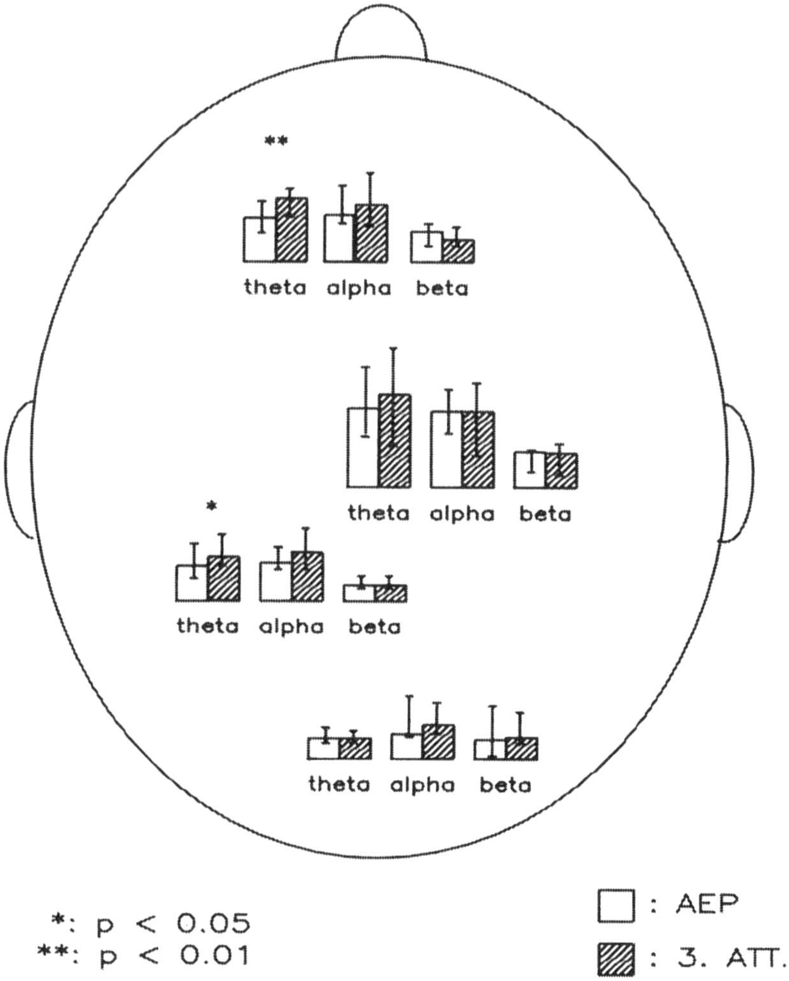

Fig. 19.9. Medians and 95% confidence intervals of the amplitudes of theta (3–6 Hz), alpha (6–13 Hz), and beta (15–30 Hz) components of standard AEPs (AEP) and responses elicited by 3rd attended tones in auditory omitted stimulus paradigm (3.ATT). (Modified from Demiralp T. and Başar E. (1992) *International Journal of Psychophysiology* 13: 147–160.)

Table 19.1. Medians of the amplitudes of theta (3–6 Hz), alpha (8–15 Hz) and beta (15–30 Hz) frequency components of standard VEPs (VEP) and responses elicited by the 3rd attended light stimuli in visual omitted stimulus paradigm (3.ATT); percent changes of amplitudes between 2 conditions (as the percent of the standard VEP amplitude) are given in the parentheses, and significant differences are marked by symbols (*: $p < 0.05$, **: $p < 0.01$). (From Demiralp T. and Başar E. (1992) *International Journal of Psychophysiology* 13: 147–160.)

		Theta (3–6 Hz)	*Alpha* (8–15 Hz)	*Beta* (15–30 Hz)
F3	VEP	4.6	3.5	3.2
	3.ATT	6.8	3.0	3.0
		(48%)**	(−13%)	(−7%)
C_z	VEP	6.7	4.8	3.9
	3.ATT	8.4	3.8	3.4
		(26%)*	(−19%)	(−12%)
P3	VEP	3.1	4.4	3.4
	3.ATT	4.5	4.2	2.5
		(45%)*	(−3%)	(−25%)
O1	VEP	4.5	5.2	2.9
	3.ATT	4.9	4.2	2.9
		(11%)*	(−20%)	(−0%)

Table 19.2. Medians of the amplitudes of theta (3–6 Hz), alpha (6–13 Hz), and beta (15–30 Hz) frequency band components of standard AEPs (AEP) and responses elicited by the 3rd attended tones in auditory omitted stimulus paradigm (3.ATT); percent changes of amplitudes between 2 conditions (as the percent of the standard AEP amplitude) are given in the parentheses, and significant differences are marked by symbols (*: $p < 0.05$, **: $p < 0.01$). (From Demiralp T, and Başar E. (1992) *International Journal of Psychophysiology* 13: 147–160.)

		Theta (3–6 Hz)	*Alpha* (6–13 Hz)	*Beta* (15–30 Hz)
F3	VEP	4.7	4.9	2.9
	3.ATT	6.7	6.1	2.4
		(44%)**	(26%)	(−17%)
C_z	VEP	8.2	7.9	3.6
	3.ATT	9.5	7.9	3.5
		(16%)	(0%)	(−3%)
P3	VEP	4.0	3.9	1.7
	3.ATT	4.4	4.9	1.6
		(10%)*	(26%)	(−5%)
O1	VEP	2.3	2.8	2.1
	3.ATT	2.3	3.5	2.2
		(0%)	(26%)	(5%)

19.4 Increase in Theta Components Is Highest in Frontal Recordings

The highest statistically significant theta increases during cognitive performance were obtained in frontal and parietal recording sites. In auditory modality, the theta increase was absolutely dominant in the frontal area (44% increase), whereas in visual modality the theta increase in the frontal recording site was only slightly higher than that in the parietal recording site (48% versus 45%). This selectivity is parallel to the results of Fuster (1991). Studies of Fuster that showed a high anticipatory activation level of frontal neurons in time-delay tasks are based on single-unit recordings in the prefrontal cortex of monkeys. Since the cognitive task in our study was also mainly based on anticipation to an expected stimulus, it is not surprising that the greatest changes are in frontal regions.

Our findings showing the strong participation of the frontal cortex in fulfilling a cognitive task are also in accordance with the results of Knight et al. (1981) on patients with frontal cortex lesions. Results of these studies indicated that the frontal lobes exhibited a modulating influence upon the endogenous negativity of ERPs produced in selective attention tasks.

19.5 In Visual Modality the Secondary Dominant Theta Increase Occurs in the Parietal Recordings

The parietal area was the secondary dominant theta center in visual modality with a percentage theta increase slightly below that obtained in the frontal area (45%), whereas in auditory modality the increase in parietal theta activity was not so prominent (10%). This property of visual ERP is in accordance with the specific functions of parietal cortex in visual information processing, as shown by means of cellular measurements on macaque monkeys (Mountcastle et al. 1975; Lynch et al. 1977). According to Mountcastle, 50% of the investigated neurons in the parietal association area 7 of the inferior parietal lobe are visual fixation cells that are active as the animal looks at visual targets that are linked to a strong drive; the rest of the neuronal elements are light-sensitive with large and bilateral receptive fields (Yin and Mountcastle 1977; Robinson et al. 1978). Lynch et al. (1977) suggest that the neurons in posterior parietal cortex are involved in selective visual attention processes. Mountcastle et al. (1981, 1984) showed that the enhanced responsiveness of light-sensitive neurons in the inferior parietal lobe does not merely occur with changes in general arousal but is more specifically related to the visual attention directed at the target light. Petersen et al. (1988) showed by means of positron emission tomography similar effects in the parietal cortex of normal humans (see also Chap. 14). Again this discussion supports the view that the distributed theta system shows a selectivity with stronger responses in frontal and parietal areas.

19.6 The Cognitive Theta Components of ERPs as a Sign of Hippocampocortical Interaction

Neuroanatomical studies show that the main thalamic input to the frontal cortex stems from the limbic system (Creutzfeld 1983). Hippocampus, which has been shown to play a key role in memory storage and learning (Zola-Morgan and Squire 1990), is an important part of the limbic system. The hippocampal theta rhythm is a well-established phenomenon (Elazar and Adey 1967; see also Chap. 25). These authors, by application of a learning paradigm, have recorded consistent and coherent patterns in the theta range in the hippocampus and subcortical structures of the cat. Başar-Eroğlu et al. (1991a,b) showed theta responses in intracranial recordings obtained from awake and freely moving cats in various layers of hippocampus. The theta responses were temporally correlated with the P300 wave of the surface-recorded ERP.

In light of these results, our finding of an increased theta response in the surface recorded ERPs of frontal recording sites during the time-prediction task might be a manifestation of a cooperative, associative activity of frontolimbic structures. These results will be further discussed in Chap. 21.

19.7 Concluding Remarks

Results of this chapter indicate once more the importance of the frequency-analysis approach in identifying various psychophysiological components of ERPs. Similar waves with similar latencies and amplitudes in the time-domain analysis can show differences and can be identified in frequency-domain presentation of the signal. Furthermore, such an analysis of surface-recorded potentials in well-designed human psychological experiments can be combined with those of subcortical recordings on animals showing characteristic rhythmic activity of deeper structures. This approach can lead to a better neurophysiological understanding of the surface-recorded potentials obtained during various cognitive performances (compare also Chap. 18 in Vol. I for a justification of component analysis in the frequency domain).

20. Topological Distribution of Oddball "P300" Responses

Erol Başar, Tamer Demiralp and Canan Başar-Eroğlu

20.1 Experimental Paradigm

A preliminary presentation of results on the dynamics of the oddball response were described in Chap. 15. As auditory stimuli, 80 dB, 1500 and/or 1600 Hz tones with 0.5 ms rise time and 800 ms duration were presented binaurally. In each recording session, firstly, the spontaneous EEG was registered for a few minutes to determine global characteristics of subjects' spontaneous EEG activity and arousal state at the beginning of the experiments. This period also helped the subjects to become familiar with experimental conditions. Thereafter, on the first group of subjects the auditory EPs, omitted stimulus paradigm and oddball paradigm were applied with short resting periods in between. The AEP experiments consisted of the presentation of 1500 Hz tones with interstimulus intervals (ISI) randomly varying between 2.5 and 4 s with a mean value of 3 s.

In the oddball paradigm the tones were presented in a pseudorandom sequence with 1600 Hz tones occurring 20% of the time and 1500 Hz tones occurring 80% of the time. The interval between tones varied randomly between 2.5 and 4 s with a mean value of 3 s as in the AEP experiments (Fig. 20.1). The subjects were instructed to keep a mental count of the number of 1600 Hz tones (infrequent target tones). With every stimulus presented a segment of EEG activity preceding and the EP or ERP following the stimulus were digitized and stored on computer disk memory. This operation was repeated approximately 100–200 times.

20.2 Topological Differences Between AFCs of AEPs

The comparison of AFCs from auditory EPs elicited in different recording sites revealed differences in the frequency contents (Fig. 20.2). Auditory EPs recorded from Cz showed a peak at 7 Hz with a shoulder at 10 Hz, whereas in the parietal region these frequencies were at the same level with an additional side peak occurring at 4 Hz. Frontal response (F3) showed characteristics similar to Cz in the alpha range, though it had a more concave form in the subalpha band. In the occipital area (O1) a residue of ongoing alpha activity and a smooth theta peak were detectable.

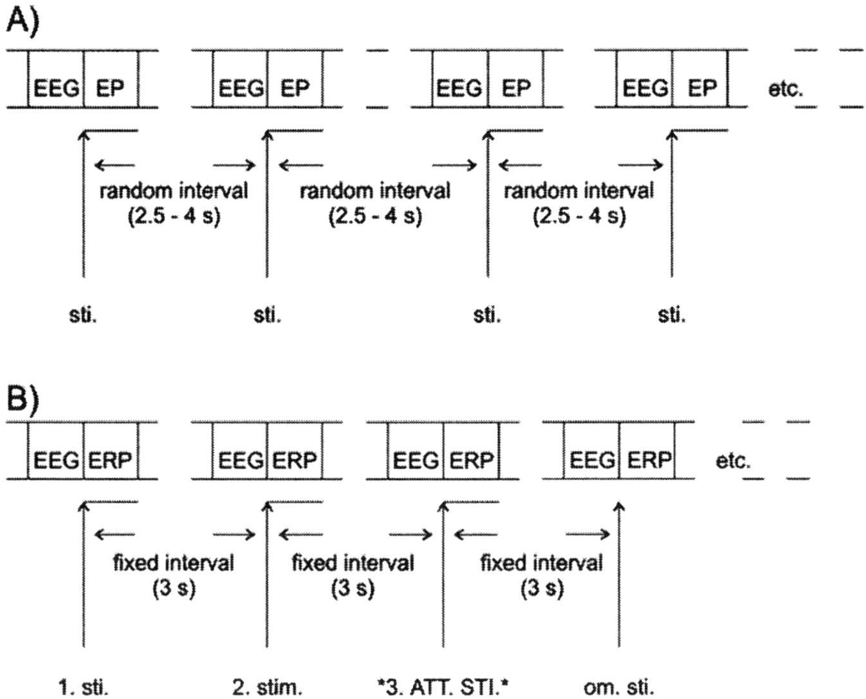

Fig. 20.1. Schematic illustration of (**a**) the oddball and (**b**) the omitted stimulus paradigms. (Modified from Başar-Eroğlu C. et al. (1992) *International Journal of Psychophysiology* 13: 161–179.)

20.3 Differences Between Time-Domain Grand Averages and AFCs of Responses in the Three Paradigms

The responses to the 3rd attended stimuli recorded from F3, Cz, and P3 locations showed marked increases in the amplitudes of N100-P200 complexes compared with the standard AEPs (Chap. 19). These increases in the amplitudes were accompanied by increases of theta-band (3–6 Hz) amplitudes in AFCs in the frequency domain. In the AFC of the occipital recording there was an increase in the theta band accompanied by a decrease in the alpha peak, though no evident response could be detected in the time domain in this location.

In the oddball experiments the target responses showed the characteristic late P300 complexes in all recording sites including the occipital area, where the earlier components were not clearly identifiable. The P300 waves in the time domain were accompanied by additional prominent delta peaks with a center frequency of 2 Hz in the frequency domain. A similar change occurred also in the AFC of the occipital area.

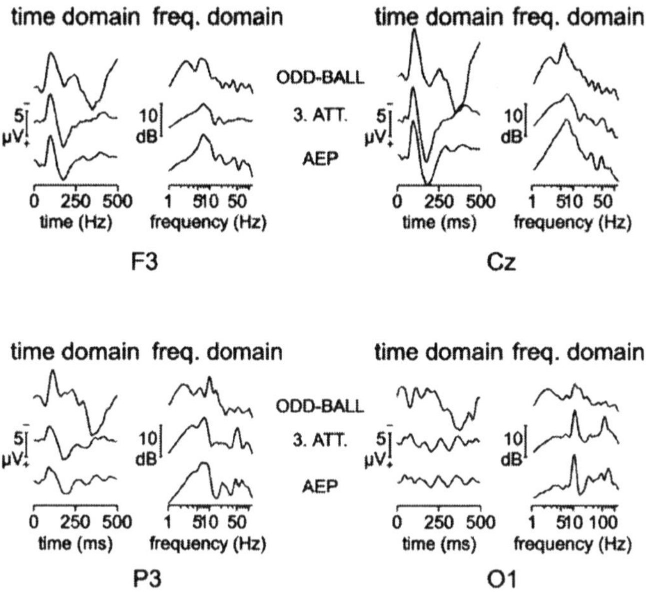

Fig. 20.2. Time-domain and frequency-domain representations of grand averages of auditory evoked potentials, responses to 3rd attended tones in the omitted stimulus paradigm (3.ATT) and responses to infrequent target tones in the oddball paradigm (ODDBALL) obtained in frontal, vertex, parietal, and occipital (F3, Cz, P3, O1) recording sites. (Modified from Başar-Eroğlu C. et al. (1992) *International Journal of Psychophysiology* 13: 161–179.)

A comparison of frequency-domain representations of responses obtained in all 4 recording sites to standard auditory EP, 3rd attended stimulus, and to oddball tones revealed a progressive increase of amplitudes of subalpha frequency components and a progressive decrease of dominant center frequency of the activity in this frequency band (Fig. 20.2).

20.4 Adaptive Filtering of the Responses

For further analysis the signals were filtered using digital filters, which caused no phase shift. The band-pass limits of the filters were selected according to the peaks in AFCs. We differentiated in the subalpha frequency range two frequency bands that were shared commonly by all conditions with some changes in center frequencies. They consisted of a delta band between 1 and 3 Hz and a theta band between 3 and 6 Hz. Figure 20.3 shows the frequency components of grand averages obtained in all 3 experiments, filtered in these frequency bands.

For statistical testing of the changes of these frequency components under different conditions of the cognitive tasks, individual responses of subjects

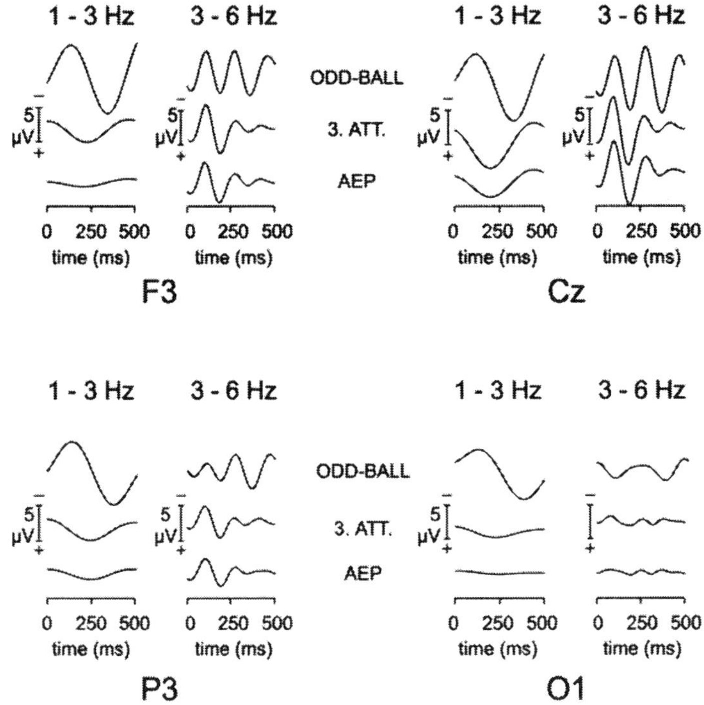

Fig. 20.3. Delta and theta frequency components of grand averages of auditory evoked potentials (AEP), responses to 3rd attended tones in the omitted stimulus paradigm (3.ATT) and responses to nonfrequent target tones in the oddball paradigm (ODDBALL) obtained in frontal, vertex, parietal, and occipital (F3, Cz, P3, O1) recording sites. (Modified from Başar-Eroğlu C. et al. (1992) *International Journal of Psychophysiology* 13: 161–179.)

were filtered with corresponding band-pass filters. The maximum peak-to-peak amplitudes of filtered frequency components in relevant time windows were measured and tested for significance of differences among the three experimental paradigms by means of the Wilcoxon–Wilcox test:

- For the delta frequency band a single time window between 0 and 500 ms was used, which is approximately equal to the period of a single delta oscillation in the frequency band 1–3 Hz (center frequency at 2 Hz).
- For theta oscillations two different time windows were used to make it possible to identify the prolonged oscillatory activity with smaller damping factors and/or delayed enhancements. The first time window covered the N100-P200 complex of auditory evoked responses (0–250 ms), and the second time window included the P300 complex (250–500 ms). (In Chap. 15 we have reported *prolonged* and *enhanced theta oscillations* in P300 responses.)

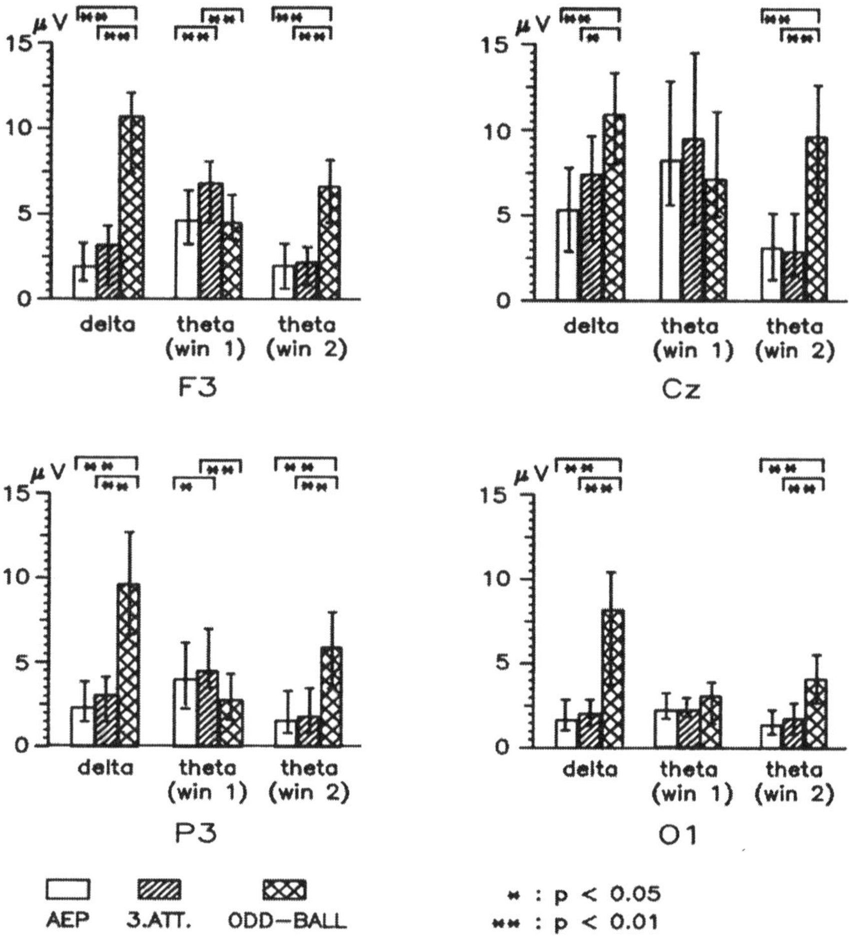

Fig. 20.4. The medians and 95% confidence intervals of maximum amplitudes of delta and theta frequency components of auditory evoked potentials (AEP), responses to 3rd attended stimuli in the omitted stimulus paradigm (3.ATT) and responses to infrequent target tones in the oddball paradigm (ODDBALL) obtained in frontal, vertex, parietal and occipital (F3, Cz, P3, O1) recording sites. The statistically significant differences are marked with symbols representing the significance levels. (Modified from Başar-Eroğlu C. et al. (1992) *International Journal of Psychophysiology* 13: 161–179.)

The medians of the amplitudes of delta, theta, and alpha components and 95% confidence intervals obtained in the 3 paradigms are shown in Fig. 20.4. The significant differences are marked with symbols representing the significance levels (see also Table 20.1).

Table 20.1. The medians of maximum amplitudes of delta, theta, and alpha frequency components of auditory evoked potentials (AEP), responses to 3rd attended stimuli in the omitted stimulus paradigm (3.ATT) and responses to infrequent target tones in the oddball paradigm (ODDBALL) obtained in frontal (F3), vertex (Cz), parietal (P3), and occipital (O1) recording sites. The percent changes of amplitudes in (3.ATT) and (ODDBALL) conditions as the percent of the standard AEP amplitudes are given in parentheses. Statistically significant differences are marked with symbols representing the significance levels (*: $p < 0.05$; **: $p < 0.01$). (From Başar-Eroğlu C. et al. (1992) *International Journal of Psychophysiology* 13: 161–179.)

		Delta (1–3 Hz)	Theta (3–6 Hz)	
			window 1 (0–250 ms)	window 2 (250–300 ms)
	AEP	1.8	4.7	2.0
F3	3.ATT	3.0 (66%)	6.7 (43%)**	2.0 (0%)
	P300	10.6 (489%)**	4.5 (−4%)	6.5 (225%)**
	AEP	5.3	8.2	3.0
Cz	3.ATT	7.3 (37%)	9.5 (16%)	2.9 (−3%)
	P300	10.9 (106%)**	7.2 (−12%)	9.7 (223%)**
	AEP	2.4	4.0	1.6
P3	3.ATT	3.2 (33%)	4.4 (10%)*	1.8 (13%)
	P300	9.7 (304%)**	2.7 (−33%)	5.8 (263%)**
	AEP	1.6	2.3	1.3
O1	3.ATT	1.9 (19%)	2.3 (0%)	1.5 (15%)
	P300	82. (413%)**	2.8 (22%)	3.9 (200%)**

The responses to the 3rd attended stimuli in the omitted stimulus paradigm depicted in all recording sites insignificant increases in the delta frequency band compared with the standard AEPs. In the oddball paradigm, there were further increases in the amplitudes of the delta components, which were statistically significant compared with the responses to 3rd attended stimuli ($p < 0.01$ at F3, P3, O1, and $p < 0.05$ at Cz) and with the AEPs ($p < 0.01$ at all locations).

In response to 3rd attended stimuli in the omitted stimulus paradigm statistically significant increases of theta oscillations occurred in the early part of the ERPs (window 1: 0–250 ms) in frontal (F3) and parietal (P3) locations compared with the standard AEPs ($p < 0.01$ and $p < 0.05$, respectively) and compared with the oddball responses ($p < 0.01$ in both locations). In the late part of ERPs no significant theta change occurred in responses to the 3rd attended stimuli. Furthermore, in oddball responses significant theta increases were registered only in the latter part of the responses (window 2: 250–500 ms). The theta changes in oddball responses were more widely distributed in comparison to those in responses to 3rd attended tones, which were localized in frontal and parietal regions. The theta increases could be observed in all 4 recording sites and were statistically significant in comparison to both the AEP and 3rd attended tone responses ($p < 0.01$ at all locations; see Table 20.1).

20.5 The "Selectively Distributed Theta-Response System in the Brain" and the Corticohippocampal Interaction

According to a number of neurophysiological, psychological, and biophysical approaches, the oscillatory phenomena in neural tissues merit nowadays considerable attention for an understanding of cognitive functions of the CNS (Bullock 1992; Başar 1992; Galambos 1992; Petsche and Rappelsberger 1992; Pfurtscheller and Klimesch 1992; Gray et al. 1992; Tononi et al. 1992). In Chap. 18 we have described results from experiments with freely moving cats where a passive P300 paradigm was used (for general remarks about the functional correlates of P300-like ERPs – focused attention, signal detection, recognitron and decision making see Chap. 14.)

Miller (1991), after his strong survey of electrophysiological and functional analysis on the corticohippocampal interplay, concludes: "Despite the lack of evidence at present, the central prediction still made is that there are cortical nodes with consistent phase relations to hippocampal theta activity." On the basis of results obtained from experiments on the P300 response of the cat brain, the responsiveness in the theta frequency range is analyzed in an extended manner below. Tentative interpretations of the results presented in this volume recently led us to postulate the existence of a *selectively distributed alpha response system* and a *selectively distributed theta response system* in the brain.

20.6 Paradigms Used

20.6.1 Oddball Paradigm

During the application of the oddball paradigm the subjects have at least two different categories of tasks to perform:

- *Focused attention* and *signal detection*
- *Matching for target recognition and decision-making*

Since the target signals were presented in a random order with random interstimulus intervals, the subjects were not able to prepare themselves for signal detection. There is no anticipation of the coming stimulus. Therefore, the signal detection task refers to the period after the stimulus delivery. The task of recognition of the target stimuli amongst the nontarget stimuli involves matching and decision-making processes. Accordingly, the involved mental processes contain at least two different groups of components as described above. (The matching mechanisms have been extensively analyzed in the recent reviews by Näätänen (1988, 1990)).

20.6.2 Paradigm with Omitted Fourth Signal and to-Be-Attended 3rd Signal

In this simpler paradigm one of the above-mentioned task categories seems mainly to be in play, namely *focused attention and signal detection.* Since the stimulation is applied repetitively with constant interstimulus intervals, a preparation takes place prior to the target. Attention is focused on the third signal (see Chap. 19). The signal is anticipated. Therefore, at the time of signal application there is no *surprise* or *novelty.* Accordingly, the difficulty that might stem from "matching for target recognition" and "decision-making" is highly reduced or not at all involved in the procedure.

Common components in ERPs of both paradigms are focused attention and signal detection. Accordingly, we tentatively assume that the delta response, the most prominent component of the oddball-ERP, is mostly involved with the signal-matching and decision-making following a novel or unexpected signal and/or partial surprise. The early theta response (window 1), which is most prominent in frontal and parietal locations during the 3rd attended stimulation, is probably due to focused attention that refers to preferential detection of selected stimuli, and not to matching for signal-recognition or decision-making.

Significant changes in theta and delta responses in different locations (as analyzed in Chap. 19 and in the present chapter) and the possible functional correlates of the applied paradigms are summarized together in the illustrative Table 20.2. During the paradigm with 3rd attended stimulus, theta response increase was recorded only immediately following the stimulation. Moreover, such increases were recorded only in frontal and parietal locations. Seemingly, for this less complicated task a theta response increase to sound stimuli in frontoparietal location is representative for the brain mechanisms in play (compare also Chap. 23).

Considering the anticipatory (time estimation) component of the task, the frontal dominance of the theta increase is in accordance with the studies based on the cellular measurements in the prefrontal cortex (Fuster 1991). Fuster showed a high anticipatory activation level of frontal neurons in time delay-tasks.

Table 20.2. Significant changes of theta and delta responses in different recording sites under both categories of paradigms and the possible functional correlates of the applied paradigms (*: $p < 0.05$, **: $p < 0.01$). (From Demiralp T. and Başar E. (1992) *International Journal of Psychophysiology* 13: 147–160.)

Paradigms	*(1) Omitted stimulus (3rd attended signals)*								*(2) Oddball (target tones)*			
Description of the task	Subject focuses his attention to the 3rd signal presendeted repetitively: probability of target occurrence = 100%, high expectancy, no surprise								Subject focuses his attention to rare oddball tones presented randomly. Probability of target occurence = 20% low expectancy, surprise, decision-making			
Functional correlates	Association, focused attention, signal detection, expectation, anticipation								Association, focused attention, signal detection, matching for target recognition, decision-making, surprise			
	3rd Attended light				3rd Attended tone				Oddball (target) tone			
	F3	Vz	P3	O1	F3	Cz	P3	O1	F3	Cz	P3	O1
Increase of delta freq. response	—	—	—	—	—	—	—	—	**	**	**	**
Increase of early theta response	**	*	*	*	**	—	*	—	—	—	—	—
Increase of late theta response	—	—	—	—	—	—	—	—	**	**	**	**

When the task additionally involves signal-matching, decision-making, and surprise, the changes in the frequency channels of ERP reach a higher degree of complexity:

1. In all locations a marked change of the delta response is recorded.
2. Increases in late theta responses (window 2) were significantly also recorded in all locations.

The existence of important delta increase suggests that processes of decision-making and surprise are reflected in this slowest EEG-response component. Increase in the delta response is probably related only to decision-making and matching, whereas the theta response seems to be involved in several tasks (focused attention, signal-detection, anticipation, and expectation) appearing either as an early or late component.

More evidence for the participation of delta response during the procedure of decision-making was provided by the experiments at the auditory threshold level. The results of Chap. 12 have shown that at the hearing threshold EPs were reduced to an almost pure delta oscillation, also detectable without frequency analysis or filter application. At the hearing threshold subjects are supposed to be involved in decision-making. Accordingly, the results of Chap. 12 attributing a decision making function at the threshold level to a delta component are in good accord with these new findings.

21. Wavelet Analysis of Oddball P300

Tamer Demiralp, Ahmet Ademoglu, and Erol Başar

21.1 Results

The wavelet analysis presented in detail in Chap. 5 of Vol. I could not be extensively applied to all experiments described in this book. Since the application of wavelet analysis is very pertinent to a dynamic approach to the study of brain signals, we treat only some preliminary results that basically confirm the results that were obtained by application of the TRFC and digital filtering methods.

The application of a 5-octave wavelet transform on the data yielded 5 sets of coefficients in the 63–125 Hz, 31–62 Hz (gamma), 16–30 Hz (beta), 8–15 Hz (alpha), and 4–7 Hz (theta) frequency ranges and the residues in the 0.5–4 Hz frequency band. Each octave contained the number of coefficients that is appropriate to display the relevant time resolution for that frequency range (Fig. 21.1, left column). The interpolation of the coefficients by the spline functions can be used to visualize the time course of each frequency band (Fig. 21.1, right column).

In order to choose the dominant feature(s) correlating with the P300 wave, the conventional task-based selection criterion for grouping the responses has been applied. Furthermore, the grand averages of the task-relevant (target) and task-irrelevant (nontarget) responses of all subjects have been computed (Fig. 21.2). The grand average of the nontarget responses contained also a P300 wave of smaller amplitude compared with the grand average of the target ones. By applying the wavelet transform (WT) to the grand average of the target and nontarget responses, the coefficients and interpolated responses corresponding to the high gamma, gamma, beta, alpha, theta, and delta ranges of the EEG have been obtained. An increased amplitude of the delta response and increased amplitudes and prolongation of the theta and alpha responses were observed in the target responses compared with the nontarget ones. This finding was in accordance with the previous results (Başar-Eroğlu et al. 1992; Stampfer and Başar 1985; Başar and Stampfer 1985; Kolev and Schümann 1992; cf. Chaps. 15 and 20).

However, the most pronounced feature of the P300 wave was the increased positivity of the 4th delta (0.5–4 Hz) coefficient after the stimulation time point, which corresponds to the time range between 300 and 430 ms. Fig-

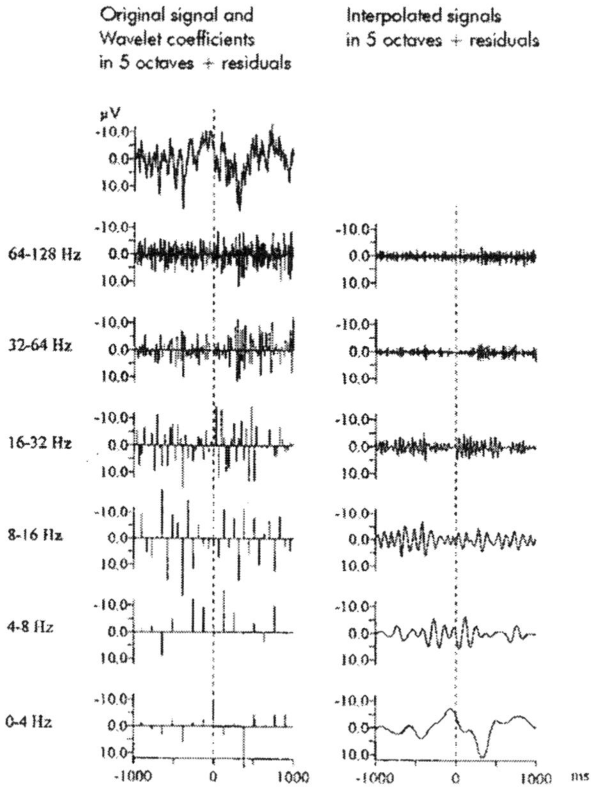

Fig. 21.1. The wavelet coefficients in 5 octaves and the residuals of a sample P300-sweep (*left column*) and the interpolated signals (*right column; dashed line*: stimulus).

ure 21.3 shows delta-band coefficients of a sample set of sweeps. It can be seen that there are some sweeps that display pronounced positive 4th delta coefficients (D4+) and some that have 4th delta coefficients of an amplitude less then equal to 0 (D4−). The *response-based selection* of the trials containing the P300 wave was carried out according to this feature. The sweeps containing positive 4th delta (D4+) coefficients were chosen as P300-sweeps.

Figure 21.4 shows the grand averages of the P300 and non-P300 responses that were chosen according to the *response-based* and *task-based* selection criteria and their delta components. It can be observed that the response-based selection of the P300 responses according to the positivity of the 4th delta coefficient (D4+) shows a P300-wave of higher amplitude compared with the target responses chosen according to the task-based selection criterion. Accordingly, the non-P300 responses obtained by the response-based classification (D4−) show smaller positive amplitudes in the P300 latency range compared with the nontarget responses.

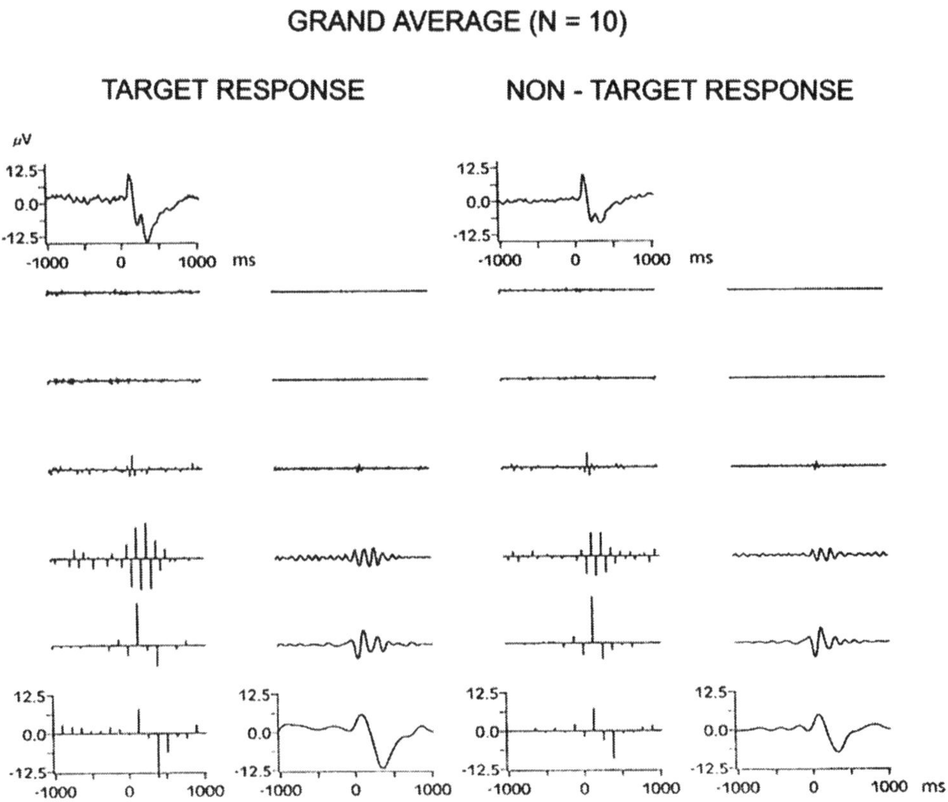

Fig. 21.2. The 5-octave spline wavelet decomposition of the grand averages of the target (*left*) and nontarget responses (*right*) of all 10 subjects. In each part the *first column* displays the original signal and the coefficients obtained in the high gamma, gamma, beta, alpha, theta, and delta bands of the EEG, and the *second column* the interpolated signals in each octave.

Figure 21.5 illustrates the averages of the nontarget responses of each subject that contain a positive 4th delta coefficient. It can be clearly seen that under this condition significant P300 waves can be obtained even if the stimuli were defined as nontarget.

Figure 21.6 shows the mean numbers of sweeps in the target, nontarget, D4+, and D4− groups after artifact elimination. The mean count of the artifact-free sweeps in the experiments was 80.9 (±8.81). With task-based classification, the mean count of the target responses without artifacts was 22.0 (±1.41), and the mean number of the nontarget responses without artifacts was 58.9 (±8.13). However, the response-based classification of the sweeps according to the occurrence of a positive delta coefficient in the P300 range (D4+) yielded a higher number of responses containing the P300 component. The mean count of the D4+ sweeps was 35.4 (±13.39) and that of

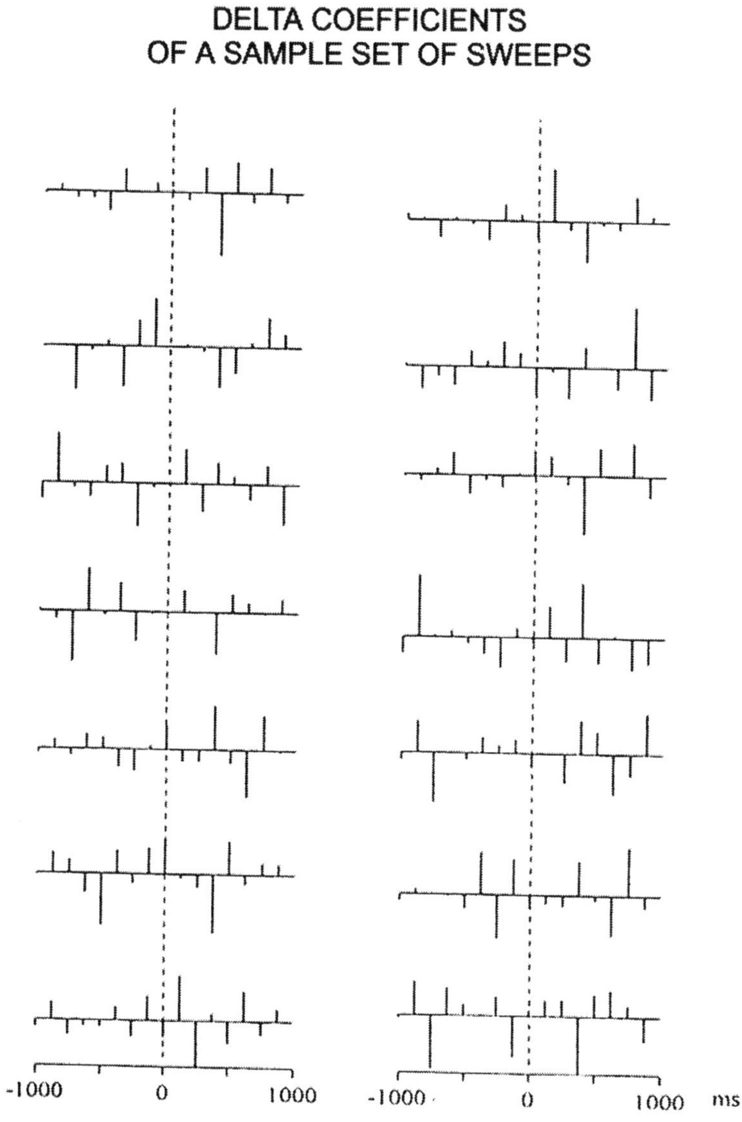

Fig. 21.3. The delta coefficients of a sample set of sweeps. The 4th coefficient after stimulation (*dashed line*) shows marked positivity in some of the sweeps, whereas it is ≤ 0 in others.

D4– sweeps was 45.5 (±9.49). In the mean, 72% of target trials were overlapping with the D4+ sweeps, whereas 28% of target trials have been identified as D4–. On the other side, only 51% of nontarget trials belonged to the group D4– whereas 49% of the nontarget trials were classified as responses containing a P300 wave.

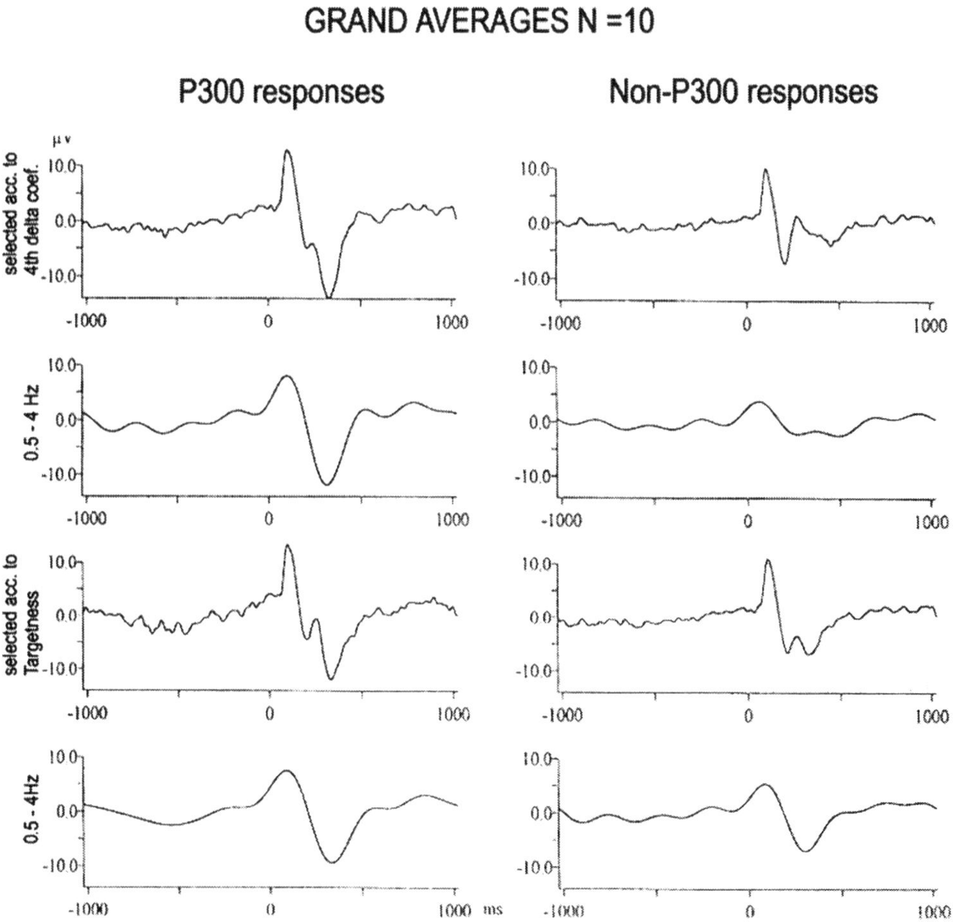

Fig. 21.4. The averages of sweeps classified as P300 responses and non-P300 responses according to the sign of their 4th delta coefficient in the wavelet decomposition (*1st row*), band-pass filtered in the delta frequency range (0.5–4 Hz) (*2nd row*); the averages of target and nontarget sweeps (*3rd row*), band-pass filtered in the delta frequency range (*4th row*).

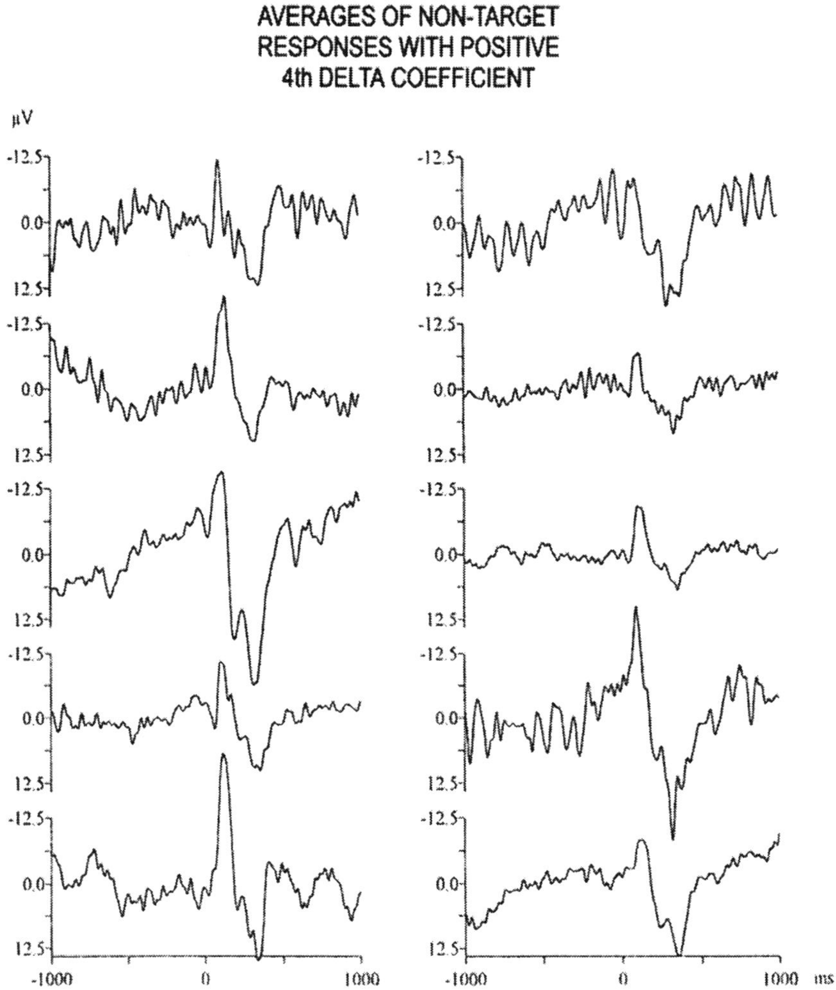

Fig. 21.5. Averages of the nontarget responses with positive 4th delta coefficients of all 10 subjects showed prominent P300 waves.

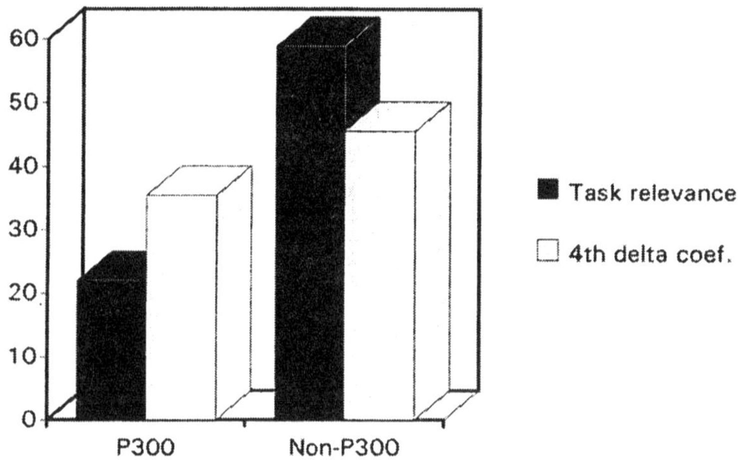

Fig. 21.6. Mean number of the sweeps belonging to the Target, Nontarget, *D*4+ and *D*4− groups.

21.2 The P300 Wave Can Be Detected in the Single-Trial ERPs by the Spline Wavelet Coefficients in the Delta Frequency Range

The sign of the 4th delta coefficient representing the delta activity in the 310–430 ms period after stimulation was used as the main feature representing the presence or absence of a P300 wave in the single trials. By averaging the sweeps with a positive delta coefficient in this time range, a clear P300 wave could be obtained, which was absent in the average of the rest of the sweeps. This result verifies our previous results on the relationship between the delta frequency component of the ERP and the P300 wave (Demiralp and Başar 1992; Başar-Eroğlu et al. 1992).

Our results show that 28% of the target sweeps did not contain a P300 component according to our classification criterion, whereas almost half of the nontarget responses contained this component, although the decision of the subjects on the "targetness" of the stimuli measured by the button-press after the target tones was 100% correct. In order to test the reliability of this classification, we averaged the target, nontarget, D4+ and D4− sweeps respectively and compared the amplitudes in the P300 latency range between target and D4+ sweeps and the nontarget and D4− sweeps (Fig. 21.7) by a paired nonparametric test (Wilcoxon–Wilcox). The statistical results show that there is an insignificant *increase* in the amplitude of the positivity in the P300 latency range in the average of the D4+ sweeps compared with the average of the target responses, whereas a significant *decrease* in the positivity is obtained in the averages of the D4− sweeps compared with the nontarget responses (Table 21.1). It shows that the method could determine

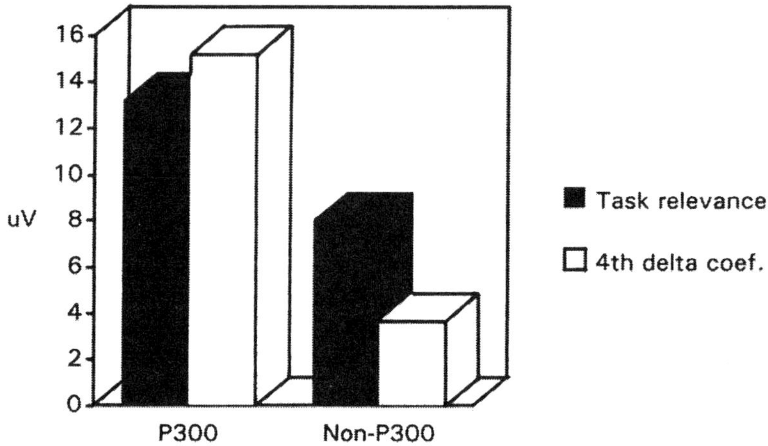

Fig. 21.7. The means of the amplitudes of the Target, Nontarget, $D4+$, and $D4-$ responses in the P300 latency range (250–400 ms) against the baseline of the average of the 1000 ms prestimulus period.

Table 21.1. The means and standard deviations of the amplitudes in the P300 latency range (250–400 ms) against the baseline of the average of the 1000 ms prestimulus period.

Target	13.20 ± 7.98	Nontarget	8.03 ± 4.89
D4+	15.18 ± 6.54	D4−	3.67 ± 3.62
p	N.S.	p	< 0.01

the nontarget sweeps containing a P300 component and eliminate the target sweeps without a significant P300 component.

21.3 The Response-Based Classification of the ERP Trials Yields Enhanced P300 Amplitudes Compared with the Average of the Target Responses

Furthermore, response-based averaging leads to an enhancement of the P300 amplitude compared with task-based averaging: 28% of the responses belonging to the target class are absent in the set of the P300 containing sweeps chosen with response-based classification. On the other hand, half of the nontarget responses are classified as P300 containing sweeps according to the response-based classification criterion. This result is in accordance with the findings of Haig et al. (1995). The authors used a method based on globally optimal vector quantization by simulated annealing for objective classification of single-trial target response subtypes of an oddball paradigm and found that only 40% of single trials had a morphology that resembled the averaged

ERP waveform. The remaining single trials had a response morphology that was different from the average. In the present chapter, we did not try to classify the subtypes of target responses, but we clustered objectively the P300 and non-P300 trials in the whole oddball recording including the target and nontarget responses. In a recent study (Schümann et al. 1995, see Chaps. 15 and 18 in Vol. I), a similar selection of the P300 sweeps has been carried out using the enhancement of the delta activity after the stimulation time-point. The average of the sweeps with enhanced poststimulus delta activity revealed P300 waves of comparable amplitude and shape as the average of the target responses.

21.4 A Functional Interpretation

Polich and Kok (1995) emphasize that although the explanations of the P300 center on the basic information-processing mechanisms of attention allocation and immediate memory, a substantial portion of P300 variation appears to be caused by factors related not only to alterations of the task structure but also to fluctuations in the arousal state of the subject. The studied factors affecting the biological state are mostly of tonic nature such as the usage of certain drugs, sleep deprivation, circadian or ultradian rhythms, or exercise (for a review see Polich and Kok 1995). However, phasic changes of the brain state during the application of the paradigm, which might be also of spontaneous nature, have rarely been studied.

The ERP analysis is based mostly on the time domain of the ERPs, where only the amplitudes and latencies of some prominent peaks and valleys are taken into consideration. Similarly, in frequency-domain representation by means of conventional power spectra, the time relations between components are lost.

In general, the time-domain and frequency-domain representations of the ERPs have given complementary information on the functional component structure. Therefore, efficient algorithms for analyzing the signal in the time–frequency plane are extremely important in extracting and relating distinct functional components. Wavelet analysis, "particularly the multiresolution representation, respects the overlapping component composition of ERPs, providing a natural way to partition them among several orthogonal functions with parallel time courses at different time scales" (Samar et al. 1995) and enables one to capture the time-dependent frequency-related information in ERPs.

21.5 The Number of Sweeps Containing a P300 Wave May Be Used as an Additional Measure in ERP Analysis

The current results might lead to another measure in the analysis of the oddball responses in the future: The number of P300 signals in different groups of subjects with different cognitive behavior, different ages, or different pathologies could be compared. Ford et al. (1994) using a P3 templating procedure showed that the diminished P3 amplitude seen in schizophrenia is partly due to a decreased number of single trials containing the P300 wave. The objective selection of the trials containing the P300 morphology among the whole data set of an oddball recording using the wavelet decomposition method could serve as a technique for establishing such a new measure in the clinical analysis of the ERP.

A number of studies have shown significant oscillatory events related to brain functions. These oscillatory components shown by frequency-domain and band-pass filtering techniques are rather band-limited signals than pure or nearly sinusoidal oscillations. The typical time courses of such oscillations show increases and decays in the amplitude in relatively short periods of time that also contribute to the bandwidth of the oscillatory signal components. By considering these properties of the oscillations contained in the ERPs and the nonoscillatory components with short lives, the attempt to obtain a perfect resolution in frequency-domain representation of the signal loses importance. Instead, the optimization of the signal representation in the time–frequency plane at the cost of losing the precise information on the frequencies gains relevance.

22. Dynamic Memory Manifested by Induced Alpha

22.1 Why Look for Internal Event-Related Oscillations?

As we have seen in the previous chapters, various neural populations in the brain can generate coherent states in which oscillatory 10 Hz activity as well as theta activity is recorded. A light flash can elicit a 10 Hz enhancement in the brain if the brain shows a disordered activity prior to stimulation. We also can evoke a 40 Hz response with sharp onset light or acoustical stimulation and so on. At this point an important question is; Can we find a way to put the brain in such coherent states of EEG activity *without* external sensory stimulation? Can we find a sensory–cognitive task to produce *coherent internal evoked potentials*, or better, *internal event-related oscillations?*

22.2 Coherent and Ordered States of EEG due to Cognitive Tasks

22.2.1 Preliminary Experiments and Method

The experiments were carried out with 16 volunteer healthy subjects, mostly students 19 to 21 years of age. The EEG was recorded in vertex, parietal, and occipital locations against the reference of the ear lobe (Cz, P3, P4, and O1 in the 10–20 system). The EEG signals were amplified by using a Schwarzer EEG machine. The subjects sat in a soundproof and echo-free room that was dimly illuminated. For stimulus preparation, evaluation of selective averaging procedure, and digital filtering, a Hewlett Packard 1000F computer was used. The filtering of EEG and of event-related potentials was carried out by using a digital filtering method. The digital filters do not create any phase shift. As auditory stimulation, 2000 Hz, 80 dB tones of 800 ms duration were applied at regular intervals of 2600 ms. Every third or fourth tone was omitted. The subjects were asked to predict and to mark mentally the time of occurrence of the omitted signals. The EEG 1 s prior to omitted stimulation was also recorded with the ERP. The light stimulator was a 20W fluorescent bulb that was electrically triggered. The duration of the light step was also 800 ms.

22.2.2 Preliminary Results in Special Cases

When subjects had learned and successfully followed the rhythmicity that is contained in the paradigm, they were usually able to increase their attention. Then, rhythmic prestimulus EEG patterns could be observed. Most of the subjects reported that at the beginning of an experimental session with repetitive signals they had difficulty in predicting the time of occurrence of the stimulus omission. Usually, in the second half of the experiment they were able to predict the time of the omitted signal. Accordingly, in our signal analyses we applied a selective averaging by grouping approximately the first ten prestimulus sweeps at the beginning of the experiment and the last ten sweeps. Figure 22.1 illustrates comparatively the averages of the first and the last ten prestimulus EEG epochs (digitally filtered between 1 and 25 Hz) that were recorded at the vertex of a subject who reported that at the beginning of the experiment he felt himself unsure and diffuse. Towards the end of the experimental session he could be more concentrated, and thus performed his task much better. The average of the ten sweeps at the end of the experiment depicted a regular rhythmic behavior with large amplitudes. The first ten sweeps tend also to the same rhythmicity, but the average is less regular and has low amplitudes.

The rhythms, which are in principle similar to those illustrated in Fig. 22.1, have been observed in all the subjects. The alignment and phase-reordering were not the same in all the subjects. The exact time of regularity and phase-reordering showed fluctuations in a time period 700 to 0 ms prior to the event.

Figure 22.2 shows approximately ten prestimulus EEG epochs from another subject, C, at the end (a) and at the beginning (b) of an experiment. Single sweeps here were digitally filtered in the frequency range between 7

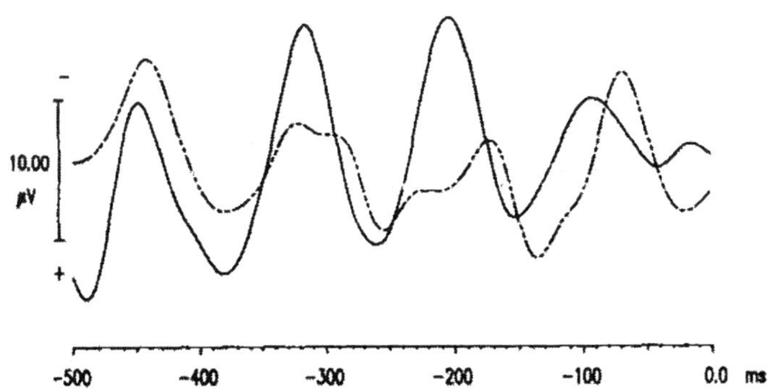

Fig. 22.1. Averages of the first (*broken line*) and last (*solid line*) ten prestimulus EEG epochs of the experiment, filtered in the 1–25 Hz frequency band. (From Başar E. et al. (1989) in: Başar E. ed. *Brain Dynamics. Progress and Perspectives.* Berlin, Heidelberg, New York: Springer.)

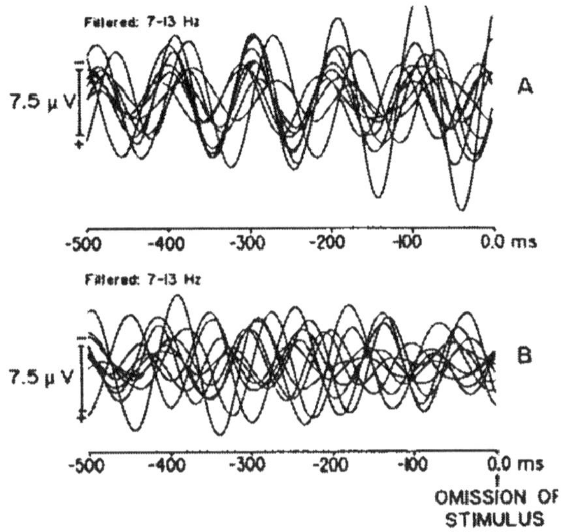

Fig. 22.2. (**a**) Approximately ten prestimulus EEG sweeps at the end of the experiment. (**b**) Approximately ten prestimulus EEG sweeps at the beginning of the experiment. (From Başar E. et al. (1989) in: Başar E. ed. *Brain Dynamics. Progress and Perspectives.* Berlin, Heidelberg, New York: Springer.)

and 13 Hz according to the rhythmicity revealed in the wide-band curve. It is easy to see that at the end of the experiment *repeatable patterns* are observed, in contrast to the beginning without such patterns.

22.2.3 Global Trends of Pretarget Event-Related Rhythms. Expectation and Reaction of Subjects; Subject Variability

In order to summarize these preliminary results we can say that subjects often decided on their own target or *mentally predicted target.* Some subjects could better predict the omitted signals after the second tones, others after the third tones. Most of the subjects showed better performance at the end of the experiment, but some were able to recognize the time of the occurrence of the target already at the beginning of the experiments. At this stage of the investigation the reliability of the results consists only in comparison of the phase-ordered EEG states with statements of the subjects as to whether or not they were able to mark the target signal mentally. It is well known that during long recording sessions the EEG can show high stationary behavior, random synchronization, and alignments. Our findings showed, however, that these phase-ordered patterns were fairly well correlated with the subjects' reports and, as will be shown below in this chapter, fluctuations could be recorded in long standing experiments for a number of subjects.

22.3 Paradigms with Increasing Probability of Occurrence

In order to reduce the possibilities of recording *randomly occurring coherent EEG signals*, we extended our paradigm as follows. In addition to paradigm 1, omitting the third or fourth stimulation (randomly), we extended our measurements to include a more difficult paradigm 2 and thus decrease the probability of the occurrence of the target signals. In paradigm 2 the omitted signal could be changed from the fourth to the seventh randomly (the occurrence of the target signal is 25% when the subject has already heard the second tone; see Table 22.1). During the same recording session the subjects had to perform another experiment, paradigm 3, which offered an easier way to mark the target signal mentally – exactly every fourth signal was omitted. This is the easiest of our paradigms, since the probability of the target occurrence is 100% when the subject has already heard the third signal.

A comparison of the experimental results showed that the subjects could emit coherent and phase-ordered pretarget EEG signals in almost all cases of the easiest paradigm, whereas the same subjects did not show the same good coherent and phase-ordered pretarget EEG during the most difficult paradigm. In this section we describe globally the results with mean-value curves from experiments with 16 subjects. Below in this chapter we will present descriptions of experiments on single subjects.

Figure 22.3 illustrates the comparison between the most difficult paradigm and the easiest one. The curves are wide-band filtered (1–100 Hz). The results are as follows:

1. Every fourth to seventh signal omitted (the most difficult paradigm):
   (a) We observed no regular rhythmicity prior to target (or omitted stimulation).
   (b) We observed an enhancement of the unfiltered EEG following the omitted stimulation (target signal).

Table 22.1. Comparison of paradigms. (From Başar E. et al. (1989) in: Başar E. and Bullock T.H., eds. *Brain Dynamics – Progress and Perspectives*. Berlin, Heidelberg, New York: Springer.)

	Paradigm	Probability of occurrence of target
1.	Every 4th or 7th stimulation (randomly) omitted (most difficult paradigm)	25% after 3rd tone
2.	Every 3rd to 4th stimulation (randomly) omitted (intermediate paradgm, less difficult)	50% after 2nd tone
3.	Every 4th stimulation (no randomness) omitted (easiest paradigm)	100% after 4rd tone

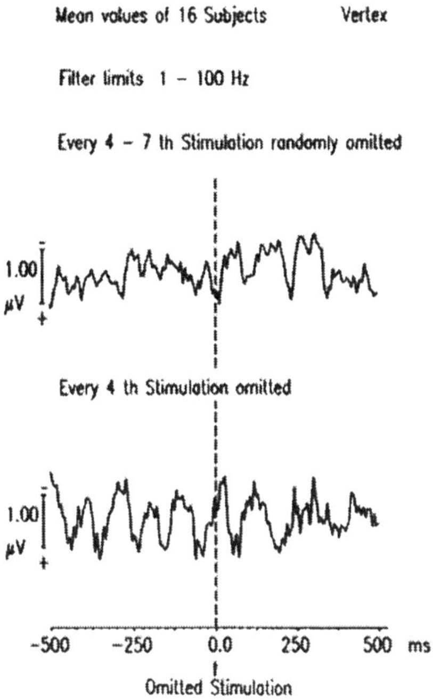

Fig. 22.3. Comparison of the most difficult (every 4th to 7th stimulation randomly omitted, *top*) and the easiest (every 4th stimulation omitted, *bottom*) paradigms as mean-value curve from 16 subjects. Filter limits, 3.5–8 Hz (theta frequency range). Target signals occur at time "0". Derivation, vertex. (From Başar E. et al. (1989) in: Başar E. ed. *Brain Dynamics. Progress and Perspectives.* Berlin, Heidelberg, New York: Springer.)

We should note here that the occurrence of target is rare. This means that upon the omitted stimulation the surprise effect should be greater. We do not analyze here the variability of the P300 wave among the subjects, which showed relevant individual fluctuations and also variabilities at the beginning and the end of the experimental session. The meaning of the mean-value curves is as follows: It can be globally stated that mean value curves from 16 subjects showed a slight EEG enhancement with peakings around 300–400 ms. The latency change of waves of the P300 family were large in our measurements (for the variability of the range of the P300-wave family, see also Galambos and Hillyard (1981)).

2. Every fourth signal omitted (the easiest paradigm):
   (a) The pretarget EEG shows a rhythmicity of around 9.5 Hz.
   (b) No EEG enhancement is observed after the omitted stimulation. The mean-value EEG signal looks like a continuation with a slightly slower rhythm.

Our results will be analyzed in three different frequency bands, as follows.

8 to 13 Hz Range

During the most difficult paradigm (every fourth to seventh signal omitted) no coherent and ordered 10 Hz activity was observed preceding the target signal (Fig. 22.4). However, following the omitted stimulation, a 10 Hz enhancement was observed.

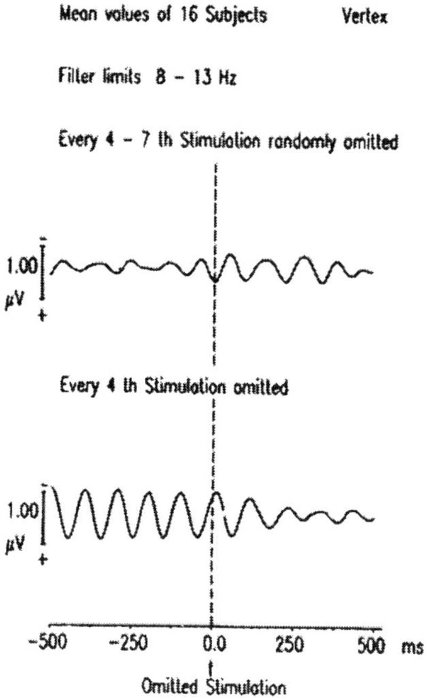

Fig. 22.4. Comparison of the most difficult (every 4th to 7th stimulation randomly omitted, *top*) and the easiest (every 4th stimulation omitted, *bottom*) paradigms as mean-value curve from 16 subjects. Filter limits, 30–50 Hz (40 Hz frequency range). Target signals occur at time "0". Derivation, vertex. (From Başar E. et al. (1989) in: Başar E. ed. *Brain Dynamics. Progress and Perspectives.* Berlin, Heidelberg, New York: Springer.)

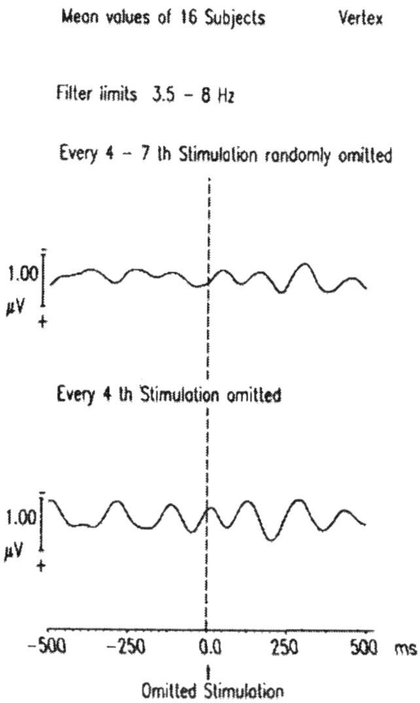

Fig. 22.5. Resting EEG of the subject J.K. *At the bottom* ten sweeps of EEG segments which were digitally filtered in the frequency range of 8–13 Hz are shown. Time "0" is arbitrarily chosen. EEG samples were recorded at the beginning (**a**), middle (**b**), and end (**c**) of the recording session. *Top*, mean value results on ten sweeps. Correlation coefficients C evaluated from three ensembles of ten sweeps. (From Başar E. et al. (1989) in: Başar E. ed. *Brain Dynamics. Progress and Perspectives.* Berlin, Heidelberg, New York: Springer.)

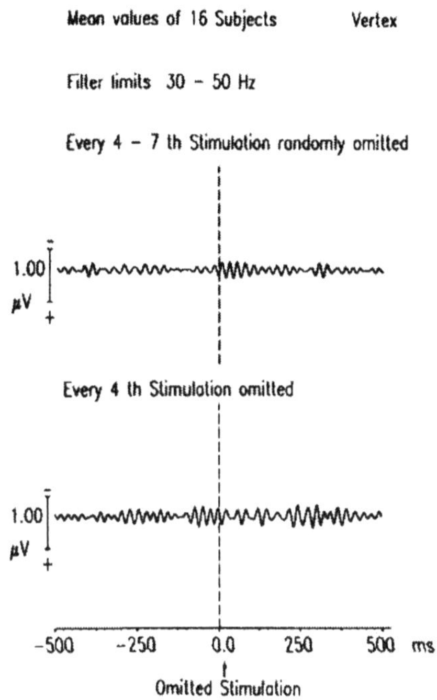

Fig. 22.6. Results of Fig. 22.3 filtered in the 30–50 Hz (40 Hz frequency) range

During the easiest paradigm (every fourth signal omitted) a coherent 9 Hz rhythmicity preceding the target was observed. On the other hand, after the omitted stimulus a blocking of the regular 10 Hz activity was observed.

3.5 to 8 Hz Range

During the most difficult paradigm, in the theta frequency range an enhancement following the omitted stimulation was observed, whereas there is no regular rhythmicity in the pretarget EEG. In other words, by this paradigm there is a reaction due to the surprise (Fig. 22.5). Compare also enhanced theta responses in hippocampus in Chap. 16.

40 Hz Range

In Fig. 22.6 the mean value curves of 16 subjects in the 40 Hz frequency range are illustrated.

During the paradigm with every fourth signal omitted we observed an increased regular rhythmicity of 40 Hz activity just prior to stimulation (50 ms prior to omitted stimulation) in the mean value curve. There is a decrease in the 40 Hz activity (or a blocking) following the omitted stimulation.

During the most difficult paradigm (every third to seventh signal omitted) there is no increased regular rhythmicity prior to target, but there is an enhancement after the omitted stimulation (approximately 250 ms after stimulation).

Results of enhancement or blocking of 40 Hz in this global analysis have the same trend as in the 10 Hz activity. Our analysis is not yet sufficient to describe whether the 10 Hz and 40 Hz enhancement (or blocking) occur simultaneously or not. The reader is refered to Chap. 17 to compare the results in the cat hippocampus.

22.4 Experiments with Light Stimulation

We will describe some experiments, during which repetitive light stimuli were used, and give more details of the analysis performed. Although a small number of subjects was used, a large number of experiments were carried out for each subject. In these series of experiments we will consider only the 8 to 13 Hz frequency range, since a detailed and exact account of all the experiments covering various frequency ranges is not feasible in this study.

Methodological Remarks

We used during these experiments the same experimental setup and procedure as described earlier in this chapter. The light stimulation consisted on light steps of 800 ms duration. The light source was a 20W fluorescent bulb that can be triggered with a short time constant. The intervals between stimuli were 2600 ms. The evaluation and interpretation of the correlation coefficient are described in the Appendix.

22.4.1 Examples of Experiments with Varied Probabilities of Stimulus Occurrence

The subject J.K. is a medical student who quickly learned the goal of the experiments and has been very cooperative during the experiments. Figure 22.7 illustrates samples of the filtered resting EEG as a control before an experiment with the cognitive task. There are three plots of the filtered EEG segments, with ten sweeps in each plot. The three plots present samples during the same recording session. The mean correlation coefficient ($\overline{C}$) of each ensemble of sweeps in a time range from −500 to 0 ms is also shown in the illustration.

The subject was instructed to be attentive to repetitive light stimuli. Every fourth light stimulation was omitted (the easiest paradigm). Subject J.K. reported that at the beginning of the experiment he could easily mark the target signal; however, after approximately ten omitted signals, or after the first 40 sensory stimulations, he could not concentrate as well, and toward the end of the measurement he had enormous difficulties in concentrating.

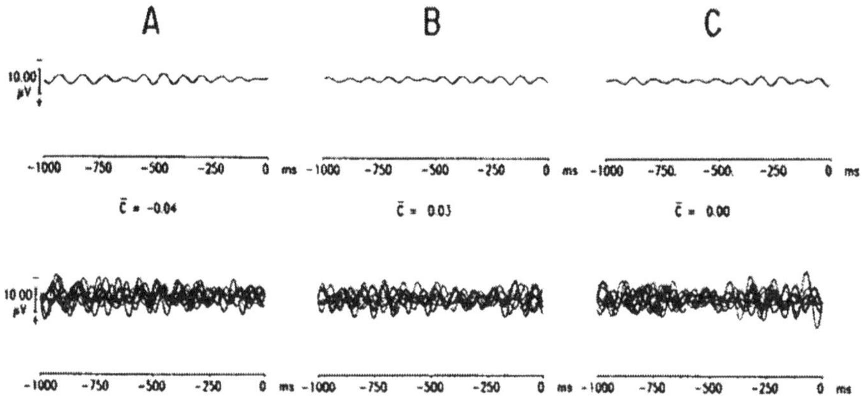

Fig. 22.7. Pretarget EEG of subject J.K. during resting EEG in: Başar E. ed. *Brain Dynamics. Progress and Perspectives.* Berlin, Heidelberg, New York: Springer.)

Figure 22.8 shows the first ten filtered sweeps together with the filtered mean value and wide-band mean curves (1–30 Hz). A clear rhythmicity and a good congruency in almost all sweeps are observed. In the following sessions of the experiments (b–c) the rhythms were less regular, and the congruency among sweeps almost disappeared. Also during this stage, 10 Hz EEG with larger amplitudes were observed in comparison with the resting EEG, which is shown in Fig. 22.7. The correlation coefficient decreased drastically against the end of the experiment: At the beginning, where the subjects reported a good performance, the correlation coefficient was high, $\overline{C} = 0.38$; later, it was lower ($\overline{C} = 0.13$ and $\overline{C} = -0.01$).

Figure 22.9 illustrates a similar experiment with the subject J.K. a few months later. Here again he reported that just at the beginning of the experiment he was able, with ease, to mark the target mentally; during the experiment he lost his ability to follow the target, but near the end he again had better control in marking the target. Figure 22.9b shows the decrease in congruency and the diminishing of the correlation coefficient. In Fig. 22.9c the congruency was better ($\overline{C}$ is $= 0.28$). On the following day we started the first experiment with the most difficult paradigm (every third to seventh stimulation omitted) and proceeded with the next-to-easiest paradigm.

During the most difficult paradigm the subject J.K. reported that at the beginning of the experiment he felt unsure whether he could follow the rhythmicity of the light signals; but in the last two-thirds of the experimental period he was able to mark a larger number of the target signals. Figure 22.10a illustrates the beginning and Fig. 22.10b the middle stage of the experiments.

The amplitudes of the EEG increased during the experiments but not the correlation coefficient. During another experiment with the easiest paradigm,

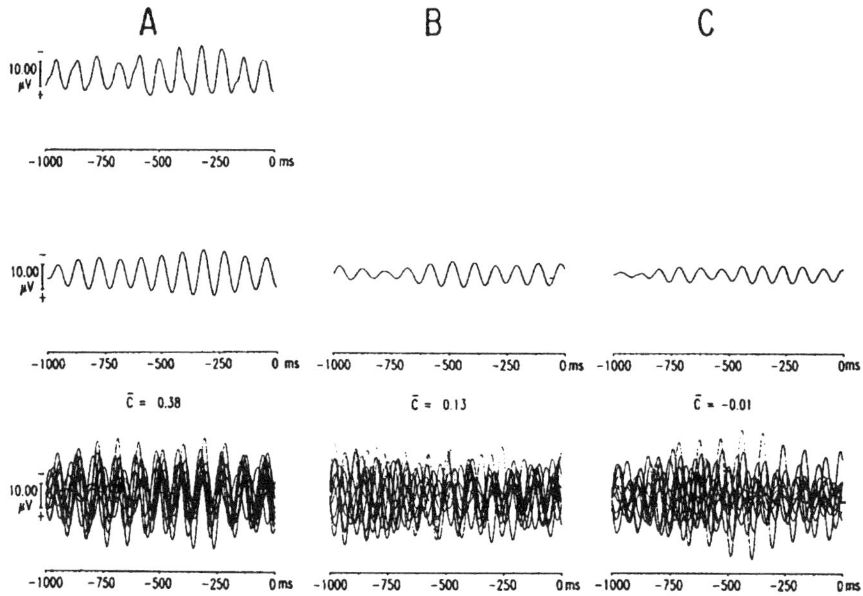

Fig. 22.8. Pretarget EEG of subject J.K. (experiment No. 3) during the easiest paradigm, every 4th signal omitted. EEG segments were filtered in the frequency range of 8–13 Hz. The time scale from −1000 ms to "0" indicates 1 s recording time prior to the target (omitted tone). (**A**) Ten single EEG samples at the beginning of the experimental session (*bottom*). Mean value curves of ten sweeps (*middle*). Broadband mean value curve from ten sweeps (*top*); filter range, 1–30 Hz. (**B**) Ten EEG samples in the middle of the experimental session (*bottom*). Mean value curve from ten sweeps (*top*). (**C**) Ten EEG samples at the end of the experimental session (*bottom*) Mean value curve from ten sweeps (*top*). The correlation coefficients $\overline{C}$ evaluated from three ensembles of ten sweeps are shown at the top of each ensemble. $\overline{C}$ here describes only the period between −500 to 0 ms. i.e., 500 ms prior to target. Subject's report: (**A**), good performance; (**B**) and (**C**), bad performance. (From Başar E. et al. (1989) in: Başar E, ed. Brain Dynamics. Progress and Perspectives. Berlin Heidelberg New York: Springer)

the subject J.K. reported that this time he had not performed well at the beginning (Fig. 22.11). However, toward the end of the experiment he definitely had better control in marking the target. Comparison of Figs. 22.8 and 22.11 shows that here an opposite effect occurred: In the experiment shown in Fig. 22.11 the congruency between single curves was better toward the end of the experiment, and $\overline{C}$ increased from 0.00 to 0.16.

In five subjects the EEG measurements during the easiest paradigm using light signals were taken after those with the most difficult paradigm. But during such an experimental session with the most difficult paradigm, congruency of single rhythms like the epochs of Figs. 22.6, 22.7, and 22.9 were not observed. Further, the correlation coefficient, calculated during the four

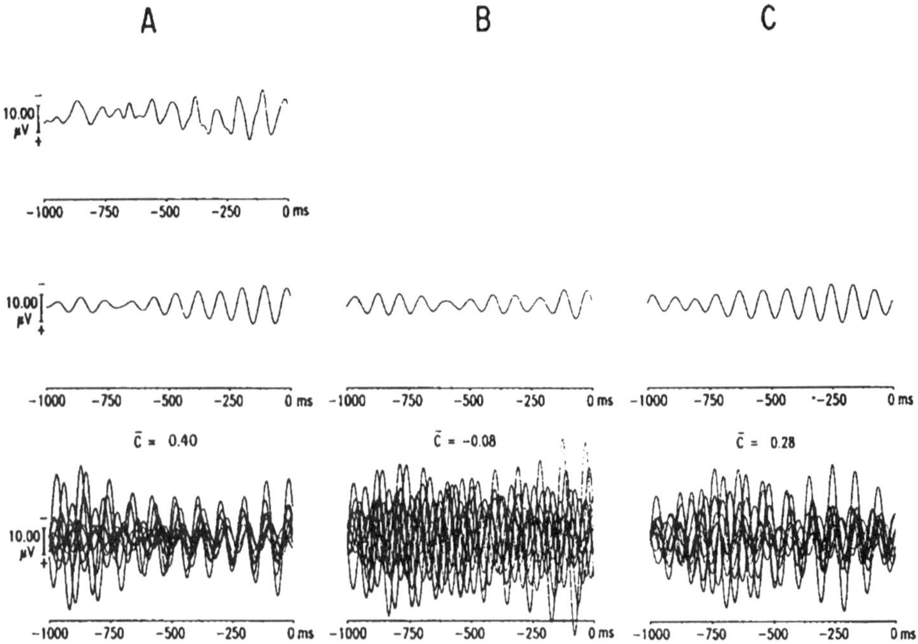

Fig. 22.9. Pretarget EEG of subject J.K. (experiment No. 15) during the same (easiest) paradigm. Explanation as for Fig. 22.8; this shows a repetition after a few months. Subject's report: (**A**) and (**C**), good performance; (**B**), bad performance. (From Başar E. et al. (1989) in: Başar E, ed. Brain Dynamics. Progress and Perspectives. Berlin Heidelberg New York: Springer)

stages of the experiment, remained in all cases around 0.05; it never reached values around 0.4.

Remarks on the Comparison of the Easiest and the Most Difficult Paradigms

We want to mention again why comparison of results using the easiest and the most difficult paradigms for the same subject is important in formulating a judgment concerning the existence of event-related pretarget rhythms. For the same subject we have the possibility to increase the probability of the occurrence of the target by up to 100%. The increase in the EEG amplitude and the tendency to regularity and phase-ordering is reflected in correlation coefficients. If the probability of the occurrence of a target is then decreased, one would expect a *less good* or even *bad performance*. In the latter case it might be expected that the phase-ordering of the EEG and the tendency to a repeatable pattern would end. On the basis of this reasoning, we applied both paradigms to five subjects on the same days, and always obtained comparable results that were similar to the results from the subject J.K.

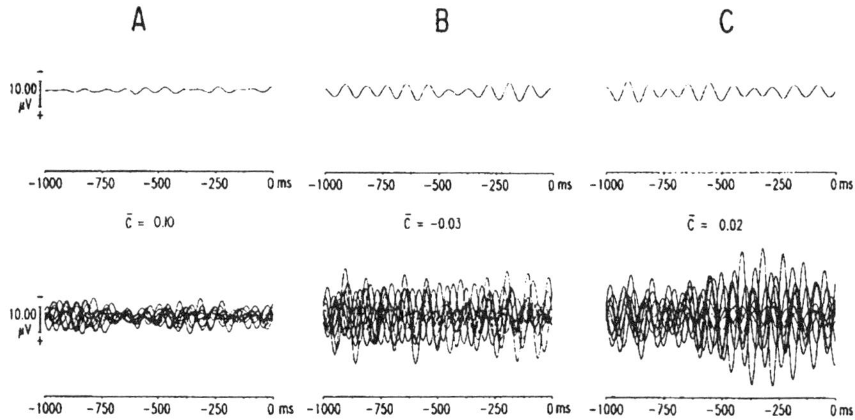

Fig. 22.10. Pretarget EEG of subject J.K. (experiment No. 16) during the most difficult paradigm, every 4th to 7th signal omitted. EEG segments were filtered in the frequency range of 8–13 Hz. The time scale from −1000 ms to "0" indicates 1 s recording time prior to the target (omitted light). (**A**) Ten single EEG samples at the beginning of the experimental session (*bottom*); mean value curve (*top*). (**B** Ten EEG samples in the middle of the experimental session (*bottom*); mean value curve from ten sweeps (*top*) (**C**) Ten EEG samples at the end of the experimental session. The correlation coefficients $\overline{C}$ here describes only the period from −500 to 0 ms. Subject's report: Tried to do well. (From Başar E. et al. (1989) in: Başar E, ed. Brain Dynamics. Progress and Perspectives. Berlin Heidelberg New York: Springer)

In the Appendix we describe an interpretation of the correlation coefficient that reflects the comparison of ten EEG signals during the same experimental period. The increase in correlation coefficient means the increase in similarity of single epochs. The fact that subjects who had reported good performance produced a mean correlation coefficient of up to $\overline{C} = 0.4$ shows that the EEG can attain good phase-ordered patterns, which is contrary to the case of recordings with less probability of occurrence. We also have to emphasize that the recording of *almost repeatable EEG* patterns during defined experiments with cognitive targets requires a large number of experiments and good cooperation with the subjects.

Different Time Windows

In Figs. 22.7 to 22.11 we have consistently considered the time window between 500 and 0 ms prior to target. Although the EEG of most of the subjects depicted a phase ordering starting 1000 to 700 ms prior to target signals, the time scale of −500 to 0 ms is the most common one for a rough preliminary evaluation. In order to avoid errors of visual inspection we started an analysis with some moving time windows prior to target. This means that we chose six time windows at various points along the time axis of −1000 to 0 ms. The

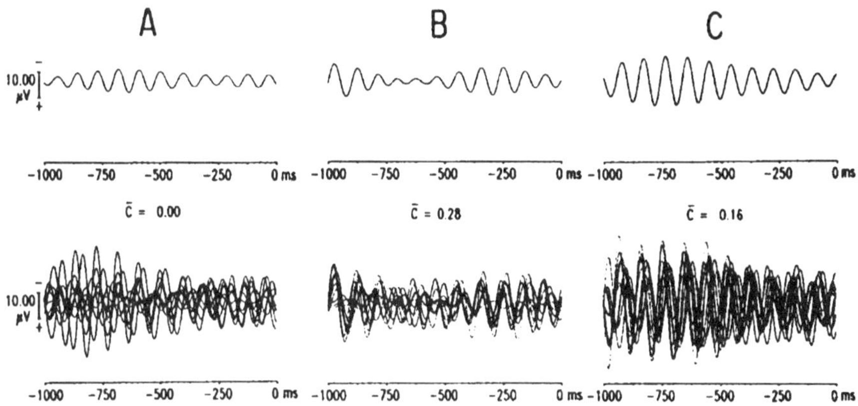

Fig. 22.11. Pretarget EEG (8–13 Hz) of subject J.K. (experiment No. 19) during the easiest paradigm, every 4th signal omitted. (**A**) EEG samples at the beginning of the experimental session (*bottom*); mean value curve (*top*). (**B**) EEG samples in the middle of the experimental session (*bottom*); mean value curve from ten sweeps (*top*) (**C**) EEG samples at the end of the experimental session. Subject's report: Performance bad at beginning (**A**), increasingly good toward the end of the experiment (**B, C**). (From Başar E. et al. (1989) in: Başar E, ed. Brain Dynamics. Progress and Perspectives. Berlin Heidelberg New York: Springer)

narrowest window had a duration of 300 ms. Larger correlation coefficients are to be expected for the smaller windows.

Let us consider the ten EEG sweeps illustrated in Fig. 22.8. For the 500 ms before stimulation, $\overline{C} = 0.40$. As Table 22.2 shows, $\overline{C}$ takes different values depending on the length and position of the time window. For a time window of -300 to 0 ms before stimulation, $\overline{C}$ has the highest value, whereas the window of -700 to -300 ms has a much lower value. The same Table 22.2 shows the correlation coefficients of control EEG sweeps of Fig. 22.7. During recording of EEG sweeps where the subject did not report a good performance, the correlation coefficients are not much higher even by choosing narrow time windows (the mean value is -0.05). The control EEG of the same subject (sweeps from Fig. 22.7) does not show significant values of $\overline{C}$ even after choosing narrow time windows.

Figure 22.10 shows the pretaget EEG and target EEG of subject J. K. during the most difficult paradigme.

For all experiments performed, the search with different time windows has been carried out; the results are similar to those in Table 22.2, indicating always highly increased mean values of correlation coefficients during good performance sessions compared with the resting EEG or bad performance sessions. Evaluation of all the subjects under study gave similar results, showing that with analysis of time the correlation coefficient is always highest during the easiest paradigm.

Table 22.2. Correlation coefficients $\overline{C}$ of the EEG of left occipital recording for the subject J.K. for several time windows. (From Başar E. et al. (1989) in: Başar E. and Bullock T. H., eds. *Brain Dynamics – Progress and Perspectives.* Berlin, Heidelberg, New York: Springer.)

Time window (ms)	$\overline{C}$ (Experiment no. 3)	$\overline{C}$ (Experiment no. 15)		$\overline{C}$ (Experiment no. 19)		$\overline{C}$ Control EEG
	Bad performance	Good performance	Bad performance	Good performance	Bad performance	
−1000 to 0	−0.07	0.15	−0.04	0.16	0.01	−0.01
−700 to 0	−0.04	0.3	−0.04	0.15	0.03	−0.05
−500 to 0	−0.1	0.4	−0.06	0.12	0	−0.04
−1000 to −500	0.1	−0.07	−0.07	0.18	0.02	−0.01
−700 to −300	−0.07	0.14	−0.06	0.14	0.04	−0.01
−300 to 0	0.03	0.46	−0.06	0.11	−0.01	−0.02
Mean value of $\overline{C}$ based on 6 time window	−0.04	0.23	−0.05	0.14	0.01	0

22.5 Long-Standing Experiments with Subject A.F.

The subject A.F. was a technical assistant in our research group. From the beginning of the experiments he had a great interest in being a subject and gave detailed reports after every measurement session. The analysis of the single sweeps by visual inspection correlated highly with his reports in most cases. As he reported that he had been able to mark the target signal mentally during a given measurement session, the single pretarget EEG curves showed, as a rule, good congruence. The agreement with his report was about 80%. Taking this degree of reliability into account, we performed seven experiments with A.F. in a period of about 3 months. The measurements and the report were made as follows:

1. The resting EEG prior to application of a paradigm was measured.
2. The easiest paradigm with repetitive light stimuli was applied.
3. After application of 30 light stimuli (and ten omitted signals) A.F. wrote a report to describe the session as one with "good performance" or with "bad performance."
4. The single sweeps of pretarget EEG were plotted, and the reliability of the subject's report was checked by means of analysis with correlation coefficients that give the degree of single-sweep congruence.

In six of the seven experiments A.F. reported that he had measurement periods with "good performance" as well as "bad performance." In one experiment he reported only "bad performance."

In Table 22.3 the correlation coefficients $\overline{C}$ in the frequency band of 8–13 Hz are listed for the resting EEG and for *good* as well as *bad performance*

Table 22.3. Correlation coefficients $\overline{C}$ of ten single epochs of EEG in left occipital recording for the subject A.F. during the easiest paradigm and of the control EEG. (From Başar E. et al. (1989) in: Başar E. and Bullock T. H., eds. *Brain Dynamics – Progress and Perspectives.* Berlin, Heidelberg, New York: Springer.)

Experiment no.	Easiest paradigm		Control EEG	
	Good performance $\overline{C}$	Bad performance $\overline{C}$	Experiment no	$\overline{C}$
24	0.16	0.01	22	0
27	0.3	−0.02	25	−0.04
32	–	−0.06	30	0.02
37	0.04	−0.05	35	0.02
42	0.13	0.01	40	−0.05
47	0.1	−0.06	45	0.01
52	0.24	0.06	50	0.03
Mean value	0.16 ($N = 6$)	−0.02 ($N = 7$)		0.00 ($N = 7$)

periods during the easiest paradigm. Experiment no. 32 is the one in which the subject could achieve a good performance in none of the measurement periods. The correlation coefficient for the resting EEG had values no higher than 0.05 and averaging < 0.002. The periods with bad performance had values averaging -0.02. The periods with good performance vary between 0.04 and 0.3, with a mean of 0.16. This mean value shows high variability for one subject. It is mainly noteworthy only as an important indicator of the trend in the experiments with A.F.

Experiments involving varied target probabilities, evaluation of correlation coefficient, and plotting of single EEG curves were performed with six subjects in similar sequences. Each subject was studied on at least two different days. All showed behavior similar to that of J.K. and partially so to that of A.F.

22.6 Quasi-deterministic EEG, Cognitive States, Dynamic Memory

The experiments described in the previous sections have shown that during cognitive tasks it is possible to measure almost reproducible EEG patterns in subjects expecting defined repetitive sensory stimuli. During such coherent states single sweeps of EEG are time-locked to a target signal, at least for periods of 10 min during which the subjects are able to mark a cognitive target mentally in a recurrent manner.[2] We use the expression "quasi-deterministic EEG" for the recurrently omitted, almost reproducible, EEG patterns. The specific brain function is related in this case to a type of short learning process and short-term memory, which we tentatively denote as *dynamic memory* (Başar 1988).

In the present chapter we have focused our emphasis on the 10 Hz frequency range, although other frequency ranges were briefly demonstrated earlier in this chapter. Tentatively, the expression "dynamic memory" could be extended to cover all relevant EEG frequency ranges and most probably combinations of several patterns. The use of the expression "quasi-deterministic EEG" finds its legitimacy in several results reported by Babloyantz (1988), Röschke and Başar (1990), and Freeman and Skarda (1985), who have shown that the EEG reflects the properties of a strange attractor (see Chap. 10 in Vol. I).

Hillyard and Picton (1979) used the term *selective attention* or simply *attention* as a construct that has a rather broad but circumscribed set of meanings, being clearly distinguished from nonselective CNS processes such

[2] A recording session with 10 target signals (10 omitted stimuli and 30 physical stimulations) requires at least 200 s or about 3 min). Some of the subjects can reach the same performances during 30–40 omitted stimuli, in other words, around 10 min.

as arousal or alertness. Attentional processes are those CNS functions that enable perceptual or motor responses to be made selectively to one stimulus category or dimension in preference to others. Irrelevant stimuli that are not being attended are either partially or completely rejected from perceptual experience, entry into long-term memory, and control over behavior. Furthermore, these authors state that attention refers to selective aspects of sensory processing. Accordingly, all experimental demonstrations of attention must measure the responsiveness of the organism to more than one category of stimulus. The initial stages of sensory processing are generally thought to proceed rather inflexibly and to consist of an initial afferent registration and feature analysis of incoming sensory data. This information persists in accessible form for about 1 s in a high-capacity sensory memory that has been termed the *buffer stores, iconic (visual) memory,* and *echoic (auditory) memory.*

During our experiments the subjects of the experiments had to pay attention to an omitted stimulus. During this process, if they were able to be attentive enough to mark the target mentally, they anticipated with 10 Hz waves that were time-locked to the target, showing almost reproducible patterns. Depending on the performance of a subject, the coherency time of such reproducible wave packets ranged usually from 300 to 1000 ms, most of it prior to the time the omitted stimulus was due.

In earlier publications Başar (1983a,b, 1990) has tentatively assumed that the evoked potentials manifest the bifurcation of the strange-attractor EEG to a limit-cycle attractor of short duration. A strange attractor is manifested by activity that appears to be random. However, the activity of a strange attractor is deterministic and reproducible if the input and initial conditions can be replicated. The experiments with human subjects described in this chapter showed that by increased certainty due to expectation of repetitive sensory signals and accordingly by increased attention stages, the subjects seem to generate internal cognitive inputs to the CNS. These cognitive inputs are probably due to repetitive similar mental efforts. According to our experimental results, if a subject cannot mentally predict the occurrence of the expected target signal (omitted stimulation), there is no averaged synchronization of the EEG in the 10 Hz frequency range (see experiments with subject J.K.) or in the 40 Hz frequency range (earlier in this chapter only global findings were shown).

Further, our results showed that the 10 Hz, 40 Hz, and 4 Hz EEGs go from disordered states to ordered coherent states during defined cognitive inputs to the CNS. Here lies a similarity to evoked potentials elicited by exogenous sensory stimuli that also show a transition from disordered to ordered coherent states.

22.6.1 What Is New in the "Dynamics of Time-Locked EEG Patterns"

Observation of changes in the amplitude of the EEG, enhancement or blocking of alpha activity during mental tasks (solving of arithmetic problems or hearing music), is not a new discovery, and relevant examples have been described in the literature (Creutzfeldt 1983; Giannitrappani 1985; Petsche et al. 1987). The new observation reported in this study is the fact that alpha waves can be time-locked to a defined cognitive target and that in well-defined experiments reproducible alpha patterns can be emitted from the human brain.

In addition to carrying out conventional analysis as described in the established P300 studies by Sutton et al. (1965) and Galambos and Hillyard (1981), we encountered developments in the analysis of ERPs in relation to missing auditory stimuli (Nakamura et al. 1986; Gauthier et al. 1986; Takasaka 1985; Friedman 1984; McCallum 1980; Simson et al. 1976). These authors' results were mostly limited to a description of latencies of potentials following the omitted stimuli and to topographic distributions of the family of P300 waves. Gauthier et al. (1986) and McCallum (1980) reported the contingent negative variation (CNV) changes during special experimental conditions. Synchronized pretarget event-related rhythms, however, were not described by these authors, since they did not use small segments of EEG or filter methods, and the subjects were not asked to perform the same mental effort to mark the target as omitted stimuli.

Lehmann (1989) stated that functional states may be recognized for relatively short epochs of EEG depicting systematic relations with types of thought processes. According to his description, the duration of spatially stable microstates covers a range of about 500 ms. Certainly, it is attractive to think of the brain microstates as being building blocks of higher information processing, which qualify as areas of consciousness only beyond a certain duration. According to our preliminary results, the microstates would have various durations depending on the basic rhythmic activity. Sensory–cognitive microstates in the alpha frequency range might have durations of about 300 ms, theta microstates about 500 ms. On the other hand, 40 Hz microstates should have shorter durations of 20 to 50 ms.

22.7 Appendix

The correlation factor ϕ of two time series $f(t)$ and $g(t)$ is defined as the mean value of their product over time $(-T, T)$ provided that $f(t)$ and $g(t)$ do not have DC components:

$$\phi = \overline{f(t)g(t)} = \lim_{T \to 1} \int_{-T}^{T} f(t)g(t)\mathrm{d}t,$$

or

$$\phi = \frac{1}{T_2 - T_1} \int_{T_1}^{T_2} f(t)g(t)\mathrm{d}t$$

if $f(t)$ and $g(t)$ are to be considered only between time points T_1 and T_2, and $f(t)$ and $g(t) \neq 0$.

It can be shown that the maximal value of the correlation function is $\phi_{\max} = \tilde{f} \cdot \tilde{g}$, where $\tilde{f}$ and $\tilde{g}$ denote the RMS values of f and g.

The value

$$C = \frac{\phi}{\phi_{\max}}$$

denotes the correlation coefficient. Its value fluctuates only within the limits -1 and $+1$. If $C = \pm 1$, it is stated that $f(t)$ and $g(t)$ are (positively or negatively) correlated. This is only the case if $f(t)$ and $g(t)$ are correlated as in the following:

$$g(t) = af(t) \quad (a \text{ a constant factor}).$$

If $C = +1 (a > 0)$, this means that f and g are completely in phase, and if, $C = -1 (a < 0)$, f and g are in counterphase. lf $C = 0$, $f(t)$ and $g(t)$ are not at all correlated. This is the case for two sine waves with different frequencies, and also for the noise signals originating from two different sources, provided the epoch analyzed is long enough; for the case of noise, C converges gradually toward 0 over many cycles of the lowest frequency in the pass-band.

In all other cases, this means that if $0 < C < 1$, then $f(t)$ and $g(t)$ are piecewise or imperfectly correlated.

The correlation coefficient is a measure to indicate the similarity of two time series. The aim of our analysis using correlation coefficients between two EEG oscillations is to try to establish the similarities in the EEG activity prior to a task. By simple superpositions of ten filtered signals, we can often recognize a similarity by visual inspection. The phase-ordering of the EEG signals during a stage with a defined cognitive load can be described in a mathematical way using the correlation function. Since we usually considered ten EEG epochs (focusing on the period just prior to the target), we used a mean value of the correlation coefficient that is the mean value of the correlations of all the ten EEG signals. In other words, for N signals,

$$\overline{C} = \frac{2}{N(N-1)} \sum_{i,j=1;i<j}^{N} C_{i}j,$$

$C_{i}j$ being thȩ correlation coefficient of EEG epochs i and j belonging to the ensemble of analyzed sweeps, and N the maximal number of analyzed sweeps.

23. Event-Related Oscillations as a Strategy in Cognition Research

23.1 Generalization of Cognitive Responses: Advantages of the Brain Dynamics Research Program and the Concept of Event-Related Oscillations

The application of the Brain Dynamics Research Program has the following advantages:

(1) The alternating application of several cognitive and sensory paradigms to all subjects brings more insight with regard to the understanding of the subject's cognitive reactivity under different tasks related to attention, memory, signal detection, and decision-making.

(2) Several systems-theoretical methods were applied in the combined analysis procedure of EEG and evoked potentials; this application brought insight to the general dynamics of processes, regardless of the paradigms and subjects used (human or cat). Because of the possibility of *selective averaging* and *single-trial analysis*, the experiment can be segmented. After the experiments one can compare, off-line, the data in different segments of the measurements. Moreover, we do not analyze only one type of wave, for example the P300, but we also consider the changes of activity in prestimulus EEG and poststimulus EEG in various frequency channels.

The number of paradigms was great, and we considered the "EEG brain codes" from different viewpoints. As we have seen in Chaps. 14, 18, 19, 20, etc., most of the important changes that occurred under learning and/or decision-making paradigms took place in the lower frequency range, between 1 to 3 Hz and/or 3 to 8 Hz. Accordingly, we are now confronted with the following question: During which type of experiments do we expect considerable changes in *theta* or *delta* frequencies, in prestimulus or poststimulus EEG segments? Returning to the main purpose of this book (namely, whether EEG and ERP carry information on brain coding), we had the relevant possibility to follow this question step by step. One of the goals was to determine whether we could discover EEG-response codes for *decision making* and *focused attention.*

(3) How does a subject make a *decision* during an experiment? The experiments on decision-making during P300 are basically similar to those where hearing at threshold level is studied. When a stimulation signal at the thresh-

old level is applied, the subject has to make a decision: *Did I hear a sound or not?* During the oddball paradigm: *Is the sound a rare one or not?* We have seen that also in cases when the subject is confronted with such a decision, the emitted oscillatory response is in the frequency range of 2–3 Hz (for single-sweep recording, see results of experiments in Chaps. 12 and 25 and Başar et al. (1992a)). In other words, we know from experiments at the threshold level that the decision-making is manifested in low-frequency induced rhythms of the subject's EEG.

In the oddball paradigm, subjects had to manage two types of tasks. First, they had to increase the attention (focused attention), and secondly, they had to make a decision following the signal detection. The decision consists in the differentiation of the *target tones* from *frequent tones*. If the subject's decisions regarding the target tones are successful, an increase in the response in the delta frequency range is recorded (see amplitude–frequency characteristics, filtered curves, and results in Chaps. 15 and 20). At this stage, the aim is a global discussion; therefore, only the grand averages of the oddball and the "3rd attended signal" (3ATT) paradigms, the respective amplitude–frequency characteristics, and filtered curves are given for the various electrode locations.

In the oddball paradigm the increase of the response in the theta frequency channel is not the only change in comparison to the ignored tones, as the delta component has a considerable time delay and increased amplitudes. In other words, if the subject detects the target, the delta response is considerably enlarged and delayed. Another important result is the difference in the theta frequency range, where the amplitudes of the response are not necessarily increased in comparison to the sensory EP, but there is a *prolongation* of the reaction in the theta frequency channel (see Chap. 20, in particular Fig. 20.2 and Table 20.1). In other words, the oscillation is less damped than is the case in responses to ignored tones or to EPs, and a second theta time window was recorded. All three components, the alpha, theta, and delta responses, are involved in the process of focused attention and decision-making.

(4) The 3ATT paradigm, where every fourth signal is omitted and the subject focuses his/her attention on the preceding 3rd signal, is useful for understanding the results discussed above. In this case the subject does not have to accomplish a "decision-making" at the time of stimulation. During the experiments or prior to experiments he has already learned that with a given stimulation interval a given sound stimulation or light stimulation will be repeatedly applied. He has only to pay attention to discover the right time of delivery of the stimulation. He does not have to perform a difficult signal detection; he knows already that the target will arrive, and he does not have to decide about the quality of the sound.

Now we will consider the theta frequency response. There is a great increase, of about 50% (see Chap. 19, in particular Fig. 19.8), in *frontal* and *parietal* recordings. It is also important to note that neither the occipital nor the vertex response showed such a statistically significant increase during this state of attention.

What conclusion can be arrived at through a comparison of these two paradigms, namely the oddball paradigm and the 3ATT paradigm? It seems that when the subject increases his focused attention, only his theta response is considerably increased, both in visual response experiments and in auditory tone experiments, i.e., independent of sensory modality. According to this consideration, and also taking into account the results of experiments at the threshold level (Chap. 12), we tentatively conclude that the *decision-making and surprise process* occurs in a lower frequency range of 2–3 Hz, whereas in cases of focused attention the theta response component shows changes and increases. The common point in both of the paradigms was the *increased attention*. Increased attention is seemingly manifested by an increase in the theta response.

(5) Now we will try to similarly discuss the paradigm concerning the alternating target with regard to Chap. 14. In this case, after the beginning of the experiment, the subject very quickly learns that every second tone will be a target: The probability of occurrence of the target is 100%. He knows exactly what to expect. He has to focus his attention to the second tone, but not for *decision-making*. In this case no delay in the delta response could be observed (see Chap. 14). This means that no large delta response occurs when the decision-making is missing.

23.2 Component Analysis Towards Functional Understanding During Cognitive Processes

The applied analysis has the advantage of comparing frequency response components. In designing the experiments we have chosen some few common aspects related to the subject's psychological states. What is a psychological state of a subject at the threshold level when he has to make a decision? If we suppose that a subject has to make a decision when similar signals at the signal threshold level are applied, the main components of the evoked response are between 1 and 4 Hz (see Chap. 12).

We emphasize results of experiments with cats, where a change or an enhancement in the theta frequency range was recorded mostly in the hippocampus. Furthermore, we mention here results of cat experiments with repetitive stimulation (repetitive auditory tones, every fourth or every fifth omitted) similar to the experiments in Chap. 19. In this case we had the most remarkable results: In the auditory cortex of the cat there are no important changes, but in the deeper layers of the hippocampus HI3 and HI4 electrodes, there is a significant increase in the theta response (see Chap. 18, in particular Figs. 18.5 to 18.7).

An explanation could be the following: The important physiological and anatomical connections between hippocampus and frontal cortex are ex-

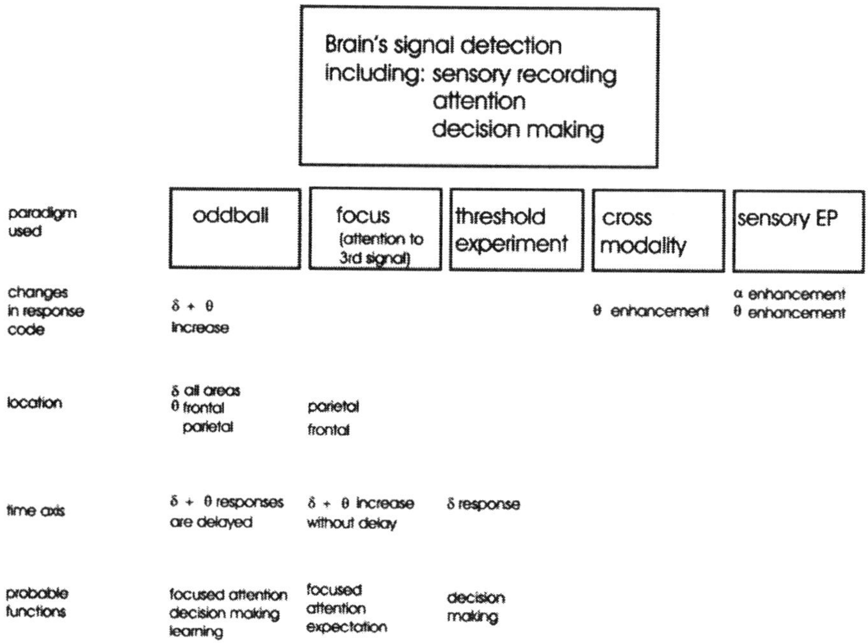

Fig. 23.1. Tentative classification of frequency responses observed with endogenous ERP paradigms and their respective functional correlates.

plained in Chap. 3 of Vol. I. Has this change or enhancement in the theta frequency range during focused attention something to do with the changes in the hippocampal rhythmicities in the area of the CA3 layer? It also has to be emphasized that usually by repetitive stimuli one observes a theta enhancement in all EP experiments reported in this book (see Chap. 20). It is interesting to note that in human experiments where the subject attends to the third stimulus for detecting the signal, the increase occurs in the theta frequency range, as in the case of hippocampus (see Table 20.1).

Figure 23.1 schematically illustrates the signal-detection and decision-making experiments that we have described in several parts of this book. The illustration shows selected examples of experiments with the help of which we were able to show the cognitive correlates of the theta or delta increases.

(1) During the oddball experiments, the subject achieved a state of *focused attention* and *decision-making.* The recorded event-related potentials depicted delta (delayed delta) plus theta response especially in the frontal and parietal locations (see Chap. 21).
(2) Concerning the experiments with repetitive stimulation where the subjects pay attention to the third signal, we noted a theta increase especially in frontal and parietal locations. There was also a delta increase,

but this occurred without delay because there was no surprise, as the subjects were in a stage of focused attention (see Chap. 19).

(3) The experiments on hearing at the threshold level showed that the response signal contained a delta wave, which was not perfectly time-locked, but the motivated subjects showed this type of emitted delta activity. In this case subjects had the task of making a decision.

(4) We have described ERP experiments where the omitted stimulus paradigm was used on freely moving cats. The hippocampus of the cat brain showed in this case an increased theta response, and the cats seemed also to be in a state of increased attention (see Chap. 18).

(5) Further experiments, during which the cats were excited with low amplitude stimulation, were described in Chap. 18. The auditory signal applied was in the range of 20 dB, so that in this case there was an increase of theta and delta activity in the cat's hippocampus. In other words, it seemed that the hippocampal theta cells were activated during the signal detection as experiments by Thompson et al. (1980) have also shown (see Chap. 18).

(6) In Chap. 22, 10 Hz induced rhythms were described that are probably related to a dynamic memory and that have characteristics of phase-locked internally induced event-related rhythms prior to the occurrence of a target.

(7) Delay and increase of the delta response are common observations in experiments with visual oddball tasks (see Chap. 18 in Vol. I and Schürmann et al. (1995). The theta increase is particularly large in parietal locations. The delay of the delta response is not dependent on the visual stimulus modality but is specific for rare stimuli (i.e., also observed with auditory stimuli).

(8) In MMN experiments delayed delta responses are observed for target tones (see Chap. 15).

(9) Bimodal stimulation in experiments with EEG measurements in humans leads to large-amplitude theta responses (see Chap. 11).

(10) Bimodal stimulation, combined with intracranial measurements in cats, shows a remarkable hippocampal response (see Chap. 1).

(11) Cross-modality measurements show that theta responses are observed for adequate as well as indeaquate sensory stimulation of primary sensory areas (see Chaps. 2 and 10).

The strategy used – several paradigms and frequency analysis – may open new avenues to quantify in frequency terms event-related brain activity involved in short-term memory processes, attention, and decision-making. The topographic patterns of the event-related activity may permit us to localize neural structures involved in these functions.

Models of brain functions related to the selectively distributed delta and theta systems are given in Chap. 28. These tentative models have been developed according to the strategies applied. Without frequency analysis and application of several sensory and cognitive paradigms the description of functional correlates as presented here could not be performed. The analysis approach using latencies and amplitudes has serious shortcomings that are clearly avoided in the frequency-domain approach.

Part IV

Integrative Systems in Brain Function

24. Functional Alphas Selectively Distributed in the Brain – A Theory

"If one understands the alpha rhythm, he will most probably understand the other EEG phenomena"

Storm van Leeuwen (1979)

24.1 The EEG 10 Hz Band Rhythms Classified

It is probably not coincidental that what is known about the EEG alpha rhythm fits rather neatly into a classification scheme similar to Galambos's scheme for the gamma band (see Chap. 26). Galambos (1992) states:

1. ***Alpha is more than just a spontaneous rhythm;*** it is the prototype of spontaneous brain rhythms.

2. ***Alpha can be induced:*** Perhaps the reader will find credible the claim that induced alpha-band rhythms are analogues of the gamma-band rhythms induced by moving stripes and odors in the animal experiments. Consider closing eyes in a lighted room: like turning off the lights, it reduces retinal illumination and initiates, or induces, an alpha-band burst in the EEG. An example of alpha-induced (by undefined) changes in internal state: the rhythm disappears the moment a subject follows the instruction "multiply 11 by 13," and it reappears as soon as the answer is delivered.

3. ***Alpha can certainly be evoked:*** Time-locking alpha-like rhythms to onset of light flashes was clearly described by M.A.B. Brazier more than 30 years ago in one of the earliest reports on human computer-averaged responses (Brazier 1960). The physiological basis of this phenomenon – a near-10-Hz periodic oscillation that continues for a second or so – remains unclear (see Mangun (1992) for a summary, Başar, Bullock 1992 eds.). Brazier's discussion of her discovery has a familiar ring:

"This finding of a rhythm phase-locked to stimulus shows that the brain now has a rhythm that has been imposed on it through a sensory system. Has this imposed rhythm supplanted "the endogenous" one; or is it the same rhythm with a shift in phase to carry the message; or is the basic rhythm still there and this one added? In the latter case the brain would have its own rhythm against which to match this one for both frequency and phase" (p. 351).

*4. **Finally alpha can be emitted:*** In the study where Başar et al. (1989) demonstrated anticipatory time-locking of gamma-band energy, a well-trained subject emitted time-locked bursts of alpha-band energy for up to a full second before the delivery of an expected target. By contrast to the modest evidence they advanced for anticipatory time-locking of waves in the gamma band, the alpha locking to the moment in future when a target will be delivered is robust and highly significant statistically.

These comparisons of alpha and gamma band prompt a possibly heuristic speculation: Because gamma-band and alpha-band rhythms behave similarly – that is, can be induced, time locked, emitted, and so on – the physiological mechanisms through which they are generated and controlled, when revealed, will have much in common.

24.2 Why the Expression "Alphas"?

The expression "alphas" (see Başar et al. 1997a) refers mainly to "alpha patterns." In general, one of the most interesting aspects of the world is that it can be considered to be made of patterns. A pattern is essentially an arrangement. It is characterized by the order of the elements of which it is made rather than by the intrinsic nature of these elements (Haken 1992).

It should be also clear that when calling the phenomenon "alpha pattern," this should be taken in a generic sense. More than 40 years ago, Grey Walter (1950) pointed out that the alpha rhythm in one subject is composed of, and is the end product of, many alpha rhythms. In 1962 he stated:

> We have managed to check the alpha-band rhythm with intracerebral electrodes in the occipital–parietal cortex; in regions which are practically adjacent and almost congruent one finds a variety of alpha rhythms, some of which are blocked by opening and closing the eyes, some are not, some are driven by flicker, some are not, some respond in some way to mental activity, some do not. What one can see on the scalp is a spatial average of a large number of components, and whether you see an alpha rhythm of a particular type or not depends on which component happens to be the most highly synchronized process over the largest superficial area; there are complex rhythms in everybody.

In several chapters it was also stated that we confront the problem of recording several types of alpha rhythms or alpha patterns. Therefore, in this chapter a concluding synopsis will be attempted in order to construct an "alpha model of the entire brain or brains."

The statements by Galambos and Walter (as given above), the results presented in the volume edited by Başar et al. (1997a), and the respective chapters of the present volume justify the expression "alphas." This expression comprises several different 10 Hz alpha activities that are related to brain functions. From the physiological viewpoint the interest in "alpha activity" disappeared in neuroscience in the late 1960s. The resonance phenomena at the alpha frequency range on the cellular level as described by Llinàs (1988), Dinse et al. (1997), and Silva et al. (1991) open a new window for the interpretation of 10 Hz oscillations. MEG recordings allow better spatial resolution (for a review see Hari 1993)). The recordings of functionally significant event-related and phase-locked 10 Hz oscillations may open new avenues, similar to the EEG research at the beginning of 1950 (see also Başar et al. 1997a).

24.3 Alphas and Alpha Activity Revisited

Since the first discovery of the alpha rhythm, research scientists have been confronted by some important questions: Where are the alpha activities generated? Do several alpha generators exist, and is it possible to find a pacemaker for the alpha rhythm in the brain? The facultative pacemaker theory of Andersen and Andersson (1968) is based on the possibility of a central pacemaker mechanism in thalamic nuclei. Our results in the preceding chapters show experimentally that resonant alpha generators are distributed selectively throughout the brain. Further, it is possible that in some structures of the brain, for example in the occipital cortex, the number of alpha generators is greater or that they are densely distributed in comparison to other structures. On the other hand, the frontal cortex of young subjects does not generate spontaneous or induced alpha activity as strongly as the occipital lobes (see Chap. 10 in Vol. I). The existence of distant alpha generators in the frontal and occipital cortex (or cerebellum) cannot be explained by volume conduction. The analysis of evoked potentials of the cat hippocampus showed that large amplitude and frequency changes were recorded in electrodes with a distance of 1 mm (see Chap. 1).

The significant and stimulus-locked enhancements of the 10 Hz response occur only when the cat has been excited by visual stimulation in the visual cortex and not by auditory stimulation. These results indicate that there is no significant volume conduction between auditory and occipital cortex. Another important experimental finding is the following: Polysensory structures such as the reticular formation and the supramodal center hippocampus generate enhanced alpha responses to visual and auditory stimulation modalities. The question of why the hippocampus generates an enhancement of 12 Hz by visual stimulation and an enhancement of 9 to 10 Hz by auditory stimulation is also a central point for understanding the functional alpha oscillations.

If the experimenter stimulates the cat brain with an auditory stimulation, only the medial geniculate nucleus reacts with an alpha enhancement

(Chap. 2). Lateral geniculate nuclei do not show enhanced, evoked 10 Hz oscillations. This is a typical and very important case to show that in the brain of a freely moving and nonanesthetized animal the enhancement of alpha activity can occur in the hippocampus independent of the modality of sensory stimulation. In other words, hippocampus is a gate par excellence for the alpha-signal transfer to several cortical locations. The electrodes were implanted in various layers of the hippocampus, and they all responded with enhanced alpha-damped oscillations. The thalamus cannot work in this way; either it does not resonate, or it responds with 10 Hz or 12 Hz (see Chap. 27). Accordingly, the thalamocortical alpha circuit strongly emphasized by several investigators cannot be the only generating site of the alphas.

24.4 Some Physiologically Based Theories on the Generation of Alpha Rhythms

24.4.1 The Facultative Pacemaker Theory

On the basis of their observations, Andersen et al. (1967a,b) and Andersen and Andersson (1968) advanced the hypothesis of facultative pacemakers in the thalamus influencing activity in the corresponding cortical areas. This idea is illustrated in Fig. 24.1. Rhythm generation is assumed to occur at many independent thalamic locations, and each nuclear group to be able to produce rhythmical activity of intraspindle frequencies. Via the specific thalamocortical fibers, the burst discharges of the thalamic rhythmical entity are imposed on a cortical area, which may correspond to a cortical column (Mountcastle 1957; Hubel and Wiesel 1962). Each thalamic burst will induce neuronal activity in the cortex, thus giving correspondence between thalamic and cortical rhythmicity. Since each thalamic rhythm generator may act as an independent unit, the same individuality in rhythmical activity will occur also at the cortical level, as illustrated in the schematic traces in Fig. 24.1.

Andersen and Andersson (1968) estimated the maximal number of rhythm generators in the cat thalamus to be on the order of 30,000–40,000, while the number of cortical columns in one hemisphere of the cat is around 25,000. These authors noticed that the high degree of synchrony of rhythmical activity usually found in thalamic and cortical areas indicates that the activity of individual generators of the rhythm is coordinated into larger units. This may be done by "distributor neurons," for example of the type described by Scheibel and Scheibel (1966) and Tömböl (1967). However, the experiments with decorticated preparations led Andersson and Manson (1971) to assume that a mechanism in the thalamus operates that gives a widespread synchronization of individual rhythm generators and that the coordination of rhythmical activity in large cortical areas is primarily an intrathalamic event.

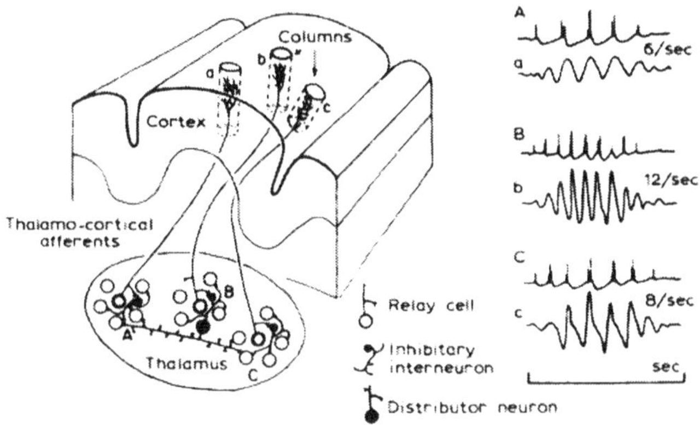

Fig. 24.1. Diagram illustrating a model of the thalamocortical rhythmic correspondence. The neurons of three thalamic nuclear groups, *A*, *B*, and *C*, send their axons to the appropriate part of the cerebral cortex, activating the columns *a*, *b* and *c*. Collaterals of these axons excite inhibitory interneurons (black), which have profusely ramifying axons that can initiate postsynaptic inhibitions simultaneously in a large number of thalamic neurons. The different thalamic nuclear groups have been given different intraspindle frequencies, times of onset and stop, and interspindle periods. The corresponding features of the cortical spindles vary accordingly, as illustrated *in the right-hand column*. The *upper lines* are imaginary spindles from the thalamic groups *A*, *B*, and *C*, whereas the lower lines show the corresponding cortical spindles as they would appear at points *a*, *b*, and *c*, respectively. (From Andersen P. et al. (1968) *Journal of Physiology* 192: 257–281.)

Başar et al. (1975a,b) and Başar (1980) commented that the idea of alpha pacemakers and a transmission possibility between thalamus and cortex is useful. However, pacemaker hypotheses should be extended to other structures including brainstem and hippocampus and also to other frequency bands.

24.4.2 The Scope of Lopes da Silva and Coworkers

Contrary to the viewpoint of Andersen and Andersson (1968), Lopes da Silva et al. (1973b) did not assume a priori that all rhythmic activities falling within the alpha range should be considered equivalent. In their investigations, the latter authors recorded the spontaneous alpha rhythms and barbiturate spindles in dogs from the same recording site. The alpha rhythms were found to differ from barbiturate spindles with respect to their cross spectra and topographic distribution. Lopes da Silva et al. (1973b) concluded that in a state of alpha rhythms, the thalamic and cortical sites act more independently of one another than under a state of light barbiturate anesthesia. Further, the same group of investigators (Lopes da Silva et al. 1973a) tested the hypothesis of Andersen and Andersson by studying particularly the relationships between alpha rhythms recorded in the lateral geniculate nucleus and visual cortex. The strength of thalamocortical relationship was measured by means of coherence functions.

Although significant thalamocortical coherences were found, cortico-cortical coherences were generally higher than the high thalamocortical coherences. The high cortico-cortical coherences were found over relatively large areas (diameters of 5 or 6 mm). The model of Andersen and Andersson, according to which the rhythmic activity of discrete cortical areas is paced by relatively small groups of neurons in thalamic nuclei, would imply the existence of low coherences between different cortical areas with high thalamocortical coherences of some specific sites. This implication is completely contradicted by the results of Lopes da Silva et al. (1973a).

The view of Lopes da Silva et al. (1973a) on the pacemaker concept is as follows: "In our view this problem cannot be solved by using a deterministic pacemaker concept. In a general sense a deterministic pacemaker has two main properties: It has auto-rhythmicity and it can impose its rhythm upon other systems. This implies absolute correlation, i.e., coherence of one, or nearly one if it is assumed that there is measurement of noise in all records."

According to Lopes da Silva et al. (1973a), alpha rhythms should be considered to be signals produced by stochastic processes. More specifically, thalamic and cortical alpha rhythms would result from the filter properties of neural networks when submitted to a random input. The networks, the alpha filters, could be identified in physiological terms in a cell population with an organization of the type described by Andersen and Andersson (1968).

Neural networks with similar design and frequency selectivity should exist in different brain areas (cortical and thalamic), where they give rise to alpha rhythms with the same frequency spectra but not necessarily with large values of coherence. Instead of thinking in terms of pacemakers, these authors proposed to think in terms of statistical correlations.

The investigation by Lopes da Silva et al. (1973a,b) in dogs with electrodes inserted through the cortex and those with a large number of electrodes spaced closely together on the surface of the cortex has elucidated that at a distance ranging from 0.5 mm to 2.5 mm on the cortex, there may be good coherence of alpha rhythm, but also that this coherence is often not absolute. In fact, a good deal of dissimilarity may be quite noticeable even from electrodes spaced closely together in different cortical layers. The waveform of alpha rhythms recorded by means of intracortical electrodes is rather different from that derived from the surface. The alpha waves may be much more spiky and present an arcade. Lopes da Silva et al. (1970a,b, 1973a,b) suggest that in the occipital cortex many small generators are present that averaged over a period of time are significantly in synchrony, though they may vary considerably from moment to moment. This holds not only vertically for the generators at different depths in one area, but also horizontally for the generators over the cortex.

24.4.3 Survey by Andersen and Andersson

A particular feature of the thalamic relay nuclei is their ability to convert a single afferent volley to a series of rhythmic discharges along the thalamocortical fibers. Adrian (1941) discovered that a single tactile stimulus elicited a series of waves in the thalamus, which he called thalamic after-discharges. Similar rhythmic activity was found by Bremer and Bonnet (1950) in the medial geniculate nucleus in response to a click. All these authors noted that the frequency of the evoked activity was around 10/sec., i.e., similar to that of the spontaneous rhythmic cortical waves. Adrian (1941) maintained that the after-discharges consisted of bursts of spikes separated by slow waves. A peripheral stimulus elicited a series of 3 to 7 such cycles. By recording from the white matter below the cortex, Adrian showed that the rhythmic discharge occurred in the thalamocortical fibers, indicating a thalamic origin of the after-discharges. Due to this rhythmic discharge in response to a single afferent volley, a series of waves are initiated in the cortex, appearing at a frequency of about 10/sec. (Bartley and Bishop 1933; Bishop 1933; Jarcho 1949; Bishop et al. 1953).

In 1941 Adrian already reported that rhythmic 10/sec activity following a single afferent volley could be recorded within or at the dorsal surface of the thalamus, even if the appropriate cortical area was removed. In other words, the thalamic nuclei contain a mechanism for the transfer of a single volley to a rhythmic 10/sec sequence without the presence of the cortical area to which the thalamocortical fibers project. In 1951 Chang advanced the hypothesis that a cortico-thalamic reverberating circuit should be the basis for the evoked rhythmic activity. The arguments for this explanation were the presence of a similar rhythmic activity in the thalamus and cortex, and the difficulty of recording thalamic rhythmic activity after removal of the appropriate cortical projection area. However, this theory is contradicted not only by the early reports of Adrian (1941) and Bremer and Bonnet (1950), but also by more recent observations by Adrian (1951), who critically tested the cortico-thalamic reverberating hypothesis, and by Galambos et al. (1952). Full support for the statement of Adrian and Bremer was given by Andersen.

24.5 Multifunctional and Selectively Distributed 10 Hz Oscillations – A New Survey

Başar (1980) reviewed further the "induced" and "evoked" alpha by stating that the rhythms are generated in distributed parts of the brain, and not in a given unique structure. Furthermore, if the brain is brought to an excited state, either by means of sensory stimulation or cognitive tasks, it can generate evoked alpha rhythms. These rhythms are not the alpha states fitting in the conventional description of the alpha activity, as stated by the rules of the EEG Society. On the contrary, if the brain is excited with a sensory stimulation, it shows short damped oscillatory behavior with a duration of approx. 300 ms upon stimulation.

Further results of Başar et al. (1979a,b) demonstrated that if a cat has been stimulated with a sensory stimulation (light or sound), an increased coherence in the 10 Hz range between all structures in the visual pathway, hippocampus, and reticular formation was observed, for all pairs of recordings (see also Chap. 1) in the visual and auditory pathways. We repeat here again: We have assumed that a central or a common mechanism could put all these various centers into a state of 10 Hz oscillatory response; i.e., upon application of sensory stimulation various structures in the cat brain generate evoked rhythms of 10 Hz in all structures at the same time (Başar 1980; see also Chap. 28 describing 10 Hz models). Furthermore, it should be emphasized that phase spectra between all pairs of studied structures did not show any phase shift in the 10 Hz frequency range. Now the survey by Başar (1980) will be extended by a survey considering the new results on alpha responses.

I. ***Alpha response at human occipital electrode*** upon sine modulated light stimulation (Van der Tweel and Verduyn Lunel 1965; Spekreijse 1966; Regan 1966b).

II. ***Strong alpha resonance*** in the alpha channel in the auditory cortex, thalamus, reticular formation, inferior colliculus, hippocampus, and cerebellar cortex of the cat upon acoustic stimulation (Başar et al. 1975b,c and 1976a; see also Chap. 1). "Strong alpha resonances" are manifold with enhanced 10 Hz band oscillation following stimulation. The duration of the oscillation is 250–300 ms, and the enhancement factors reach values higher than 1.5.

III. ***Strong resonant alpha responses of 12 Hz upon light stimulation*** observed in the visual pathway, reticular formation, and hippocampus of the cat (Başar et al. 1977a; see also Chap. 1).

IV. ***Perfect time locking*** in the alpha channel during spindle sleep in the cat cortex following acoustic stimulation (Başar et al. 1975c and 1976b).

V. ***Time locking in the alpha frequency range at scalp electrodes*** of human subjects with closed eyes upon auditory stimulation (Başar et al.

1976c). This is the time-locked, but not enhanced, response to inadequate stimulation.

VI. ***Enhancement and perfect time locking*** in the alpha frequency channel at scalp electrodes upon acoustic stimulation of human subjects with open eyes (Başar et al. 1976b; see also Chap. 9).

VII. ***Enhancement and time-locked alpha response*** upon photic stimulation with 50–150 ms latency at scalp electrodes of human subjects sitting in a dimly illuminated room (Başar et al. 1977b). This response is to be differentiated from the alpha blocking seen in spontaneous EEG when a subject opens his or her eyes.

VIII. ***Alpha after-discharge.*** Enhanced alpha component with 500–600 ms latency (following alpha-blocking) at human scalp electrodes (Barlow and Estrin 1971; Lansing and Barlow 1972; Nogawa et al. 1976).

IX. ***Strong resonant alpha responses*** in various brain structures such as the auditory cortex, medial geniculate nucleus, inferior colliculus, reticular formation, and hippocampus of the cat during slow wave sleep stage (Başar et al. 1975c and 1979b; Özesmi and Başar 1974). The frequency is slightly higher (see Chap. 5).

X. ***Induced 10 Hz rhythm without a physical stimulation.*** Anticipatory 10 Hz oscillations that are phase-locked to the moment of a cognitive target (for a review see Başar et al. 1989 and Chap. 22).

XI. ***Prolonged alpha-response component of the P300 response.*** In experiments with certain cognitive tasks, induced EEG rhythmicities show an event-related prolongation (Başar-Eroğlu et al. 1992; Stampfer and Başar 1985, 1988; Chap. 15). This was observed by comparing EPs obtained in the following paradigms:

1. "Sensory EP" sessions consisted of standard EP measurements obtained in response to tone bursts of 2000 Hz frequency.
2. In "cognitive" sessions, the same stimuli were applied, but subjects had been instructed to wait for deviant tones of 1950 Hz frequency.

For data analysis, we combined digital filtering and a single sweep wave identification method (Kolev and Daskalova 1990).

The time locking of single sweeps was prolonged in "cognitive" sessions. Prolonged responses were observed both in the theta (4–7 Hz) and in the alpha (7–13 Hz) ranges. Examples of prolonged alpha oscillations are given in Fig. 24.2, which shows averaged EPs (solid lines) and histograms (bars) obtained by the single sweep wave identification method in four subjects. The left column refers to "sensory EPs" sessions, whereas the right column refers to "cognitive" sessions. The histogram bars show a longer time locking of sweeps in "cognitive" sessions than in "sensory EP" sessions, resulting in a prolonged oscillation visible in the averaged EP.

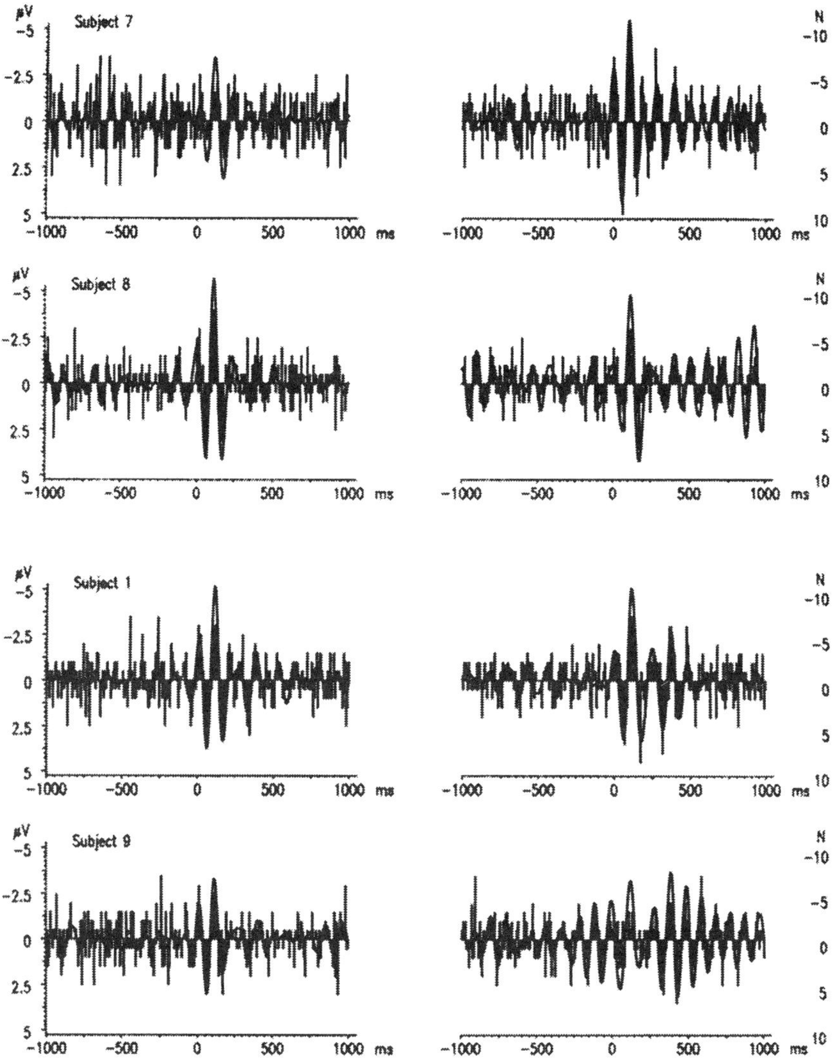

Fig. 24.2. Filtered averaged EEG-EPs (lead: Cz) and corresponding SSWI histograms from all single sweeps for 4 subjects in experimental session without (*left column*) and with task (*right column*). Frequency limits are 7 to 13 Hz. *Along the left y-axis*: amplitudes of EPs in μV (referring to curve), *along the right y-axis*: sum of counts (referring to SSWI histogram bars; for computation, see Kolev V. and Schürmann M. (1992) *International Journal of Neuroscience* 67: 199–213). Time interval for histogram plotting: 8 ms.

It was hypothesized that the prolonged endogenous ERP components may be due to induced rhythmicities triggered by additional information processing when a high degree of certainty has to be resolved during the "cognitive" session (Kolev and Schürmann 1992; for the applied methods with histograms see Chap. 6 in Vol. I).

XII. ***Event-related synchronization and desynchronization of alpha in relation to movement and memory.*** The terms "event-related desynchronization" (ERD) and "event-related synchronization" (ERS) are used by Pfurtscheller and coworkers to describe the generation of more or less coherent oscillating potentials. In item XIIa, the role of alpha and mu ERD in the framework of movement-related oscillations will be discussed. Item XIIb is about the relation between ERD/ERS and memory processes.

XIIa. ***Movement-related changes of mu and central beta rhythms (Fig. 24.3a,b).*** Voluntary self-paced movement can be divided into 3 phases: preparation and planning, execution, and recovery from movement. Each of these phases has its characteristic spatiotemporal patterns of EEG reactivity. As an example, preparation for finger movement is accompanied by a contralateral mu (beta) desynchronization starting about 2 sec prior to movement onset. This event-related desynchronization (ERD) becomes bilaterally symmetric during movement and recovers slowly within some seconds after movement onset. The mu ERD is focused to electrodes C3 and C4, both overlaying primary sensorimotor hand areas. This premovement ERD has a similar time course with brisk, preprogrammed and slow, feedback-controlled movement (Stancak and Pfurtscheller 1996) and is interpreted as a correlate of the readiness of all sensorimotor structures needed to perform a motor task.

Embedded in the bilateral mu ERD during and after movement are short-lasting beta bursts (postmovement beta synchronization; Pfurtscheller et al. 1996) that display their maximum about 0.6 s after movement offset. These beta bursts have a clear contralateral dominance and are more circumscribed than the mu ERD and are focused to the primary motor area. Finger movement results in a postmovement beta synchronization over the primary hand area; foot movement results in a beta synchronization over the foot area close to the vertex. The interesting observation is that finger movement results in beta oscillations mainly in the frequency band between 16 and 20 Hz, whereas foot movement induces beta oscillations between 18 and 25 Hz. This means that the postmovement beta synchronization is highly movement-specific, not only in its localization but also in its frequency. Higher frequencies can be explained by a smaller cortical circuitry involved in coherent activity.

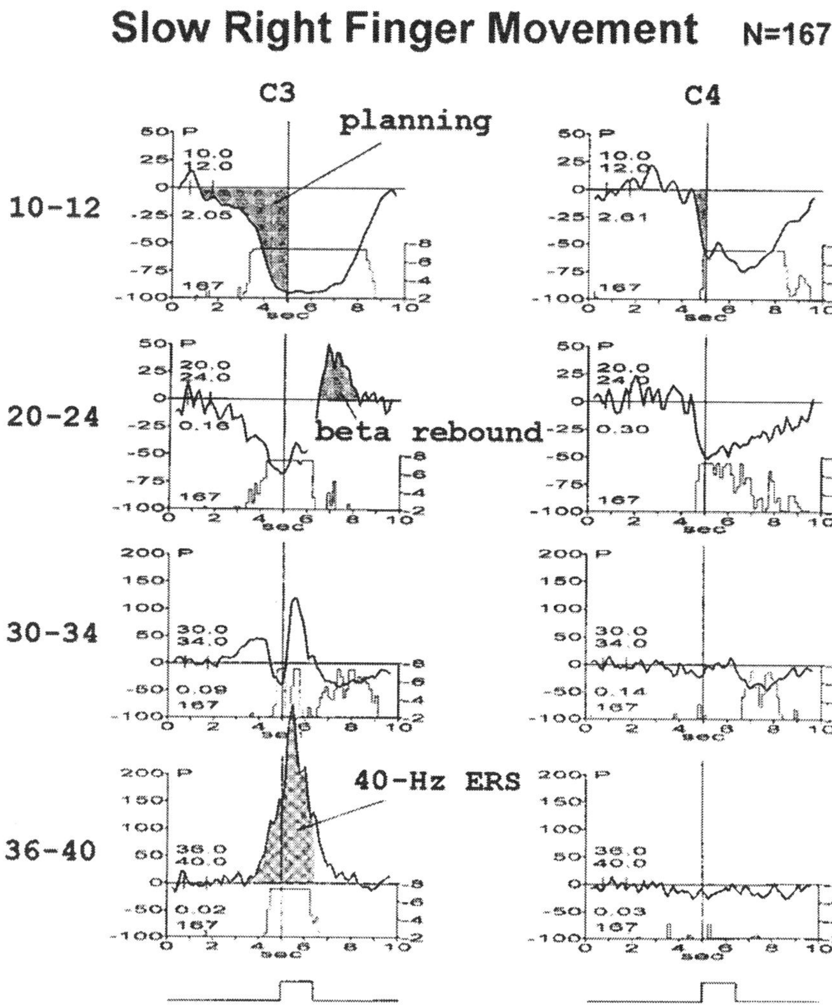

Fig. 24.3a. Band power versus time courses for frequency bands as indicated calculated for EEG signals recorded at electrodes C3 and C4 during slow (duration about 1.5 s) self-paced index-finger extensions; inter-movement interval $> 10s$. The data show the contralaterally starting mu and beta desynchronization (premovement mu ERD), the bilateral mu and beta ERD during execution of movement, the contralateral postmovement beta synchronization (beta rebound) and gamma activity (around 40 Hz) during execution of movement. Note that the postmovement beta synchronization and the gamma bursts are embedded in the desynchronization mu rhythm and the beta bursts, starting clearly after the gamma burst declined. (Courtesy of Dr. G. Pfurtscheller, Graz.)

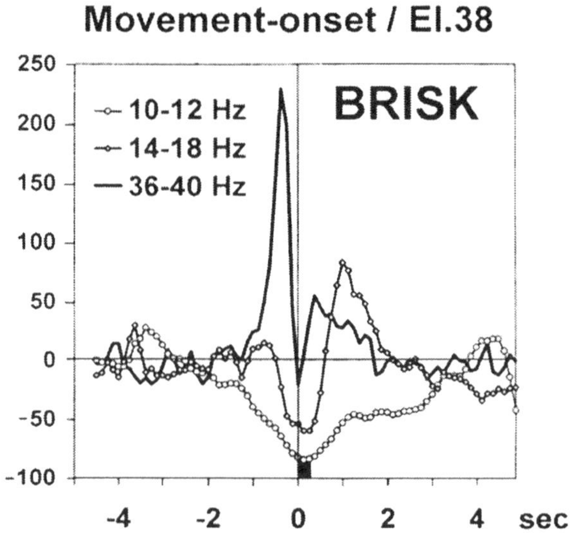

Fig. 24.3b. Superimposed band power versus time diagram for brisk, self-paced finger movements. It can be seen that the mu (10–12 Hz) displays ERD over more than 4 seconds, starting about 2 seconds prior to movement onset. Shortly before movement onset, gamma bursts (36–40 Hz) are found, and after termination of movement beta bursts (14–16 Hz) are found. Both beta and gamma bursts are embedded in the desynchronized mu rhythm. (Courtesy of Dr. G. Pfurtscheller, Graz.)

Execution of movement is characterized by maximal desynchronization of mu and central beta rhythms in both hemispheres. Recovery from movement is accompanied by a contralateral beta synchronization.
Besides postmovement bursts of beta activity with frequencies below 30 Hz, bursts of gamma activity around 40 Hz before and/or during execution of movement can also be found (Pfurtscheller and Neuper 1992): These 40 Hz oscillations are characteristic for Rolandic areas during sensorimotor attention and/or motor preparation and are found during information processing. Large-amplitude gamma oscillations have, therefore, a completely different meaning than large alpha waves, which are characteristic of "idling" cortical areas. In other words, excited cortical structures can show a *desynchronized alpha-band activity* and embedded bursts of gamma oscillations. Beta oscillations between 13 and 30 Hz are in between the alpha and gamma frequency ranges, and are found after a motor act is terminated. Whether at this time the motor cortex is "idling" or is actively processing (e.g., reafferent information from the receptors involved in the movement) or is involved in cancellation of the motor program can only be speculated upon and needs further investigation.

It is interesting to note that bursts of movement-related gamma oscillations over primary motor and supplementary motor areas display a high linear coupling (Andrew and Pfurtscheller 1996), whereas the gamma and mu oscillations of the sensorimotor areas of the two hemispheres do not display bilateral coupling.

It should be noted that induced sensory alpha responses and anticipatory alpha activity emitted by the subject expecting auditory or visual targets in the work of Başar et al. (see Chap. 22) are not in contradiction to results of the Pfurtscheller group. Sensory alpha enhancements can be elicited only if the amplitude of the prestimulus alpha activity is low and the stimulation signal is adequate to the recording site. The enhanced or sensory alpha responses usually have a duration of 250–300 ms and are time-locked to stimulation. The alpha oscillation prior to targets in cognitive tasks (see Chap. 22) are also different from the phenomena described by Pfurtscheller's group. According to the work of this group, there are several ways to induce different degrees of synchronization in the 10 Hz activity. This phenomenon was also described by Grey Walter (1950, 1964) in an explicit fashion.

XIIb. ***Memory-related alpha desynchronization:*** Klimesch's experimental findings (Klimesch 1996, 1997) support the following hypothesis about the relation between brain oscillations and memory processes: Short-term (episodic) memory demands lead to a synchronization in the theta band, whereas long-term (semantic) memory demands lead to a task-specific desynchronization in the upper alpha band. Klimesch further suggests that short-term (episodic) memory processes are reflected by oscillations in an anterior limbic system, whereas long-term (semantic) memory processes are reflected by oscillations in a posterior-thalamic system (see also Chap. 15 for Klimesch's interpretation).

XIII. ***Spontaneous and evoked alphas in invertebrate ganglia.*** Further evidence showing that the alpha activity cannot be explained through generators only in the thalamus or cortex can be found in the experiments performed on the cerebral ganglion of aplysia (Chap. 6) and also with isolated ganglia of *Helix pomatia.* These studies provided evidence that spontaneous or electrically induced 10 Hz activity can be recorded in vitro in these small neural populations each consisting only of approximately 2000 neurons. Although the amplitudes of 10 Hz activity are fairly low and cannot reach the higher-amplitude synchronized states of the mammalian cortex, there is some good synchrony. Furthermore, amplitudes of spontaneous and induced alpha rhythms can be modulated by using pharmacological agents (Schütt et al. 1992; Schütt and Başar 1992; see Chap. 6).

The classification of sensory or cognitive evoked or induced alphas given in this section is a restricted one, since more examples can be found in the literature concerning alpha oscillatory components. This only serves to give the idea of richness of functionally related alpha components

24.6 Secondary Alpha Response or Alpha Response with Delay

In Chaps. 2 and 10 it was shown that inadequate stimuli could not generate significant and time-locked cortical alpha enhancements in the first 300 ms upon stimulation. The occipital cortex of the cat brain does not respond with enhanced 10 Hz if the stimulation is an auditory one (i.e., inadequate stimulation). Figure 2.8 shows single sweeps and the averaged potentials filtered upon inadequate (auditory) and adequate stimulation in the alpha frequency range (right and left columns, respectively).

In these types of potentials in the poststimulus interval between 250 and 300 ms, a small 10 Hz enhancement can be recognized. The time locking is weak in comparison to results on adequate stimuli. Therefore, although single sweeps may contain fairly high 10 Hz amplitudes, the responses are not perfectly phase-locked, and furthermore, they are delayed (see Chap. 2). These types of responses are also obtained in the auditory cortex by means of visual stimulation (inadequate stimulation) as well as emitted 10 Hz oscillations following the omitted stimuli in the cortex of the cat brain (see Chap. 16).

These "secondary," or "delayed," 10 Hz responses were also recorded from human brains during cross-modality experiments with EEG (see Chap. 10) as well as with MEG recordings (see Chap. 13). What can be the cause of this delayed oscillatory 10 Hz activity, which we call secondary alpha enhancements?

In Chap. 27, a gating mechanism related to the thalamic nuclei for the alpha oscillatory response is discussed. Thalamic 10 Hz enhancement is recorded only if the stimulation is an adequate one. This means that the medial geniculate nucleus depicts 10 Hz alpha enhancement only if auditory stimulation has been applied, but it does not react, upon visual stimulation, with an immediate 10 Hz enhancement (i.e., within the first 200 ms). In contrast, hippocampus always reacts with an ample 10 Hz oscillatory behavior upon stimulation. This oscillatory response of hippocampus is around 9 to 10 Hz for auditory stimulation and approx. 12 Hz for visual stimulation. It seems that gates open to every type of sensory stimulation in the hippocampus and that this structure merits the expression "supramodal gate par excellence" as described by Swansson (1981; see Chap. 3 in Vol. I) also with the analysis of EPs.

We focus our attention now on the model concerning the generation of resonance loops between hippocampus and cortex as described by Miller (1991; see Fig. 24.4). Miller has estimated the time of signal transfer between hippocampal formation, prefrontal cortex, and nonlimbic association cortex using conduction velocities. The schematic presentation by Miller (Fig. 24.4) plausibly shows delay times in the various parts of a cortical hippocampal loop. The total loop time may be more than 200 ms, which means that one way for conveying information from the hippocampal formation to the nonlimbic association cortex may take around 120 to 150 and up to 200 ms.

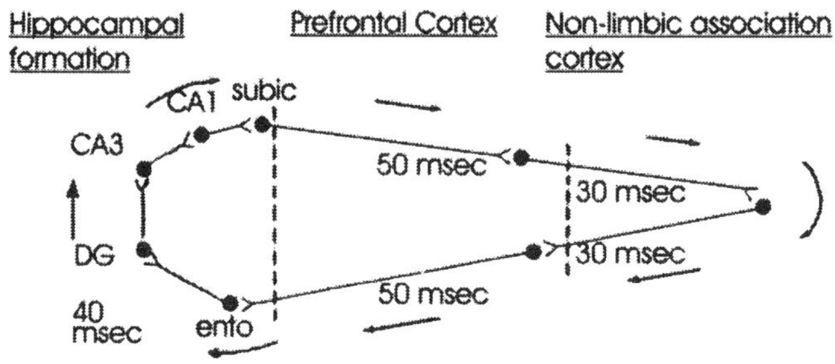

Fig. 24.4. Schematical representation of plausible delay times in the various parts of a corticohippocampal loop; the 40 ms delay within the hippocampus is derived from data presented by Miller (1991). ento; Entorhinal cortex; DG; denated gyrus; subic; subiculum. Total loop time may be in excess of 200 ms. (From Miller R. (1989) *Psychobiology* 17: 115–128.)

According to the experimental results presented in Chaps. 1 and 27, all sensory stimulations evoke a 10 Hz oscillation with a duration of 250–300 ms in the hippocampus. The expression *every sensory stimulation* has to be pointed out here. The sensory stimulation may be an auditory one, in which case it will evoke 10 Hz oscillatory waves, or it can be a visual one, which then triggers a 12 Hz oscillation immediately following this stimulation. This sensory information arises upon stimulation, most probably conveyed to the association cortex and to frontal nonlimbic association cortex as well to other parts of the cortex, and also to the primary sensory areas. A current spread or volume conduction can also exist, but most probably the volume conduction is negligible, because of the long distances between auditory and visual cortices. Now, when recording a response with our electrodes in a primary sensory area we would have to add 20 to 50 ms.

In other words, if the hippocampus generates an oscillatory 10 Hz pattern following every stimulation mode and if the hippocampus transmits this signal to other association areas, then the primary sensory areas would receive information approx. 150 to 200 ms following the application of the stimulation, i.e., later. Accordingly, it can be hypothesized that the secondary 10 Hz oscillations recorded in the primary visual or auditory areas have their pacemaker source in the hippocampus or in the brainstem. At least our data may explain or show that such a transfer of information in the 10 Hz frequency range from hippocampus to cortex is possible. Hippocampus itself receives signals from brainstem structures such as the reticular formation (see Chap. 3 in Vol. I).

If an auditory stimulation generates 10 Hz activity in the reticular formation, this signal can be hypothetically transferred to hippocampus and from there to prefrontal cortex and nonlimbic association cortex. In this case there is the possibility that the signal can reach the primary sensory areas, too, via the nonlimbic association cortex, as recorded in the cortical potentials (Fig. 24.5).

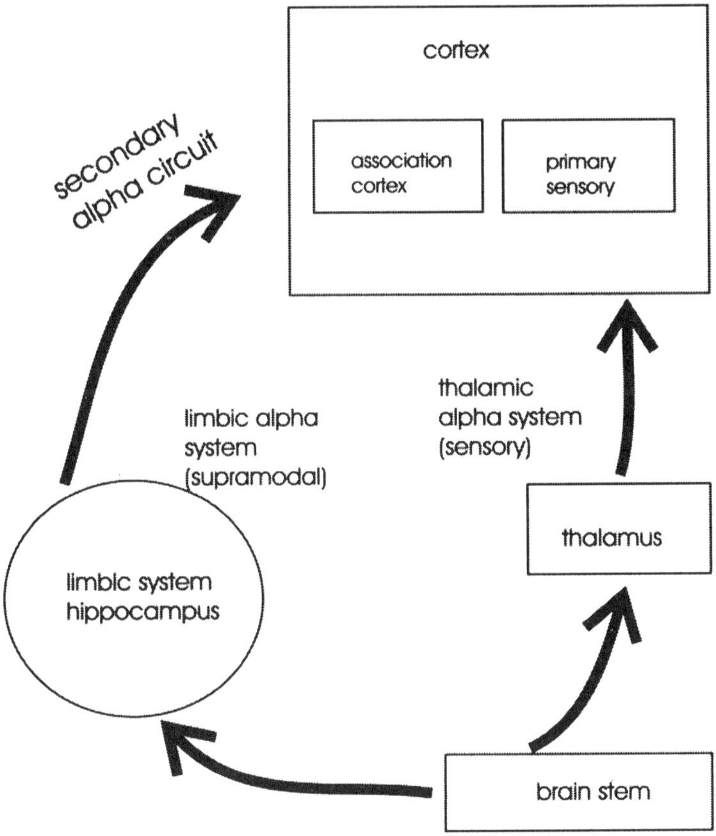

Fig. 24.5. Hypothetical diagram of the primary and secondary alpha circuit.

In this way not only do specific afferents from sense organs not reach the primary cortical areas, but the same sensory information will be partly sent to association areas of the cortex via the limbic system (hippocampus). On the other hand, the sensory input reaches the mesencephalic reticular formation. As stated by the theory of Hernandez-Peón (1961), the reticular formation acts as a filter and is also a gate structure par excellence, controlling the information flow to the primary sensory areas and the limbic system as well as to polymodal association areas of the cortex (see Chap. 3 of Vol. I).

The 10 Hz response recorded in the reticular formation has also marked enhancements similar to those in the hippocampus. The information flow over the mesencephalic reticular formation also produces "delayed signals." This signal transmission over the reticular formation can also be considered as an additional system acting parallel to the prolongation of the 10 Hz activity. If we follow the same scheme used for the hippocampus, the delay of signals over the reticular formation would again be in the range of 200–300 ms.

24.7 Synopsis on Multiple Functions of "Alphas"

We will group results and interpretations of functions of alpha generators into 4 categories:

24.7.1 Memory Mechanisms and Alpha

Klimesch (1993, 1996, 1997) claims that the vast majority of experiments correlating alpha with cognitive performance are not concerned with memory processes. The general view here is that in comparison to a resting period, task demands tend to attenuate or desynchronize alpha rhythms (Berger 1929).

As compared to the abundance of experiments dealing with alpha power measurements, relatively few studies focused on task-related shifts in alpha frequency. The experiments by Osaka et al. (1984) showed that only for difficult but not for easy tasks, alpha frequency increases selectively in the hemisphere that is dominant for a particular task. Clear evidence of a positive relationship between dynamic memory performance and increase and synchronized alpha activity was indicated by Başar et al. (1985, 1987, 1989). Again, clear evidence for a positive relationship between memory performance and mean alpha was reported by Klimesch et al. (1990).

EEG signals were recorded by Klimesch et al. (1993) from subjects performing a modified version of Schneider's and Shiffrin's memory-search paradigm in order to test whether individual alpha frequency (IAF) is related to memory performance and/or attentional demands. The results of their investigations showed that memory performance has a strong effect on individual alpha frequency. As compared to a resting period, the difference in individual alpha frequency between age-matched *good* and *bad memory performers* reached a maximum when subjects were actually retrieving information from their memory. During retrieval, the alpha frequency of good performers is 1.25 Hz higher than for bad performers. The results of amplitude analyses demonstrate further that during retrieval, alpha desynchronization is more pronounced for bad performers than for good performers.

Results of Klimesch (1996) and of Başar et al. (1989; see Chap. 22) on dynamic memory demonstrate that alpha activity is strongly correlated with working memory and probably with long-term memory engrams (for Klimesch's interpretation see also Chap. 15).

24.7.2 Alpha as Sensory Response

Since the days of Adrian, "evoked alpha" was regarded as a sign of the reactiveness of the CNS to sensory stimuli. Sensory evoked 10 Hz responses were recorded in several cortical and intracortical structures (see Chaps. 1 and 9). The sensory stimulation can evoke 10 Hz oscillatory behavior also at the cellular level (Creutzfeldt et al. 1966; Dinse et al. 1997; Silva et al. 1993 and Chap. 9 in Vol. I).

24.7.3 Alpha and Motor Processes

Earlier in this chapter the results of Pfurtscheller and coworkers on motor correlates of alpha activity were explained in detail. Here we should again mention the statements of Llinàs (1989), who assigned to oscillations and resonances in the central nervous system diverse functional roles. This author assumed that oscillations play a role in determining global functional states (for example, sleep–wakefulness or attention, and timing in motor coordination). Both these examples and especially the results of Pfurtscheller et al. (1996) show how the alpha activity is involved in movements.

24.7.4 Association Mechanisms and Attention

In a series of publications Norbert Wiener (1955, 1956, 1957) advanced the concept that the alpha rhythm might give a reflection of a relatively stable clocking mechanism of the brain, to serve a gating function. Wiener ascribed the hypothesized clocking mechanism to mutual entrainment of an ensemble of nonlinear alpha-rhythm generators such that the frequencies of individual members of the ensemble would have a tendency to be pulled towards a central frequency, in a manner analogous to that in which large numbers of some species of fireflies flash synchronously.

Several results in distributed cortical and intracranial structures underlined in the present chapter lead definitely to the conclusion of correlation of alpha with movements, memory, sensory processing, attention, and last, but not least, with general association mechanisms. The immediate question that now arises is as follows: *Are different alphas, as selectively distributed generators involved in general mechanisms of the brain, deterministic signals coming from hidden or undefined sources that work in a random manner?* The answer, based on strong empiricism, is yes.

As will be described in Chap. 29, a unique frequency component is, in general, not sufficient to find the exact electrical correlates of a brain function. The experimenter has to pay attention to the 10 Hz components (or to the several alphas ranging from 7 to 13 Hz) in his experimental strategy or design.

As several results in Chap. 12 in Vol. I also emphasize, the amplitude, time course, and frequency contents of evoked potentials (especially N100-P200 wave complexes) strongly depend on the amplitude of alpha activity prior to a sensory stimulation. Recent publications by Brandt and Jansen (1991) and Brandt (1997a) confirm the findings of Başar (1980) regarding the inverse relationship between evoked potentials and prestimulus EEG. In addition, these authors describe a model presenting evidence that the generation of spontaneous EEG activity can acquire the waveforms of EPs when pulse-like signals serve as input. In particular, these authors describe a marked correlation between N100 amplitude and prestimulus EEG. The data of Jansen and coworkers provide support to the hypothesis that the spontaneous 10 Hz activity and visual evoked potentials are generated by some of the same neural structures and that the visual evoked potentials are due to distributed action rather than dipolar sources.

Assuming that several 10 Hz generators are distributed in the brain and further supposing that these activities act as nonlinear clocks (as hypothesized by Norbert Wiener), then we are confronted with the question whether 10 Hz processes may facilitate, *overall*, association mechanisms in the brain. When a sensory or cognitive input elicits "10 Hz wave trains" in several brain structures, then it can be expected that this general activity can serve as *a communication signal par excellence* between different structures (see also results by Başar et al. (1997b)).

24.8 "Alphas" Selectively Distributed in the Brain

The expression *distributed alpha system* points out that there are various structures in the brain that are able to produce 10 Hz activity either spontaneously or by a deterministic sensory or cognitive event that may be either external or internal. The determinism may be such that the occurrence of the oscillation is directly related to the stimulation condition; in such a case we speak about *event-related oscillations.* Or the determinism may be such that the occurrence does not directly follow the stimulation; in such a case we speak of induced oscillations (see Chap. 22).

In this chapter we have discussed at least 20 types of functional correlates of the alpha activity or of the alpha response. In accordance with these considerations it is plausible that alpha networks are distributed in the brain in a selectively distributed way. These networks exist in the brainstem, in cerebellum, in thalamus, in the limbic system, in sensory and association and motor areas of the cortex. The alpha oscillatory behavior is involved in responding to sensory or cognitive inputs and is dependent on stimulus-adequate circuits of the brain. These experimentally established working types of the alphas might play an essential role in sensory and cognitive communication in the brain. We assume that the combination of these alpha networks is a powerful dynamic property of the brain.

24.9 An Integrative Theory of Alphas

(1) Different 10 Hz oscillations (alphas) can be recorded from various areas of the human scalp and intracranial structures of the cat brain. They are related to diverse sensory and/or cognitive functions.

(2) *"Spontaneous" alphas are not noise but probably quasideterministic signals."* 10 Hz oscillations of the brain are not pure noise as demonstrated by several results: Best examples are the analysis of correlation dimension and phase-locked and reproducible 10 Hz rhythms preceding a cognitive target.

(3a) *Evoked 10 Hz oscillations* can be generated in several structures of the brain simultaneously. The damped oscillations (of approximately 200–300 ms duration) after sensory stimulation are sensitive to the modality of stimulation and the recording site.

(3b) When subjects are given a cognitive task, *event-related alpha rhythms* may be prolonged (cognitive components).

(4a) *Alpha does not necessarily reflect "passive states" or "idling" of the brain.* This is a consequence of items 3a,b. Moreover, it is possible that passive states or idling are components of general states of the brain, their causality being yet unknown (hidden causality).

(4b) *Alphas have multiple functional correlates.* 10 Hz rhythms have important functional correlates including sensory, motor, and memory functions.

(5) *Alpha rhythms at the cellular level.* Spontaneous or evoked 10 Hz oscillations are also measured at the cellular level; thus 10 Hz activity reflects basic physiological properties of the brain.

(6) *Distributed alpha system.* Rather than trying to locate a unique alpha generator, it is preferable to formulate the existence of a "diffuse and distributed alpha system" (Başar and Schürmann 1996). The following physiological results concerning *hippocampal* alpha responses illustrate this proposition:
   (i) Auditory and visual stimulations elicit in cat hippocampus strong and stable alpha responses (10 Hz oscillations of approx. 300 ms), which are visible without filtering.
   (ii) Cortical and thalamic 10 Hz responses can be elicited only by stimulations that are adequate for the respective area. In contrast, hippocampal 10 Hz responses are present in all types of stimulations.
   (iii) In coherence functions computed from visual evoked responses, the hippocampocortical coherence is significantly larger than the thalamocortical coherence (the latter being extremely low following inadequate stimulation).

   Thus, thalamocortical circuits are not unique in generating alpha responses; the hippocampus may even have a more general significance.

(7) *Alpha as a universal code or universal operator.* The results mentioned above underline the major physiological meaning of 10 Hz oscillations, which may be comparable to the putative universal role of gamma responses in brain signaling.

25. Theta Rhythms in Integrative Brain Function

25.1 Functional Importance of Theta Rhythms

Possibly, theta rhythms are the most intensively investigated phenomenon in neurophysiological studies aiming at correlating rhythms with cognitive functions. In the present chapter, we will emphasize the recent theory developed by Miller (1992) about the role of theta rhythm in integrative brain function. We will further pursue functional correlates and integrative aspects of theta oscillations and underline the concept of the "selectively distributed integrative theta system" in the brain. This means that there are various structures in the brain that are able to produce theta activity either spontaneously or by a deterministic sensory or cognitive event that may be either external or internal. The determinism may be such that the occurrence of the oscillation is directly related to the stimulation condition; in such a case we speak about *evoked oscillations*. Or the determinism may be such that the occurrence does not directly follow the stimulation; in such a case we speak of *induced oscillations*.

In several earlier chapters of this book it was stated that event-related theta oscillations are highly correlated with mechanisms of associative learning and attention as well as retrieval (see Chaps. 14, 18, 19, and 21). Theta oscillations are not the only ones that are involved in these cognitive processes. However, the distributed theta system of the brain may be the system with the most influence on integrative cognition and association. Results and discussions on the functional correlates of induced 10 Hz and 40 Hz rhythms show that event-related oscillations in higher frequency ranges are also involved in cognitive processes. Further, the ERPs describing complex functions do not represent homogeneous oscillations but a superposition of oscillations over a wide frequency range (see Chaps. 29 and Chap. 14 in Vol. I).

25.1.1 A Summary of Theta Rhythms in the Limbic System

Hippocampal theta activity has been extensively analyzed since the pioneering work of Grastyán et al. (1959) and Adey et al. (1960). In Table 25.1 there are a few examples of cat experiments during exploration and searching behavior as well as during motor behavior. *Orienting* is a coordinated response

Table 25.1. Correlations between orienting behavior and rhyhmic slow activity (RSA) in cats (From Miller R. (1991) *Cortico-Hippocampal Interplay and Representation of Contexts in the Brain.* Berlin, Heidelberg, New York: Springer.)

Reference	Circumstances of experiment	Findings
(A) Spontaneous orienting		
Radulovacki and Adey (1965); Whishaw and Vanderwolf (1973)	Spontaneous orienting	4–5 Hz RSA
Brown (1968); Kemp and Daada (1975)	Active visual search and fixed	RSA present (frequency higher during active visual search)
Bennett (1970	Periodic orienting, after exploratory behavior had habituated	RSA present
Bennett and Gottfried (1970)	Noncued DRL performance	NO RSA except during orienting
Berrett et al. (1973)	Continuously reinforced lever pressing	No RSA except during orienting
Kemp and Kaada (1975)	Active visual search	RSA correlated with pupillary dilatation and degreee of eye opening
Coleman and Lindsley (1975)	Alerts vs. relaxted posture	RSA present during alertness
Radil-Weiss et al. (1976)	Acute pretrigeminal cat	RSA correlated with ocular and EEG signs of orienting
Gralewicz (1981)	Shuttlebox avoidance	RSA during attentive postures
(B) Orienting acquired during conditioning		
Grastyan et al. (1959); Lissak and Grastyan (1960)	Appetitive and excape tasks	CS came to elicit both RSA and orienting during training. With overtraining both disappear together
Brown and Shryne (1964)	Repeated tones	RSA and orienting develop together
Brown (1968)	Repeated tones	RSA initially absent, despite orienting, but desvelops later at times of orienting
Bennett (1970)	Clicker signaling food delivery	Orienting and RSA develop together, and disappear together at al later stage of training
Grastyan and Vereckei (1974)	Appetitive training, CS and goal box spatially seperate	Orienting and RSA develop tohether, and disappear together at a later stage of training
Coleman and Lindsley (1977)	Continuously reinforced lever pressing; or alternating periods of reinforcement and nonreinvorcement	RSA develops in association with attentive postures and orienting
Buszaki et al. (1981)	GO/NO training	RSA frequency increased as orienting developed
Buszaki et al. (1981)	Unavoidable shock signaled by tone CS	Orienting declined although RSA at CS persisted

that appears to indicate *alertness, arousal*, or *readiness* to process information. Orienting is, of course, closely linked to attentive states and learning. This table contains only examples of the cat hippocampus. An extended review of functionally related theta activity is given by O'Keefe and Nadel (1978).

The examples shown in the table make clear that it is usually not possible to relate spontaneous and induced rhythmicities only to a single function (see also Chap. 24). At least 20 papers indicate that hippocampal theta activity can be controlled in various brainstem nuclei (Vertes et al. 1982). Accordingly, the concept of distributed theta networks in the brain and the multifunctional character of theta activities should be considered when trying to understand integrative brain function.

According to Lopes da Silva (1990, 1992) the rhythmic slow activity (RSA), or theta rhythm, may be considered the *fingerprint* of all limbic structures, although it is most prominent in the hippocampal formation. Lopes da Silva suggests three types of functional correlates of RSA:

1. RSA may serve a *gating function* on the flow of information through the hippocampus.
2. RSA may facilitate the *matching* of the *resonance* between the hippocampal formation outputs and the circuits of target structures.
3. RSA can have a *facilitating* role in the *induction/enhancement* of long-term potentiation (LTP) in the circuits of the hippocampus and of target structures.

Further, this author puts forward the hypothesis that *oscillatory states* in neuronal networks may constitute a mechanism used by the nervous system to regulate changes of state in these networks.

Swanson (1983) concludes that the *hippocampal formation* may be thought of as a class of complex *supramodal association cortex* that receives and integrates information from each of the sensory modalities and projects back to complex polymodal and other supramodal association areas (to visceromotor control systems, particularly in the hypothalamus) and to the striatum. The conclusion of Swanson is based on several physiological and anatomical investigations (e.g., Jones and Powell 1970; Mesulam et al. 1977). Besides theta resonances, 40 Hz resonances to acoustical stimuli and the delayed P300 40 Hz complex during cognitive tasks in the cat (see Başar-Eroğlu and Başar 1991 and Chap. 17) are also observed in the hippocampus (see Table 25.1). The hippocampus, as a supramodal association structure, is involved in several spontaneous and induced rhythmicities. Accordingly, it is to be emphasized that the supramodal character of hippocampus cannot be described alone with its marked theta activity (Horowitz et al. 1973, Leung et al. 1992).

25.2 Earlier Experiments on Induced or Evoked Theta Oscillations

Adey et al. (1960) undertook early, pioneering work on theta rhythms of the limbic system of the cat brain during conditioning where they used spectral and coherence functions (see also Elazar and Adey 1967 and Table 25.1). These investigators described relevant experiments demonstrating that rhythmic field potentials of the cat brain are related to behavior.

The use of the coherence function comparing EEG activity in various nuclei of the brain was useful in refuting the view that "EEG is an epiphenomenon" (Adey 1989). The induced theta rhythm and the *task-relevant coherence* in the limbic system of the cat brain is a milestone in EEG research. During these experiments the cat hippocampal activity exhibits a transition from irregular activity to coherent induced rhythms. Such results have encouraged Başar and Özesmi (1972) and Başar and Ungan (1973) to choose the hippocampus as a model for a possible resonance theory of the brain. The experiments performed by this group showed that such an explanatory model was feasible.

The reader is referred to Chap. 1 for an illustration of a hippocampal EP of a waking cat, recorded after an auditory stimulation, and for the respective amplitude–frequency characteristics obtained by means of Fourier transform. A relevant theta response dominates the frequency response of the hippocampus.

Hippocampus has two remarkable types of regularities that may be relevant in the present context:

1. The anatomical organization of the hippocampus is not diffuse. It is morphologically precisely defined in an arrangement of pyramidal cells and neural populations called regions CA1, CA2, and CA3 (see Chap. 3 in Vol. I).
2. Hippocampus generates at least two types of regular theta rhythms depending on the behavioral state of the animal and also on the task used (see Table 25.1).

Başar (1980) assumed that if the brain receives a sensory stimulation and if a structure has its own intrinsic activity, then the structure would respond with its natural frequencies (see also Chap. 14 in Vol. I). Since hippocampus is a supramodal structure par excellence having the theta rhythm as a major natural frequency, a sensory stimulation should bring this structure to a type of theta oscillation manifested in EPs (see below in this chapter for a description of hippocampal theta cells). The hippocampal theta resonance was considered by Başar and Özesmi (1972) as an important model to start a resonance theory of the brain. It was stated that the evoked potentials of hippocampus reflect ample theta resonances. This view was also one of the core ideas in "EEG-Brain Dynamics" (1980).

25.3 Correlating with Orienting: Review by Miller

A theory of how the hippocampal theta rhythm can entrain populations of neurons in the cerebral cortex into resonant neural activity is extensively discussed by Miller (1991).

Miller summarized that hippocampal theta activity occurs in cats in association with locomotion and other body movements, in a manner similar to that seen in rats and other small mammals. However, there are a great many exceptions to these correlations, where theta activity can occur during immobility, or movement can occurs without theta activity. Many of these exceptions occur in the course of learning tasks. There are suggestions that theta activity occur at the time when performance of a learned task is improving most rapidly and declines as tasks become familiar. Apart from these results, there is a great deal of evidence that both spontaneous and conditioned orienting is accompanied by low-frequency theta in the cat.

Petsche et al. (1962) followed up the early evidence from lesion experiments that showed a septal involvement in the theta rhythm by recording from units in the septum of curarized rabbits. Theta activity was elicited by sensory stimulation or electrical stimulation of the *reticular formation.* A proportion of septal units showed regular rhythmic bursts of impulses, in phase with the hippocampal theta rhythms (see Fig. 25.1).

These findings have been confirmed many times: Vinogradova and Zolotukhina (1972), studying alert rabbits, found that only one-third of neurons in both medial and lateral septum showed "theta bursts." Apostol and Creutzfeldt (1974), using curarized rabbits, recorded simultaneously the septal and hippocampal EEG and the spike activity of individual septal units during spontaneous rhythmic activity or rhythms evoked by sensory stimuli.

Brazier (1968) recorded from the hippocampus in 30 patients prior to surgery (28 cases of temporal lobe epilepsy, 2 nonepileptics). The hippocampal EEG showed peak power in the 2–4 Hz range, but peak of coherence with EEG from other sites (e.g., parahippocampal gyrus) was in the 4–8 Hz range. Lieb et al. (1974) recorded from the hippocampus, amygdala, and parahippocampal gyrus. A small peak was seen at 8–10 Hz, variable between patients, with a broad peak at slower frequencies. Halgren et al. (1978) made some behavioral observations in a single-case study. The hippocampus was synchronized at 5–6 Hz during quiet resting, when the patient tensed all his muscles or made simple alternating movements. Increased synchrony was also seen during testing of patients on a series of word meanings or a tapping sequence. Desynchronization occurred during speech or rapid breathing, tying a bow, imitating movements of the experimenter, or while the patient was giving a verbal description of a word.

Miller describes rhythmic activity in the hippocampus as not depending on a rhythmic input from the septal regions (see Chap. 3 in Vol. I). The isocortex also influences the hippocampal theta rhythm, but it can promote such a rhythm only in so far as the septal neurons are already rhythmically

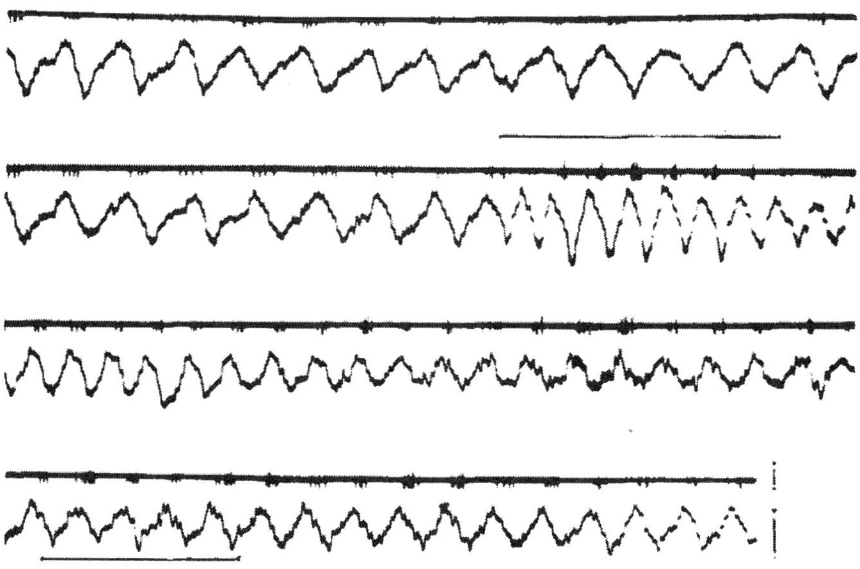

Fig. 25.1. Single-unit recording from a medial septal neuron, showing rhythmic bursting of the neuron (*upper trace*) phase-locked to the hippocampal theta rhythm (*lower trace*). Theta activity is induced by esserine. Sensory stimulation (*horizontal bar*) accelerates the rhythm. Calibration: time, 1 s; voltage; 1 mV for unit, 500 μV for EEG. (From Petsche H. et al. (1962) *Electroencephalography and Clinical Neurophysiology*, 14: 202–211.)

active. The entorhinal cortex appears to have a role that may sometimes be of greater quantitative importance than the rest of the isocortex. The influence of the entorhinal cortex can be explained on the basis of a direct relay of rhythmic signals from septum to hippocampus via the entorhinal cortex. Resonance between hippocampus and entorhinal cortex may also play a role. It is, however, difficult to explain the role of most of the isocortex on the basis of a pathway from septum, above the corpus callosum, to the hippocampus. This leaves Miller with the alternative hypothesis that there is a component of the overall theta rhythm that is maintained by recursive loops reciprocally connecting the hippocampus and cortex via a direct posterior route behind the corpus callosum.

25.4 Theta Activity in the Prefrontal Cortex

There is some recent evidence that the midline prefrontal region of the cortex can generate theta activity in certain cognitive states. This was reported by Mizuki et al. (1980): EEG rhythms of 5–5.5 Hz frequency appeared with some regularity during the performance of simple repetitive mental-arithmetic tasks. It also has been reported that such frontal theta activity is even more common during REM sleep (Hayashi et al. 1986). In addition, Lange et al. (1978), using spectral analysis of frontal EEGs, showed that theta frequencies (3–7 Hz) increased during motor or verbal learning tasks.

Miller (1991) argues that all these data are compatible with the thesis that theta activity in frontal regions is associated with a theta activity in the hippocampus. It might be expected that the prefrontal cortex should show theta activity at the same time as the hippocampus, in view of the strong connections between the two structures. If such a relation could be established, one would have a method of determining the behavioral correlates of at least one aspect of hippocampal theta activity in man, which would be of general importance. Here, we mention the importance of theta activity controlling the reactivity of human frontal lobes to sensory stimulation (see Chap. 12, Vol. I and Chaps. 18 and 19 in the present volume for a description of frontal and hippocampal theta activity).

25.5 Miller's Description of the Relation of Intracellular Potentials to EEG Activity in the Theta EEG Activity

Fox et al. (1983) recorded intracellularly from several pyramidal cells in CA1 and CA3 in urethanized rats and found during spontaneous theta activity that the maximum hyperpolarization occurred just before the positive peak of dentate gyrus theta, i.e., just before the negative trough of CA1 (local) theta activity. Leung and Yim (1986) had similar findings in this preparation. In quantitative terms, negativity of the rhythmic membrane potential lagged behind the negativity of the extracellular theta rhythm by about 260 degrees (or in alternative terms was 100 degrees in phase advance over the extracellular EEG). A third paper (Nunez et al. 1987), using urethanized, curarized rats, showed, for rhythmic potentials, that the negative phase of the extracellular theta activity was always locked to the positive phase or the rising depolarization phase of the intracellular rhythm. These three results for pyramidal cells in CA fields are in rough agreement, as illustrated in Fig. 25.2.

The theoretical phase relation between intracellular and extracellular theta oscillations merits discussion. In theory, if currents flowing into or out of a neuronal element are the sole source of the local extracellular fields, the time of maximum extracellular negativity should coincide with the time of maximum inward flow of positively charged ions. In other words, extracellular negativity should be 90 degrees phase advanced compared with intracellular depolarization.

Miller (1992) draws conclusions on hippocampal sources generating theta rhythms in freely moving rats. According to this author, although a phase reversal may occur in the theta band, the upper part of CA1 is in relation

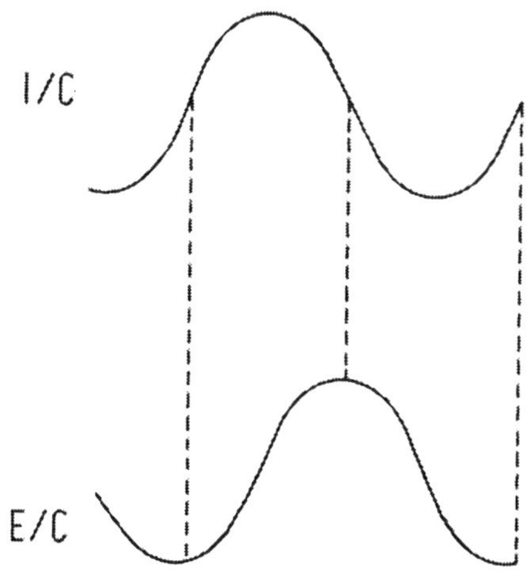

Fig. 25.2. Schematic drawing of the relative phase of sinusoidal potential waves, as recorded intracellularly (*above*) and extracellularly (*below*). Intracellular potentials are 90 degrees phase advanced over extracellular ones (or equivalently, intracellular depolarization lags extracellular negativity by 90 degrees). (From Miller R. (1991) *Cortico-Hippocampal Interplay and Representation of Contexts in the Brain.* Berlin, Heidelberg, New York: Springer.)

to the deeper part of CA1 or the dentate gyrus, to which it appears to be related by a constant delay.

In the freely moving rat, however, theta activity in the superficial part of field CA1, though often approximately phase reversed compared with that in the deeper part of CA1 or the dentate gyrus, actually appears to be related to this by a constant delay. A schematic representation of this relationship is presented in Fig. 25.3. The evidence is consistent with the idea of theta-modulated signals reaching the dentate gyrus and distal apical dendrites of CA1 along the perforant path, these signals having a phase advance (constant duration of lead) over the signals generated in the superficial part of CA1. In this condition there appears also to be theta activity generated in fields CA3 and CA2, which lie on the transverse pathway between the dentate gyrus and field CA1. It is therefore plausible that in the free-moving preparations, the entorhinal cortical input controls theta activity in the dentate gyrus, which then controls CA1 theta activity (Miller 1991).

In Chap. 1 it is shown that the cat hippocampal theta response has greater amplitudes in comparison to all other recordings. Furthermore, Başar-Eroğlu et al. (1991a,b) demonstrated that the theta response to omitted stimuli gave rise to greatest theta enhancements in the CA3 region of the hippocampus. Demiralp and Başar (1992) and Demiralp et al. (1994) further demonstrated that by application of repetitive auditory stimuli, the frontal areas as well as the hippocampal CA3 gave distinct and prominent theta responses with ample enhancements (see also Chaps. 18 and 19).

What we have seen in the above description of Miller's concept is the fact that the limbic system, including hippocampus and septum, and parts of the cortex as enthorhinal cortex and parietal cortex are strongly involved in the resonance loop in the theta frequency range. Although there are at least two different frequency components in the hippocampus, there is some evidence

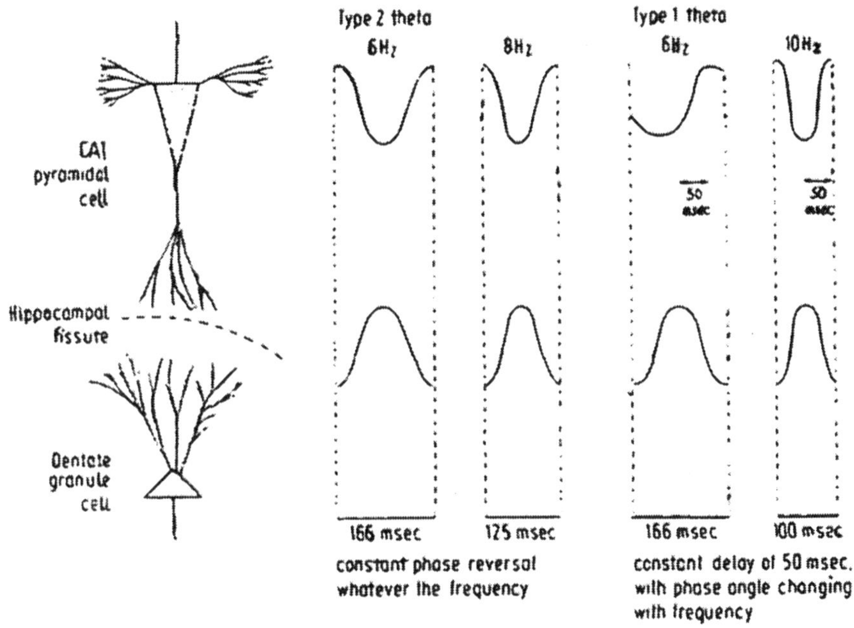

Fig. 25.3. Schematic illustration of the phase relations between the superficial generator (pyramidal cell layer of CA1) and the deep generator (deep stratum radiatum, and putatively the dentate gyrus) of theta rhythms in the hippocampal complex. In type 2 theta activity, there is phase reversal that stays constant despite variations in rhythm frequency. In type 1 theta activity, the superficial generator lags behind the deep one by an interval of about 50 ms, which stays constant with varying frequency, while the phase angle between the two varies. Thus at 6 Hz, the superficial theta rhythm generator lags the deep one by 108 degrees, while at 10 Hz it lags by 180 degrees. (From Miller R. (1989) *Psychobiology* 17: 115–128.)

that the brain can be put in a "theta state" by impulses coming either from external stimulation, or these radius areas can be put together in theta states also driven by the reticular formation or other parts of the brainstem. The model described by Miller (1991) is not only of high importance for the present chapter, but also for several other parts of this book.

The first premise of Miller's theory is that there are reciprocal connections between the hippocampus and a large portion of the cortex. This theory emphasizes the fact that direct afferent and efferent connections between hippocampus and neocortex appear to involve a limited extent of cortex, particularly some of the *parietotemporal association areas*, and also the *prefrontal* and *limbic areas* of the cortex. The theta modulation signals are likely to influence limbic and prefrontal areas and also other areas of the cortex (mainly association cortex).

The second premise of Miller's theory concerns the conduction delay in the reciprocal pathway between association cortex and limbic system. Figure 25.4 illustrates a schematic presentation of plausible delay times in various parts of the corticohippocampal loop. Miller estimates in rats a corticohippocampal total loop time in the range of 120 to 200 ms. He furthermore argues that such slowly conducting loops between hippocampus and cortex take on significance when it is realized that the values given above (120–200 ms) are in the range of the period for the theta rhythm in the rat (frequency of 5–8 Hz). He further states: "When an animal is in a particular environment, a collection of widely dispersed neurons throughout the cortical mantle may be set into a tonic state of elevated activation, as a result of signals reaching the cortex via any of its sensory systems. As yet, however, the activation of these neurons is not coordinated as a "cell assembly," so one cannot speak of this collection as categorically recognizing the environment. Consider now the relation we find between hippocampus and cortex: There is a series of recursive loops, with some temporal divergence/convergence in the pathway in each direction, and there is the possibility of regular oscillation of neural activity imposed at

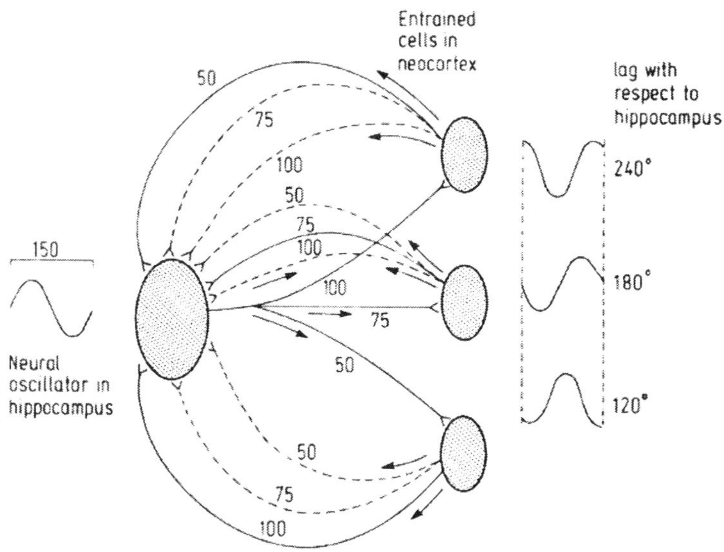

Fig. 25.4. Scheme for self-organization of phase-locked loops. The master oscillator (hippocampus) has afferent and efferent links with a variety of neocortical loci. All links include lines with a range of delays. (Conduction delays (ms) for each axonal link are shown by numbers adjacent to each axon). By strengthening of selected synaptic connections of axons with appropriate delays (*solid lines*, depiced here as having total loop times of 150 ms), the neocortical loci can become entrained to the hippocampal rhythm. Different neocortical loci will have different phase relations with the hippocampal rhythm (*right*) (Modified from Miller R. (1989) *Psychobiology* 17: 115–128.)

the hippocampal end. In this situation, the "Hebbian mechanism" favors the strengthening of connections which permit resonance.

Such resonance between hippocampus and cortex is envisaged to have an important functional role in the registration and retrieval of information in the cortex. Each pattern of resonant loops will raise the activation of a specific collection of senses which are widely dispersed across the cortex."

Miller also developed a model in which neural oscillators in the hippocampus may entrain cells in the neurocortex with different phase lags with respect to hippocampus. This model for self-organization is illustrated in Fig. 25.4. Accordingly, Miller envisages the hippocampus operating in two quite different modes, which he calls *baseline oscillator* and *cortical entrainment* modes.

By further descriptions of the corticohippocampal interaction, Miller predicts that memory retrieval and performance may depend on theta frequency. Miller argues also that resonant loops are to be seen as playing a most important part in information processing in the forebrain. An important part of Miller's theory consists in his design for a prototype cerebral cortex. This model is based on the concept of a modification of synapsis as a basis for learning. Here, Hebb's rule (1949) has again to be taken into consideration:

> *When an axon of cell A is near enough to excite a cell B and repeatedly or persistently takes part in firing it, some growth process or metabolic change takes place in one or both cells, such that A's efficiency as one of the cells firing B is increased.*

Hebb's rule implies that groups of synapses converging on a single neuron and having the tendency to fire together will become strengthened as a group. This is the principle of cooperativity.

Miller (1991) describes several submodels in which *sensory cortex, association cortex,* and *motor cortex* are interrelated and work in cooperation. He also describes the concept of distributed memory based on resonance phenomena of the assembly between various parts of the sensory association cortex and motor cortex. Figure 25.5 explains sensory signals giving rise to unimodal association that are able to converge or diverge on multimodal association areas.

We summarize in the next section that theta activity and theta responses are involved in learning and memory processes as shown by several experiments described in this book and are supported by the corticohippocampal interplay theory of Miller. We also have to point out that in humans the important theta increases in frontal and parietal areas are related to large theta enhancements during selective attention processes. Furthermore, the experiments on P300 responses of the cat brain described in Chap. 16 are also in the same streamline. We will argue in Chap. 28 that the "distributed theta system" and the "distributed alpha system" of the brain are interrelated and may work in a parallel way.

In Chap. 12 in Vol. I it was shown that theta activity in the frontal part of the human brain controls the amplitudes of the frontal evoked potentials.

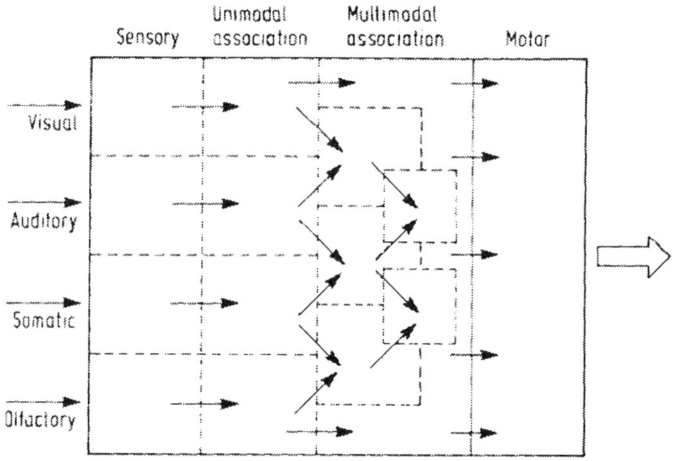

Fig. 25.5. A more realistic plan for a cortex that can represent the environemnt as a whole. (From Miller R. (1991) *Cortico-Hippocampal Interplay and Representation of Contexts in the Brain.* Berlin, Heidelberg, New York: Springer.)

In other words, the reaction of frontal areas of the human brain depend on its theta state. We will also describe (in Chap. 28) a selectively distributed parallel processing theta system. This distributed system might have the strongest theta components in limbic structures or in cortical ones working strongly together with the limbic system (frontal cortex, parietal cortex). Theta frequency components are either spontaneous patterns or induced ones obtained by the utilization of Fourier analysis or "single-trial theta epochs" of evoked potentials according to our combined analysis procedure. When we report on a *selectively distributed parallel processing theta system on the entire brain,* we want to emphasize that the theta generators must exist in several areas other than the limbic system or frontal cortex. The existence of the marked theta responses in the visual cortex of the human brain is an example of the ditributed nature of the parallel processing theta system (see Chaps. 9 and 10).

In Chaps. 2 and 10, where we have described the cross-modality experiments in the human and cat visual cortex, we have seen that neither the acoustical nor the other three stimulations generate in the occipital areas induced rhythms in the alpha frequency range, but there are enhancements of theta response in the visual cortex in response to an auditory stimulation. We have concluded that in this case the theta response is an association response par excellence.

25.6 Selectively Distributed and Induced Theta Oscillations in the Brain; A Theory

We summarize the theta oscillations described in this book.

1. Theta responses were recorded in the hippocampus of the cat brain following randomly applied auditory stimulation. The theta responses have different amplitudes in different layers of the hippocampus (see Chap. 1).
2. If the auditory stimulation is applied to the cat repetitively with interstimulus intervals of 2.5 s, significant changes in the amplitude of the theta response by each consecutive stimulus are observed. The cat is seemingly in a *stage of expectancy* if the stimulations are repetitive. The fourth stimulation, i.e., the last one prior to omitted stimulation or prior to the target, increases the amplitude in an average range of 40%. This is a type of short-term potentiation due to the repetitive stimulation.
3. An oscillation or induced rhythm to a target is also registered by a cognitive target in the form of omitted stimuli. The P300 of the cat hippocampus is a delayed and prolonged theta oscillation par excellence. There is only an omitted stimulation, i.e., a pure cognitive task, and this nonphysical event evokes a theta oscillation.
4. Similar to the results in the hippocampus of the cat brain, repetitive stimulation (auditory or visual) puts the brain into a state of increased attention, as shown in detail in Chap. 19 (cf. Demiralp and Başar 1992). The frontal human EP contains a high theta component. One can say as well that the response is reduced almost to a homogeneous theta oscillation in the frontal and parietal locations of the human brain. One can recognize in unfiltered single epochs that the response is an almost pure theta oscillation.
5. In the oddball paradigm we have to deal with a prolonged theta reaction (Chaps. 15 and 20).

It is not simple to give a working hypothesis concerning theta oscillations. To be completely efficient one should try to describe all types of theta generators or possible induced rhythmicities. The experiments of Adey (1960, 1969) were the first to indicate that induced theta rhythms in the limbic system and in hippocampus can be described as important functional correlates for cognitive processes in the central nervous system. *Learning theta* and *increased coherence in the theta frequency* range gave the first hints for correlating the EEG activity with functional states. From such experiments we have arrived at the following conclusion: If electrodes are near pyramidal cells of CA3, the evoked potentials contain large theta components. Assuming that (1) the EP globally describes the rhythmic firing properties of a cell group and that (2) the physiological properties of potentials emanating from the group are directly reflected in its EPs, the theta response can be *decoded* from this compound potential (i.e., from the EP; Başar 1980).

Now a new, important approach has been introduced by Miller (1991), who carefully described well-chosen references from the literature concerning theta activity in the limbic system and cortex and the interaction of cortex and limbic system. We argue that Miller's theory is another type of approach for the concept of theta resonances in the limbic system established by Başar and Özesmi (1972) and Başar and Ungan (1973) and extended by Başar-Eroğlu et al. (1992).

According to the data presented it can be assumed that theta rhythms, spontaneous or induced, can be considered as important building blocks of functional signaling in the nervous system. However, as we will try to explain in Chap. 29, a unique rhythm or frequency response component in EPs cannot be the only functional processor for a given function. *Complex functions are certainly interwoven with several oscillations, which may occur singly or in combination.*

Please compare also Chapter 12 in Volume I and recently published report by Başar et al. (Electroencephalogr Clin Neurophysiol **108**: (1998): 101–109).

26. Gamma-Band Responses in the Brain: Functional Significance

Canan Başar-Eroğlu, Erol Başar, Martin Schürmann, Atsuko Schütt, Daniel Strüber, Michael Stadler, and Sirel Karakaş

Gamma band (40 Hz) activities exist in a number of brain structures of different species, with seemingly different functional/behavioral correlates. This rhythm also has different dynamics in various structures and under different experimental conditions: It exists spontaneously and/or can be evoked, induced, or emitted with different latencies and relations to sensory–cognitive events.

Recent measurements of gamma-band activity at the *cellular level* stress high-level functional correlates such as binding of features (for alpha-like activity at the cellular level, see Dinse et al. 1997 and Chap. 24). In contrast, field potential measurements in both humans and animals demonstrate that gamma-band responses may have multifold functions both in "obligatory" sensory and in cognitive processing.

With respect to the generators of gamma activity, experimental data hint at the existence of a distributed gamma system in the brain. Furthermore, an interpretation of gamma rhythms as universal functional building blocks is suggested.

Even the neurophysiologist who is not working in this field is nowadays familiar with gamma rhythms. Studies include conventional spike analysis in anaesthetized animals, sensory-evoked potentials, and, most markedly, higher nervous activity, i.e., electrophysiology of cognitive processes. The functional correlates proposed so far include linking of perceptual information ("binding") and vibrations in the memory leading to the question whether "grandmother is an oscillation" (Stryker 1989). However, the functional correlates of the gamma response and its origin, now being investigated in several outstanding laboratories, often lead to controversies. These controversies are encountered (1) in descriptions related to localization or origin of gamma generators (visual cortex, frontal cortex, thalamus), (2) functional properties of gamma rhythms (obligatory sensory processes, cognitive processes), and (3) in temporal properties of this signal: Are gamma band responses phase-locked or time-locked? Does the response occur immediately, or does a long latency exist between excitation and response? For a survey of these findings see Başar and Bullock (1992) and Pantev et al. (1994).

Therefore, the aim of this chapter is to review the literature on gamma activity in the human and animal brain at different levels of complexity. It covers (1) observations at the cellular level, (2) EEG-evoked potential recordings from different species, and (3) from different brain structures. The multifold functional interpretations of the 40 Hz responses are discussed.

26.1 Historical Note: Four Phases of Pioneering Studies Related to the Gamma Band

We propose the following historical staging of four more or less separable phases: The **first phase** of studies on gamma-band activity was initiated by Adrian (1942), who reported that the application of odorous substances to the olfactory mucosa gave rise to a train of sinusoidal oscillations lasting for the duration of the stimulus. The "induced activity" was usually between 30 and 60 Hz. Later, Lavin et al. (1959) and Hernandez-Peon et al. (1960) stated that the 40 Hz activity could be elicited by a wide range of nonolfactory stimuli.

During the **second phase**, roughly 1960–1980, the **"40 Hz induced rhythms"** remained obscure to the majority of electrophysiologists with some notable exceptions:

1. Freeman undertook experiments to establish that the 40 Hz activity is a basic functional signal of the brain. Freeman (1975) emphasized that the 40 Hz wave packet has a key function in perceptual models of the olfactory bulb of the rabbit. Freeman (1975) and Freeman and Skarda (1985) have shown that the EEG of the olfactory bulb and cortex in awake and motivated rabbits and cats shows a characteristic temporal pattern consisting of bursts of 40–80 Hz oscillations, superimposed on a surface negative baseline potential shift synchronized to each inspiration. According to Freeman, the neural activity that is induced by an odor during a period of learning provides the specification for a neural template of strength connections between the neurons made active by that odor. Subsequently, when the animal is placed in the appropriate setting, the template may be activated in order to serve as a selective filter for search and detection of the expected odor (Freeman 1979).
2. Sheer (1976) pioneered psychological studies by means of scalp recordings. Later he interpreted the ongoing 40 Hz rhythm as a sign of focused cortical arousal (Sheer 1989).
3. Van der Tweel and Spekreijse (1969) found a visual high-frequency response after sinusoidal light stimulation in the occipital recordings of the human brain.
4. Başar and Özesmi (1972) used for the first time the expression "gamma response" in the cat hippocampus. Başar and Ungan (1973) introduced the first analysis by means of adaptive digital filtering to report a 40 Hz oscillatory response in the cat hippocampus (confirmed later by Leung (1992)). These investigations were probably the first studies to show physiological evidence of the gamma response as a component of evoked potentials.

Başar et al. (1975a,b) published evidence that the 40 Hz response belongs to a general brain response that can be recorded in the whole brain.
5. Başar et al. (1976b) published the first results related to the human auditory 40 Hz response.

The **third phase** is marked by the increasing popularity of the gamma-band response following the publication by Galambos et al. (1981). The gamma band response thus found general acceptance in human studies on sensory and cognitive processes (see below).

The **fourth phase** was initiated by the discovery of the 40 Hz oscillations at the cellular level (Gray and Singer 1987; Eckhorn et al. 1988), attracting the attention of all neuroscientists. Even investigators who had never worked with EEG and/or rhythms in the brain recognized the functional importance of this response.

26.2 A Classification of Gamma-Band Activities by Galambos

Galambos (1992) classified the gamma rhythms recorded in different species into four categories: spontaneous, induced, evoked, and emitted.

1. The **spontaneous gamma rhythm** contributes at any given moment a fraction of the total energy a brain is generating.
2. The **induced gamma rhythm** is a series of oscillations following sensory stimulation that are not time-locked to the stimulus. This type of gamma oscillation is recorded in the olfactory bulb of the rabbit (Freeman 1979) and visual cortex of the cat during stimulation with moving stripes (Gray and Singer 1987; Eckhorn et al. 1988).
3. **Evoked gamma** is tightly time-locked to an external stimulus. This type of gamma activity is recorded from the human scalp during stimulation with auditory clicks or flashing lights (Başar 1980; Başar et al. 1987; Galambos et al. 1981; Galambos and Makeig 1988; Bressler 1990; Pantev et al. 1991).
4. **Emitted gamma rhythms** are oscillations that follow the omitted stimuli in a train of regular stimuli in fish, cat, and man (Bullock et al. 1990; Başar-Eroğlu and Başar 1991; Başar et al. 1989).

For the purpose of the present chapter, we will not precisely follow Galambos's categories, but we will distinguish between 40 Hz activity at the cellular level and 40 Hz activity in field potential recordings, i.e., in EPs and ERPs.

26.3 40 Hz Responses at the Cellular Level

Gray and Singer (1987, 1989) and Gray et al. (1989) have reported that neurons in the cat visual cortex exhibit oscillatory responses in the frequency range of 40–60 Hz. These oscillations occur in synchrony with cells located within a functional column and are tightly correlated with oscillatory field potentials. Additionally, Gray and Singer (1989) provided evidence that neurons in spatially separate columns synchronize their oscillatory responses depending on the spatial separation and orientation preference of the cells. Based on these results, Gray and Singer proposed that the synchronization of oscillatory responses of spatially distributed, feature-selective cells might be a way to establish relations between features in different parts of the visual field. Eckhorn et al. (1988, 1989a,b) also found stimulus-evoked resonances of 35–85 Hz throughout the visual cortex when primary coding channels were activated by their specific stimuli and raised the question of what neuronal/cortical activity the gamma band is correlated with.

26.4 40 Hz Responses in Field-Potential Recordings: Sensory and/or Cognitive Processes?

The 40 Hz evoked response can be elicited by using auditory and visual stimuli (Başar et al. 1987; Pantev and Elbert 1994). According to these studies, it can be argued that the 40 Hz activity might represent an important general information processing of the brain similar to the 10 Hz activity. As already mentioned above, we have shown that the so-called middle latency response of human auditory EP reflects the 40 Hz wave packet elicited in the first 100 ms following stimulus onset (Başar et al. 1987, 1991). The 40 Hz activity has small amplitudes in comparison to alpha or theta activity and is therefore almost masked by these large-magnitude activities. The use of digital filtering, however, makes it possible to analyze 40 Hz activity separately.

26.4.1 Measurements in Animals

Evoked-Potential Measurements in Cats. Başar and Ungan (1973) demonstrated the existence of a distinct auditory gamma response in the hippocampus of the cat brain by using for the first time adaptive filtering techniques The relation of gamma responses to obligatory sensory responses is supported by the following findings:

1. The 40 Hz response is a robust, almost stable component of the sensory-evoked potential. It can be recorded in cortex, hippocampus, and cerebellum of the cat brain. It is phase-locked and occurs in the first 100 ms after sensory stimulation. The response has also a secondary oscillation with a latency of approx. 300 msec (Başar and Özesmi 1972; Başar and Ungan 1973; Başar 1980; Başar et al. 1995).
2. A similar phase-locked oscillation is also a component of the human auditory and visual response (Başar et al. 1976b, 1987)

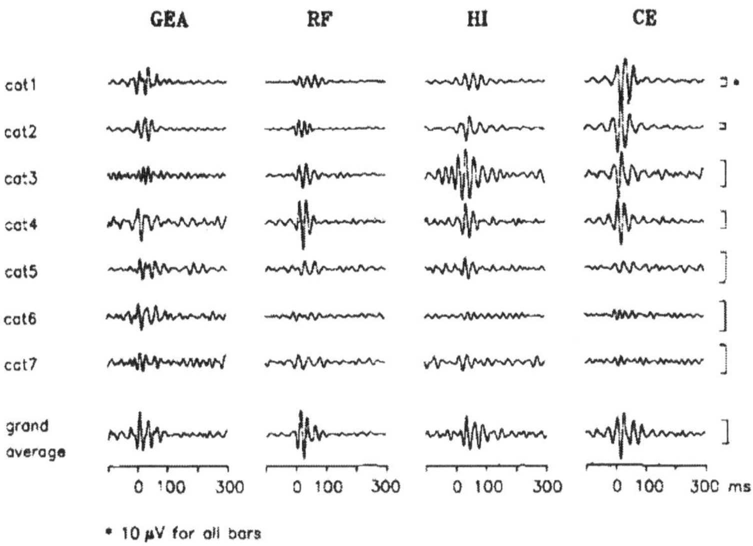

Fig. 26.1. Sensory-evoked responses in different brain structures of the cat. Gamma-band (30–70 Hz) filtered auditory responses of single cats in auditory cortex (GEA), reticular formation (RF), hippocampus (HI), and cerebellum (CE). *Bottom row*: gamma-band filtered average ($n = 7$) of responses. (Modified from Başar E. et al. (1995) *IEEE Engineering in Medicine and Biology* 14: 400–410.)

Figure 26.1 shows the averaged evoked responses of seven cats in four brain structures: gyrus ectosylvianus anterior (GEA, auditory cortex), hippocampus (HI), formatio reticularis (RF), and acoustical vermis of the cerebellum (CE). The EPs are filtered in the gamma frequency range. All of the averaged responses demonstrate an enhancement in the gamma-band activity following the stimulation, which shows the time-locked nature of the gamma-band response. The grand averages of all the experiments filtered in the gamma range are presented at the bottom of the illustration.

The data were recorded from intracranial structures with chronically implanted electrodes. During recording the cats were unrestrained and unanaesthetized (freely moving). As auditory stimulation, tone bursts of 2000 Hz and 80 dB (SPL) were randomly applied. The single trials of the evoked potentials were then selectively averaged by rejecting movement and sleep artifacts. We have used an adaptive digital filtering: Firstly, the amplitude–frequency characteristics (AFC) were computed by application of fast Fourier transform (FFT) to the averaged EPs. The existence of a gamma response in the AFC was a prerequisite for digital filtering (see Başar 1980). The filters used in our investigations do not introduce any phase shift.

Furthermore, an important point demonstrated by the results described above is that the gamma-band response may exist in parallel (simultaneously) in many subsystems of the brain (Başar et al. 1995; Demiralp et al. 1995b).

Therefore, any interpretation of the functional significance and generators of this activity, based on observations in any one of these subsystems, would not be general enough. An integrative approach to diffuse gamma-band activities of the brain seems to be more explanatory.

Event-Related Potential Measurements in Cats. In earlier studies we have shown that the so-called P300-40 Hz responses occur 250–400 ms following task-relevant stimuli. The appearance of the P300 response is usually combined with a wave packet of 40 Hz responses in hippocampus (Başar-Eroğlu and Başar 1991), reticular formation, and auditory cortex (Başar-Eroğlu and Schürmann 1994). Similar results were also observed from human experiments by means of different paradigms (Başar et al. 1993).

Leung (1992) confirmed the existence of the gamma response in hippocampus. The author claimed that spontaneous or evoked brain activity in the hippocampus shows a 20–70 Hz beta rhythm under some conditions, typically during behavioral activation and accompanied by a theta rhythm. Vanderwolf et al. (1978) have shown a fast activity of 20–70 Hz in hippocampus that increases in power during voluntary behavior as compared with automatic behavior.

Cognitive Processes Cause a Delay: Long Latency 40 Hz Induced Rhythms in Cats. A P300 40 Hz component has been recorded in the CA3 layer of the cat hippocampus using the omitted stimulus paradigm, in which every fifth tone of a repetitive auditory stimulation was omitted. A dominant and significant 40 Hz time-locked wave packet occurs approx. 300 ms after the omitted stimuli and not in the first 100 ms. The 40 Hz response is superimposed with a slow wave after 270 ms latency (Başar-Eroğlu and Başar 1991). The P300 40 Hz response is also recorded in cortex and reticular formation; i.e., similar to the sensory-evoked 40 Hz rhythm, it is diffusely distributed in the brain (Başar-Eroğlu and Schürmann 1994).

The oddball paradigm applied to human subjects also revealed the existence of a significant time-locked component with a large latency (Başar-Eroğlu and Schürmann 1994).

Evoked Potential Measurements in Snail Ganglia. Evoked field potentials in the *Helix pomatia* visceral ganglion were measured using an electrical impulse as stimulation. A 40 Hz response was recorded in the first 100 ms after the stimulation (Schütt et al. 1992; Schütt and Başar 1992). The 40 Hz oscillatory response is phase-locked and ample; it is also visible without application of filters. The amplitude of the gamma responses can be modulated by pharmacological agents in arthropods. Kirschfeld (1992) emphasized that nervous systems as simple as those of insects also show gamma response. This author further states that in such a simple system the existence of gamma response cannot be explained by high cognitive processes. Our results from snail ganglia led us to the same interpretation. Furthermore, the model of snail ganglia is less complex, and the electrical stimulation is a more direct one in comparison to light stimulation.

26.4.2 Measurements in Humans

Evoked Potentials. The 40 Hz human auditory response was first reported by our research group (Başar et al. 1976b). Later, we published the relation between 40 Hz spontaneous activity in human EEG recordings and the middle latency response of the auditory evoked potential (Başar et al. 1987). By means of single trial analysis of EEG-EP epochs, we demonstrated the inverse relation between amplitudes of the prestimulus 40 Hz EEG activity and the 40 Hz response of the evoked potential (EP). In other words, if the spontaneous 40 Hz activity had large amplitudes, there was less enhancement in the response, and vice versa.

Figures 26.2a and 26.2b show this inverse relationship between evoked and spontaneous activities. Figure 26.2a shows ten randomly selected single trials (note the low-amplitude 40 Hz response in the averaged EP). For Fig. 26.2b, only single trials with marked amplitude enhancement after stimulation were selected; in these curves, the responses are perfectly time-locked to stimulation (please note the small amplitudes of the 40 Hz prestimulus EEG).

Figure 26.2c shows three selectively averaged EPs (averages of 40 single trials each). In Fig. 26.2c (a) a conventional averaged EP is presented, without any selection of trials with respect to amplitude or phase-angle criteria of prestimulus EEG or 40 Hz response; (b) illustrates the same curve, filtered 30–50 Hz; (c) presents an EP averaged from single trials selected with specific criteria (marked amplitude enhancement after stimulation); (d) is the 30–50 Hz filtered component of (c); (e) shows an averaged EP derived from 40 Hz responses with low enhancement factor; and (f) is the respective 40 Hz component. In the averaged EP of (a), the peaks P30 and P60 are clearly recognized, whereas the same peaks are more ample in the selective average of (c). However, in (e) P30 and P60/N50 are almost abolished. The result of applying a 30–50 Hz stop-band filter (which theoretically rejects the 40 Hz response) to the conventional averaged EP of (a) is shown in (h). Note that (h), obtained with stop-band filtering, is very similar to (g), obtained with low enhancement trials ((g) is identical with (e)).

These results were interpreted as follows: The existence of high-amplitude 40 Hz activity preceding the auditory stimulation is an obstacle for the occurrence of a 40 Hz response (Başar 1980; Başar et al. 1987). By averaging selected EP epochs with ample spontaneous 40 Hz activity (with small enhancement), the averaged EPs obtained showed no middle-latency response. The same picture is obtained when the 40 Hz response is "rejected" by the use of a stop-band filter. Pantev et al. (1991) applied the same digital filtering technique and found similar results by analyzing magnetoencephalographic recordings in human subjects.

Attention-Related 40 Hz Activity. Several papers have handled the importance of 40 Hz activity in states of attention and motivation. It has been shown for gamma waves that their amplitude increases as a subject becomes

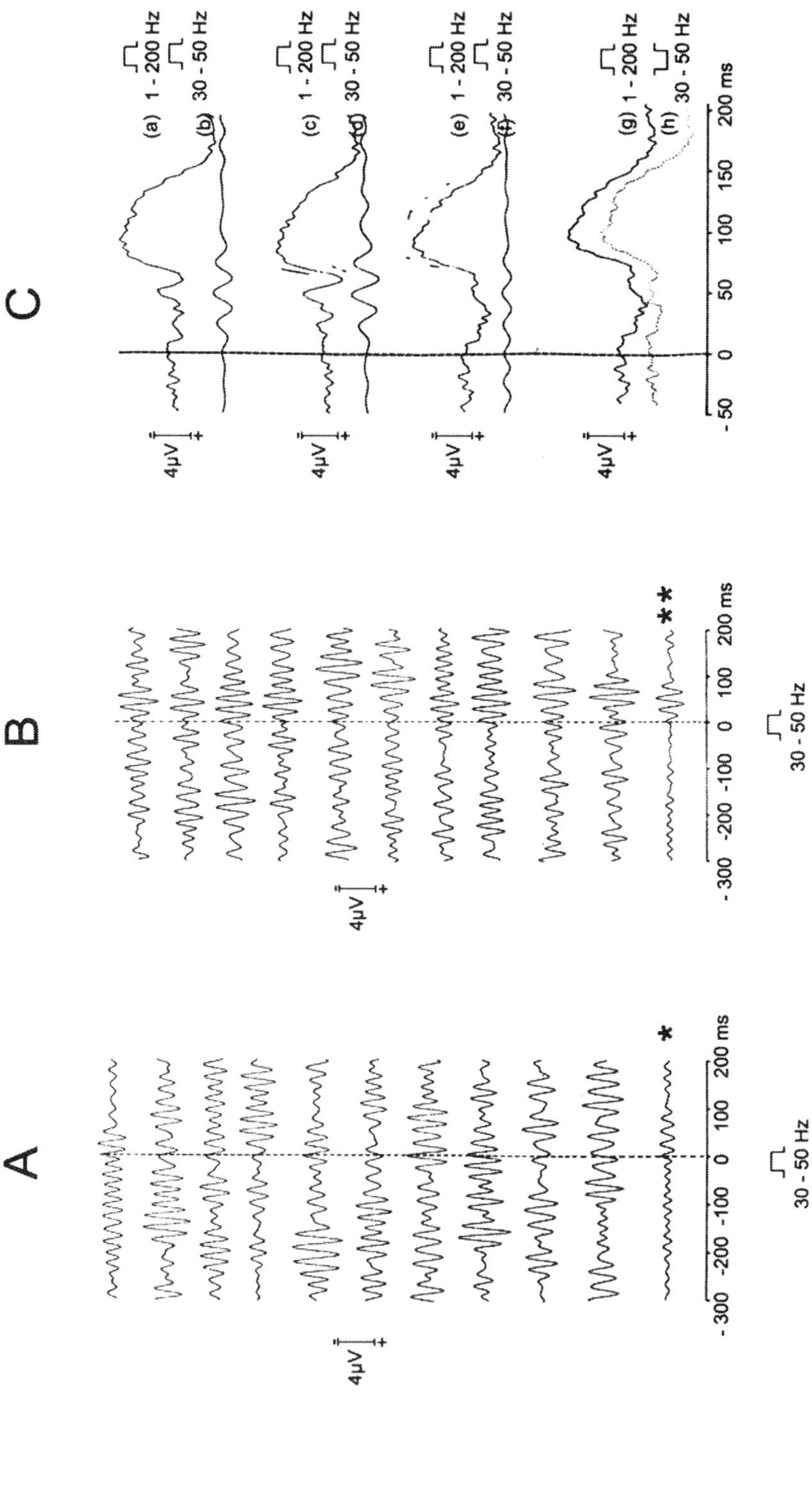
A
B
C
4µV
- 300 -200 -100 0 100 200 ms
30 - 50 Hz
(a) 1 - 200 Hz
(b) 30 - 50 Hz
(c) 1 - 200 Hz
(d) 30 - 50 Hz
(e) 1 - 200 Hz
(f) 30 - 50 Hz
(g) 1 - 200 Hz
(h) 30 - 50 Hz
- 50 0 50 100 150 200 ms
*
**

more attentive (see Bouyer et al. (1987) for comparable cat experiments). Recently, Tiitinen et al. (1993) reported that selective attention enhances the auditory 40 Hz transient response in humans, especially over the frontal and central areas. The 40 Hz response was larger when subjects paid attention to stimuli rather than ignoring them, so that these authors proposed a physiological correlate of selective attention in the 40 Hz transient response in humans. This interpretation is in accordance with our results during multistable perception (see below). However, it is to be emphasized that the 40 Hz activity recorded during perceptual switching is an *endogenously* induced activity.

Jokeit and Makeig (1994) have measured different event-related patterns of gamma-band power of fast- and slow-reacting subjects by means of FFT. Llinas and Ribary (1993) found reduced activity during delta sleep, whereas oscillatory activity was stronger during both wakefulness and REM sleep. Recently, Joliot et al. (1994) have suggested that oscillatory 40 Hz activity is related not only to primary sensory processing but also could serve a broad cognitive binding function. Pulvermüller et al. (1994) suggested stronger gamma responses to words compared to pseudowords both in EEG and MEG recordings. Goertz et al. (1994) presented results showing that dynamics of 40 Hz activity are related to stimulus evaluation in sensorimotor processing during a visual discrimination task.

During selective somatic attention, Desmedt and Tomberg (1994) demonstrated a transient phase locking of the gamma waves generated in the contralateral prefrontal and parietal cortical areas in spite of their wide separation. The phase locking started about 130 ms after the somatosensory stimulus, stayed phase-locked for about 125 ms, and then desynchronized until after the reaction time and P300 closure.

Fig. 26.2. (**a**). Ten randomly selected single EEG-EP trials filtered with digital filters of 30–50 Hz. *, average of these trials. Stimulation is applied at time "0 ms". (**b**). Ten single EEG-EP trials, digitally filtered (30–50 Hz) and selected for high enhancement, i.e., high-amplitude increase after stimulus (in comparison with prestimulus EEG amplitude). **, average of these trials. (**c**). Selectively averaged evoked potentials (EPs); averages of 40 single trials each, with different selection of trials (human vertex recordings). (**a**) EP averaged from **randomly selected** single trials ("conventional" averaged EP); (**b**) same EP, filtered 30–50 Hz. (**c**) EP averaged from single trials selected for specific criteria (**marked amplitude enhancement** in the 40 Hz range after stimulation); (**d**) same EP, filtered with band limits 30–50 Hz. (**e**) EP averaged from single trials with **low amplitude enhancement**; (**f**) same EP, filtered with band limits 30–50 Hz. (**g**) identical with (**e**) – average of low enhancement trials; (**h**) result of applying a 30–50 Hz stop-band filter (which theoretically rejects the 40 Hz response) to the conventional averaged EP of (**a**). Note the similarity of (**h**), obtained with stop-band filtering, and (**g**), obtained with low enhancement trials. (Modified from Başar E. et al. (1987) *International Journal of Neuroscience* 33: 103–117.)

Perception-Related 40 Hz Activity. Lutzenberger et al. (1995) wrote that visual stimulation alters local 40 Hz responses in the EEG. Their results show that area-specific 40 Hz responses are a correlate of perception of coherent visual patterns in humans. These authors came to this interpretation in the light of studies by Eckhorn et al. (1988) and Engel et al. (1992).

Tallon et al. (1995) tested the hypothesis that synchronized oscillatory activity in the gamma range plays a role in visual feature binding of coherent static objects. The authors used two coherent Kanizsa triangles (one with real contours and one with illusory contours) and a noncoherent one. They found a specific 30 Hz power increase only in the case of coherent stimulation, no matter whether the real or the illusory triangle was used. This finding supports the hypothesis of a temporal code for coherency representation. However, the authors emphasize that it is not yet clear whether their results can be directly related to those obtained with intracortical electrodes in the anaesthetized cat by Gray et al. (1989).

Kojo et al. (1993) reported results of psychophysical experiments concerning the spatial and temporal properties of illusory figures. They demonstrated that the illusory Kanizsa triangle could be seen, although only one inductive element was present at any time. This spatiotemporal window – within which the illusion is visible – is discussed in connection with the synchronized gamma waves reported by Eckhorn et al. (1988). The authors suggest that two different brain locations where gamma waves are synchronized would activate the "neurons" between them and cause the illusory contours. This assumption is in correspondence to an earlier formulation of a coactivation theory of illusory contours put forward by Stadler and Dieker (1972).

The above-cited results of Desmedt and Tomberg (1994) are also in line with the microphysiological evidence obtained on mammalian visual cortex. The selective functional synchronization between prefrontal and parietal human brain areas reflects the transient binding of discrete cognitive features.

The results of Başar-Eroğlu et al. (1996b,c) show that viewing multistable patterns also leads to an enhancement of 40 Hz activity. Although the 40 Hz increase is highest at fronto-central locations, it could also be observed in parietal and occipital areas. This means that the 40 Hz activity is enhanced over widely spread regions of the cortex, which is compatible with the concept of a selectively distributed gamma response system in the brain.

Evoked Gamma-Band Response and Its Relationship to the Middle Latency Response. In various studies of Başar et al. (1976a,b, 1987) and in a recent study by Galambos (1992) it is shown that the auditory and visual transient evoked responses, recorded from the human scalp, contain a train of short latency wavelets in the gamma frequency range, precisely time-locked to the stimulus and lasting approx. 100 ms. Başar (1980) has extended earlier studies on the cat hippocampus and demonstrated that the most ample hippocampal evoked gamma can be recorded at the CA1 layer. Pantev et al. (1991) discovered a gamma-band transient magnetic oscillatory response, evoked in the human brain by the onset of auditory stimuli, consisting of four or more cycles phase-locked to stimulus onset in approx. the 20 ms to 130 ms poststimulus intervals, similar to results of Başar et al. (1987), indicating a close relationship between the gamma-band activity of the brain and the middle latency auditory-evoked response: "The wave complex P25, N50, P60 seems to represent modulation of the 40 Hz mechanism. The wavelets of this complex are minima and maxima of the same oscillation, namely of the resonating 40 Hz wave packet." The same viewpoint is supported by Galambos (1992).

Evoked Gamma Rhythm as a Rostrocaudal Scan in the Human Brain. The studies by Llinas and Ribary (1992) on the auditory system in humans, using the MEG technique, indicate synchronized 40 Hz oscillatory activities over large cortical areas that are highly organized in space and time. According to Llinas and Ribary (1992), the spatiotemporal magnetic-field pattern of gamma-band activity suggests the presence of a coherent rostrocaudal sweep of activity repeating every 12.5 ms due to a continuous phase shift over the hemisphere. "The findings indicate the presence of a scanning-like process most probably of intrinsic origin that covers much of the brain surface with a focus on the activated sensory area." The authors propose that this scanning occurs on account of thalamocortical-resonant synaptic interactions.

26.5 Functional Interpretation of 40 Hz Responses in Light of Comparative Data

26.5.1 The Binding Problem: Gamma-Band-Induced Rhythm as a Mechanism of Feature-Linking in the Visual Cortex

Parallel distributed networks in the brain propagate any event into separately processed stimulus attributes, storing the memory of the corresponding information in different regions. Retrieval of this information from memory is mediated by multiple brain substructures. Recovery and cognitive evaluation of the event when it reoccurs engages anatomical areas dispersed throughout the brain. This statement is strongly based on cognitive experiments employing single and multiple unit activity and field potentials in animals, event-related potentials and EEG in humans, and a great variety of clinical studies using modern imaging techniques.

Distributed gamma responses (as shown in Fig. 26.1; see Sect. 6.2 for details) may be electrophysiological correlates of such distributed networks in the brain and of the information processing in these networks. Such responses have been observed by our research group and by John et al. (1975) in the cat brain, in accordance with the concept of synchronized selectivities as proposed by Başar (1980). We suggest that approaches at solving the *binding problem* (discussed by several neuroscientists) should be based not only on the search for single generators: Distributed gamma responses should be taken into account.

Recently, Jürgens et al. (1995; Jürgens and Rösler 1995) have argued that what appears as gamma-band responses obtained from EEG or MEG signals may in fact be an artifact of the analysis of spectral responses. In particular, changes observed in gamma-band power may be harmonics of changes in alpha activity. In a series of articles Başar et al. (1976a,b, 1995) have shown that the gamma oscillations are not harmonics of alpha activity. Also, the independent existence of both rhythms has been shown (Pfurtscheller and Neuper 1992). In a recent study by Pulvermüller et al. (1995) it is also indicated that the gamma oscillations are not, as a rule, harmonics of alpha activity.

26.5.2 The Diffuse and Selectively Distributed Gamma System of the Brain

The Diffuse Gamma System as a Generalized Concept of the Brain. We have already introduced the concepts of distributed alpha, theta, and gamma response systems as a consequence of our long-standing studies on the 40 Hz response (Başar et al. 1991; Başar-Eroğlu et al. 1992).

Başar (1980) stated that the 40 Hz response occurs simultaneously in all areas of the brain. The term "diffuse" is used in order to describe the distributed nature of the gamma response in the brain. At this level of investigation, it is not possible to define the connections between the elements of these systems exactly from one neuron to another one or to define the directions of signal flow and exact boundaries of neuronal populations involved. However, this description is necessary to emphasize that rhythmic phenomena in these frequency ranges are not unique features of the observed single subsystem of the brain, and that their simultaneous existence in distant brain structures may be a relevant and important point in understanding the cooperative activities of distinct brain structures.

With a generalized concept of the brain as a distributed parallel processing system, in a wide framework of results with various species, brain structures, and experimental conditions reviewed, we can state tentatively that the gamma-band rhythm is an important, universal operator in brain function, which is distributed to many subsystems of the brain. Edelman (1979) states: "In such systems, integrative functions may emerge from the dynamics of the entire network rather than from linear computations performed at each nodal point in the circuit." Goldman-Rakic (1988), in search of a topography of cognition, concludes: "If subdivisions of limbic, motor, sensory, and associative cortex exist in developmentally linked and functionally unified networks, as the anatomical, physiological, and behavioral evidence suggests, it may be more useful to study the cortex in terms of information processing functions and systems rather than traditional but artificially segregated sensory, motor, or limbic components and individual neurons within only one of these components."

According to our approach, the resonating loops between different brain structures and the rhythms common to several subsystems of the brain serve as an important tool in understanding this link between anatomically distant but functionally complementary structures. The diffuse gamma system seems to be one important example of this type of linking parallel activities of the brain.

Beyond this general integrative concept, finer analysis and functional interpretation of specific examples may be necessary. However, it then should be considered as a part of the *distributed gamma response system* of the brain. While a loosely time-locked 40 Hz response is recorded in the sensory cortex or a tightly time-locked gamma-band activity is recorded over the whole neocortex, then it should be asked; What is the function of the thalamic nuclei (Steriade et al. 1990), hippocampus, reticular formation, and other brain structures including the cerebellar cortex (Başar 1980; Başar and Özesmi 1972; Başar and Ungan 1973; Başar-Eroğlu and Başar 1991) having the ability to respond in this frequency range at the same time? The introduced paradigm to measure 40 Hz activity changes during perceptual multistability, provides an important support to the concept of a diffuse gamma-band system with various functional correlates and different cortical generators.

Multiple Generators and Multiple Functional Correlates of the Gamma System. For the human or cat brain it is impossible to define a unique source or a unique 40 Hz generating structure. According to the results there is (some) evidence that a great number of 40 Hz generators are diffusely distributed in the whole brain (parallel processing). Some examples are illustrated in Fig. 26.1. 40 Hz responses were recorded in cortex, hippocampus, and cerebellum with different latencies. Due to the long distance between hippocampus and cerebellum, hippocampus and auditory cortex volume conduction of 40 Hz signals can be ruled out. It was shown that with only 1 mm distant electrodes, the amplitudes of the gamma signal are altered considerably (Başar-Eroğlu and Başar 1991). Accordingly, the illustrated phase-locked hippocampal, cortical, and cerebellar potentials stem from different generators without phase lag.

Alone, the results presented in this chapter confirm that 40 Hz responses are involved with all types of cognitive and sensory processes, ranging from simple signal detection to complex CNS activity such as recognition of ambiguous figures. In event-related measurements (omitted tones in the cat brain and oddball experiments in the human brain) 40 Hz responses are recorded with a long latency, whereas simple sensory stimulation also elicits a 40 Hz response without time lag (Fig. 26.2.)

Gamma Rhythms as Functional Building Blocks. The evaluation of the above statements leads to the following generalization: 40 Hz signals of the brain (spontaneous, induced, or evoked) belong to important functional building blocks of brain electrical activity. 40 Hz wave packets may occur in different and distant structures, may act in parallel or separately, depending on the modality of sensory or cognitive stimulation. They may show phase locking, time locking, or weak time locking. Moreover, the fact that spontaneous, ample, and distinct evoked 40 Hz rhythms can be measured upon simple electrical stimulation in isolated snail ganglia demonstrates that this signal presents a universal code of the basic properties of CNS signaling. In an invertebrate ganglion at must 2000 neurons exist; the helix gamma response is certainly not a result of high-level cognitive processes as presented in Chap. 6; Sec. 6.6.

26.5.3 Conclusion

It can be seen from the above examples and the results of the present chapter that gamma-band activities can be recorded in widely distributed brain structures and in brains or ganglia of different species with seemingly different functional and behavioral correlates. This rhythm also has different dynamics in various structures and under different experimental conditions: It exists spontaneously and/or can be evoked, induced, or emitted with different latencies and relations to sensory–cognitive events.

In conclusion, we tentatively propose that the gamma-band activity is a part of the common language elements of the brain, which also might play an important role in the mutual information transfer between subcomponents of the parallel processing network of the brain, as is the case with a number of other oscillatory activities such as theta, alpha, and beta rhythms.

Although a number of experimental findings demonstrating the importance of gamma-band oscillatory activities in brain function exist, a further series of complementary studies is necessary to show the analogous and parallel activities of the gamma-band rhythms studied until now in different species and brain structures under comparable experimental conditions.

For further general conclusions the reader is recorded to Chap. 29.

27. Structures, Brain Waves, and Their Functions

27.1 Parallel Processing – A Principle of Brain Function Accessible to Investigation by Means of Field Potentials and EEG

In the present chapter we begin a difficult task. We will explore functions of various brain structures by considering experiments on brain waves. The expression "brain wave" includes spontaneous activity, evoked responses, induced rhythms, event-related oscillations, etc. This approach has the well-known weakness that when analyzing the field potentials, it is difficult to define boundaries of brain nuclei and their electrical activity. The approach has, however, important usefulness, since a great quantity of data can be collected and interpreted from several electrodes distributed in the brain. Furthermore, it is possible to perform measurements during continuously changing cognitive states. This way, EPs or EEG segments are recorded in the cortex, limbic system, thalamus, and cerebellum. They can be compared in waking and freely behaving animals.

This type of recording during behavioral states cannot possibly be managed with recordings of single-cell electrodes. Studies on functional correlates of structures like sensory cortices, hippocampus, and thalamic relay nuclei are based mostly on experiments using unit recordings. A major difficulty concerning the interpretation of experiments with single-unit recordings (for example experiments on corticothalamic information transfer) is the limitation of the results from a few neurons that can be impaled. Accordingly, the author assumes that every hypothesis on the localization of "thalamocortical circuit as 10 Hz generator" is restricted and not acceptable with regard to the results of experiments described in this book: The alpha, theta, and gamma generators are selectively distributed in the brain.

An important result was the following: Upon visual stimulation the auditory thalamic and cortical structures did not react with a significant 10 Hz enhancement (see Chaps. 1 and 2), and conversely, visual cortical areas and lateral geniculate nucleus did not react with 12 Hz enhancement to auditory stimuli (Chap. 2). Theta responses were, however, existent as significant enhancements mostly in all the structures, also after application of inadequate stimuli. On the other hand, hippocampus showed a completely different quantitative behavior upon application of sensory stimulation. Although sensory-specific thalamic and cortical centers do not react with a 10 Hz enhancement, hippocampus does respond strongly with 10 Hz responses after auditory and visual stimuli as the results given in Chap. 1 clearly show.

Although in the conventional neurophysiology literature it is accepted that thalamocortical circuits are the only alpha-generating pacemaker structures, the findings presented in this book clearly demonstrate that the hippocampal alpha generators also exist and may probably play a major role in *integrative and associative brain function.*

Chapter 1 also shows phase spectra indicating a zero phase in the 10 Hz range among thalamus, cortex, reticular formation, hippocampus, inferior colliculus, etc. This means that alpha and theta enhancements occur as parallel processing without phase lag in all studied structures. These last findings include an important physiological statement: Probably, some higher-frequency impulses reach reticular formation, thalamus, and hippocampus at the same time. After this preliminary excitement, theta and alpha generators are excited almost at the same time in these centers.

27.2 The Basic Cortical Circuit and Cortical Oscillatory Responses

As in every other region of the brain, the synaptic circuits in the cortex are built up out of the triad of neural elements: input fibers, output neurons, and intrinsic neurons. Insight into the complexities of neocortex can be gained by primarily by considering the simple types of cortex. Careful consideration of the organization of these regions shows that each cortex consists of a basic circuit consisting of three types of connections:

1. Afferent input fibers (IN) make excitatory synapses onto distal apical dendritic spines of pyramidal neurons.
2. Intrinsic axon collaterals are reexcitatory (RE) to pyramidal neurons over long distances.
3. Local interneurons are activated by afferents to give feedforward inhibition (FI) and/or axon collaterals to give lateral inhibition (LI). These connections can be identified in the basic circuit diagram of Fig. 27.1, which is reproduced from Shepherd (1988).

As shown in Fig. 27.1b, granular association cortex is very similar to simple cortex. The primary input (IN), by definition, is from other parts of cortex, and this terminates mainly in the most superficial layer. Pyramidal cells mediate reexcitation of pyramidal cells through direct connection to their axon collaterals (RE). Feedforward inhibition (FI) and lateral inhibition (LI) are also mediated in much the same way as in simple cortex.

In primary sensory cortex, such as visual cortex, the afferent input ends mainly in a specific population of stellate cells that define a new layer, lamina

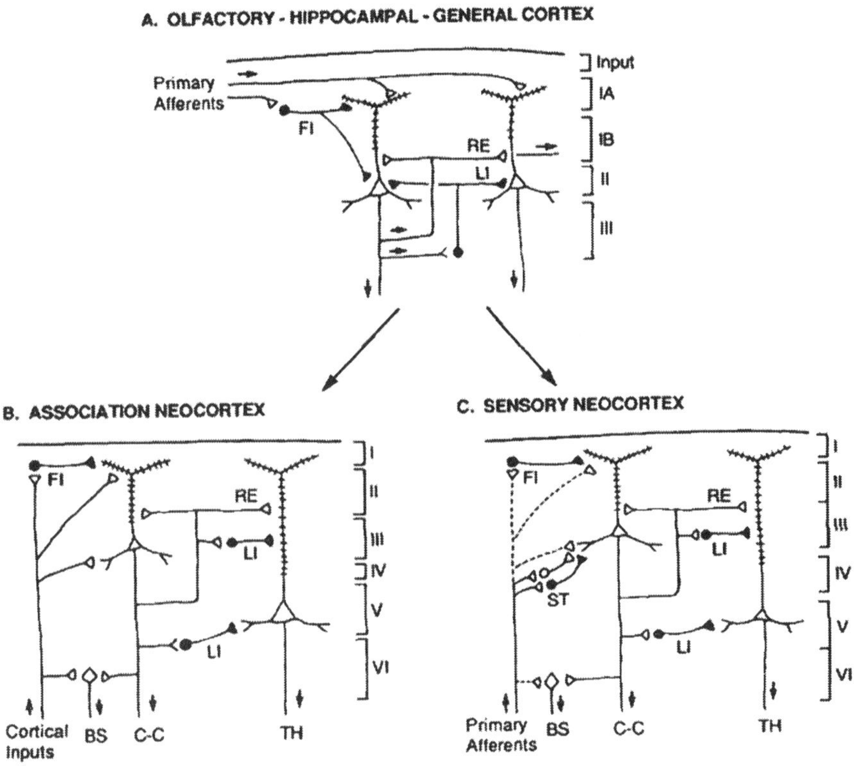

Fig. 27.1. Principles of local circuit organization in the cerebral cortex. (**a**). Simple cortex, as represented by olfactory cortex, hippocampus, and reptilian dorsal (general) cortex. Layers indicated on *right*. Abbreviations: FI, feedforward inhibitions; LI, lateral (and feedback) inhibition; RE, recurrent (and lateral) excitation. (**b**). Agranular association areas of neocortex. Layers on *right*. Output sites below: BS, brainstem; C-C, corticocortical; TH, thalamus. (**c**). Primary sensory areas of neocortex. Note that the main difference from the circuit in **b**) is the predominance of primary afferents from the thalamus that project to an intracortical relay through stellate (ST) cells in layer IV. Connections to other layers are still present, but less prominent (*dashed lines*). Excitatory connections shown by open profiles; inhibitory, by closed profiles. (Modified from Shepherd G.M. (1987) in: Gazzaniga M.S., ed. *Perspectives in Memory Research*. Cambridge: MIT Press.)

IV, midway through the cortex (see Fig. 27.1c). This population of cells can be seen to function as a kind of intracortical relay, receiving the thalamic input and transferring it to the pyramidal neurons, providing thereby a more complex preprocessing of information prior to its input to the pyramidal neurons.

Shepherd poses the following important question: "What makes neocortical circuits unique?" and he gives a perspective on the properties that are specially related to the neocortex.

1. It is placed where it is accessible to every major sensory input, arriving either directly (from the olfactory cortex) or relayed from below through the brainstem and thalamus.
2. It is a layered structure doubled back on itself, so that inputs arrive from the depths and outputs exit through the depths. This leaves the cells in every layer potentially accessible to every input. When the local circuits through collaterals and interneurons are added to the picture, the potential ways by which information can be integrated, stored, and recombined become enormous. Such a structure is no longer dominated by the operational sequence demanded from a particular input or output.
3. Rather than there being one type of output cell, as is common in many of the centers, there are several types. In fact, each layer is the source of output fibers; some fibers go only to other layers (such as outputs from layers I and IV); others go to different distant targets (such as outputs from Layers II, III, V, and VI; see Fig. 27.1a–c). As a result, each layer, in effect, acts as a semi-independent unit, defined by its own particular inputs, outputs, intrinsic connections, and relations to its neighboring layers.
4. If this structure is given the chance to expand, through some property of its not-quite-rigid brain case, then one has the opportunity to continue enlarging individual areas, or adding new ones, in order to combine information from new combinations of output targets.

Accordingly, the following thesis can be hypothetized: After a *light stimulation*, a special sensory thalamic center, such as the *lateral geniculate nucleus*, does not only respond with a 12 Hz evoked oscillation, but it transmits the 12 Hz oscillations to the visual cortex. If a cat is stimulated with auditory signals of 2000 Hz and 80 dB, the neural afferent inputs to visual cortex cannot be excited directly (see Fig. 27.1). But still, there exists a neural input to visual cortex from other primary areas as well as association areas of the cortex. As shown in Chap. 2, evoked responses to acoustic stimulation recorded in the visual cortex do not have a 10 Hz response but mostly a dominant theta response. The major utility of the evoked potential recording is the fact that there are still large responses despite the inadequate stimuli. This type of response it not recorded in single neurons of the primary sensory areas. However, it is a dominant theta response. In the filtered response, we observe a slight alpha enhancement, too, but with a remarkable latency (between 150 and 300 msec).

We underline our tentative approach, assuming that theta enhancements originate mostly from associative responses (see also Chaps. 1, 18, 19, and 25). These, in turn, result from interplays between association areas of the cortex, limbic system, and several parts of the frontal cortex, but without the necessity of integrative activity of special sensory thalamic nuclei. In the case of auditory transmission, we would assume that lateral geniculate nucleus is not activated in the vicinity of 10 Hz. The delayed 10 Hz enhancement in the cortex might come from transmission via hippocampus and frontal cortex to visual associative areas and then to parietal cortex and primary visual cortex (see Chap. 24).

We make similar statements concerning the auditory cortex and the medial geniculate nucleus. Medial geniculate nucleus also does not show any primary alpha enhancements during the first 100 msec after a stimulation in the auditory cortex if the sensory stimulation is not adequate (see Chap. 2). However, large theta enhancements are observed in thalamus and cortex after an auditory stimulation. It is also important to state that the primary cortex does not react to sensory stimulation with significant alpha enhancements, as explained in Chaps. 2, 9 and 10.

Recordings of sensory EPs in the thalamus and cortex confirm the theory of the existence of thalamocortical 10 Hz circuits when adequate stimulation is applied. But our findings have pointed to an important distinction between responses of the visual and auditory cortices. These two primary areas show a difference in their response codes: We record 12 Hz from the visual pathway and 10 Hz from the auditory system.

27.3 Thalamus: Sensory Gate for the Alpha Response

Such thalamic nuclei as the lateral geniculate and medial geniculate nuclei are centers, that convey information coming from receptive fields via brainstem to primary sensory areas. One fundamental principle of thalamic organization that has gained widespread acceptance only in recent decades is that all dorsal thalamic nuclei project to the cerebral cortex. There appear to be no exceptions to this rule, and previous ideas of relay nuclei, intrinsic or associational nuclei, and of different cortically and striatally projecting nuclei are no longer tenable. However, it is by no means permissible to think of each dorsal thalamic nucleus as having the same type of relationship with the cortex as its neighbors (Macchi 1983; Herkenham 1986).

27.3.1 Classical Thalamocortical Projection

There is a traditional belief that all thalamocortical axons arising within a particular thalamic relay nucleus terminate in one or at most a limited number of cortical areas and that within these areas they end in layer IV with a highly discrete topographic ordering. Such a belief underlies not only the concept of functional area specificity in the cortex but also the fine-grain, topographic organization of the sensory cortical areas and the idea of a sequential interlaminar flow of intracortical activity, starting in layer IV (according to Steriade et al. 1990). There is little reason to doubt that areal specificity and topographic order are fundamental organizing principles in the thalamocortical projections of most cells in major thalamic relay nuclei such as the ventral posterior, the medial geniculate, and the dorsal lateral

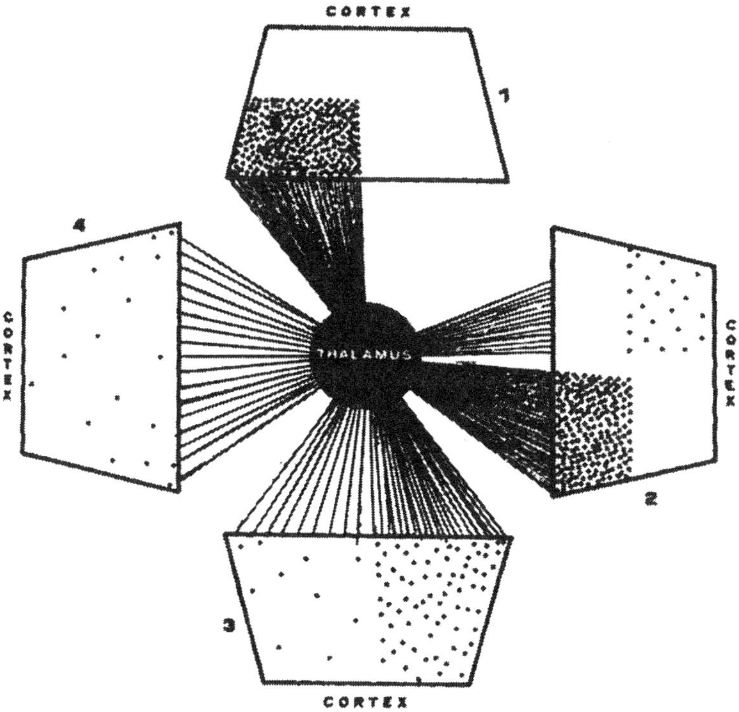

Fig. 27.2. Scheme of the patterns of thalamocortical projection in the cat. 1: Nuclei projecting densely on single cortical fields. 2: Nuclei projecting densely on one area and diffusely on another. 3: Nuclei projecting diffusely on several cortical fields but with regional concentrations. 4: Nuclei projecting diffusely over widespread areas. (From Macchi G. (1983) in: Macchi G., Rustioni A., Spreafico R., eds. *Somatosensory Integration in the Thalamus*. Amsterdam: Elsevier.)

geniculate. However, many thalamic nuclei display more diffuse, seemingly less highly organized thalamocortical connections, and most nuclei, including the principal relay nuclei cited, have projections terminating in cortical layers outside layer IV (Fig. 27.2).

Among the nuclei projecting only to the cortex, Macchi (1983) has distinguished four major types (Fig. 27.2) based upon the diffuseness or nondiffuseness of their cortical projections. This represents a useful scheme that combines the old nucleus-to-field pattern of organization with the new knowledge of cells with differential projections in individual nuclei.

In the experiments described in the Chap. 1 we found that thalamic and cortical frequency characteristics are in good accord if dominant peaks along the frequency axis are compound. Furthermore, we have seen in several experimental descriptions that evoked responses were measured in the auditory system as 9–10 Hz and 12 Hz in the visual system.

In single EEG-EP epochs filtered between 8 Hz and 15 Hz we observed that cortical or thalamic structures react with enhanced alpha responses if the sensory stimulation is an adequate one, whereas no reactions could be observed in the 10 Hz frequency range to inadequate stimuli. Here again we discuss these results together with regard to visual stimulation applied simultaneously lateral geniculate nuclei, visual cortex (area 17), and hippocampus of the cat brain (only for prototype sweeps) (see Fig 27.3 in next section). The striking difference to those responses obtained by auditory stimulation is illustrated in Chaps. 1 and 2. There are no reactions in LG and visual cortex, but there are large 10 Hz enhancements in the hippocampal response. What does this mean? In Chaps. 2, 10, and 24 we outlined a working hypothesis stating that 10 Hz responses are mostly a sign of adequate sensory excitement. This is, however, seemingly the case for structures like thalamic relay nuclei and primary cortices. In lower structures, such as the reticular formation, large alpha enhancements are observed for all types of stimuli. If the outlined working hypothesis indicating "a special transmission of sensory signals in the 10 Hz frequency range" is true, then one can suppose that thalamus acts as a "gating structure" for the 10 Hz transmission.

The information transfer in slower frequency ranges, i.e., in the theta frequency channel, does not have the same functional nature. Large enhancements (theta enhancements, shown in Chap. 1 and Chap. 2) are also observed. From the anatomical description presented by Macchi (Fig. 27.2) we have learned that several types of projections and diffuse bundles of projections exist between thalamus and cortex. Accordingly, it is possible that theta oscillations are also conveyed to cortex when the stimulation is not an adequate one.

27.4 Hippocampus

27.4.1 Hippocampus as "Supramodal Structure"

According to Mesulam and his colleagues (1977), each primary sensory area projects first to unimodal association areas, which primarily underlie sensory-specific discrimination functions. These projections are essential because different parts of a primary sensory field are usually not interrelated by association fibers. Unimodal association areas, in turn, project to sites of sensory convergence, which the authors called *polymodal associated areas*. In the primate, these areas are found, for example, in parts of the prefrontal region and near the superior temporal sulcus. Finally, certain regions in the inferior parietal lobule and in other parts of the cingulate gyrus and the prefrontal region appear to receive their major cortical inputs from *polymodal association areas*, and these were called *supramodal association areas (hippocampus)*. Functional studies of the inferior parietal lobule in the monkey (Motter and Mountcastle 1981) suggest that *supramodal cortex* may be involved in very *complex functions associated with visual attention, spatial orientation and perception*, and the integration of visual and manual operations (see also Chap. 14).

The concept and nomenclature regarding the polymodal and supramodal areas are presented in the more extended illustrations of Flohr (1991), as well in Chap. 28.

27.4.2 Cross-Modality Experiments

Again we take into consideration the presented single-EEG epochs of hippocampus, which are compared with those of cortex and thalamus with regard to inadequate stimuli. Although inadequate stimulation does not elicit any 10 Hz enhancement in thalamus and cortex, ample 10 Hz enhancements are recorded in hippocampus also following inadequate stimulation. Illustrations shown in Fig. 27.3 can be indicated as prototypes for all the experiments performed with 11 cats. Now we shall try to follow the signal transmission from receptive fields to upper centers of the brain. Every sensation gives rise to neural impulses that reach the brainstem and, later on, thalamic relay centers and cortex (see Chap. 3 in Vol. I). On the other hand, we now see that several nuclei of the limbic system and hippocampus receive inputs from brainstem and especially from reticular information.

We have an important experimental situation, in which hippocampus reacts to sensation (adequate or inadequate) with relevant alpha response and, according to the findings shown in Chap. 1, being completely independent from thalamus. This reaction clearly demonstrates the existence of an alpha circuit (or an alpha transmission line or alpha transmission neural pathway), which also includes hippocampal formation.

Furthermore, these results teach us that we can use both the sensory and event-related potentials as an important tool for further neurophysiological and functional correlates. Visual sensory information, coming from retina, is certainly also relevant for the auditory centers or all types of associative neural populations of the brain. How could such information reach the cortex, in which visual and auditory sensations must be evaluated together? If the light evokes 12 Hz oscillations in lower parts of the brain and if 12 Hz oscillation is related to a light stimulation, then hippocampus processes this information and probably conveys it further to primary sensory or association areas of the cortex.

The 12 Hz code is processed not only in primary visual areas, but probably in other association areas of the cortex, too. The 10 Hz oscillation of the auditory cortex will be also processed in auditory association areas. Upon an inadequate stimulation, in most cases a secondary 10 Hz wave with a latency of 150–300 msec can be observed (Fig. 2.2). Miller (1992) has calculated that in the rat brain the neural conduction time is around 120 ms from hippocampus to cortex. Is there any possibility that hippocampal 10 Hz oscillations propagate to several areas of the cortex as well as to other limbic structures?

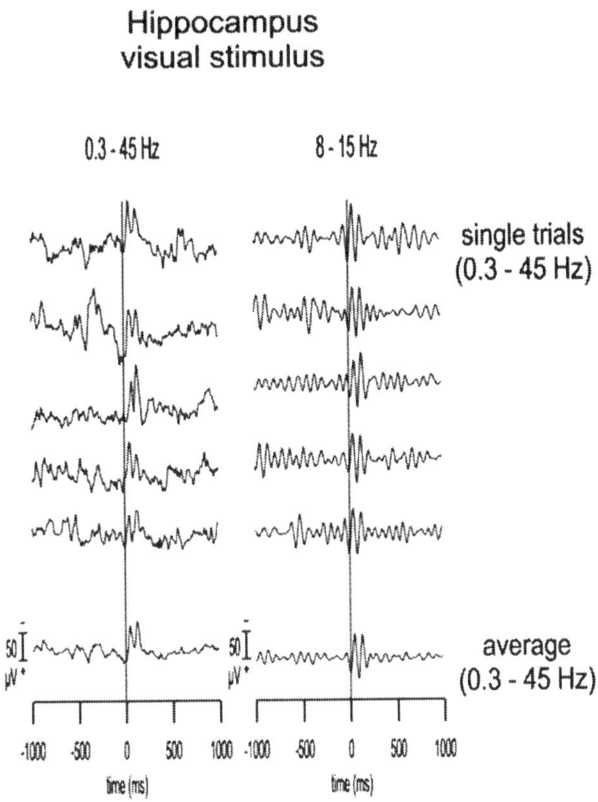

Fig. 27.3. Hippocampal alpha responses in the cat hippocampus recorded with intracranial electrodes as described in Chap. 1. *Left column*, wide-band filtered curves. *Right column*, curves filtered in the alpha range (8–15 Hz). *Top*, single epochs. *Bottom*, averaged EPs.

In order to establish such a theory, one should be able to track the oscillations with multiple electrodes in several parts of the cortex.

27.5 Frontal Cortex

ERPs obtained with paradigms inducing focused attention, P300, and with stimuli giving rise to high expectancy states have shown marked electrophysiological changes in the frontal cortex, parietal cortex, and the limbic system. It was shown that the frontal areas of the human cortex react with enormous theta enhancements to cognitive stimulation requiring states of focused attention and short-term memory (see Chaps. 19 and 20). In the human frontal cortex a theta enhancement increase of 50% was recorded while a subject paid attention to a target that was expected 100%. Similar experiments with cats demonstrated that also in the CA3 layer of the hippocampus an increase of 40% theta enhancement is to be seen.

In P300 types of experiments, again learning tasks led to a theta increase with a time delay in frontal and parietal recordings (see Chaps. 18 and 19). These results clearly demonstrate that tasks requiring attention give rise to marked theta increases in evoked-potential components. When comparing the results of experiments with simple light or sound stimulation, in which the EPs contain dominant alpha responses, we are inclined to state that cognitive loading increases the weight of theta components in comparison to alpha components. Furthermore, the increase in theta responses takes place mostly in frontal hippocampal or parietal structures. Even the omitted stimuli, which give rise to a P300 response in the cat's hippocampus, have a dominant theta component, again with the largest component in the CA3 layer. Under the impression of the results described in the previous sections, we again are confronted here with the idea that association processes are accompanied by theta enhancements (see Chap. 16).

We go further, taking into account one of the major functions of the frontal cortex: As Fuster (1991) extensively stated, the frontal cortex is highly involved in processes of anticipation. When we apply repeated light stimulation to a subject and ask him to focus his attention on the third signal applied (Chap. 19), then the subject certainly anticipates the presentation of the target. Since the anticipation is one of the major functions of frontal lobes, and since a significant theta increase is observed in the evoked potentials, we can assign to frontal cortex, also in linkage with hippocampus, a dominant localization of theta processes (see Chap. 25).

Maintaining this idea in our scope, we take another step. In Chap. 12 in Vol. I experiments showed clearly that frontal EPs are strongly controlled or influenced by the theta activity of frontal cortex. Accordingly, it is assumed that theta rhythm is a major operating rhythm of the frontal cortex. As stated elsewhere (Brandt et al. 1991; Rahn and Başar 1992; Chap. 22), the major operating rhythm of occipital areas is alpha. It is certain that alpha responses or theta responses can be detected in various intracranial and cortical structures. The existence of a significant difference in the major operating rhythms in occipital or frontal areas gives strong support to the possibility that theta (spontaneous, evoked, induced) and alpha (spontaneous, evoked, induced) rhythms have fundamentally different functional operations. But during some functional states, major operating rhythms can change their functional roles; the nature of the experiment, i.e., tasks, can influence the weight of these functional components in brain rhythms.

In Chap. 28 we will return again to the cognitive correlates of these various oscillatory components: We will put forward a hypothesis where all the brain rhythms will be brought under a selectively distributed delta–theta–alpha–gamma system of the brain.

27.6 Cerebellum

Although functional correlates of the cerebellum have been described in all textbooks as being related to movements and motor processes in general, this view is now changing because of accumulated new data on cognitive cerebellar functions. Leiner et al. (1989) described new evoked connections between the cerebellum and cerebral cortex, as illustrated in Fig. 27.4. Furthermore, the same authors reviewed results of published data concerning cerebellar participation in human neural functions. Among these functions are:

1. Cerebellar activation during mental imagery
2. Cerebellar activation during word processing
3. Cognitive deficit in rapidly shifting attention between sensory modalities
4. Cognitive planning and cognitive operation in three-dimensional space
5. Verbal and nonverbal intelligence and verbal associative learning

The review of Leiner et al. (1993) includes data on measurements of PET, SPECT, and computer tomography.

In the present book the following important information emerged from studies with the cat cerebellum:

(1) The cerebellar cortex has similar dynamics in comparison to other structures of the brain: The frequency characteristics contain, for example,

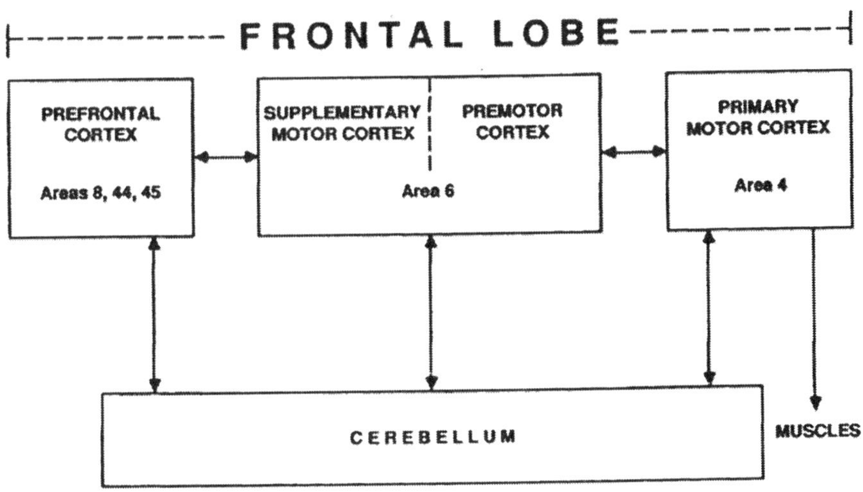

Fig. 27.4. Parallel cerebellofrontal connections in the human brain. Different parts of the cerebellum are linked to different areas of the frontal lobe. Because these connections enable the frontal areas to send signals to the cerebellum and receive signals from it, the cerebellum can serve as an information-processing adjunct to all of these frontal areas. Information can be processed rapidly in such cerebrocerebellar loops during the waking hours of life. (From Leiner C.H. et al. (1989) *Behavioral Neuroscience* 103: 998–1008.)

maxima in 10 Hz, 40 Hz, and other known frequency bands. Since we have used auditory stimulation to analyze evoked potentials, we can tentatively assume that after the stimulation, auditory associative reactions are processed also at the cerebellar level. At least one can definitely state that the cerebellar dynamics are perfectly coordinated (correlated) with the integrative dynamics of the entire brain.

(2) An important aspect of cerebellar electrical activity is its spontaneous activity in the frequency range between 200 Hz and 300 Hz. This very high frequency activity of the cerebellum was described in Chap. 10 of Vol. I. Furthermore, it was demonstrated that at least 25% of the data recorded during the waking stage of the cats were chaotic signals and not "noise." Furthermore, we also have recorded phase-locked very high frequency response components in the analysis of dynamics by means of frequency characteristics and filtered responses (see Chap. 1).

(3) After the application of pharmacological agents the relevant finding was the following: Upon the injection of cerulein into cats, the cerebellar dynamics to auditory stimuli did not change at all, whereas cortical and hippocampal responses were completely changed (see Chap. 8). This means that by use of pharmacological agents the electrical responses and entire dynamics of the cerebellum can be electrically discorrelated and separated from the excitability of other parts of the brain; i.e., the dynamic integration of the cerebellar response has a very important functional correlation that cannot at all be explained through volume-conducted activities of other centers.

This example once more shows that under the given conditions, some parts of the brain can keep their own dynamics that are not necessarily reflected in other structures. For this implication the reader is also referred to Chap. 18 in Vol. I.

In this book we cannot describe functional brain waves in all studied structures. These few examples describing new functional developments in cortex, cerebellum, hippocampus, and thalamus are given in order to point out the utility and importance of applying the concept of event-related oscillations and systems-theoretical methods in integrative neurophysiology.

28. Brain Functioning: Integrative Models

Erol Başar and Martin Schürmann

In previous chapters related to event-related oscillations and cognitive processes, methods and approaches of the Brain Dynamics Research Program were utilized. Special use was made of the concept *going out of the system.* The strategies used in understanding the cognitive correlates of the event-related oscillations had the aim of investigating the common components in results obtained by application of diverse psychophysiological paradigms.

28.1 EEG Frequencies as General Operators

We have often used the expression *general operator* for various EEG frequencies. What is an operator? An operator does not only perform a function. It also helps in the transmission of information from one location to another. When we say that 10 Hz activity can be considered as an operating signal, this means not only that a given structure of the brain generates a 10 Hz pattern, but that it sends 10 Hz-signals to other structures to elicit excitation on the receptive neural populations. This means that the signal "EEG oscillation" acts also as an electrical message to elicit activation in another brain structure.

When considering this, we do not assume that 10 Hz EEG signals in the range of 20–100 μV are necessarily related to action potentials, but the 10 Hz oscillations at the neural level (spikes with oscillatory behavior) are probably the acting electrical signals that change the excitatory level. We assume similar mechanisms for 40 Hz, 4 Hz, and all other oscillations.

As we have previously shown, these oscillations or *wave packets*, can be recorded in the entire brain by being *in phase*, *prolonged* or *delayed.* Probably, not only the 2 Hz, 4 Hz, 10 Hz, and 40 Hz activities are manifestations of excited structures, but they are involved in integrative functions communicating and interacting with a great number of neural populations distributed in the brain.

28.2 Do EEG Frequencies Reflect Repertoires of Higher Brain Function?

The remarkable diversity of the nervous systems of various animals and their exquisite capacity for adaptive function is considered by neurobiologists to be an intriguing subject. G. Edelman (1978) points out that despite their complexity, however, all nervous systems appear to obey similar general principles at the level of morphological expression of neuronal structures and their mechanisms of signal transmission. Edelman claims that at the functional level, fundamental and methodological confusion still reigns. Therefore, he poses more general questions on the evolutionary development and function of higher brains.

1. Does the brain operate according to a single principle to perform its higher-order and cognitive functions?
2. Can one discern a general mechanism that is required for the performance of cognitive processes despite the many differences in brain substructures?
3. At what level does such a mechanism operate? At the level of cells, molecules, or circuits?

Since the properties of membrane potentials at the cellular level are known to be invariant in various brain structures or brain systems, common neurophysiological mechanisms seem to exist. But it is also evident that these cellular properties cannot explain complicated brain functions. Mountcastle (1976) stated, "The center problem of brain physiology is now to understand the actions of large populations of neurons, actions that may not be wholly predictable after properties of subjects. The main problem of brain physiology is to understand the nature of repertoire building by populations of cell groups" (compare also Fessard's (1961) viewpoint as given in Chap. 31).

When assuming that the higher cognitive functions of the brain can be analyzed more accurately by the activity of neuron populations, we are confronted with two basic questions:

1. Can the EEG frequencies reflect basic functional properties of brain function?
2. Can transition of the EEG from one state to another be explained as "regime change" during the performance of a given function?

The basic properties of the functioning of the CNS require an exact description of attention, learning, and memory. As Edelman analyzed, the memory states reflect parallel processing of the signals and asynchrony that should be taken into account to discover the initial function of the time period assumed for the phasic states. Hippocampal rhythms with frequencies of 3–8 Hz have been detected, and it is conceivable that cycles could be as long as 300 msec. In other words, we are confronted with some basic frequency signals in the theta frequency range during cognitive processes in the human brain and also the cat brain. Again, the frequency range of 10 Hz plays an important role during high states of expectancy (the observed 10 Hz induced rhythms related to memory processes were explained in Chaps. 22 and 24; see also Chap. 25).

In the last part of this volume we describe step by step a theory on the existence of the selectively distributed alpha and theta and 40 Hz systems of the brain. We propose now the existence of EEG modules that can be active in parallel, being selectively (diffusely) distributed in the brain. EEG modules, in this context, are neural populations giving rise to coherent EEG activities in different frequency channels.

We assume that the EEG frequencies that can attain coherent states belong to repertoires of the higher brain function. The superposition of several of these signals makes it possible to connect behavioral events with an ensemble of event-related oscillations. Since the event-related potentials are now very popular for scientists of psychophysiology who use a diversity of psychological paradigms, consideration of EEG dynamics can bring a deeper understanding to the correlation between cognitive performance and physiological states.

28.3 From "Functional EEG Modules" to "Selectively Distributed Frequency Systems" in the Brain

Regan (1989) discussed in an interesting way the single-neuron doctrine and alternative ideas for understanding the physiology and functional organization of the brain. Regan pointed out *pontifical neurons, grandmother neurons,* the *Barlow neuron doctrine* for perception, and Mountcastle's ideas about cortical modules. Sherrington introduced the notion of *one ultimate pontifical nerve cell*, which integrates CNS function. The grandmother neuron is a neuron that responds to nothing else but the face of one's grandmother. Barlow (1972) stated that the sensory system is organized to achieve as completely as possible a representation of the sensory stimulus with a minimum of active neurons. According to his concept, perception corresponds to making a selection from the very numerous high-level neurons that corresponds to a pattern of external events of the order of complexity of the event that is symbolized by the word "grandmother."

Szenthagothai's (1983) well-known illustration of a 300 μm diameter cortical module (Fig. 28.1) is one of the important examples of neural modules and local nerve circuits, an ensemble that plays a significant functional role. According to Szenthagothai, modular organization allows a higher degree of specific connectivity to be achieved with a minimum of genetic instructions.

Mountcastle (1976) defined the basic function unit as a *minicolumn* approx. 30 μm in diameter containing 100–300 neurons. Larger processing units called *macrocolumns* contain up to several hundred *minicolumns*.

The idea of *distributed system* plays an important role in the statements or theories of all these scientists. In Mountcastle's words, *"prominent among*

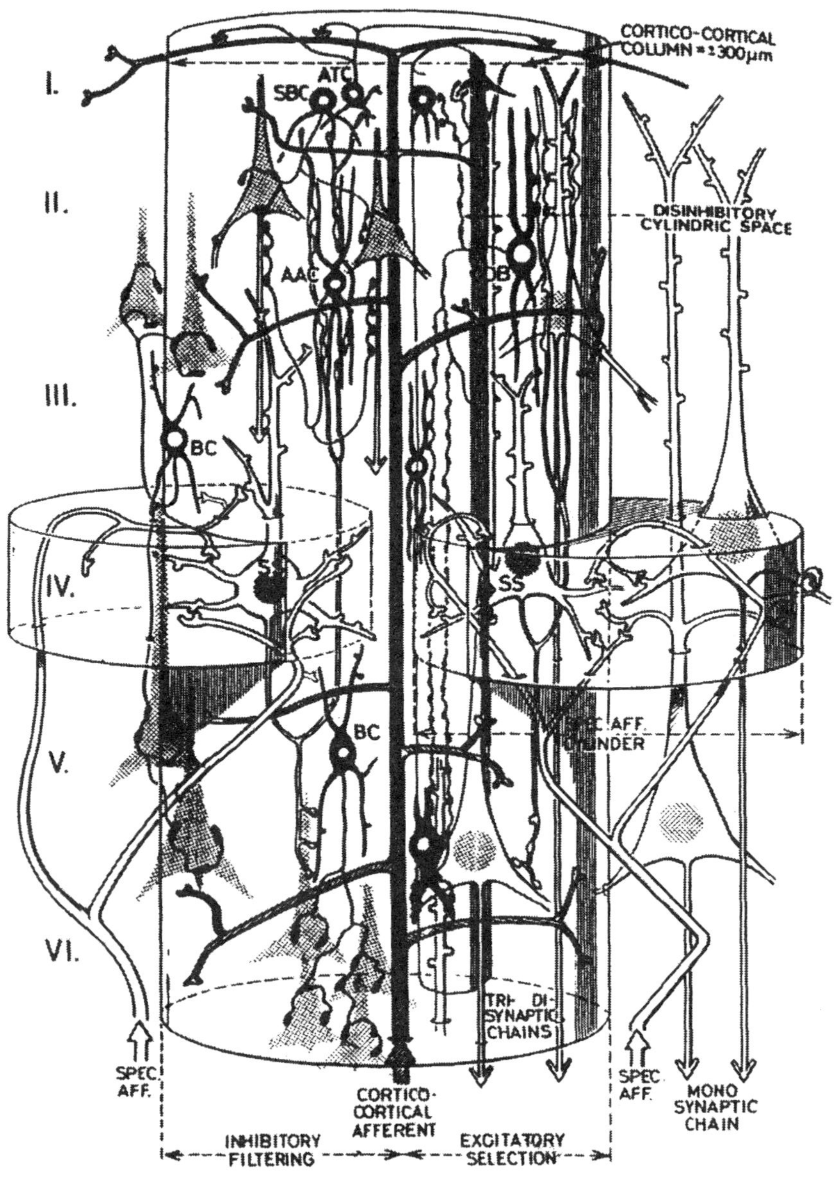

Fig. 28.1. Neuron circuits of a representative corticocortical column. Cells in black are known or assumed to be inhibitory. ACT: axonal tuft cell. SC: small basket cell. AAC axoaxoric (chandelier) cell. BC: basket cell. The following pyramidal cells and excitatory interneurons are drawn in outline. SS: spiny stellate cell. CDB: double bouquet cell, assumed to be predominantly disinhibitory, acting in a very narrow vertical cylindric space due to its narrow vertical axon strands. (From Szentágothai J. (1983) *Reviews in Physiology, Biochemistry and Pharmacology* 98: 35–61.)

them is the concept that the brain is a complex of widely and reciprocally interconnected systems and that the dynamic interplay of neural activity within and between these systems is the very essence of brain function." Again coming back to those statements, the large entities of the brain are themselves composed of replicated modules. The linked sets of modules of the various brain entities are defined as a *distributed system.*

Freeman (1975) has named the theory of using dynamics of neural masses the *new Sherrington doctrine*, in which neural populations play the significant functional role. Roy John (1988) described a *hyperneuron* consisting, again, of neural populations as a functional important entity in the brain.

According to the experiments described in this book, the gamma activity is not recorded only in a small given brain population but in widely distributed neural populations (see Chaps. 3 and 26). The 10 Hz resonances are also widely distributed and depend on functions processed in the brain. Depending on *regimes* or *states* of the brain, the limbic system, brainstem, thalamus, and cortex are all involved with 2 Hz, 4 Hz, 10 Hz, and 40 Hz firing or with all of them. The distribution of coherently working EEG modules does not take place only in the cortex; we assume that they are selectively distributed in the entire brain.

One of the examples is the frontal–hippocampal–parietal system, which gains a highly organized theta state during attentive processes. By combining examples and concepts arising from the analysis of EEG and evoked potential responses, we aim at an extension and generalization from modular concepts to modular frequency systems operating in the entire brain. This is a new hypothesis that has emerged from the results of the present book. This hypothesis is based on experiments in which brain functions are explained on the basis of measurement of field potentials that reflect the activity of large neural populations. These neural populations can also be interpreted as functionally active modules. These explanations are based on the different frequency systems of the brain, which are not defined in space and time precisely. But even then, such explanations seem more efficient in explaining the dynamic aspect of the percepts.

With the neuron doctrine alone, as it was originally proposed by Sherrington (1948), it is not possible to interpret the functional contributions of alpha, theta, delta, and gamma responses. The generators giving rise to these frequency responses are extremely sensitive to the modality of sensory and cognitive inputs. The existence of alpha, theta, delta, and gamma generators that act in a parallel fashion and are found in distributed areas of the brain can be yet explored only with macroelectrodes that are placed with an adequate physical separation. Tracking of properties of functionally related distant single neurons is not yet possible because of technical limitations. At the first step, approaching the higher functions of the brain is easier when we assume the existence of modules that generate EEG-like signals. The foregoing reasoning is an attempt at extending the Sherringtonian neuron doctrine to neural populations with oscillatory firing properties. At this point, it should again be remembered that potentials measured from human subjects reflect the properties of a large number of neurons and, possibly, multiple neural populations.

During a P300-type of experiment, the brain goes over to a regime in which distributed theta and delta neural networks are mostly in play. During pure sensory stimulations several structures of the brain go over to a 10 Hz regime (see the example in Chap. 1). During the slow-sleep stage the brain does not only show slow wave activity in the delta frequency range, but all EPs from intracranial and cortical structures of the cat brain and cortex of the human brain depict a dominance of delta responsiveness.

28.3.1 Does Cortico-Cortical communication between EEG Modules in the Distant Parts of the Cortex Exist?

According to Szentagothai (1983), connections are established between vertical columns and modules (Fig. 28.1). These connections can also link frontal areas to occipital ones. Now the question is whether the hypothesized EEG generating modules in the frontal and occipital areas are also connected. If event-related 10 Hz oscillations are measured in occipital and frontal areas, then it is certainly legitimate to define such modules, and thus connections between these modules. The same possibility also exists in theta, delta, and gamma frequency bands. With the knowledge presented in this book it is not possible to clearly indicate that the modules described above also serve for processing information in the EEG frequency channels. Volume conductions between distant modules (for example, occipital and frontal cortices) are impossible. Accordingly, the high-amplitude delta responses following target signals in P300 experiments must have different local generators. Another probability consists in the fact that neural impulses can reach several parts of the cortex following an event and generate in cortical areas oscillatory responses in various frequency channels depending on the sensory information or task.

The schematic illustrations describing selectively and in parallel working EEG systems, to be explained in the next section, are based on the hypothetical existence of large populations of modules in the cortex.

28.4 Tentative Conclusions

The concept of selectively distributed alpha, theta, delta, and gamma systems is difficult to explain, because these systems have no precise spatial boundaries. Further, no exact specification can be stated with regard to the time domain, since these systems do not always occur at the same point along the time axis. Therefore, we supply some schematic illustrations that are not precise but are based on experimental results as well as on physiological and anatomical evidence. When these facts are combined, a new configuration

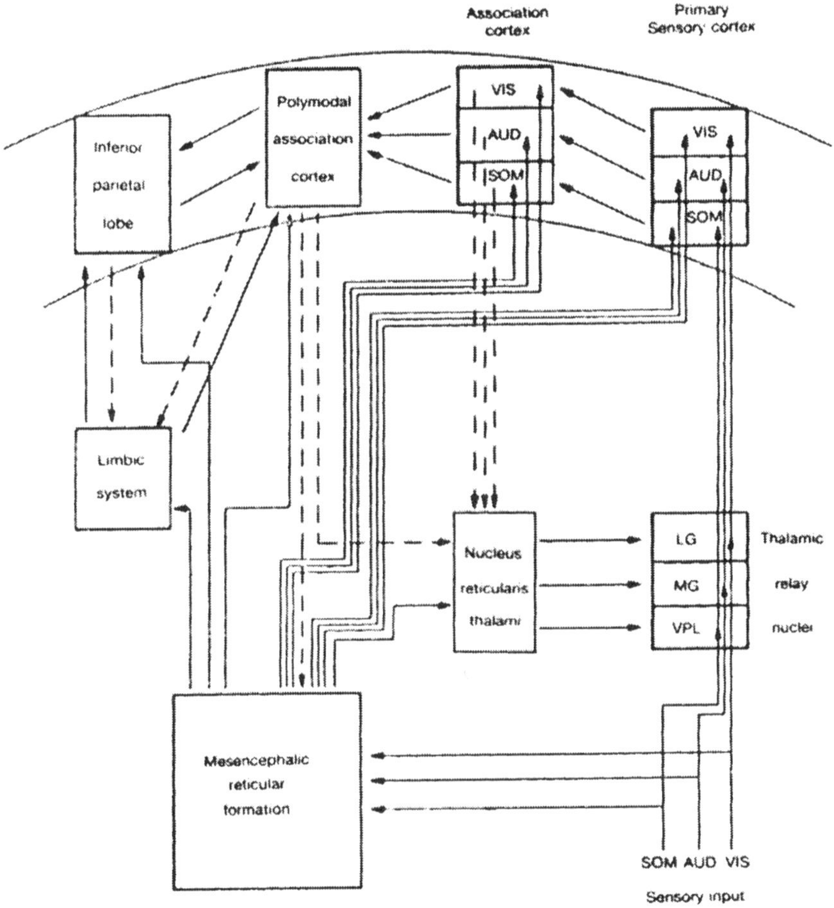

Fig. 28.2. Flow of information in the auditory, somatosensory, and visual pathways, reticular formation, limbic system, and association areas of the cortex. (Sensory and cognitive neural pathways in this figure modified from Flohr H. (1991) *Theory and Psychology* 1: 245–262.)

emerges. However, the processes in this configuration are frozen in the time domain; i.e., they are limited to durations of a maximum of 1 s.

Certainly, the brain is working in a more complicated manner, but the model ideas are presented so as to expose strategies that could be applied productively in brain research.

Flohr (1991) described the physiological and anatomical connections in the brain in a simplified and transparent manner (see Fig. 28.2).

1. *Specific afferents from sense organs reach specific thalamic nuclei* before going to primary cortical areas (see Chap. 3 of Vol. I). For instance, au-

ditory information is transmitted through the medial geniculate nucleus to the primary auditory area; visual afferents are transmitted through the lateral geniculate nucleus to area 17 of the occipital cortex.

2. *Nonspecific afferents reach the cortex from the mesencephalic formation.* It has now been established that the reticular formation is connected to different nuclei with specific afferent connections. There is a second site where the reticular formation influences the processing of primary afferents: the thalamic relay nuclei. The nucleus reticularis thalami, a thin sheet of neurons, surrounds the dorsal thalamus and inhibits the thalamic relay nuclei. Its control function is, in turn, affected by collaterals of thalamocortical pathways, by collaterals from corticothalamic projections, and by inhibitory afferents from the mesencephalic reticular formation. As we have already described in Chap. 3 of Vol. I, there are important connections within the cerebral cortex involving the association areas. Primary auditory, somatosensory, and visual fields each project to adjacent unimodal association areas, which, in turn, project to secondary unimodal association fields. The unimodal association areas project to a number of polymodal sensory areas, lying in the cingulate gyrus and the parietal, temporal, and frontal lobes. The functions of these areas are vaguely described as a cross-modal association and synthesis (the cross-modality experiments and the concept of cross-modal association are described extensively in Chaps. 2 and 10).

The polymodal association areas project to the inferior parietal lobe, which has been termed a *supramodal* area. Polymodal and supramodal regions have connections to the limbic system; these connections provide the anatomical substrate by which motivational states influence cortical processing of sensory stimuli (see Chap. 3 in Vol. I).

The advantage of using this simplified illustration is tremendously important for understanding additive cognitive information processing of cortical areas. Following sensory and event-related stimuli, every sensation induces in our brain also cognitive loading, at least for matching processes. Furthermore, all presented cognitive targets evoke sensations; the respective neural processes are interwoven and require, for final processing, at least three neural loops next to purely sensory connections in the simple sensory systems. These loops are:

1. *secondary connections* to the cortex over the reticular formation
2. *secondary connections* over the limbic system
3. The connections within the cortex between association areas

28.4.1 Activation of Alpha System with Light

When a subject (human or cat) has been visually stimulated, large alpha enhancements are recorded in the mesencephalic reticular formation, lateral geniculate nucleus, and, in parallel, in the hippocampus. There are also large

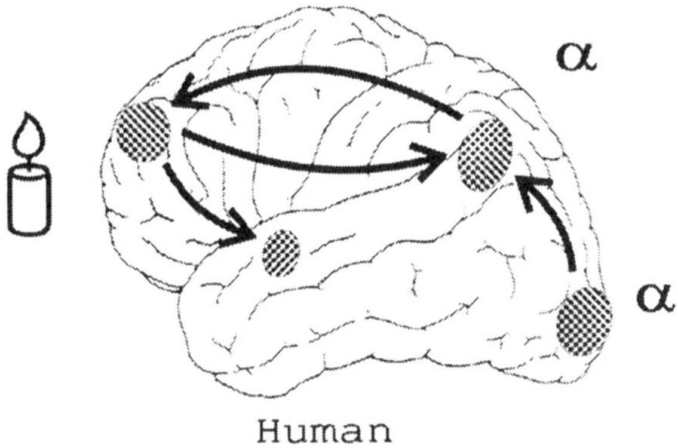

Fig. 28.3. Activation of alpha response "modules" upon visual stimuli. Frontal and parietal responses are not parallel, but secondary delayed and smaller responses.

alpha enhancements in the visual and association cortices. The information flows to the polymodal association cortex and to the limbic system (results from Chaps. 1 and 9). This information transfer is presented in Chap. 1. The results have also shown that large theta enhancements were recorded in the thalamus, hippocampus, primary cortex, and association cortices including the frontal lobes. The same figure illustrates that alpha enhancements cannot be observed in the cortical auditory areas and in the medial geniculate nucleus upon visual stimulation. However, theta enhancements are present in these structures.

Figure 28.3 shows the hypothetical information flow in the human cortex during the first second following a stimulus. There are large and dominant alpha responses in the occipital cortex. The alpha information then reaches the parietal cortex, and via the known association areas the frontal cortex, but not the auditory areas.

The neurophysiological processes in subcortical structures are demonstrated in Fig. 28.4. Large alpha and theta enhancements are observed in the limbic system, reticular formation, and lateral geniculate nucleus. No alpha enhancements are recorded in nuclei of the auditory pathways. Although the theta enhancements are a major response component also in thalamus and cortex, the alpha response was not recorded at all in these structures.

With regard to EPs, we have recorded large enhancements in the occipital and parietal cortices, but generally only small and delayed alpha responses. The hypothetical schema in Figs. 28.3 and 28.4 is also supported by neuroanatomical findings about connections that travel a long distance from the occipital cortex to the frontal cortex and from there to the temporal cortex.

28.4.2 Activation of the Alpha System with Auditory Stimulation

Figures 28.5 and 28.6 explain a similar hypothetical flow of information upon an auditory stimulation. In this case alpha enhancements dominate the structure in the auditory pathways, i.e., reticular formation and hippocampus.

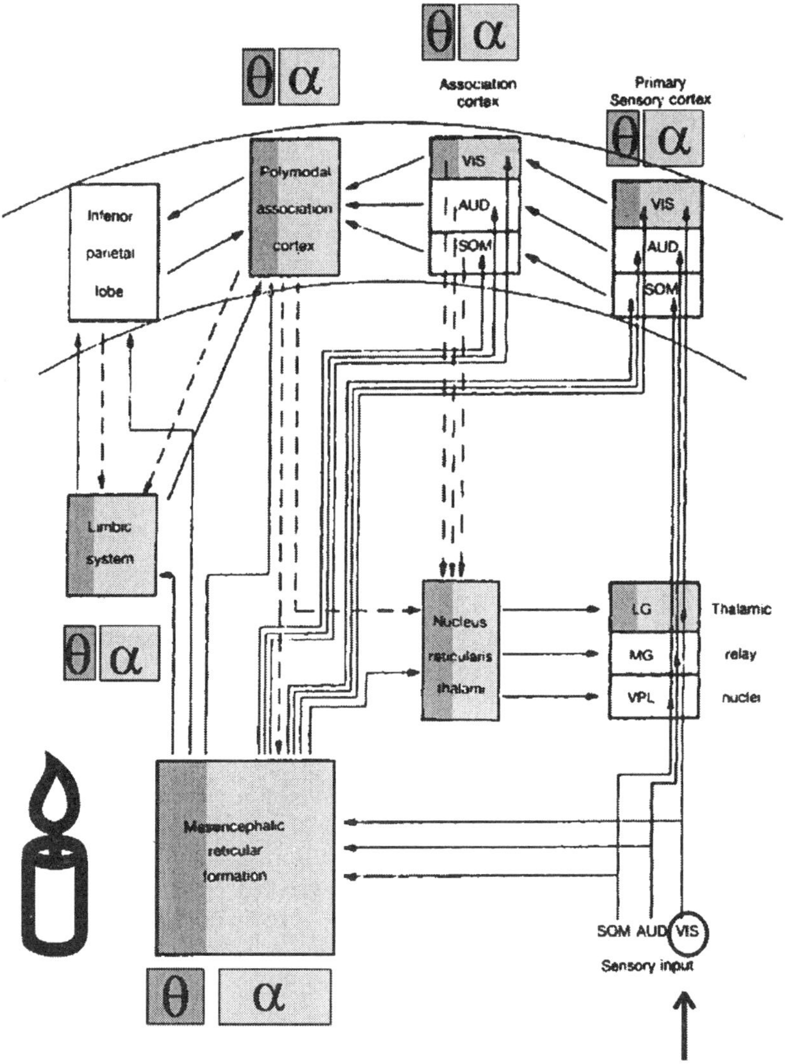

Fig. 28.4. Flow of oscillatory information in alpha and theta frequency channels in the visual pathway, reticular formation, limbic system, and association areas of the cortex. Letters alpha and theta indicate the existence of strong enhancements. (Sensory and cognitive neural pathways in this figure modified from Flohr H. (1991) *Theory and Psychology* 1: 245–262.)

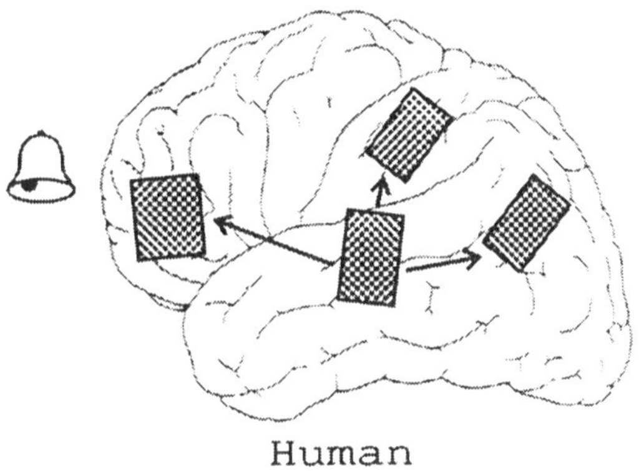

Fig. 28.5. Activation of alpha response modules upon auditory stimuli. Frontal responses probably delayed and smaller.

The neurophysiological processes in subcortical structures depict large alpha-response components in reticular formation, medial geniculate nucleus, and auditory cortex. It is noteworthy that reticular formation and hippocampus show large alpha responses, whereas lateral geniculate nucleus and visual cortex depict only theta responses. Theta enhancements are present in all structures, regardless of whether the stimulus is inadequate or adequate. The way of thinking used for auditory stimulation can be applied to visual stimulation also, but now the largest alpha enhancements are marked in temporal, parietal, and occipital areas.

28.4.3 Activation of Theta and Delta Systems

We again use the same anatomical pictures of Flohr to explain the electrical responses occurring in the region of 300 msec upon a cognitive target. For the sake of simplicity, we do not consider the delayed and prolonged 10 Hz responses and take into consideration only delta and theta responses. Experiments described in Chaps. 14, 15, 18, and 21 confirmed large delta and theta enhancements around 300 msec following the stimuli. Figure 28.7 explains the process limited to approximately 1 s following a cognitive (i.e., target) stimulus.

If the stimulation applied contained an event-related target, the hypothetical flow of information would shows large theta and delta enhancements in all areas of the cortex, but probably also in the substructures. Large theta and delta responses and delayed alpha responses would only be observed at 300 ms after stimulation (see also Chap. 16).

28.4.4 Experiments with Focused Attention

We can also present similar pictures for the experiments that we have performed where auditory stimulation was used during which the subject had to pay attention to the third signal (for the paradigm, see Chap. 19). With

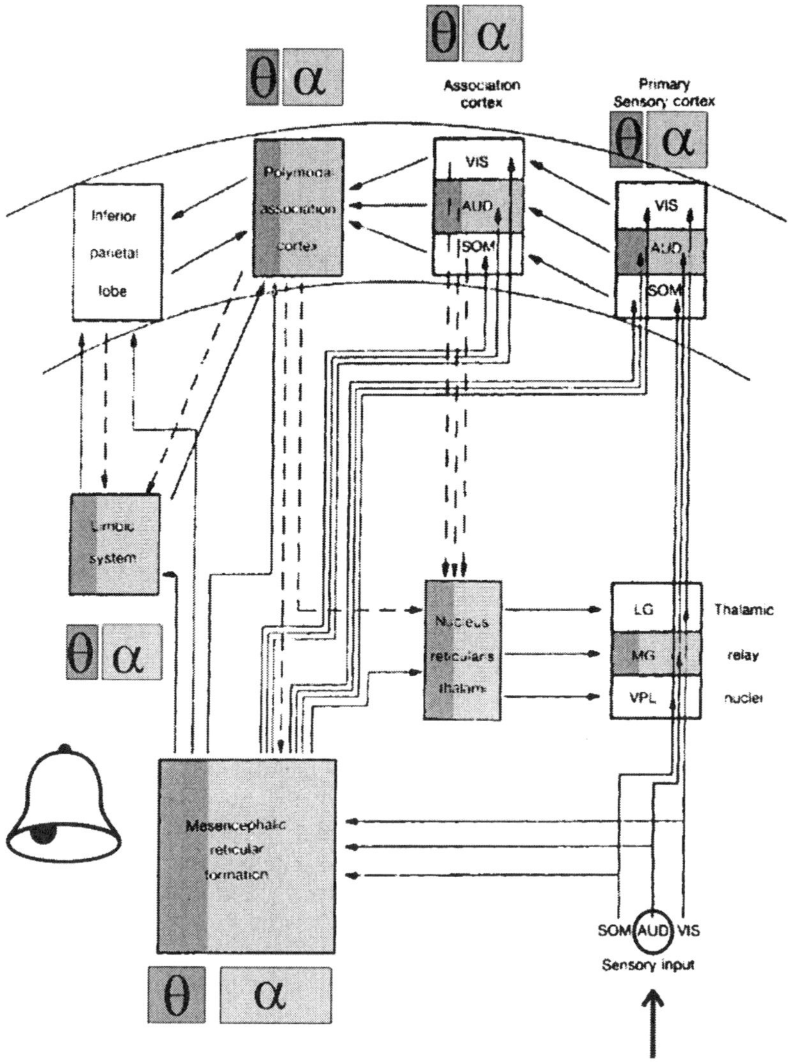

Fig. 28.6. Flow of oscillatory information in alpha and theta frequency channels in the auditory pathway, reticular formation, limbic system, and association areas of the cortex. Letters alpha and theta indicate the existence of strong enhancements. (Sensory and cognitive neural pathways in this figure modified from Flohr H. (1991) *Theory and Psychology* 1: 245–262.)

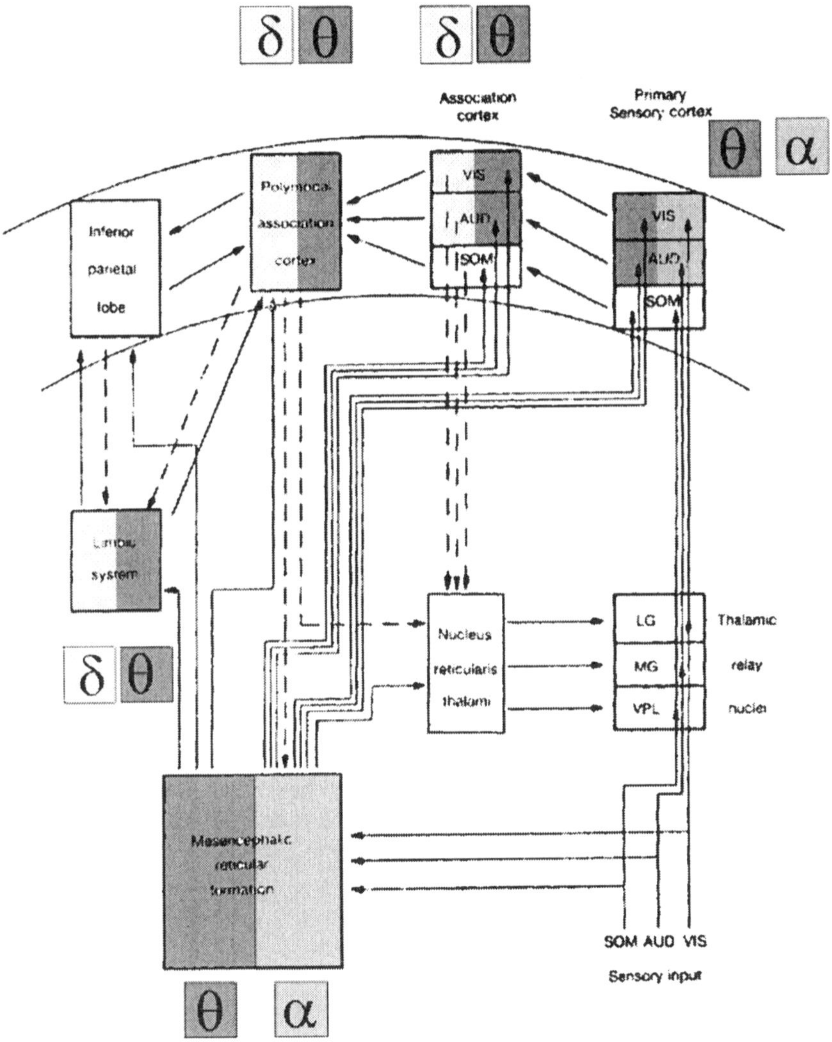

Fig. 28.7. Flow of oscillatory information in theta and delta frequency channels in the visual and auditory pathways (VIS, AUD) and limbic system. Letters theta and delta indicate the existence of strong enhancements. (Sensory and cognitive neural pathways in this figure modified from Flohr H. (1991) *Theory and Psychology* 1: 245–262.)

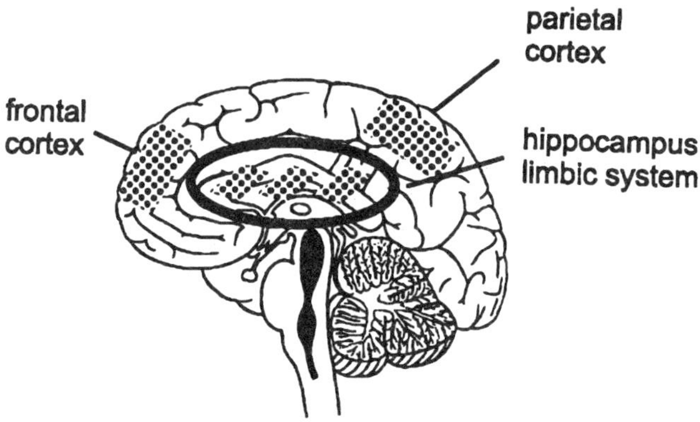

Fig. 28.8. A hypothetical diagram of the selectively distributed theta response system.

regard to the experiments with humans, the third stimulation elicited large enhancements in the theta frequency range, especially in the frontal and parietal recordings. However, these large theta enhancements occurred with a delay of approx. 300 ms after the stimulation. On the other hand, the changes in hippocampal responses occurred without any delay.

Although for this paradigm there is no information on changes in theta activity in the human hippocampal recordings, we hypothesize the existence of a theta-response circuit including hippocampus, frontal, and parietal areas of the cortex. In all experiments described in this book, tasks that have involved focused attention have given rise to theta enhancements. Accordingly, we propose relations between the hippocampal, frontal, and parietal systems as given in Fig. 28.8.

29. EEG and Event-Related Oscillations as Brain Alphabet

29.1 The Integrative Character of the Compound Potential EP

The results of this book show that oscillatory generators are selectively distributed. Thus, upon excitation (sensory and/or cognitive), *diffuse areas* are activated. Accordingly, the existence of diffusely distributed parallel-processing alpha, theta, and gamma systems in the brain were hypothesized in previous chapters. As we have also often pointed out, the 10 Hz, 40 Hz, and 4 Hz enhancements in several distant locations cannot be explained with the phenomenon of volume conduction (see especially Chap. 17).

According to the working hypothesis presented in Chap. 11 of Vol. I, the EEG consists of the activity of an ensemble of generators producing oscillatory activity in several frequency ranges. These oscillators are usually active in a random manner. However, by application of a sensory stimulation, all these generators are probably coupled, or they act together in the same phase space, and the phase locking gives rise to an enhanced rhythm, which we call *enhanced EEG oscillations* or *enhanced oscillations*. The ensemble of responses in these various frequency ranges then gives rise to the compound EPs that are the superposition of several event-related or induced oscillations in the channels of 4–5 Hz, 10 Hz, 20 Hz, and 40 Hz (see Chap. 14 in Vol. I). One can here raise the following criticism: We have dissected the EP into frequency components – alpha-response, theta-response, gamma-response, etc. Does the brain work in this way? Indeed, we can find functional or cognitive states in the brain in which one component gains a dominating responsiveness, thus making only that type of response visible (see, for example, Chap. 19).

Further, in the ensemble of experiments outlined in this book, we have had a number of strategies to demonstrate the superposition principle. These various strategies were:

1. Change the position of the recording electrode (Chaps. 1 and 2)
2. Ablation or pathological lesions (Chap. 16 in Vol. I)
3. Use of phârmacological agents (Chap. 8)
4. Reduce or change the intensity of stimulation or use several stimulation paradigms (Chap. 12)
5. Increase the level of cognitive input (Chaps. 19 and 18)

6. Compare with recordings in single units (Chap. 9 in Vol. I)
7. Change the cognitive input (Chaps. 15, 21, 14)

All these important points or strategies have demonstrated that the event-related oscillations are often independent components and that different functional states (induction, evocation, stimulation) can cause causal changes of the event-related oscillations. In Chap. 18 of Vol. I it has clearly been shown that event-related oscillations in diverse frequency bands are real physiological and/or psychological components.

29.2 Compound P300 Potential

We will emphasize here as a prototype the P300 in order to show that we can start to use a new type of nomenclature regarding EPs. This nomenclature is based on the concept that EPs are superpositions of several event-related oscillations:

1. The major change in the P300 response in comparison to sensory EPs consists in a significant increase in the delta response distributed in all locations of the skull. The dominating response to rare stimuli is a delta response, which, however, occurs with a considerable latency.
2. Another important change in the P300 response consists in an increase in the theta component, especially in parietal and frontal locations in the time window around 300 ms following the rare stimulus. In other words, increased theta wave packets do not occur immediately, but they are delayed and contribute to the composition of the late component of P300, which is the characteristic of this response.
3. In the alpha frequency range, no increase in the 10 Hz oscillation can be observed. However, the 10 Hz oscillation is prolonged. It seems that the usual sensory component that we call *adequate 10 Hz enhancement* requires more time for reaction.
4. We have in this compound P300 potential another important change in the 40 Hz frequency band in the region of 300 ms following the stimulation, and a significant increase in 40 Hz can be observed (Chap. 21).

These findings relate several functional correlates, of which we will mention two:

1. The analyses with filtering techniques show that the 2 Hz, 4 Hz, 10 Hz, and 40 Hz wave packets can react independently and move around the time axis without being related to each other. However, a superposition of all these wave packets in the vicinity of 300 ms following the stimulation is observed. From the functional viewpoint, theta increases following the focused attention to a target can immediately elicit a theta increase upon stimulation (Chap. 19). The design of such an experiment shows that the

experimenter can operate with these wave packets in order to relate them to defined functions.

2. The compound potential, P300, teaches us that if several functions, such as *surprise, decision-making, focused attention, short-term memory,* and *learning*, are involved, then not only analyses of single-frequency components are important. Instead, we need the combination of all the components in order to judge the function contained in such a compound potential.

We pointed out the effect of learning during recording of such P300 potentials: At the beginning of an experiment subjects are not yet comfortable with the cognitive task, but once they learn to differentiate the rare tones (target tones) from frequent tones (nontarget tones), then theta-response waves are observed after approx. 300 ms following the stimulation, as is usually analyzed in the P300 potential. In several parts of this book we have described the correlation of theta waves in a number of learning experiments, as first demonstrated in the pioneering work by Ross Adey (1960, 1969) (see also Chap. 25, particularly Table 25.1). One would say, *What is new in this description? Do we have new statements about learning and theta waves if these processes have already been described previously?* The answer is, *Definitely yes.* All types of experiments correlating theta activity with behavior and learning are designed in such a way that the analyzer looks only at the theta activity. Here, we have the compound potential that we call *evoked potential* or *event-related potential*, and we can dissect this potential into frequency components and try to correlate those with functions. Correlation of theta-response increase with increased attention provides one such example.

29.3 A Cognitive Input Reduces the Compound Potentials to Almost Homogeneous Oscillatory Responses

In Chap. 19 we have used the strategy of increasing the expectancy of human subjects in a special task. The subjects were stimulated with repetitive light stimulation, where their attention was focused on the third "attended" signal. In frontal locations of the subjects, EPs were recorded indicating a significant increase in the theta frequency band. Additionally, it was observed that these EPs were reduced to homogeneous theta oscillations that can be observed without digital filtering. Here single-trial EPs from frontal recordings were analyzed during such an experiment (see Chap. 19 for examples of EPs resembling homogeneous theta oscillations).

29.4 How Is a Compound EP Almost Reduced to Homogeneous Oscillatory Response Potential in the Delta Frequency Range?

In Chap. 12 it was shown that by application of auditory signals with different intensities, major changes were observed not only in the EPs but also in the amplitude–frequency characteristics. By decreasing stimulation intensity, the frequency components higher than 7 Hz were attenuated progressively, and at the hearing threshold the evoked potentials were reduced to a homogeneous oscillation in the region of 2 Hz. In other words, the compound potential can be reduced by the application of psychophysiological strategies only to one component. In this case, we tend to say that in such situations we do not have compound potentials, but induced rhythms only in a given frequency band (see Chap. 21).

29.5 Event-Related Rhythms in 5 Hz and 10 Hz: Reduction of the Compound Potential by Topological Differentiation

In previous sections, we have seen the effects of specific cognitive tasks on theta responses in the frontal cortex. According to MEG measurements one can find homogeneous 10 Hz or 5 Hz evoked responses upon sensory stimulation in areas separated by small distances.

Since the spatial resolution of MEG measurements is significantly higher than that of scalp-recorded EPs, there should also be a possibility of observing evoked rhythms in at least the *alpha* and/or *theta* frequency ranges in time recordings of auditory evoked MEG responses. In order to examine this, a series of experiments on 15 subjects was performed. We emphasize only one experiment, to be considered as prototype. In Chap. 13 we have mentioned a type of alpha enhancement observed at a temporal location and, at the same time, within a spatial distance of only 2 cm, a type of *alpha blocking*.

The advantage of using magnetic recordings (instead of electrical recordings) in the computation of amplitude–frequency characteristics is clearly demonstrated in Chap. 13. The very pronounced change in the frequency distribution – as a consequence of a spatial displacement of only a few centimeters – cannot be seen on the basis of electrical recordings from the scalp. How does the evoked response appear in the time domain for two such different frequency characteristics? For further examples of how homogeneous oscillatory responses (theta and alpha) can be recorded from the skull with the help of the MEG, see Chap. 13.

29.6 Brain Codes: Brain Alphabet EEG?

In Chap. 1 of Vol. I the question had been raised whether EEG signals represent *neural codes* related to CNS functions. In previous chapters, treatments of EEG codes and EPs as coding signals were explained in detail. In the following a speculative thought formulated by Başar (1989) is considered.

Let us consider three different brain structures, or networks, A, B, and C, which have excitabilities in the four frequency ranges of 4 Hz, 10 Hz, 20 Hz, and 40 Hz, which we will call theta, alpha, beta, and gamma. Let us assume that structure A can be excited and can excite structure B, which in turn can excite structure C. When structure A excites structure B, there will be a transmission of signals in one of these frequency channels from structure A to structure B; and structure B, if not already excited in a different frequency, will then excite structure C in the same frequency range (see Fig. 29.1).

Let us now give another name to the specific resonant modes, calling them *passletters*, because we have seen that a brain structure into which activity in the 10 Hz frequency range is entering will, by an excitability rule, resonate in that range. In this case the passletter would be excitability at 10 Hz. A second passletter is at 20 Hz, and the third passletter is at 40 Hz. The term *password*, or *pass characteristic*, has already been used by Barlow (1961), who tried to develop a theory of passwords and pass characteristics in neural networks. Our theory is somewhat different from his, but we use his expressions.

According to empirical results, theta, alpha, beta, and gamma resonances do occur in the brain. Therefore, we can suppose that the passletters represent resonant modes that help to optimize signal transfer between various brain

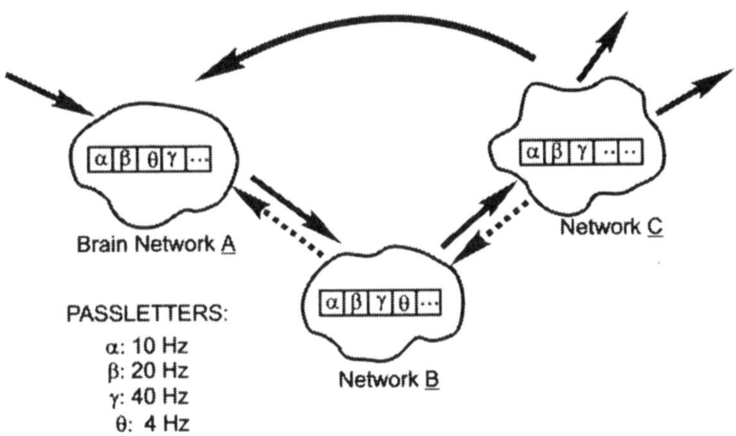

Fig. 29.1. Schematic explanation of passletters. (From Başar E. (1988) in: Başar E., ed. *Dynamics of Sensory and Cognitive Processing by the Brain*. Berlin, Heidelberg, New York: Springer.)

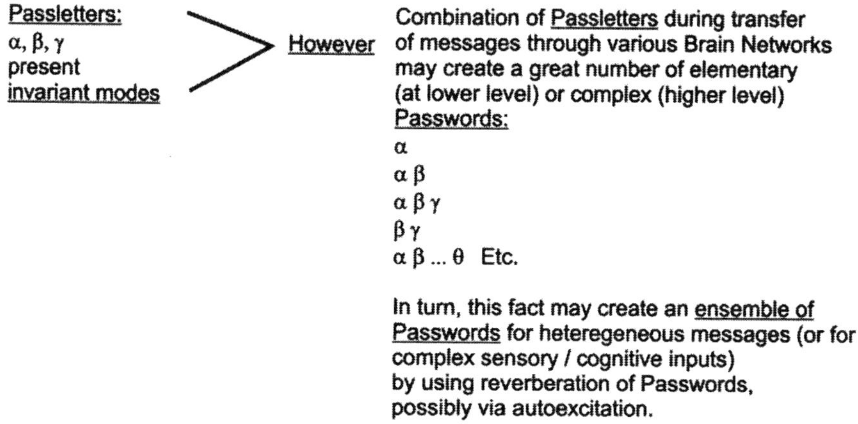

Fig. 29.2. From passletters to passwords. (From Başar E. (1988) in: Başar E., ed. *Dynamics of Sensory and Cognitive Processing by the Brain*. Berlin, Heidelberg, New York: Springer.)

structures. This idea of a way in which brains may work was suggested by the excitability rule derived from experiments.

Now we will go further and speculate that the passletters alpha, beta, gamma, and theta represent resonant modes that are almost invariant. However, passletters may perhaps be combined during the transfer of messages through various brain networks to create a great number of less elementary or complex combinations that we may call passwords. For example, there can be only alpha transmission in all three of the structures A, B, and C, or there can be a combination of alpha and beta (passletters α, β). A still more complicated password would be alpha–beta–gamma (α, β, γ) and so on (Fig. 29.1). Note that these passletters may constitute an *alphabet* (Fig. 29.2).

The information capacity of a structure can be important, especially when it is busy, for if the information capacity of structure B is limited, or if structure B is busy at the time, the transmission in one or another of these special channels will not be optimal. This fact, in turn, may lead to the creation of an ensemble of passwords for complex sensory–cognitive inputs and other heterogeneous messages by using reverberations of passwords, possibly via autoexcitation. How might such autoexcitation occur? One possibility is that structure A could excite structure B again, as in limbic system networks, for example. In such a case, it is possible that as a consequence of B, C could start to reverberate. Such reverberation would be possible only if all the structures had the ability to resonate in the same frequency channel.

Finally, one can speculate further about why it would be advantageous for a brain to transmit by means of resonance phenomena. We have thought of one possibility. The simplest invariant transfer functions are probably represented in the brain by resonant networks, but this fixed "hardware" may create a much richer array of useful "software." That is, it could be an economical principle of the brain to achieve good internal communication first by using a small number of similarly structured networks as channels for a small number of frequencies, and then to combine the elementary passletters into a large number of compound passwords that could transmit a wide variety of complex patterns. (Suchs structured networks ca be alpha modules or gamma modules described in Chaps. 17 and 28)

The building of such patterns is, as stated above, probably due to cooperative phenomena of activity in large neural populations. We therefore tentatively assume that the use of a synergetic concept is of basic importance for understanding the EEG phenomena related to brain function.

29.7 Examples of the Brain Alphabet EEG

In this section we assigned the role of letters to natural EEG frequencies. However, as we know from experimental results, the response frequencies are not always at the same position on the time scale following a stimulation. There are immediately occurring alpha oscillations, or delayed alpha, as well as delayed theta and delta activities. Phase-locked frequencies and prolongations of oscillations must also be considered as features of passwords. Not only the frequency, but also the sequential repetition of oscillations has to be considered as part of the passwords. The amplitude of the responses and the weighting of amplitudes of different EEG codes are also contributions to the formation of passwords.

In the following, we illustrate four speculative cases describing what such passwords might look like:

1. Target P300
2. Nontarget P300
3. Experiments with increased attention to the 3rd stimulation
4. Threshold experiments
5. Occipital sensory EPs

(see illustration in Fig. 29.3).

29.8 The Concept of "EEG Codes" as an Important Step Towards the New Integrative Neurophysiology

Signals such as the EEG and event-related oscillations can be measured simultaneously from several parts of the brain. There is no need to restrain the animals or to give them pharmacological agents to be able to carry out the experiments. In contrast, measurements of single-unit activity require experimental setups that involve artificial paradigms. In most cases experimental

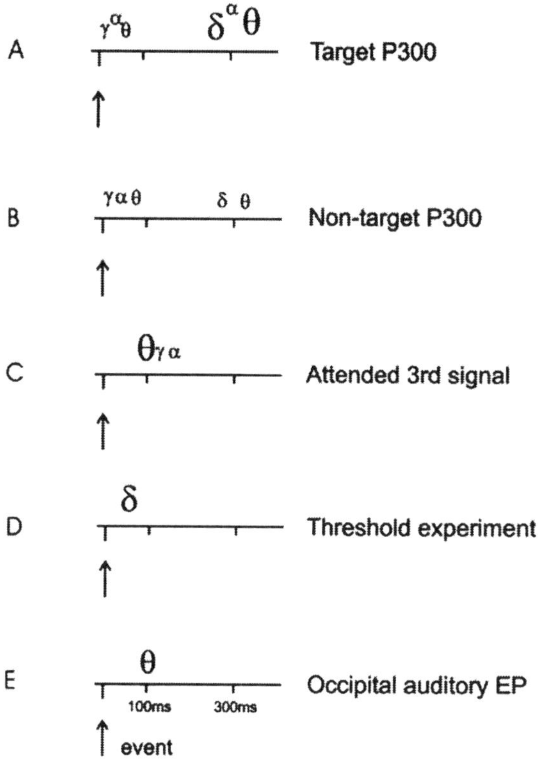

Fig. 29.3. (**a**). Oddball P300 response is characterized by a delayed large delta component, prolongation of alpha and theta. (**b**). Non-target P300 response has less ample alpha, delta, and theta responses and shows no delays or prolongations. (**c**). When a subject focuses his/her attention on the "attended 3rd signal," a theta response dominates the frontal ERP. (**d**). During experiments with near-threshold stimulation, only delta responses are observed. (**e**). In occipital locations a theta response is recorded upon auditory stimulation (see "cross-modality" experiments in Chaps. 2 and 10).

animals are anaesthetized. As a rule it is not possible to perform simultaneous recordings from distant parts of the brain with multiple electrodes, whereas this is possible using field potentials (i.e., EEG).

Another important advantage of using the EEG is recent developments in cognitive brain electrophysiology (see Picton 1992; Heinze et al. 1994b; Başar and Bullock 1992). With the help of new computer methodologies, in the last decade it has become possible to design experiments that utilize sophisticated psychophysiological paradigms and also to analyze quickly the data (see Part III of the present volume). All these outlined aspects support the contention that the EEG and its manifestations such as EP and ERP will in the near future provide important keys to understanding integrative brain functions.

29.9 Thoughts Concerning the So-Called Grandmother Cell

In earlier debates Stryker (1989) raised the question whether the so-called grandmother cell is an oscillation or oscillatory behavior of cells in the visual cortex firing in the gamma frequency band (see also Chap. 9 in Vol. I). This chapter argued that a function is now usually related to a unique oscillatory response, but moreover that the superposition of oscillations gives rise to the compound potentials representing complex functions. To look at a picture of an old woman (who is not one's grandmother but has similarities to her) and a picture of one's grandmother would again evoke similar types of oscillating behavior.

Accordingly, the author would suggest that there are no typical grandmother cells in the brain, but that the oscillatory actions from the limbic systems and association cortices would respond with various frequency amplitudes and oscillation durations. Such a reaction is a result of this *alphabet*, and resulting *passwords*.

Furthermore, it is assumed that no special neurons, but several neural populations that are diffusely distributed in the entire brain, would contribute to the sensation or recognition of "grandmother's face".

29.10 Possible Operator Properties of EEG Frequencies

As stated at the beginning of Chap. 28, an *operator* not only transmits information from one location to another but also performs functions. Considering 10 Hz activity as an operating signal implies that a given structure of the brain does not only show a 10 Hz pattern, but that the 10 Hz patterns, which are sent to other structures, start new physiological excitations on the receptive neural populations as well. Oscillations in several frequency ranges, or *wave packets*, can circulate in the entire brain, being *in phase, prolonged*, or *delayed*. In conclusion, the 10 Hz, 4 Hz, and 40 Hz activities are not only manifestations of excited structures; they are involved in *integrative functions* of communication and interaction between distributed neural populations in the brain.

30. Event-Related Oscillations in Brain Function

Most functions require the integrated action of neurons located in many regions. Localization of function means that certain areas of the brain are more concerned with one kind of function than with others. The existence of a significant difference in the major operating rhythms in occipital or frontal areas gives strong support to the possibility that theta (spontaneous, evoked, induced) and alpha rhythms (spontaneous, evoked, induced) have fundamentally different functional operations. But during some functional states, major operating rhythms can change their functional roles; the nature of the experiment, i.e., tasks, can influence the weight of these functional components on brain rhythms.

(from Chap. 27)

At the turn of the twentieth century Karl Lashley believed that various parts of the brain are equipotential and that for many functions, virtually any part of the brain could substitute for any other. In contradiction to this hypothesis, subsequent experiments have shown that even highly complex brain functions can be attributed to specific brain areas. Localization does not imply, however, that any specific function is exclusively mediated by only one region of the brain.

In the same way of thinking, several event-related oscillations in various frequency bands are, in the following synopsis, assigned to multifold brain functions.

30.1 Selectively Distributed Theta Oscillations: Properties, Functions, and Hypotheses

30.1.1 Properties

1. Event-related theta oscillations are usually masked by alpha oscillations.
2. Theta oscillations have larger amplitudes and more regular waveforms in frontal and central areas (Fz, Cz, Pz association areas).

3. Theta oscillations in cortex are most probably coupled with *hippocampal theta generators* during cognitive processes as results and discussions in Chaps. 18 and 25 showed.
4. Theta frequency components are either *spontaneous patterns* or *induced oscillations*. These components were manifested by the utilization of "single-trial theta epochs" of EPs according to our combined analysis procedure. By describing a *diffusely distributed parallel processing theta system on the entire brain*, we emphasize that the theta generators must also exist in several areas other than only in the limbic system or frontal cortex. The existence of the marked theta responses in the visual cortex of the human brain is an example of the distributed nature of the parallel-processing theta system (see Chaps. 9 and 10).
5. **Theta activity controls frontal and parietal responses** (see below for major operating rhythms). If prestimulus theta oscillations have low amplitudes, the N100/P200 amplitudes of frontal EPs are increased, as was demonstrated in Chap. 12 in Vol. I: Theta activity in the frontal part of the human brain controls the amplitudes of frontal EPs. In other words, the reaction of frontal areas of the human brain depends on its theta state. This *selectively distributed system* might have the strongest theta components in limbic structures, or in cortical ones working strongly together with the limbic system (frontal cortex, parietal cortex; see also Chap. 18).

30.1.2 Functions

1. Theta response is the most stable component of the cat P300 response. This response has its highest amplitude in the CA3 layer. According to Miller's theory, **corticohippocampal interaction** and resonance play a basic role in cognitive processing (see Chap. 25).
2. Bimodal sensory stimulation induces great increases in frontal theta response, thus demonstrating that complex events increase the **frontal processing** in the theta range.
3. Event-related oscillations in the theta band are prolonged and/or have a second time window approx. 300 ms after the presentation of the target in oddball experiments. The prolongation of event-related theta oscillations contributes to the N200/P300 waveform (see Chaps. 19 and 20). Prolongation of theta is interpreted as being correlated with **selective attention.**
4. **Association.** Event-related theta oscillations are also observed after an *inadequate stimulation*, whereas event-related alpha oscillations are not present if the stimulation is an inadequate one. Accordingly, the associative character for event-related theta oscillations is more pronounced than for higher-frequency event-related oscillations.
5. **Multiple functions of major operating oscillations (rhythms) in theta band.** Event-related potentials obtained with paradigms inducing focused attention, P300, and with stimuli giving rise to high-expectancy

states have shown marked electrophysiological changes in the frontal cortex, parietal cortex, and the limbic system. It was shown (Chaps. 20 and 21) that the frontal areas of the human cortex reacted with enormous theta enhancements to cognitive stimulation requiring states of focused attention and short-term memory. In the human frontal cortex a theta increase of 50% was recorded while the subject paid attention to a target that was expected (probability 100%). Similar experiments with cats demonstrated that also in the CA3 layer of the hippocampus a theta increase of 40% was to be seen. In P300 types of experiments, again learning tasks led to a theta increase with a time delay in frontal and parietal recordings. These results clearly demonstrate that cognitive tasks give rise to marked theta increases in evoked potential components. When comparing the results of experiments with simple light or sound stimulation, in which the evoked potentials contain dominant alpha responses, we are inclined to state that cognitive task increases the weight of theta components in comparison to alpha components. Furthermore, the increase in theta responses mostly takes place in frontal, hippocampal, or parietal structures. Even the omitted stimuli that give rise to a P300 response in the cat hippocampus induce a dominant component, again with the largest component in the CA3 layer.

6. **Orientation.** In Table 25.1 there are a few examples of cat experiments during exploration and searching behavior as well as during motor behavior. *Orienting* is a coordinated response that appears to indicate *alertness, arousal*, or *readiness* to process information.

30.1.3 Hypotheses

1. **Complex functions.** According to the statements above and the data presented in several Chaps. (15, 19, 20, and 25), it is clear that event-related theta oscillations can be considered as important building blocks of the functional signaling in the nervous system. However, as stated in Chap. 29, a unique rhythm or event-related oscillation cannot be the only functional processor for a given function. *Complex functions are certainly interwoven with several oscillations, which may occur singly or in combination.*
2. **Thalamus gate.** Seemingly, if a sensory stimulation is interrupted already in the *gate thalamus* in the 10 Hz frequency channel, the theta component of the same sensory message is not blocked in thalamus but conveyed further to cortex. These working hypotheses can be better understood or find more support from analysis of electrophysiological behavior of hippocampus where single EEG-EP epochs are used. (See Chap. 27)
3. **Carrier signal for cognition?** Postulated that different types of projections might carry either theta or alpha information, the theta information would be probably transferred diffusely to several parts of the cortex as well as to association areas, which in turn may indirectly reach primary

sensory areas of the cortex. This chain of thoughts may even be developed so far to assign to the theta responses a function as a carrier signal for cognitive association processing.

This working hypothesis does not only find support with the experiments described here, but the description of all experiments with association tasks indicated that theta enhancements occur regardless of the type of experiments and of subjects used:

- Theta enhancements are very high in the hippocampus of the cat (Chaps. 16 and 18).
- Theta enhancements are observed in frontal and parietal recordings in experiments during states of focused attention and high expectancy (Chap. 19).
- Results of Miller on corticohippocampal signal processing support the functional role of theta transmission in all cognitive states related to association.

However, the existence of a significant difference in the major operating oscillations in occipital or frontal areas gives strong support to the possibility that theta (spontaneous, evoked, induced) and alpha (spontaneous, evoked, induced) rhythms have fundamentally different functional operations. But during some functional states, major operating rhythms can change their functional roles; the nature of the experiment, i.e., the task, can influence the weight of these functional components on brain rhythms.

30.2 Selectively Distributed Alpha Oscillations: Properties, Functions, and Hypotheses

30.2.1 Properties and Functions

1. **Alpha is not noise – it is a quasi-deterministic Signal** (see Chap. 10 in Vol. I).
2. **Alpha as sensory response in the human and cat brain.** In Chaps. 2 and 10 it was shown that inadequate stimuli could not generate significant and time-locked cortical alpha enhancements in the first 300 ms upon stimulation. The occipital cortex of the cat brain does not respond with enhanced 10 Hz if the stimulation is an auditory one (i.e., inadequate stimulation).
3. **Break of the sensory alpha response upon injury in the optic nerve.** Multiple sclerosis patients with opticus neuritis do not show alpha response after visual stimulation, thus clearly demonstrating the strong relation to sensory functions of the alpha response.

   According to the outlined working hypothesis indicating "a *special transmission* of sensory signals in the 10 Hz frequency range," one might suppose that *thalamus* acts as a *gating structure* for the 10 Hz transmission (see Chap. 27).

4. **Long-latency alpha oscillations.** In these types of potentials in the poststimulus interval between 250 and 300 ms, a small 10 Hz enhancement can be recognized. The time locking is weak in comparison to results on adequate stimuli. Therefore, although single sweeps may contain fairly high 10 Hz amplitudes, the responses are not perfectly phase-locked, and furthermore, they are delayed. These types of responses are also obtained in the auditory cortex by means of visual stimulation (inadequate stimulation; see Chaps. 2 and 10).
5. **Distributed alpha response to sensory stimuli in the cat brain.** Since the days of Adrian, *evoked alpha* has been a sign of the reactiveness of the CNS to sensory stimuli. Sensory-evoked 10 Hz responses were recorded in several cortical and intracortical structures (see Chaps. 1, 2, 9, and 10).
   Alpha oscillations with a duration of approx. 250 ms are recorded in the cerebellum, reticular formation, and inferior colliculus after auditory stimulation of 80 dB and 2000 Hz. Alpha oscillations with a duration of approx. 250 ms are recorded in the visual cortex, lateral geniculate nucleus, hippocampus, and reticular formation after visual stimulation.
6. **Prolonged alpha-response component of the P300 response.** Following cognitive targets, the event-related alpha oscillations are *prolonged.* During oddball experiments the response to the target has a duration of up to 380 ms. This prolonged oscillation contributes to the 200/380 ms peaks. Our unpublished results showed also a prolonged 10 Hz response to visual targets (see Chaps. 15, 20, and 24).
7. The sensory stimulation can **evoke 10 Hz oscillatory behavior also at the cellular level** (Creutzfeldt et al. 1966; Dinse et al. 1997; Silva et al. 1993; see Chap. 9 in Vol. I).
8. **Alpha can be induced.** Perhaps the reader will find credible the claim that induced alpha-band rhythms are analogues of the gamma-band rhythms induced by moving stripes and odors in animal experiments. Consider closing the eyes in a lighted room: As with turning off the lights, this reduces retinal illumination and initiates, or induces, an alpha-band burst in the EEG. As an example of alpha induced by (undefined) changes in internal state, the rhythm disappears the moment a subject follows the instruction "multiply 11 by 13," and it reappears as soon as the answer has been delivered.
   This finding of a rhythm phase-locked to stimulus shows that the brain now has a *rhythm that has been imposed on it through a sensory system.* Has this imposed rhythm supplanted the *endogenous* one? Or is it the same rhythm with a shift in phase to carry the message? Or is the basic rhythm still there and this one added? In the latter case the brain would have its own rhythm against which to match this one for both frequency and phase.
9. **Event-related synchronization and desynchronization: Movement-related Alpha.** Pfurtscheller and Klimesch (1992) use the terms

"event-related desynchronization," or ERD, and "event-related synchronization," or ERS, to describe the ability of neural structures to generate more-or-less coherent oscillating potentials. ERD describes the attenuation or blocking, and ERS is the enhancement of oscillating potentials within the alpha and beta frequency bands.

10. **Spontaneous and evoked alphas in invertebrate ganglia.** Further evidence showing that the alpha activity cannot be explained through generators in only the thalamus or cortex is the recordings of the cerebral ganglion of *Aplysia* (see Chap. 6) and also with isolated ganglia of *Helix pomatia*. Electrically induced 10 Hz activity can be recorded in vitro in these small neural populations. Furthermore, amplitudes of spontaneous and induced alpha rhythms can be modulated using pharmacological agents (see Chap. 8).
11. **Memory-related event-related 10 Hz oscillations – alpha can be emitted.** Başar et al. (1989) demonstrated that a well-trained subject emitted time-locked bursts of alpha-band energy for up to a full second before the delivery of an expected target. By contrast to the modest evidence they advanced for anticipatory time locking of waves in the gamma band, the alpha locking to the moment in the future when a target will be delivered is robust and highly significant statistically.
    Results of Klimesch (1996) and of Başar et al. (1989, Chap. 22) on dynamic memory demonstrate that alpha activity is strongly correlated with working memory and probably with long-term memory engrams. According to Başar and coworkers, induced alpha rhythms can be considered to be internally evoked potentials during periods of expectation of a light or auditory target.
12. **Hippocampal event-related alpha oscillations.** According to the experimental results presented in Chaps. 1 and 27, all sensory stimulations evoke a 10 Hz oscillation with a duration of 250–300 ms in the hippocampus. The expression "every sensory stimulation" has to be pointed out here. The sensory stimulation may be an auditory one, and then it will evoke 10 Hz oscillatory waves, or it can be a visual one, which then triggers a 12 Hz oscillation immediately following this stimulation. This sensory information arises upon stimulation and is most probably conveyed to the association cortex and to frontal nonlimbic association cortex as well as to parts of the cortex other than primary sensory areas.

30.2.2 Hypotheses

1. **Alpha and communication processes.** The 10 Hz processes may facilitate, *overall* association mechanisms in the brain. When a sensory or cognitive input elicits "10 Hz wave trains" in several brain structures, then it can be expected that this general activity can serve as a *communication signal par excellence* between different structures (see Chap. 24).

2. **Alpha activity controls EPs.** As several results in Chap. 12 of Vol. I also emphasize, the amplitude, time course, and frequency contents of evoked potentials (especially N100-P200 wave complexes) strongly depend on the amplitude of alpha activity prior to a sensory stimulation.
3. **Selectively distributed 10 Hz oscillations.** Rather than trying to locate a unique alpha generator, it is preferable to formulate the existence of a "selectively distributed parallel processing alpha system" (Chap. 24). The physiological results concerning *hippocampal* alpha responses support this proposition:
   - Auditory and visual stimulations elicit in cat hippocampus strong and stable alpha responses (10 Hz oscillations of approx. 300 ms), which are visible without filtering.
   - Cortical and thalamic 10 Hz responses can be elicited only by stimulations that are adequate for the respective area. In contrast, hippocampal 10 Hz responses are present in all types of stimulations.
   - In coherence functions computed from visual evoked responses, the hippocampocortical coherence is significantly larger than the thalamocortical coherence (the latter one being extremely low following inadequate stimulation).
   - Thus, thalamocortical circuits are not unique in generating alpha responses; the *hippocampus* and *formatio reticularis* may have an even more general significance.
   - In case the selectively distributed alpha system is not developed, no sensory 10 Hz response is registered: Three-year-old children show a 10 Hz response neither to auditory nor to visual stimulation ("brain response susceptibility": see Chap. 31 of the present volume and Chaps. 13 and 14 of Vol. I).
4. The statements above lead to the tentative interpretation of *alpha as a universal code or universal operator in the brain.* The major physiological meaning of 10 Hz oscillations is comparable to the putative universal role of gamma responses in brain signaling.

30.3 Functions and Hyphotheses Related to the Selectively Distributed Gamma Oscillations

The functional correlates of the gamma oscillations described in the results presented in this book so far include linking of perceptual information ("binding"; Gray and Singer 1987; Gray et al. 1989; Eckhorn et al. 1988) and "vibrations in the memory" (Douglas and Martin 1995), leading to the question whether "grandmother is an oscillation" (Stryker 1989). Below we will deal with the diverse functional correlates of gamma activity, followed by a general view of gamma function.

1. Freeman (1975) and Freeman and Skarda (1985) have shown that the EEG of the **olfactory bulb** and cortex in awake and motivated rabbits and cats shows a characteristic temporal pattern consisting of bursts of 40–80 Hz oscillations, superimposed on a surface negative baseline potential shift coupled to each inspiration.
2. Since 1972 Başar's laboratories have published around 30 papers directly related to 40 Hz responses or indicating 40 Hz in complex response rhythms, finally proposing a working hypothesis assuming the existence of a **selectively distributed parallel-processing gamma system.**
   (a) **Auditory gamma-band responses** are selectively distributed in different brain structures (not only cortical but subcortical). The 40 Hz response is a robust, almost stable component of the **auditory EP** in cats. It can be recorded in cortex, hippocampus, and cerebellum. It is *phase-locked* and occurs in the first 100 ms after the sensory stimulation. The response has also a secondary oscillation with a latency of approx. 300 ms (see Chaps. 3 and 26).
   (b) In a similar way, the 40 Hz response is observed as a component of the **visual EP** in cats (Schürmann et al. 1997; see Chap. 3).
   (c) A similar *phase-locked oscillation* is also a component of the **human auditory and visual response.**
   (d) The 40 Hz response 100 ms following the applied stimulation is a pure sensory component. Karakaş and Başar (in press) recently showed that in oddball, MMN, and single-stimulation paradigms after Polich et al. (1994) the 40 Hz response in the first 100 ms remained unchanged; i.e, it is not related to cognitive tasks.
3. A **P300-40 Hz component** has been recorded in the CA3 layer of the cat hippocampus by means of an ERP paradigm by using omitted stimuli following repetitive auditory stimulation as a target. A dominant and significant 40 Hz time-locked wave packet occurs approx. 300 ms after the stimulation and not in the first 100 ms.
4. **Attention-related 40 Hz activity.** Several papers have handled the importance of 40 Hz activity in states of attention and motivation. Tiitinen et al. (1993) reported that selective attention enhances the auditory 40 Hz transient response in humans especially over the frontal and central areas.
5. A significant gamma-band (30–50 Hz) activity increase in EEG during states of **"perceptual switching"** was measured induced by an ambiguous stimulus pattern. The most significant 40 Hz enhancements were measured in frontal areas and can reach increases of 40 to 50% in states of naive and active observations (Başar-Eroğlu et al. 1996b).

30.3.1 Hypotheses

1. Neuroscientists discuss the **"binding problem":** How is the spatially distributed but temporally coherent (simultaneous) electrical activity stemming from a large number of elementary neural components integrated to

functional activity? Gamma rhythms have been suggested to be associated with binding phenomena:

- Gray and Singer (1987, 1989) and Gray et al. (1989) have reported that neurons in the cat visual cortex exhibit oscillatory responses in the frequency range of 40–60 Hz. These oscillations occur in synchrony with cells located within a functional column and are tightly correlated with oscillatory field potentials. Gray and Singer proposed that the synchronization of oscillatory responses of spatially distributed, feature-selective cells might be a way to establish relations between features in different parts of the visual field.
- Eckhorn et al. (1988, 1989a,b) also found stimulus-evoked resonances of 35–85 Hz throughout the visual cortex when primary coding channels were activated by their specific stimuli, and raised the question whether coherent oscillations do reflect a mechanism of feature-linking in the visual cortex.
- Despite its current popularity, this hypothesis is probably insufficient to explain the "ubiquity of gamma rhythms" (Desmedt and Tomberg 1994).

2. According to Llinás and Ribary (1992), the spatiotemporal **magnetic field pattern of gamma-band activity** suggests the presence of a coherent rostrocaudal sweep of activity repeating every 12.5 ms due to a continuous phase shift over the hemisphere. The authors propose that this scanning occurs on account of thalamocortical resonant synaptic interactions.
3. **General view on gamma function.** These results should be taken into account when interpreting gamma rhythms, which may rather be *universal building blocks* than specific correlates of perceptual binding or higher cognitive processes. We propose that diffusely distributed gamma generators are activated by different sensory and cognitive events and that they react with different time delays. This property of brain structures was called *synchronized selectivities* or *synchronized resonances* to point out that upon stimulation, generators of oscillatory potentials in the brain act in a coherent way. They may facilitate signal transmission by means of resonance. (Başar, 1980)

The statements listed above underline that gamma activity has *multiple functional correlates:* The results presented in this book alone confirm that 40 Hz responses are involved with all types of cognitive and sensory processes ranging from simple signal detection to complex CNS activity such as recognition of ambiguous figures. The evaluation of the above statements leads to the following generalization: 40 Hz signals of the brain (spontaneous, induced, or evoked) belong to important **functional building blocks** of brain electrical activity. 40 Hz wave packets may occur in different and distant structures and may act in parallel or separately depending on the modality of sensory or cognitive stimulation. They may show *phase locking, time locking,* or *weak time locking.*

30.4 Selectively Distributed Delta Oscillations: Functions and Hyphotheses

1. **Sensory function.** The delta response oscillation (the delta response) is recorded in auditory and visual EPs in all scalp recordings of the human brain and also in cortical and intracortical structures of the cat brain. This means that sensations evoke a delta response; i.e., delta response oscillations are encountered as sensory components in brain responses (Chaps. 1, 2, 9, and 10).
2. **Cognitive functions.** The amplitude of the delta response is considerably increased during oddball experiments. The delta response to the target signals is delayed and prolonged. Accordingly, it was concluded that the delta response is related to *signal detection* and *decision-making* (see Chaps. 20 and 21).
3. **Cognitive delta response is selectively distributed in humans.** The response to visual targets have their greatest response amplitude in parietal locations, whereas the greatest delta response amplitudes are observed in central and frontal areas following auditory target signals (see Chaps. 20 and 18 in Vol. I and Chap. 28).
4. **Signal detection at hearing threshold.** Human subjects do respond with delta oscillations at the hearing threshold. Accordingly, the role of the delta response in signal detection and decision-making is demonstrated in different types of cognitive experiments (see Chap. 12).
5. MMN is shaped by a delayed delta response superimposed upon a significant theta response.
6. Sleep delta responses are probably the major processing signals in the sleeping brain.

30.5 Conclusion: Multiple Functions

In this chapter we have described approximately 50 types of oscillatory activities with definitive or tentative explanations of their functional relations (there are more examples in this book; however, we do not summarize all of them here).

Our results have clearly demonstrated that it is not possible to assign to a given type of oscillatory activity only one function. These oscillations have *multiple functions* and act as *universal operators*, or *codes*, of brain functional activity. Moreover, besides the frequency and site of the activity, some other parameters are also involved (or interwoven) in brain functioning. These parameters are: *enhancement, time locking, phase locking, delay of the oscillation*, and *prolongation of oscillations.*

Complex and integrative brain functions are manifested in the superposition of several oscillations and frequency stabilization, but also in the following parameters: degree of prolongation, enhancement, delay, time locking, and phase locking in several time windows.

Part V

Conclusion

31. An Integrative Neurophysiology Based on Brain Oscillations

Erol Başar and Sirel Karakaş

An *integrative neurophysiology* should describe activity of brains of different species and their manifold functions from a global viewpoint. Further, it must rely on extensive experimental work: The EEG oscillations permit the analysis of sensory and cognitive functions together – both in freely behaving animals and in the thinking and feeling human brain. *Processing of sensation and cognitive functions* are interwoven; the type of approach utilized in the present book and its companion volume makes the analysis of integrative brain functions feasible. An *integrative neurophysiology* must embrace both physiological and cognitive functions. The bridging of both types of functions is possible only by common approaches. Brain oscillations seem to be basic events for establishing such an integrative discipline. Since the *natural frequencies* of the CNS are manifested in *EEG oscillations*, such an approach is probably one of the most fruitful or promising approaches to creating an integrative neuroscience in the future.

In this chapter we will summarize principles and theories that emerged from the present book and Vol. I. Chapter 30 of the present volume summarizes our findings about functional correlates of event-related oscillations, thus completing the present volume. Prior to making concluding remarks, we will treat an important question posed by Fessard 40 years ago. We think that in concluding our books we are able to find a satisfactory approach to the main claim of Fessard. A well-functioning system must obey some general rules, or laws, even if it is a very complex one. The brain is the most complex system known in our near universe. Is the brain not a well-functioning system?

31.1 Oscillations Govern the General Transfer Functions in Neural Tissues of the Brain

Fessard (1961) emphasized the role of neuronal networks in the brain:

> The brain, even when studied from the restricted point of view of sensory communications, must not be considered simply as a juxtaposition of private lines, leading to a mosaic of independent cortical territories, one for each sense modality, with internal subdivisions

corresponding to topical differentiations. The track of a single-unit message is doomed to be rapidly lost when one tries to follow it through a neuronal field endowed with network properties, within which the elementary message readily interacts with many others. Unfortunately, we still lack principles that would help us describe and master such operations in which heterosensory communications are involved. These principles may gradually emerge in the future from an extensive use of multiple microelectrode recordings, together with a systematic treatment of data by modern electronic computers, so that pattern-to-pattern transformation matrices can be established and possibly generalized. For the time being, it seems that we should do better to try to clear up such principles as seem to govern the *most general transformations – or transfer functions* – of multiunit homogeneous messages during their progressions through neuronal networks.

The transfer function measures the ability of a network (here, neural networks of the brain) to increase (facilitate) or impede (inhibit) transmission of signals in given frequency channels. Thus the properties of the *transfer function* constitute the *main* framework for *signal processing and communication*. The existence of general transfer functions would then be interpreted as the existence of networks distributed in the brain that show similar frequency characteristics or facilitate or even increase the signal transmission in given frequency channels. In an electric system optimal transmission of signals is often reached when distributed subsystems of the system are tuned to the same frequency range. Does the brain have such subsystems tuned in similar frequency ranges, or are there common frequency modes in the brain? The results of our books provide a real, definitive approach to this question. The described frequency characteristics in all brain tissues that embrace the resonant oscillatory processes or selectively distributed oscillatory systems of the brain (*delta, theta, alpha, beta, gamma*) constitute and govern mathematically the general transfer functions of the brain (see the definition of frequency characteristics in Chap. 4 of Vol. I).

All brain tissues in both the animal and human brains (including isolated ganglia of invertebrates and low vertebrates, and the human brain) react to sensitive and cognitive inputs with oscillatory activity within *almost invariant and general governing frequency channels*. Experimental results have shown that the degree of synchrony, amplitudes, durations, and phase lags are continuously varying, but similar oscillations are in general present in the activated brain tissues.

31.2 Brain Oscillatory Theory and Functional Interpretations

31.2.1 Spontaneous Oscillations

1. **Natural frequencies of the brain.** The brain has several types of natural oscillatory activities in various frequency ranges: delta, theta, alpha, gamma, and highest frequencies. These oscillations may occur spontaneously, or they can be *emitted, induced,* or *evoked.*
2. **The EEG is not noise: chaotic signals.** The natural oscillations are not always noise but probably reflect properties of chaotic attractors: We use the expression quasi-deterministic to describe this oscillatory behavior.
3. **Natural frequencies (EEG frequencies) are recorded at the cellular level.** Spontaneous and evoked or induced rhythms are also observed in 10 Hz, 40 Hz, beta, and delta frequencies at the cellular level. The descriptions in Chap. 9 of Vol. I clearly show that one can correlate the spikes at the cellular level with field potentials (EEG-EPs). Although the relation between EEG and single-cell activity was also demonstrated earlier, new results of Llinás (1988), Singer et al. (Gray and Singer 1987; Gray et al. 1989), Eckhorn et al. (1988), and Dinse et al. (1997) using multiple electrodes extended this approach, now almost reaching perfection. Together with the nonlinear approach, electroencephalography will probably gain importance for describing the states of the brain.
4. **Major operating rhythms.** Experimental results presented in this book demonstrate that the EPs and the EEG immediately prior to stimulation build an entity with the following EP. The results described in Chap. 12 of Vol. I further demonstrate that these are *major operating rhythms* in several areas of the cortex. These major operating rhythms also influence the shape of each potential in different frequencies. A good example is that the theta activity is the major operating rhythm in the frontal cortex. On the other hand, the major operating rhythm of the occipital cortex is 12 Hz oscillation.

31.2.2 The Origin of Event-Related Oscillations

1. **The properties of the transfer functions** interwoven with *resonances* and *natural frequencies* constitute the main framework for *signal processing* and *communication: Alpha, theta, gamma,* and *delta responses* provide the main electrical information processing in the brain.
2. **Transitions of EEG from disordered to ordered states.** The frequency-domain description of EPs in the cortex, thalamus, reticular formation, hippocampus, and cerebellum show a similar overall frequency structure to the ongoing activity (EEG), indicating a resonating universal mechanism: The sensory stimulus brings the brain into a more *coherent*

state. In response to the stimulus, the frequency bands in various structures of the brain become much sharper and narrower, and they become coherent in phase and frequency. In a given frequency channel the magnitude of the response is enhanced against the magnitude of the prestimulus activity (see Chap. 11 in Vol. I).

3. **Response susceptibility of the brain:** The results have led us to derive a *rule of excitability,* which can be stated as follows: If a brain structure shows spontaneous rhythmic activity in a given frequency channel, then this structure is also in the same frequency channel, and moreover, it will produce *internal evoked potentials* to internal afferent impulses originating in the CNS or respond in the form of EPs to external sensory stimuli with patterns similar to those of internal evoked potentials (see Chap. 14 in Vol. I).
4. **Superposition principle:** EPs are a superposition of delta, theta, alpha, beta, and gamma oscillations that are enhanced or phase-locked depending on the nature of the stimuli (sensory or cognitive).
5. **Brain real responses.** Using several physiological, psychological, or biophysical methods and strategies, it was shown that the evoked rhythms ranging from delta to gamma frequencies are real brain responses that are related to functions (see Chap. 18 in Vol. I).
6. **Internal Evoked Potentials.** The findings on stereodynamics of brain potentials showed that the evoked responses in all the nuclei and in all the frequencies are strongly dependent on the spontaneous activities just prior to stimulus. There are cases in which the filtered EEG-EP epochs already depict, in the EEG portion, ample potentials similar to the filtered EP signals taking place immediately after stimulation. The resemblance in the shapes of EPs and such EEG bursts leads us to use the expression *internal EP* for the description of large-amplitude and synchronized EEG recordings. When the relevant internal EPs are recorded before the stimulation, usually the EPs induced by stimulation do not have large amplitudes. Accordingly, we assume that the research on single-trial EPs will also be useful in contributing to our understanding of EEG population dynamics.

31.2.3 Functional Interpretation

1. **Multiple functions related to oscillations.** On the one hand, spontaneous and event-related oscillations are correlated with several brain functions. On the other hand, functions are related to a superposition of oscillatory responses. The amplitude and duration of the oscillation are also correlated with function and are modified according to functional changes. EP components that we called *alpha, theta, delta, beta,* and *gamma responses* are correlated to various functions, as demonstrated in several parts of our books and summarized in Chap. 30. *Superposition* of all these *oscillatory signals* can be related to complex functions. The complexity

of an event is not reflected only by EPs. In a number of experiments the spontaneous EEG can be also considered as *internally evoked potentials*, where the inputs are coming from as yet hidden sources.

2. **EEG as a functional brain code, oscillations as brain alphabet.** The core concept of the monograph "EEG-Brain Dynamics" (1980) was based on the statement that the EEG is not simply a noise but that it is in all probability one of the most useful signals of the brain related to EPs. This core concept has been supported and extended to include brain function related to EEG by also taking into account results of this book. The EEG itself is considered here not only as a quasi-deterministic signal but as the most useful activity in its capacity as *brain code for function.*
3. According to the findings in this book, all the **EEG frequency generators are selectively distributed** within the entire brain. Sensory or cognitive inputs bring these EEG generators into a resonating state. The described *selectively distributed oscillatory systems* govern the *excitability* and *communication* of all brain structures. Not only in the cortex but also in deeper structures of the brain, for example in the brainstem, 10 Hz, 4 Hz, and other resonances can be registered. During such activity, the brain reaches coherent states with regard to time and space. Not only the cortex, but the entire brain is involved in complex responses, such as the P300 response. In some of the functional reactions only a few EEG components are involved, and one can sometimes encounter simple oscillatory responses, for example the delta response at the hearing threshold. Another typical example is the experiments during which the subjects pay attention to repetitively applied third signals. In this case the frontal event-related potentials are reduced almost to a unique and homogeneous theta oscillation.
   It is further proposed that *event-related oscillations* behave like **letters of an alphabet** and that complete brain functions are reflected in "words" constituted of letters such as *alpha, beta, gamma.* **Selectively distributed oscillatory networks** in frequency bands of delta, theta, alpha, beta, and gamma play a major role within the brain and for sensory and cognitive brain functions. Not only the *type* but the *size* and *combinations* of the letters describe these functions (see also Chap. 29). *Sensations* and *cognitive events* evoke superimposed oscillations, which are transmitted to all brain tissues with various degrees of intensity, synchronization, duration, and delay almost in parallel. Depending on the sensory and cognitive nature of the inputs and site of recordings, oscillations in various frequency channels follow the input. The various types of oscillations that contribute to or are manifestations of brain functioning are described in Chap. 30.
4. **Real advantages – an example. Hippocampal alpha networks** also control the alpha response. The widely held theory on alpha generators is based on the existence of thalamocortical circuits (see Chap. 24). The results presented in this book, however, demonstrate the existence of *hip-*

pocampal alpha generators that control the alpha enhancements to all types of sensory modalities. These results have been achieved with the help of cross-modality experiments in the cat brain as well as through the utilization of a coherence function (see Chap. 1). These results contradict the mainstream theories for alpha generators such as the facultative pacemaker theory explained in Chap. 24. It is important to point out that only an integrative analysis using data from the entire brain can enable the establishment of theories regarding generators. Such an analysis can be made only by using semimicroelectrodes, as was the case with our analysis, and not by single-unit studies.

31.2.4 The Approaches of Relevance

1. **Single-sweep analysis.** In order to understand the physiological or cognitive contents of evoked potentials it is suggested to perform single-sweep analysis. The methods introduced are (a) *pass-band filtering* performed by means of adaptive digital filters; (b) *wavelet analysis*, a method of time–frequency analysis; (c) *SSWI* for the assessment of phase-locking single sweeps in a way independent of amplitude (see Chaps. 4, 5, and 6 of Vol. I).
2. **The voyage through brains.** A comparative analysis of brain oscillatory responses in various brains (invertebrates, fish, cat brain, human brain) are presented in Chaps. 6 and 7. Similar patterns of oscillatory response behavior were recorded upon auditory stimulation, in electroception and also in invertebrate ganglia to direct electrical stimulation. Accordingly, we assume that oscillatory behavior reflects basic network properties of brains. It would be erroneous to assign, for example to the gamma band, specific cognitive functions, as some authors emphasize.

31.2.5 Final Conclusion

Integrative brain activity is manifested in EEG and event-related oscillations. The examples presented in these books for describing cognitive potentials in the cat and human brains were obtained under specific experimental paradigms. The functional activities of the central nervous system are certainly not limited to the experiments and results covered by the present book. The core concept is related to the utilization of the EEG code as an important key for understanding the *integrative activity* of the brain.

32. A "Neurons–Brain" Doctrine: New Thoughts

There are several trends extending or renewing Sherrington's "neuron doctrine." Freeman has proposed a "neo-Sherringtonian view" of integrative brain function, whereas Barlow and Mountcastle have proposed modern views on the neuron doctrine (see Chap. 28).

Relying on the results of the two volumes, we present here a new doctrine to describe a basic framework for signaling of communication and functional operating in the brain. This new doctrine cannot be perfect and cannot cover all results and principles accumulated in the last decades. However, it provides a plausible and progressive framework that might extend the old doctrine. The latter should be considered as a special case of the more general new *neurons doctrine.*

1. The *neuron* is the basic signaling element of the brain.
2. *Oscillatory neural activity* is considered as a basic signal reflecting natural frequencies of the brain (this thesis relies on work of Verzeano (1971), Singer (1987), Silva (1991), Eckhorn et al. (1988), Dinse et al. (1997); see Chap. 9 in Vol. I).
3. *Neural assemblies* replace the *neuron* in the description of complex brain functions. This view diverges from Sherrington's "neuron doctrine." Taking a metaphor from the physical sciences, neurons can be considered as *atoms*, neural assemblies that contribute to a function as *molecules.* Accordingly, the metaphor is similar to *statistical mechanics* and *gas laws* (see Sect. 4.2 in Vol. I).
4. *Oscillatory activities (event-related, induced,* or *spontaneous)* govern the most general *transfer functions* in the brain (frequency characteristics and power spectra are governed by *alpha, beta, gamma, theta, delta,* etc. oscillations – which is confirmed by the wavelet approach). Further, as stated in Chap. 31, the general transfer functions provide a framework for electrical information processing in the brain.
5. Oscillations in different frequency ranges are a property of the neurons (see #2 above). *Selectively distributed oscillatory neural populations,* however, behave *with "molecular properties"* (alpha, beta, gamma, delta, theta being *atomic properties*). These oscillatory networks are activated upon sensory stimulation or event-related tasks by manifestation of
   (a) synchronization of neural activity,

(b) partial synchrony,
(c) enhancements, or
(d) blocking or desynchronization of oscillations,

depending on the nature of the sensation or event, and accordingly depending on the function performed. These selectively distributed networks are *operators* of general brain functions including *communication and association* and data retrieval (molecular properties; see the previous item).

6. *Major operating rhythms* play a key role in association and communication. Topological distribution of oscillations is heterogeneous, and their functions are manifold. Accordingly, parallel processing is not perfect among distributed populations since the major operating rhythms are selectively distributed. *Examples:* Alpha responses are not recorded in the medial geniculate nucleus and in the auditory cortex in response to light stimulation, whereas the lateral geniculate nucleus and the primary visual cortex respond with large alpha enhancements (more examples are summarized in Chap. 30).
7. *Types of neurons* do not play a major role for frequency tuning of oscillatory networks. The neural architectonics of the cerebral cortex, cerebellum, and hippocampus are completely different. In spite of this, all these structures behave with almost similar frequency responses.
8. Distributed oscillatory networks react selectively upon application of *pharmacological agents. Examples:* Cerulein causes a great change in hippocampal evoked response that takes the form of a homogeneous 3-4 Hz response, whereas the frequency response of the cerebellum remains completely unchanged, since this agent does not have any effect on the cerebellum. Acetylcholine activates the 4 Hz response of the hippocampus enormously, whereas other structures are less influenced (see Chap. 8).
9. *Functions in the brain* are manifested by varied degrees of *superpositions* of oscillations in EEG frequency ranges. There are varied degrees of responsiveness depending on the strength of the stimulation or the event presented to the CNS. Accordingly, neuronal assemblies do not react with a type of all-or-none behavior as in the single neuron doctrine.
10. There exists a strong inverse relation between prestimulus oscillations and brain responses. Spontaneous oscillations control the amplitude and shape of population responses (see Chaps. 12 and 14 in Vol. I and Arieli et al. 1996).

33. Epilogue: EEG Oscillations in Integrative–Cognitive Neurophysiology

My "Brain Dynamics" concept of 1980 was based strongly on oscillatory dynamics and resonances in brain tissues. At the end of 1978 we had completed the experiments that made possible the monograph *EEG Brain Dynamics* (Başar 1980). A few years later the extension of nonlinear approaches (showing that the EEG signal is quasi-deterministic and not a simple noise) brought to EEG research a new viewpoint. Later, discoveries related to gamma oscillations on the cellular level opened completely new horizons in the neurosciences, and the concept of *oscillatory brain dynamics* gained great importance in all research laboratories.

Nowadays, an increasing number of brain scientists are employing the approach of oscillatory components of event-related potentials, electroencephalography, and magnetoencephalography. When we also think of the important remark made by Mountcastle about a possible turnaround in the analysis of brain field potentials and their importance, we can imagine that this field will grow in an astonishingly fast and fruitful way. I also think that we will discover and rediscover Hans Berger's ideas as a breakthrough in the brain research of the twentieth century. The development of modern computers, of interesting software, nonlinear analysis, and the possibility of using magnetoencephalography records from the whole cortex will open new horizons. Today we may not be able to imagine all the new territory that is to be discovered. I think that we are at the beginning of a very important stage in brain research. In the year 2000 it may be that we will be witness to a new area in brain research in which probably the EEG oscillations together with complementary brain imaging techniques will be a focal point.

This book and its companion volume present collections of experiments made over the course of a nearly 15-year voyage in brain research, during which we have collected results ranging from invertebrate brains to the highest cognitive functions in the human brain. The methods applied ranged from simple statistics to chaos analysis and wavelet approaches. I think that this ambitious approach may contribute to a newly emerging discipline of neuroscience in which dynamics and oscillations play the major role. Another important fact we have learned from the results presented in this book is that neuroscience researchers have to consider the brain as an *integrative system* and not depend on the results from a special structure or application of a special paradigm only. From the results we have obtained within the last 15 years, a prognosis for the coming 10 or 15 years can be easily formulated: A renaissance of EEG oscillations has now begun. This was possible mostly through the utilization of powerful computers and new algorithms in the analysis of EEG events. A new and very important step can now be expected by the parallel use of whole-cortex magnetoencephalography and functional magnetic resonance imaging. The experimenter who chooses these algorithms to explore higher nervous activity should essentially accept the EEG components as the most important functional building blocks of the brain. My opinion is that we have more than a "paradigm shift" in the neurosciences, as was mentioned by Mountcastle (1992; see Vol. I). Neuroscientists can finally attack one of the core problems of the brain: the understanding of brain functioning by means of its natural frequencies, namely, the *EEG oscillations.*

The interesting area of research opened by Hans Berger and enriched by Lord Adrian is again poised for a brilliant lift-off.

References

Adey, W.R. (1966) Neurophysiological correlates of information transaction and storage in brain tissue. In: Stellar, E. and Spraque, J.M. (eds.) Progress in Physiological Psychology. Academic Press, New York London, 1–43

Adey, W.R. (1989) Cell membranes, electromagnetic fields, and intercellular communication. In: Başar, E., Bullock, T.H. (eds.) Brain Dynamics – Progress and Perspectives. Springer, Berlin Heidelberg New York, pp. 26–42

Adey, W.R., Dunlop, C.W., Hendrix, C.E. (1960) Hippocampal slow waves. Distribution and phase relationships in the course of approach learning. Arch. Neurol. 3: 74–90

Adey, W.R, Walter, D.O., Lindsley, D.F. (1962) Subthalamic lesions: effects on learned behaviour and correlated hippocampal and subcortical slow-wave activity. Arch. Neurol. 6: 194–207

Adler, G., Adler, J. (1989) Influence of stimulus intensity on AEP components in the 80- to 200-millisecond latency range. Audiol. 28: 316–324

Adrian, E.D. (1941) Afferent discharges to the cerebral cortex from peripheral sense organs. J. Physiol. (Lond.) 100: 159–191

Adrian, E.D. (1942) Olfactory reactions in the brain of the hedgehog. J. Physiol. (Lond.) 100: 459–473

Adrian, E.D. (1951) Rhythmic discharges from the thalamus. J. Physiol. (Lond.) 113: 9–108

Adrian, E.D. (1931) Potential changes in the isolated nervous system of *Dytiscus marginalis*. J. Physiol. (Lond.) 72: 132–151

Adrian, E.D. (1937) Synchronized reactions in the optic ganglion of *Dytiscus*. J. Physiol. (Lond.) 91: 66–89

Albus, M. (1988) Cholecystokinin. Progr. Neuro-Psychopharmacol. Biol. Psychiatr. 12: 5–22

Aleksanov, S.N., Vainstein I.I., Preobrahenskaya, L.A. (1986) Relationship between electrical potentials of the hippocampus, amygdala and neocortex during instrumental conditioned reflexes. Neurosci. Behav. Physiol. 16: 199–207

Allison, T., Matsumiya, Y., Goff, G.D., Goff, W.R. (1977) The scalp topography of human visual evoked potentials. Electroencephalogr. Clin. Neurophysiol. 42: 185–197

Andersen, P., Andersson, S.A. (1968) Physiological Basis of the Alpha Rhythm. Appleton-Century-Crofts, New York

Andersen, P., Andersson, S.A., Lomo, T. (1967a) Some factors involved in the thalamic control of spontaneous barbiturate spindles. J. Physiol. (Lond.) 192: 257–281

Andersen, P., Andersson, S.A., Lomo, T. (1967b) Nature of thalamocortical relations during spontaneous barbiturate spindle activity. J. Physiol. (Lond.) 192: 283–397

Andersson, S.A., Manson, J.R. (1971) Rhythmic activity in the thalamus of the unanesthetized decorticate cat. Electroenceph. Clin. Neurophysiol. 31: 21–34

Andrew, C., Pfurtscheller, G. (1996) Event-related coherence as a tool for studying dynamic interaction of brain regions. Electroencephalogr. Clin. Neurophysiol. 98: 144–148

Apostol, G., Creutzfeldt, O.D. (1974) Crosscorrelation between the activity of septal units and hippocampal EEG during arousal. Brain Res. 67: 65–75

Arieli, A., Sterkin, A., Grinvald, A., Aertsen, A. (1996) Dynamics of ongoing activity: explanation of the large variability in evoked cortical responses. Science 273: 1868–1871

Arlinger, S.D. (1975) Threshold extrapolation for N1-latencies in auditory evoked responses. Rev. Laryng. (Bordeaux) 96: 147–152

Arellano, A., Schwab, R. (1950) Scalp and basal recording during mental activity. In Proceedings of the 1st International Congress of Psychiatry, Paris

Babloyantz, A., Nicolis C., Salazar M. (1985) Evidence of chaotic dynamics of brain activity during the sleep cycle. Phys. Lett. (A) 111: 152–156

Baddeley, A. (1986) Working Memory. Clarendon Press, Oxford

Baddeley, A. (1988) Cognitive psychology and human memory. Trends Neurosci. 11: 176–81

Baddeley, A., Hitch, G. (1993) The recency effect: implicit learning with explicit retrieval? Mem. Cognit. 21: 146–55

Baddeley, A. (1996) The fractionation of working memory. Proc. Natl. Acad. Sci. USA 93: 13468–72

Baddeley, A. (1992) Working memory. Science 255: 556–559

Baddeley, A. (1986) Modularity, mass-action and memory [editorial]. Q. J. Exp. Psychol. A. 38: 527–33

Baile, C.A., McLaughlin C.L., Della-Fera, M.A. (1986) Role of cholecystokinin and opioid peptides in control of food intake. Physiol. Rev. 66: 172–234

Baleydier, C., Maugiere, F. (1980) The duality of the cingulate cortex in monkey. Neuroanatomical study and fuctional hypothesis. Brain 103: 525–554

Ball, G.J., Gloor, P., Schaul, N. (1977) The cortical electromicrophysiology of pathological delta waves in the electroencephalogram of cats. Electroencephalogr. Clin. Neurophysiol. 43: 346–61

Barlow HB (1961) Possible principles underlying the transformation of sensory messages. In: Rosenblith WA, (ed.) Sensory Communication. MIT Press, Cambridge, MA

Barlow, J.S., Estrin, T. (1971) Comparative phase characteristics of induced and intrinsic alpha activity. Electroencephalogr. Clin. Neurophysiol. 30: 1–9

Bartley, S.H., Bishop, G.H. (1933) The cortical response to stimulation of the optic nerve in the rabbit. Amer. J. Physiol. 103: 159–172

Bartnik, E.A., Blinowska, K.J., Durka, P.J. (1992) Single evoked potential reconstruction by means of wavelet transform. Biol. Cybern. 67: 175–181

Başar, E. (1972) A study of the time and frequency characteristics of the potentials evoked in the acoustical cortex. Kybernetik, 10: 61–64

Başar, E. (1976) Biophysical and physiological systems analysis. Addison-Wesley, Reading, MA

Başar, E. (1980) EEG-Brain Dynamics. Relation between EEG and Brain Evoked Potentials. Amsterdam: Elsevier

Başar, E. (1983a) Synergetics of neuronal populations. A survey on experiments. (With an appendix by Başar, E. and Röschke, J.). In: Başar, E., Flohr, H., Haken, H., Mandell, A.J. (eds.) Synergetics of the Brain. Berlin Heidelberg New York, Springer, pp. 395–406

Başar, E. (1983b) Toward a physical approach to integrative physiology. I. Brain Dynamics and Physical Causality. Am. J. Physiol. 245: R510–R533.

Başar, E. (1988) EEG–dynamics and evoked potentials in sensory and cognitive processing by the brain. In: Başar E. (ed.) Dynamics of sensory and cognitive processing by the brain. Springer, Berlin Heidelberg New York, pp. 30–55

Başar, E., ed. (1990) Chaos in Brain Function. Springer, Berlin Heidelberg New York

Başar, E. (1992) Brain natural frequencies are causal factors for resonances and induced rhythms [Epilogue]. In: Başar, E., Bullock T.H. (eds.) Induced Rhythms in the Brain. Birkhäuser, Boston, MA, pp. 425–467

Başar, E., Bullock, T.H., eds. (1992) Induced Rhythms in the Brain. Birkhäuser, Boston, MA

Başar, E., Özesmi, C. (1972) The hippocampal EEG activity and a systems analytical interpretation of averaged evoked potentials of the brain. Kybernetik, 12: 45–54

Başar, E., Schürmann, M. (1994) Functional aspects of evoked alpha and theta responses in humans and cats. Biol. Cybern. 72: 175–183

Başar, E., Schürmann, M. (1996) Alpha rhythms in the brain–functional correlates. News Physiol. Sci. 11: 90–96

Başar, E., Stampfer, H.G. (1985) Important associations among EEG-dynamics, event-related potentials, short-term memory and learning. Int. J. Neurosci. 26: 161–180

Başar, E., Ungan, P. (1973) A component analysis and principles derived for the understanding of evoked potentials of the brain: Studies in the hippocampus. Kybernetik, 12: 133–140

Başar, E., Gönder, A., Özesmi, C., Ungan, P. (1975a) Dynamics of brain rhythmic and evoked potentials. I. Some computer methods for the analysis of electrical signals from the brain. Biol. Cybern. 20: 137–145

Başar, E., Gönder, A., Özesmi, C., Ungan, P. (1975b) Dynamics of brain rhythmic and evoked potentials. II. Studies in the auditory pathway, reticular formation and hippocampus during the waking stage. Biol. Cybern. 20: 145–160

Başar, E., Gönder, A., Özesmi, C., Ungan, P. (1975c) Dynamics of brain rhythmic and evoked potentials. III. Studies in the auditory pathway, reticular formation, and hippocampus during sleep. Biol. Cybern. 20: 161–169

Başar, E., Gönder, A., Ungan, P. (1976a) Important relation between EEG and brain evoked potentials. I. Resonance phenomena in subdural structures of the cat brain. Biol. Cybern. 25: 27–40

Başar, E., Gönder, A., Ungan, P. (1976b) Important relation between EEG and brain evoked potentials. II. A system analysis of electrical signals from the human brain. Biol. Cybern. 25: 41–48

Basar, E., Demir N., Durusan, R., Ungan, P. (1977a) Synchrony and coupling between simultaneously recorded brain potentials. Electroencephalogr. Clin. Neurophysiol. 43: 500–501

Basar, E., Durusan, R., Gönder, A., Ungan, P. (1977b) Human alpha response to photic stimuli. Electroencephalogr. Clin. Neurophysiol. 43: 500

Basar, E., Gönder, A., Ungan, P. (1978) Comparative frequency analysis of single EEG-evoked potentials records. International Evoked Potentials Symposium, Nottingham, September 4–6

Basar, E., Demir, N., Gönder, A., Ungan, P. (1979a) Combined dynamics of EEG and evoked potentials. I. Studies of simultaneously recorded EEG-EPograms in the auditory pathway, reticular formation and hippocampus of the cat brain during the waking stage. Biol. Cybern. 34: 1–19

Basar, E., Durusan, R., Gönder, A., Ungan, P. (1979b) Combined dynamics of EEG and evoked potentials. II. Studies of simultaneously recorded EEG-Programs in the auditory pathway, reticular formation and hippocampus of the cat brain during sleep. Biol. Cybern. 34: 21–30

Başar, E., Başar-Eroğlu, C., Rosen, B, Schütt, A. (1984) A new approach to endogenous event-related potentials in man: Relation between EEG and P300-wave. Int. J. Neurosci. 24: 1–21

Başar, E., Rosen, B., Başar-Eroğlu, C, Greitschus, F. (1987) The associations between 40 Hz-EEG and the middle latency response of the auditory evoked potential. Int. J. Neurosci. 33: 103–117

Başar, E., Başar-Eroğlu, C., Röschke, J., Schütt, A. (1989) The EEG is a quasi-deterministic signal anticipating sensory-cognitive tasks. In Başar, E., Bullock, T.H. (eds.) Brain Dynamics, Springer, Berlin Heidelberg New York, pp. 43–71

Başar, E., Başar-Eroğlu, C., Rahn, E., Schürmann, M. (1991) Sensory and cognitive components of brain resonance responses: An analysis of responsiveness in human and cat brain upon visual and auditory stimulation. Acta Otolaryngol. (Stockh.) Suppl. 491: 25–35

Başar, E., Başar-Eroğlu, C., Parnefjord, R., Rahn, E, Schürmann, M. (1992a) Evoked potentials – ensembles of brain induced rhythmicities in the alpha, theta and gamma ranges. In: Başar, E., Bullock, T.H. (eds.) Induced Rhythms in the Brain. Birkhäuser, Boston, MA, pp. 155–181

Başar, E., Dössel, O., Fuchs, M., Rahn, E., Saermark, K., Schürmann, M. (1992b) Evoked alpha responses from frontal-temporal areas in multichannel SQUID systems. Proceedings of IEEE Satellite Symposium on Neuroscience and Technology, pp. 28–33

Başar, E., Başar-Eroğlu, C., Demiralp, T. Schürmann, M. (1993) The compound P300–40 Hz response of the human brain. Electroencephalogr. Clin. Neurophysiol. 87: 14P

Başar, E., Bassar-Eroğlu, C., Demiralp, T., Schürmann, M. (1995) Time and frequency analysis of the brain's distributed gamma-band system. IEEE Eng. Med. Biol. 14: 400–410

Başar, E., Hari, R., Lopes da Silva, F.H., Schürmann, M., eds. (1997a) Brain Alpha Activity – New Aspects and Functional Correlates. Int. J. Psychophysiol. 26: 1–482

Başar, E., Schürmann, M., Başar-Eroglu, C., Karakaş, S. (1997b) Alpha Oscillations in brain functioning: an integrative theory. In: Başar E, Hari R, Lopes da Silva F.H., Schürmann M. (eds.) Brain Alpha Activity – New Aspects and Functional Correlates. Int. J. Psychophysiol. 26: 5–29

Başar, E., Kolev, V., Yordanova, J., Başar-Eroglu, C. (1997c) Is the alpha rhythm a control parameter for brain responses? Biol. Cybern. 76: 471–480

Başar, E., Başar-Eroglu, C., Demiralp, T., Ademoglu, A., Schürmann, M. (in press a) Oscillatory brain dynamics, wavelet analysis and cognition. Brain Cogn.

Başar, E., Schürmann, M., Başar-Eroğlu, C. (in press b) Potentials, rhythmic slow, behavioral correlates. In: Adelman, G., Smith, B.S. (eds.) Encyclopedia of Neuroscience. Birkhäuser, Boston, MA

Başar-Eroğlu, C. (1990) Eine vergleichende Studie corticaler und subcorticaler ereigniskorrelierter Potentiale des Katzengehirns. Thesis (Habilitationsschrift), Medizinische Universität zu Lübeck

Başar-Eroğlu, C, Başar, E. (1987) Endogenous components of event-related potentials in hippocampus: an analysis with freely moving cats. In Johnson, R.Jr., Rohrbaugh, J.W., Parasuraman, R. (eds.) Current Trends in Event-Related Potential Research. Electroencephalogr. Clin. Neurophysiol. 40: 440–444

Başar-Eroğlu, C, Başar, E. (1991) A compound P300–40 Hz response of the cat hippocampus. Int. J. Neurosci., 60: 227–237

Başar-Eroğlu, C., Schürmann, M. (1994) Subcortical P300–40 Hz response of the cat brain. In: Pantev, C., Elbert, T., Lütkenhöner, B. (eds.) Oscillatory Event Related Brain Dynamics. Plenum, New York, pp. 367–382 (NATO ASI Series; Series A: Life Sciences Vol. 271)

Basar-Eroglu, C., Basar, E., Schmielau, F. (1989) P300 response in cortical and subcortical structures of the cat brain. Ninth International Conference on event related potentials of the brain (EPIC IX), Nortwijk

Başar-Eroğlu, C., Başar, E., Schmielau, F. (1991a) P300 in freely moving cats with intracranial electrodes. Int. J. Neurosci. 60: 215–226

Başar-Eroğlu, C., Schmielau, F., Schramm, U, Schult, J. (1991b) P300 response of hippocampus with multielectrodes in cats. Int. J. Neurosci., 60: 239–248

Başar-Eroğlu, C., Başar, E., Demiralp, T, Schürmann, M. (1992) P300 response: possible psychophysiological correlates in delta and theta frequency channels. Int. J. Psychophysiol. 13: 161–179

Başar-Eroğlu, C., Strüber, D., Stadler, M., Kruse, P., Başar, E. (1993a) Multistable visual perception induces a slow positive EEG wave. Int. J. Neurosci. 37: 139–151

Basşar-Eroğlu, C., Warecka, K., Schürmann, M., Basşar, E. (1993b) Visual evoked potentials in multiple sclerosis: frequency domain analysis shows reduced alpha response. Int. J. Neurosci. 73: 259–276

Başar-Eroğlu, C., Strüber, D., Stadler, M., Kruse, P., Greitschus, F. (1995) Slow positive potentials in the EEG during multistable visual perception. In: Kruse, P., Stadler, M. (eds.) Ambiguity in Mind and Nature. Multistable Cognitive Phenomena. Springer, Berlin Heidelberg New York, pp. 391–408.

Basşar-Eroğlu, C., Basşar, E., Zetler, G. (1996a) Effects of Ceruletide and Haloperidol on auditory evoked potentials in the cat brain. Int. J. Neurosci. 85: 131–146

Başar-Eroğlu, C., Strüber, D., Kruse, P., Başar, E., Stadler, M. (1996b) Frontal gamma band enhancement during multistable visual perception. Int. J. Psychophysiol. 24: 113–125

Başar-Eroğlu, C., Strüber, D., Schürmann, M., Stadler, M., Başar, E. (1996c) Gamma band responses in the brain: a short review of psychophysiological correlates and functional significance. Int. J. Psychophysiol. 24: 101–112

Beagley, H.A., Sayers, B.M., Ross, J.A. (1979) Fully objective ERA by phase spectral analysis. Acta Otolaryngol. (Stockh.) 87: 270–278

Beagley, H.A., Kellog, A. (1970) Amplitude of the auditory evoked response at high intensities. Sound 4: 86–90

Bennett, T.L. (1970) Hippocampal EEG correlates of behaviour. Electroencephalogr. Clin. Neurophysiol. 28: 17–23

Bennett, T.L., Gottfried, J. (1970) Hippocampal theta activity and response inhibition. Electroencephalogr. Clin. Neurophysiol. 29: 196–200

Bennett, T.L., Herbert, P.N., Moss, D.S. (1973) Hippocampal theta activity and the attention component of discrimination learning. Behav. Biol. 8: 173–181

Berger, H. (1929) Über das Elektrenkephalogramm des Menschen. I. Bericht. Arch. Psychiat. Nervenkr. 87: 527–570

Berger, H. (1938) Das Elektrenkephalogramm des Menschen. Nova Acta Leopold. 6: 38

Berger, T.W, Thompson, R.F. (1977) Limbic system interrelations: Functional division among hippocampal septal connections. Science 197: 587–589

Berger, T.W., Thompson, R.F., (1978a) Neuronal plasticity in the limbic system during classical conditioning of the rabbit nictitating membrane response. I. The hippocampus. Brain Res. 145: 323–346

Berger, T.W., Thompson, R.F. (1978b) Neuronal plasticity in the limbic system during classical conditioning of the rabbit nictitating membrane response. II: Septum and mammillary bodies. Brain Res. 156: 293–314

Berger, T.W., Thompson, R.F. (1978c) Identification of pyramidal cells as the critical elements in hippocampal neuronal plasticity during learning. Proc. Natl. Acad. Sci. USA 75: 1572–1576

Berger, T.W., Alger, B., Thompson, R.F. (1976) Neuronal substrate of classical conditioning in the hippocampus. Science 192: 483–485

Best, P.J., Ranck, J.B.Jr. (1982) Reliability of the relationship between hippocampal unit activity and sensory-behavioral events in the rat. Exp. Neurol. 75: 652–64

Birbaumer, N., Elbert, T., Canavan, A.G.M., Rockstroh, B. (1990) Slow potentials of the cerebral cortex and behavior. Physiol. Rev. 70: 1–41

Bishop, G.H. (1933) Cyclic changes in excitability of the optic pathway of the rabbit. Am. J. Physiol. 103: 213–224

Bishop, P.O., Jeremy, D., McLeod, J.G. (1953) Phenomenon of repetitive firing in lateral geniculate of cat. J. Neurophysiol. 16: 443–447

Black, A.H. (1975) Hippocampal electrical activity and behaviour. In: Isaacson, R.I., Pribram, K.H. (eds.) The Hippocampus, Vol. II. Plenum, New York, pp. 129–168

Bleckmann, H., Weiss, O. and Bullock, T. H. (1989) Physiology of lateral line mechanoreceptive regions in the elasmobranch brain. J. Comp. Physiol. A 164: 459–474.

Boden, P.R., Hill, R.G. (1988) Effects of cholecystokinin and pentagastrin on rat hippocampal neurones maintained in vitro. Neuropeptides 12: 95–103

Böhmig, L. (1883) Beiträge zur Kenntnis des Zentralnervensystems einiger pulmonaten Gastropoden: Helix pomatia und Limnaea stagnalis. Inauguraldiss., Leipzig

Botte, M.G., Bujas, Z., Chocholle, R. (1975) Comparison between the growth of the averaged electroencephalic response and direct loudness estimation. J. Acoust. Soc. Amer. 58: 208–213

Bouyer, J.J., Montarom, M.F., Vahnee, J.M., Albert, M.P., Rougeul, A. (1987) Anatomical localization of cortical beta rhythms in cat. Neurosci. 22: 863–869

Brandt, M.E. (1997) Visual and auditory evoked phase resetting of the alpha EEG. In: Başar, E., Hari, R., Lopes da Silva, F.H., Schürmann, M. (eds.) Brain Alpha Activity – New Aspects and Functional Correlates. Int. J. Psychophysiol. 26: 285–298

Brandt, M.E., Jansen, B.H. (1991) The relationship between prestimulus-alpha amplitude and visual evoked potentials amplitude. Int. J. Neurosci. 61: 261–8

Brandt, M.E, Jansen, B.H., Carbonari, J.P. (1991) Prestimulus spectral EEG patterns and the visual evoked response. Electroencephalogr. Clin. Neurophysiol. 80: 16–20

Brazier, M.A.B. (1960) Long-persisting electrical traces in the brain of man and their possible relationship to higher nervous activity. In: Jasper, H.H., Smirnov, G.D. (eds.) The Moscow Colloquium on Electroencephalography of Higher Nervous Activity, The EEG Journal, Montreal, pp. 347–358

Brazier, M.A. (1968) Studies of the EEG activity of limbic structures in man. Electroencephalogr. Clin. Neurophysiol. 25: 309–318.

Bremer, F., Bonnet, V. (1950) Interprétation des réactions, rhythmiques prologuées des aires sensorielles de l'écorce cérébrale. Electroencephalogr. Clin. Neurophysiol. 2: 384–400

Bressler, S.L. (1990) The gamma wave: a cortical information carrier? Trends Neurosci. 13: 161–162

Brown, B.B. (1968) Frequency and phase of hippocampal theta activity in the spontaneously behaving cat. Electroencephalogr. Clin. Neurophysiol. 24: 53–62

Brown, B.B., Shryne, J.E. (1964) EEG theta activity and fast activity sleep in cats as related to behavioural traits. Neuropsychologia 2:311–326

Buzsaki, G., Haubenrauser, J., Grastyan, E., Czopf, J., Kellenyi, L. (1981) Hippocampal slow wave activity during appetitive and aversive conditioning in the cat. Electroencephalogr. Clin. Neurophysiol. 51:276–290

Buchwald, J.S., Squires, N.S. (1982) Endogenous Auditory Potentials in the Cat. In: Woody, C.D. (ed.) Conditioning: Representation of Involved Neural Function. Plenum Press, New York, pp. 503–515

Bullock, T.H. (1945) Problems in the comparative study of brain waves. Yale J. Biol. Med., 17 657–679

Bullock, T.H. (1974) Comparison between vertebrates and invertebrates in nervous organization, In: Schmitt, F.O., Worden, F.W. (eds.) The Neurosciences. Third Study Program. MIT Press, Cambridge, pp. 343–346

Bullock, T. H. (1979) Processing of ampullary input in the brain: comparison of sensitivity and evoked responses among elesmobranchs and siluriform fishes, J. Physiol. (Paris) 75: 397–407.

Bullock, T. H. (1981) Neuroethology deserves more study of evoked responses. Neurosci. 6 (1981) 1203–1215.

Bullock, T. H. (1982) Electroreception. Ann. Rev. Neurosci. 5: 121–170.

Bullock, T.H. (1983) Electrical signs of activity in assemblies of neurons: compound field potentials as objects of study in their own right, Acta Morphol. Acad. Sci. Hung., 31: 39–62

Bullock, T.H. (1984a) Physiology of the tectum mesencephali in elasmobranchs. In: Vanegas, H. (ed.) Comparative Neurology of the Optic Tectum. Plenum Press, New York, pp. 47–68.

Bullock, T.H. (1984b) Ongoing compound field potentials from octopus brain are labile and vertebrate-like, Electroencephalogr. Clin. Neurophysiol. 57: 473–483

Bullock, T.H. (1988a) Compound potentials of the brain, ongoing and evoked: perspectives from comparative neurology. In: Başar, E. (ed.) Dynamics of sensory and cognitive processing by the brain. Springer, Berlin Heidelberg New York, pp. 3–18 (Springer Series in Brain Dynamics, Vol. 1)

Bullock, T.H. (1988b) The comparative neurology of expectation: stimulus acquisition and neurobiology of anticipated and unanticipated input. In: Atema, J., Popper, P.R.A.N., Tavolga, W.N. (eds.) Sensory Biology of Aquatic Animals. Berlin Heidelberg New York, Springer, pp. 195–222

Bullock, T. H. (1989) The micro-EEG represents varied degrees of cooperativity among wide-band generators: spatial and temporal microstructure of field potentials, in: Başar, E. Bullock, T.H. (eds.) Brain Dynamics: Progress and Perspectives. Springer, Berlin Heidelberg New York, pp. 5–12

Bullock, T.H. (1992) Introduction to induced rhythms: a widespread, heterogeneous class of oscillations. In Başar, E., Bullock, T.H. (eds.), Induced Rhythms in the Brain. Birkhäuser, Boston, MA, pp. 1–26

Bullock, T.H. (1993) How Brains May Work. Boston, MA: Birkhäuser

Bullock, T.H., Başar, E. (1988) Comparison of ongoing compound field potentials in the brains of invertebrates and vertebrates. Brain Res. Rev. 13: 57–75

Bullock, T. H., Corwin, J. T. (1979) Acoustic evoked activity in the brain of sharks. J. Comp. Physiol. 223–234.

Bullock, T. H., Heiligenberg, W. (1986) Electroreception. Wiley, New York

Bullock, T.H., Horridge, G.A. (1965) Structure and Function in the Nervous Systems of Invertebrates, Vol. 1, pp. 22–351 and Vol. 2, pp. 801–1386. Freeman, San Francisco, CA

Bullock, T. H., Orkland, R. Grinnel, A. D. (1977) Introduction to Nervous Systems. Freeman, San Francisco, CA

Bullock, T. H., Bodznick, D. A., Northcutt, R. G. (1983) The phylogenetic distribution of electroreception: evidence for convergent evolution of a primitive vertebrate sense modality. Brain Res. Rev. 6: 25–46

Bullock, T.H., Hofmann, M.F., Nahm, F.K., New, J.G., Prechtl, J.C. (1990) Event-related potentials in the retina and optic tectum of fish. J. Neurophysiol. 64: 903–914

Bunney, B.S. (1988) Effects of acute and chronic neuroleptic treatment on the activity of midbrain dopamine neurons. Ann. N.Y. Acad. Sci. 537: 77–85

Butler, R.A., Keidel, W.D., Spreng, M. (1969) An investigation of the human cortical evoked potential under conditions of monaural and binaural stimulation. Acta Oto-Laryngol. 68: 317–326

Buzsáki, G. (1985) Electroanatomy of the hippocampal rhythmic slow activity (RSA) in the behaving cat. In: Buzsáki, G., Vanderwolf, C.H. (eds.) Electrical Activity of the Archicortex. Budapest, Akademiai Kiado, pp. 143–164

Buzsáki, G. (1992) Network properties of the thalamic clock: role of oscillatory behaviour in mood disorders. In: Başar, E., Bullock, T.H. (eds.), Induced Rhythms in the Brain. Boston, MA: Birkhäuser, pp. 235–250

Chang, H.T. (1950) The repetitive discharges of corticothalamic reverberating circuit. J. Neurophysiol. 13: 235–257

Cavada, C., Llamas, A., Reinoso-Suarez, F. (1983) Allocortical afferent connections of the prefrontal cortex in the cat. Brain Res., 260: 117–120

Chapman, R.M., Bragdon, H.R. (1964) Evoked responses to numerical and nonnumerical visual stimuli while problem solving. Nature 203: 115–1157

Chapman, R.M. (1973) Evoked potentials of the brain related to thinking. In: McGuigan, F.J., Schoonover, R.A. (eds.) Psychophysiology of Thinking: Studies of Covert Processes. Academic Press, New York, pp. 69–108

Chui, C.K. (1992) An Introduction to Wavelets: Wavelet Analysis and its Applications. Academic Press, Boston, MA

Cohen, L. (1989) Time-Frequency Distribution – A Review. Proc. IEEE: 77

Cohen, D. (1968) Magnetoencephalographic evidence of magnetic fields produced by alpha-rhythm currents. Science 161: 784–786

Coleman J.R., Lindsley, D.B. (1975) Hippocampal electrical correlates of free behaviour and behaviour induced by stimulation of two hypothalamic-hippocampal systems in the cat. Exp. Neurol. 49: 506–528

Courchesne, E. (1978) Changes in P3 waves with event repetition: Long-term effects on scalp distribution and amplitude. Electroencephalogr. Clin. Neurophysiol. 45: 754–766

Creutzfeldt, O.D. (1974) The neuronal generation of the EEG. In: Rémond, A. (ed.) Handbook of Electroencephalography and Clinical Neurophysiology. Elsevier, Amsterdam

Creutzfeldt, O.D. (1983) Cortex Cerebri. Springer, Berlin Heidelberg New York

Creutzfeldt., O.D., Watanabe, S., Lux, H.D. (1966) Relationship between EEG-phenomena and potentials of single cortical cells. I. Evoked responses after thalamic and epicortical stimulation. Electroencephalogr. Clin. Neurophysiol. 20: 1–18

Curio, G., Mackert, B.M., Burghoff, M., Koetitz, R., Abraham-Fuchs, K., Harer, W. (1994) Localization of evoked neuromagnetic 600 Hz activity in the cerebral somatosensory system. Electroencephalogr. Clin. Neurophysiol. 91: 483–487

Curio, G., Drung, D, Koch, H., Muller, W., Steinhoff, U., Trahms, L., Shen, YQ., Vase, P., Freltoft, T. (1996) Magnetometry of evoked fields from human peripheral nerve, brachial plexus and primary somatosensory cortex using a liquid nitrogen cooled superconducting quantum interference divice. Neurosci. Lett. 206: 204–206

Dafny, N., Jacob, R.H., Jacobson, E.D. (1975) Gastrointestinal hormones and neural interaction within the central nervous system. Experientia (Basel) 31: 658–660

Dahl, D. (1987) Systemically administered cholecystokinin affects an evoked potential in the hippocampal dentate gyrus. Neuropeptides 10: 165–173

Dahl, D. (1988) Central administration of cholecystokinin potentiates evoked potential amplitude in the hippocampal dentate gyrus. Neuropeptides 11: 147–151

Daubechies, I. (1990) The Wavelet Transform, Time-Frequency Localization and Signal Analysis. IEEE Trans. Inf. Theory 36: 961–1005

Davis, H. (1964) Enhancement of evoked cortical potentials in humans related to a task requiring a decision. Science 145: 182–183

Davis, H., Zerlin, S. (1966) Acoustic relations of the human vertex potential. J. Acoust. Soc. Amer. 39: 109–116

De Nabias, B. (1894) Recherches histologiques et orgnologiques sur les centres nerveux des gastropodes. Actes Soc. linn. Bordeaux 47: 11–202

Demiralp, T., Başar, E. (1992) Theta rhythmicities following expected visual and auditory targets. Int. J. Psychophysiol. 13: 147–160

Demiralp, T., Basşar-Eroğlu, C., Rahn, E., Basşar, E. (1994) Event-related theta rhythms in cat hippocampus and prefrontal cortex during an omitted stimulus paradigm. Int. J. Psychophysiol. 18: 35–48

Demiralp, T., Başar-Eroğlu, C., Başar, E. (1996) Distributed gamma band responses in the brain studied in cortex, reticular formation, hippocampus, and cerebellum. Int. J. Neurosci. 84: 1–13

Desmedt, J. (ed.) (1979) Cognitive Compenents in Cerebral Event-Related Potentials and Selective Attention. Karger, Basel (Progress in Clinical Neurophysiology, Vol. 6)

Desmedt, J.E., Tomberg, C. (1994) Transient phase-locking of 40 Hz electrical oscillations in prefrontal and parietal human cortex reflects the process of conscious somatic perception. Neurosci. Lett. 168: 126–129

Desmedt, J.E., Debecker, J. (1979) Wave form and neural mechanism of the decision P350 elicited without prestimulus CNV or readiness potential in random sequences of near threshold auditory clicks and finger stimuli. Electroencephalogr. Clin. Neurophysiol. 47: 648–670

Desmedt, J.E., Huy, N.T., Bourguet, M. (1983) The cognitive P40, N60 and P100 components of somatosensory evoked potentials and the earliest electrical signs of sensory processing in man. Electroencephalogr. Clin. Neurophysiol. 56: 272–282

Dietl, M.M., Probst, A., Palacios, J.M. (1987) On the distribution of cholecystokinin receptor binding sites in the human brain: an autoradiographic study. Synapse 1: 169–183

Dinse, H.R., Krüger, K., Best, J. (1990) A temporal structure of cortical information processing. Concepts Neurosci. 1: 199–238

Dinse, H.R., Krüger, K., Akhavan, A.C., Spengler, F., Schöner, G., Schreiner, C.E. (1997) Low-frequency oscillations of visual, auditory and somatosensory cortical neurons evoked by sensory stimulation. In: Başar E, Hari R, Lopes da Silva F.H., Schürmann M. (eds.) Brain Alpha Activity – New Aspects and Functional Correlates. Int. J. Psychophysiol. 26: 205–227

Dolecek, R., Jezek, M., Adamkova, M., Polivka, P., Kubis, M., Sajnar, J., Doleckova, D., Kurkova, J. Zavada, M. (1983) Endorphin releasers: a new possible approach to the treatment of pain after burns. A preliminary report. Burns Incl. Therm. Inj. 10: 41–44

Domesick, V.B. (1988) Neuroanatomical organization of dopamine neurons in the ventral tegmental area. Ann. N.Y. Acad. Sci. 537: 10–26

Donald, N.W. (1983) Neural selectivity in auditory attention: Sketch of a Theory. In: Gaillard, A.W.K., Ritter, W. (eds.) Tutorials in Event Related Potential Research. Elsevier, Amsterdam

Donchin, E. (1979) Event-related potentials. A tool in the study of human information processing. In: H. Begleiter (ed.) Evoked Brain Potentials and Behaviour. Plenum Press, New York

Donchin, E. (1981) Surprise! ... Surprise! Psychophysiol. 18: 493–513

Donchin, E., Cohen, L. (1967) Averaged evoked potentials and intramodality selective attention. Electroencephalogr. Clin. Neurophysiol. 22: 537–546

Donchin, E., Ritter, W., McCallum, W. (1978) Cognitive psychophysiology: the endogenous components of the ERP. In: Callaway, E., Tueting, P., Koslov, S. (eds.) Event-related Brain Potentials in Man. Academic Press, New York

Donchin, E., Heffley, E., Hillyard, S.A., Loveless, N., Maltzman, I., Ohman, A.D., Rösler, F., Ruchkin, D., Siddle, D. (1984) Cognition and event-related potentials. II. The orienting reflex and P300. Ann. N. Y. Acad. Sci. 425: 39–57

Douglas, R.J., Martin, K.A.C. (1995) Neural networks. Vibrations in the memory. Nature 373: 563–564

Duffy, F.H., Denkla, M.B., Sandini, G. (1980) Dyslexia: regional differences in brain electrical activity by topographic mapping. Ann. Neurol. 7: 412–420

Dudkin, K.N., Glezer, V.D., Gauselman, V.E., Panin, A.I. (1978) Types of receptive fields in the lateral geniculate body and their functional model. Biol. Cybern. 29: 37–47

Durusan, R. (1976) Analysis of the cerebellar cortex electrical activity by methods of systems theory. Thesis, Hacettepe University, Ankara

Echteler, S. M. and Saidel, W. M. (1980) Some connections of the teleost telencephalon. Soc. Neurosci. Abstr. 6: 629.

Eckhorn, R., Bauer, R., Jordan, W., Brosch, M., Kruse, W., Munk, M. Reitboeck, H.J. (1988) Coherent oscillations: a mechanism of feature linking in the visual cortex? Biol. Cybern. 60: 121–130

Eckhorn, R., Reitboeck, H.J., Arndt, M., Dicke, P. (1989a) Feature linking via stimulus-evoked oscillations: experimental results from cat visual cortex and functional implications from a network model. Conference on Neural Networks, Washington

Eckhorn, R., Bauer, R., Reitboeck, H.J. (1989b) Discontinuities in visual cortex and possible functional implications: relating cortical structure and function with multielectrode/correlation techniques. In: Başar, E., Bullock, T.H. (eds.) Brain Dynamics – Progress and Perspectives. Springer, Berlin Heidelberg New York, pp. 267–278 (Springer Series in Brain Dynamics, Vol. 2)

Edelman, G.M. (1978) Group selection and phasic reentrant signaling: a theory of higher brain functions. In: Edelman, G.M., Mountcastle, V.B. (eds.) The Mindful Brain. MIT Press, Cambridge, MA, pp. 51–100.
Elazar, Z., Adey, W.R. (1967) Spectral analysis of low frequency components in the electrical activity of the hippocampus during learning. Electroencephalogr. Clin. Neurophysiol. 23: 225–240
Ellis, H.C., Hunt, R.R. (1993) Fundamentals of Cognitive Psychology. Brown and Benchmark, Oxford
Engel, A.K., König, P., Kreiter, A.K., Schillen, T.B., Singer, W. (1992) Temporal coding in the visual cortex: new vistas on integration in the nervous system. Trends Neurosci. 15: 218–226
Erspamer, V. Melchiorri, P. (1983) Actions of amphibian skin peptides on the central nervous system and the anterior pituitary. In: Müller, E.E., MacLeod, R.M. (eds.) Neuroendocrine Perspectives, Vol. 2. Amsterdam, Elsevier, pp. 37–106

Fairchild, M.D., Jenden, D.J., Mickey, M.R, Yale, C. (1980) The quantitative measurement of changes in EEG frequency spectra produced in the cat by sedative-hypnotics and neuroleptics. Electroencephalogr. Clin. Neurophysiol. 49: 382–390
Fessard, A. (1961) The role of neuronal networks in sensory communications within the brain. In: Rosenblith, W.A. (ed.) Sensory Communication. MIT Press, Cambridge, MA
Fessard, A., Szabo, T. (1974) Physiology of receptors. II. Peripheral mechanisms of electroreceptors in teleosts. In: Fessard, A. (ed.) Handbook of Sensory Physiology III/3. New York: Springer, pp. 64–95.
Flohr, H. (1991) Brain processes and phenomenal consciousness: a new and specific hypothesis. Theory and Psychology 1: 245–262
Ford, J.M., White, P., Lim, K.O., Pfefferbaum, A. (1994) Schizophrenics have fewer and smaller P300s: a single-trial analysis. Biol. Psychiat. 35: 96–103
Fox, S.E., Wolfson, S. Ranck J.B, (1983) Investigating the mechanisms of hippocampal theta rhythms: Approaches and progress. In: Seifert, W. (ed.) Neurobiology of the Hippocampus. Academic Press, London, pp. 303–319
Franaszczuk, P.J., Blinowska, K.J. (1985) Linear model of brain electrical activity, EEG as a superposition of damped oscillatory modes. Biol. Cybern. 53: 19–25
Freeman, W.J. (1975) Mass Action in the Nervous System. Academic Press, New York
Freeman, W.J. (1979) Nonlinear dynamics of paleocortex manifested in the olfactory EEG. Biol. Cybern. 35: 21–37
Freeman, W.J. Skarda, C.A. (1985) Spatial EEG-patterns, non-linear dynamics and perception: the neo-Sherringtonian view. Brain Res. Rev. 10: 147–175
Friedman, D. (1984) P300 and slow wave: the effects of reaction time quartile. Biol. Psychol. 18: 49–71
Fuster, J.M. (1989) The Prefrontal Cortex: Anatomy, Physiology, and Neuropsychology of the Frontal Lobe. Raven Press, New York

Galambos, R. (1989) Electrogenesis of evoked potentials. In: Başar, E., Bullock, T.H. (eds.) Brain Dynamics – Progress and Perspectives. Springer, Berlin Heidelberg New York, pp. 13–25
Galambos, R. (1992) A comparison of certain gamma band (40 Hz) brain rhythms in cat and man. In: Başar, E., Bullock, T.H. (eds.) Induced Rhythms in the Brain, pp. 201–216

Galambos, R., Hillyard, S.A. (1981) Electrophysiological approaches to human cognitive processing. Neurosci. Res. Prog. Bull. 20: 141–265

Galambos, R., Makeig, S. (1988) Dynamic changes in steady-state responses. In: Başar, E. (ed.) Dynamics of Sensory and Cognitive Processing by the Brain. Springer, Berlin, pp. 103–122

Galambos, R., Rose, J.E., Bromiley, R.B., Hughes, J.R. (1952) Microelectrode studies on medial geniculate body of the cat. II. Response to clicks. J. Neurophysiol. 15: 359–380

Galambos, R., Makeig, S., Talmachoff, P. (1981) A 40 Hz auditory potential recorded from the human scalp. Proc. Natl. Acad. Sci. USA 78: 2643–2647

Gaudereau, P., St.-Pierre, S., Pert, C.B., Quirion, R. (1985) Cholecystokinin receptors in mammalian brain. A comparative characterization and visualization. Ann. N. Y. Acad. Sci. 448: 198–219

Gauthier, P., Hamon, J.F., Gottesmann, C. (1986) Etude de la variation contingente négative (VCN) et des composantes tardives des potentiels évoqués au cours du sommeil chez l'homme. Rev. Electroencephalogr. Neurophysiol. Clin. 15: 323–330

Gelperin, A. (1989) Perspectives in Neural Systems and Behavior. Liss, New York, pp. 121–136

Gerschenfeld, H.M., Ascher, P., Tauc, L. (1967) Two different excitatory transmitters acting on a single molluscan neurone. Nature 213: 358–359

Gevins, A.S. (1984) Analysis of electromagnetic signals of the human brain: milestones, obstacles and goals. IEEE Trans BME, 31: 833–850

Giannitrapani, D. (1985) The electrophysiology of intellectual functions. Karger, Basel

Gille, H.G., Ullsperger, P. (1987) Influence of irrelevant acoustic input upon individual activation regulation indicated by parameters of evoked brain potentials (EP). Neurosci. 22: S48

Glaser, E.M., Ruchkin, D.S. (1976) Principles of Neurobiological Signal Analysis. Academic Press, New York

Goertz, R., Jokeit, H., Küchler, E. (1994) Event related dynamics of 40 Hz electroencephalogram during visual discrimination task. Int. J. Neurosci., 79: 267–273

Goff, E.R., Allison, T., Vaughan, H.G.Jr. (1978) The functional neuroanatomy of event-related potentials. In: Tuetin, E.P., Koslow, S.H. (ed.) Event-Related Potentials in Man. Academic Press, New York, pp. 1–79

Goff, G.D., Matsumiya, Y., Allison, T., Goff. W.R., (1977) The scalp topography of human somatosensory and auditory evoked potentials. Electroencephalogr. Clin. Neurophysiol. 42: 57–76

Goodin, D.S., Squires, K.C., Hederson, B.H., Starr, A. (1978) An early event-related cortical potential. Psychophysiol. 15: 360–365

Goldman-Rakic, P.S. (1988) Topography of cognition: parallel distributed networks in primate association cortex. Ann. Rev. Neurosci., 11: 137–156

Gralewicz, C. (1981) Hippocampal electrophysiological correlates of an avoidance behaviour in the cat. Electroencephalogr. Clin. Neurophysiol. 52: 306–315

Grastyan, E., Vereckei, L. (1974) Effects of spatial separation of the conditioned signal from the reinforcement: a demonstration of the conditioned character of the orienting response or the orientational character of conditioning. Beh. Biol. 10: 121–146

Grandori, F., Hoke, M., Romani, G.L., eds. (1990) Auditory Evoked Magnetic Fields and Potentials. Karger, Basel (Advances in Audiology, Vol. 6)

Grastyan, E., Lissak, K., Madarasz, I., Donnhoffer, H. (1959) Hippocampal electrical activity during the development of conditioned reflexes. Electroencephalogr. Clin. Neurophysiol. 11: 409–430

Gray, C.M., Singer, W. (1987) Stimulus specific neuronal oscillations in the cat visual cortex: a cortical function unit. Soc. Neurosci. 404: 3

Gray, C.M., Singer, W. (1989) Stimulus-specific neuronal oscillations in orientation columns of cat visual cortex. Proc. Natl. Acad. Sci. USA. 86: 1698–1702

Gray, C.M., König, P., Engel, A.K., Singer, W. (1989) Oscillatory response in the cat visual cortex exhibits intercolumnar synchronization which reflects global stimulus properties. Nature 338: 334–337

Gray, C.M., Engel, A.K., König, P., Singer, W. (1992) Mechanisms underlying the generation of neuronal oscillations in cat visual cortex. In: Başar, E., Bullock, T.H. (eds.) Induced Rhythms in the Brain. Birkhäuser, Boston, MA, pp. 29–45

Greenstein, Y.J., Pavlides, C., Winson, J. (1988) Long-term potentiation in the dentate gyrus is preferentially induced at theta rhythm periodicity. Brain Res. 438: 331–334

Green, R.L. (1992) Human Memory: Paradigms and Paradoxes. Lawrence Erlbaum, Hillsdale, NJ

Gruzelier J, ed. (1996) New Advances in EEG and Cognition. Int. J. Psychophysiol. 24: 1–187

Hagino, Y., Moroji, T., Iizuka, R. (1989) A behavioural pharmacological study on intracerebroventricularly administered CCK-8 related peptides in mice. Neuropeptides 13: 107–113

Haig, A.R., Gordon, E., Rogers, G., Anderson, J. (1995) Classification of single-trial ERP sub-types: application of globally optimal vector quantization using simulated annealing. Electroencephalogr. Clin. Neurophysiol., 94: 288–297

Haken, H. (1977) Synergetics. An introduction. Springer: Berlin Heidelberg New York

Haken, H. (1983) Introduction. In: Başar, E., Flohr, H., Haken, H., Mandell, A.J. (eds.) Synergetics of the Brain. Springer, Berlin Heidelberg New York, pp. 3–25

Halgren E, Babb TL, Crandall PH (1978) Human hippocampal formation EEG desynchronization during attentiveness and movement. Electroencephalogr. Clin. Neurophysiol. 44: 778–781

Halgren, E., Stapleton, J.M., Smith, M., Altafullah, I. (1986) Generators of the Human Scalp P3. In: Cracco, R.Q., Bodis-Wollner, I. (eds.) Evoked Potentials. Liss, New York, pp. 269–284

Haller, B. (1913) Die Intelligenzsphären des Molluskengehirns. Arch. mikr. Anat. (I) 81: 233–322

Halliday, A.M., McDonald, W.I., Mushin, J. (1972) Delayed visual evoked response in optic neuritis. Lancet i, 982–985

Hari, R., Kaukoranta, E. (1985) Neuromagnetic studies of somatosensory system: principles and examples. Prog. Neurobiol. 24: 233–56

Hari, R., Lounasmaa, O.V. (1989) Recording and interpretation of cerebral magnetic fields. Science 244: 432–6

Hari, R. (1990) Magnetic evoked fields of the human brain: basic principles and applications. Electroencephalogr. Clin. Neurophysiol. Suppl. 41: 3–12

Hari, R. (1993) Magnetoencephalography as a Tool of Clinical Neurophysiology. In: Niedermeyer, E., Lopes da Silva, F. (eds.) Electroencephalography – Basic Principles, Clinical Applications, and Related Fields. Williams and Wilkins, Baltimore, pp. 1035–1061

Hari, R. (1991) On brain's magnetic responses to sensory stimuli. J. Clin. Neurophysiol. 8: 157–169

Hari, R., Salmelin, R. (1997). Human cortical oscillations. Trends Neurosci. 20: 44–49

Harrison, J.B., Buchwald, J.S., Kaga, K. (1986) Cat P300 present after primary auditory cortex ablation. Electroencephalogr. Clin. Neurophysiol. 63: 180–187

Harrison, J.B., Dickerson, L.W., Song, S., Buchwald, J.S. (1990) Cat-P300 present after association cortex ablation. Brain Res. Bull. 24: 551–560

Hartline, P.H. (1987) Multisensory Convergence. In: Adelman, G. (eds.) Encyclopedia of Neuroscience. Birkhäuser, Boston, MA Basel Stuttgart, pp. 706–709

Hashimoto, T., Yanagisawa, N. (1990) Acute reduction and long-term improvement of chorea with ceruletide (cholecystokinin analogue). J. Neurol. Sci. 100: 178–185

Hayashi H, Iijima S, Sugita Y, Teshiwa Y, Tashiro T, Matsuo R, Yasochima A, Hishikawa Y, Ishihara T (1986) Appearance of frontal mid-line theta rhythm during sleep and its relation to mental activity. Electroenecephalogr. Clin. Neurophysiol. 66: 66–70

Hebb, D.O. (1949) The organization of behaviour. Wiley, New York

Heiligenberg, W. (1977) Principles of electrolocation and jamming avoidance in electric fish. A neuroethological approach. In: Braitenberg, V (ed.) Studies of Brain Function. Springer, New York, pp. 1–85

Heiligenberg, W. , Kawasaki, M. (1992) An internal current source yields immunity of electrosensory information processing to unusually strong jamming in electric fish. J. Comp. Physiol. A, 171, 309–316.

Heinrich, H., Gaus, F., Dickhaus, H. (1991) Zeit-Frequenz-Charakteristika evozierter Potentiale. Biomed. Technik, 36 (Ergänzungsband), 165–166

Heinze, H.J., Mangun, G.R., Burchert, W., Hinrichs, H., Scholz, M., Münte, TF., Goas, A., Scherg, M., Johannes, S., Hundeshagen, H. (1994a) Combined spatial and temporal imaging of brain activity during visual selective attention in humans. Nature 372: 543–6

Heinze, H.J., Münte, T.F., Mangun, G.R. (1994b) Cognitive Electrophysiology. Birkhäuser, Boston, MA

Herkenham, M. (1986) New perspectives on the organization and evolution of non-specific thalamo-cortical projections. In: Jones, E.G., Peters, A. (eds.) Cerebral Cortex, Vol. 5: Sensory-Motor Areas and Aspects of Cortical Connectivity. Plenum, New York, pp. 403–445

Hernandez-Péon R (1961) Reticular mechanisms of sensory control. In: Rosenblith, W.A. (ed.) Sensory Communication. MIT Press, Cambridge

Hernandez-Péon, R., Lavin, A., Alcocer-Cuaron, C., Marcelin, J.R. (1960) Electrical activity of the olfactory bulb during wakefulness and sleep. Electroencephalogr. Clin. Neurophysiol. 12: 41–58

Hillyard, S.A. (1981) Selective auditory attention and early event-related potentials: a rejoiner. Can. J. Psychol. 35: 159–174

Hillyard, S.A. Picton, T.W. (1979) Event related brain potentials and selective information processing in man. In: Desmedt, J.E. (ed.) Cognitive Components in Cerebral Event-Related Potentials and Selective Attention. Karger, Basel, pp.1–50 (Progress in Clinical Neurophysiology, Vol. 6)

Hillyard, S.A., Hink, R.F., Schwent, V.L., Picton, T.W. (1973) Electrical signs of selective attention in the human brain. Science 182: 177–180

Hiripi, L., Salánki, J. (1973) Role of monoamines in the central regulation of periodic activity in *Anodonta cygenea* L. (Pelecypoda). In: Salánki, J. (ed.) Neurobiology of Invertebrates: Mechanisms of Rhythm Regulation. Budapest, Akademiai Kiado, pp. 391–401

Hoke, M., Lehnertz, K., Pantev, C., Lütkenhöner, B. (1989). Spatiotemporal aspects of synergetic processes in the auditory coretex as revealed by the magnetoencephalogram. In: Başar, E. (ed.) Brain Dynamics. Springer, Berlin, pp. 84–105

Horowitz, J.M., Freeman, W.J., Stoll, P.J. (1973) A neuronal network with a background level of excitation in the cat hippocampus. Int. J. Neurosci. 5: 113–123

Irle, E., Markowitsch, H.J. (1982a) Connections of the hippocampal formation, mamillary bodies, anterior thalamus and cingulate cortex. Exp. Brain Res., 47: 79–94

Irle, E., Markowitsch, H.J. (1982b) Widespread cortical projections of the hippocampal formation in the cat. Neurosci. 7: 2637–2647

Ishihara, T., Yoshii, N (1972). Multivariate analytic study of EEG and mental activity in juvenile delinquents. Electroencephalogr. Clin. Neurophysiol. 33: 71–80

Itoh, S., Lal, H. (1990) Influences of choleocystokinin and analogues on memory processes. Drug Dev. Res. 21: 257–276

Jaffe, D.B., Aitken, P.G. Nadler, J.V. (1987) The effects of cholecystokinin and cholecystokinin antagonists on synaptic function in the CA1 region of the rat hippocampal slice. Brain Res. 415: 197–203

James, W. (1890) The Principles of Psychology. New York: Holt

Jarcho, L.W. (1949) Excitability of cortical afferent systems during barbiturate anesthesia. J. Neurophysiol. 12: 447–457

Jeftinija, S., Miletic, V. Randic, M. (1981) Cholecystokinin octapeptide excites dorsal horn neurons both in vivo and in vitro. Brain Res. 213: 231–236

John, E.R., Bartlett, R., Shimokochi, M. (1975) Electrophysiological signs of the readout from memory. I. Raw data observation. Behav. Biol., 14: 247–282

John, E.R. (1967) Mechanisms of Memory. Academic Press, New York

John, E.R. (1988) Electrophysiological and radiographic evidence for the mediation of memory by an anatomically distributed system. In: Başar, E. (ed.) Dynamics of Sensory and Cognitive Processing by the Brain. Springer, Berlin Heidelberg New York, pp. 56–87

John, E.R. (ed.) (1990) Machinery of the Mind. Birkhäuser, Boston, MA

John, E.R., Prichep, L.S., Chabot, R.J. (1989) Quantitative electrophysiological maps of mental activity. In: Başar, E., Bullock, T.H. (eds.) Brain Dynamics – Progress and Perspectives. Springer, Berlin Heidelberg New York, pp. 56–87

Johnson, R. Jr. (1988) Scalp recorded P300 activity in patients following unilateral temporal lobectomy. Brain 111: 1517–1529

Johnson, R. Jr., J.W. Rohrbaugh, Parasuraman, R., eds. (1990) Event-Related Brain Potentials. Basic Issues and Applications. Oxford University Press

Jokeit, H., Makeig, S. (1994) Different event related patterns of gamma-band power in brain waves of fast- and slow-reacting subjects. Proc. Natl. Acad. Sci. USA 91: 6339–6343

Joliot, M., Ribary, U., Llinas, R. (1994) Human oscillatory brain activity near 40 Hz coexists with cognitive temporal binding. Proc. Natl. Acad. Sci. USA 91: 11748–11751

Jones, E.G., Powell, T.P.S. (1970) An anatomical study of converging sensory pathways within the cerebral cortex of the monkey. Brain 93: 793–820

Jouvet, M. (1964) Les différents états de sommeil: leurs mécanismes neurophysiologiques. In: Die Nervenphysiologie in gegenwärtiger Sicht. Barth, Leipzig (Nova Acta Leopold. 28)

Jürgens, E., Rösler, F. (1995) How do alpha oscillations influence gamma activity. psycoloquy.95.6.09.brain-rhythms.10

Jürgens, E., Rösler, F., Henninghausen, E., Heil, M. (1995) Stimulus-induced gamma oscillations: harmonics of alpha activity? Neuroreport 6: 813–816

Kandel ER (1976) The Cellular Basis of Behavior. Freeman, San Francisco

Kapas, L., Obal, F. Jr., Alfoldi, P., Rubicsek, G., Penke, B. Obal, F. (1988) Effects of nocturnal intraperitoneal administration of cholecystokinin in rats: Simultaneous increase in sleep, increase in EEG slow-wave activity, reduction of motor activity, suppression of eating, and decrease in brain temperature. Brain Res. 438: 155–164

Karakaş, S. (1997) A descriptive framework for information processing: an integrative approach. Int. J. Psychophysiol. 26: 353–368

Karakaş, S., Başar, E. (1983) High-frequency components of human visual evoked potentials. Int. J. Neurosci. 19: 161–172

Karakaş S, Başar, E. (in press) Early gamma response is sensory in origin: a conclusion based on cross-comparison of results from multiple experimental paradigms. Int. J. Psychophysiol.

Kemp I.R., Kaada, B.R. (1975) The relation of hippocampal theta activity to arousal, attentive behaviour and somato-motor movements in unrestrained cats. Brain Res. 95: 323–343

Kettner, R.E., Shannon, R.V., Nguyen, T.M., Thompson, R.F. (1980) Simultaneous behavioral and neural (cochlear nucleus) measurement during signal detection in the rabbit. Percept. Psychophys. 28: 504–513

Kettner, R.E., Thompson, R.F. (1982) Auditory signal detection and decision processes in the nervous system. J. Comp. Physiol. Psychol. 96: 328–331

Kerkut, G.A., Lambert, J.D.C., Gayton, R.J., Loker, J.E., Walker, R.J. (1975) Mapping of nerve cells in the subesophageal ganglia of Helix aspersa. Comp. Biochem. Physiol. 50A: 1–25

Kirschfeld, K. (1992) Oscillations in the insect brain: do they correspond to the cortical gamma waves of vertebrates? Proc. Natl. Acad. Sci. USA 89: 4764–4768

Klatzky, R.L. (1980) Human Memory: Structures and Processes. Freeman, New York

Klatzky, R.L. (1984) Memory and Awareness: An Information Processing Perspective. Freeman, New York.

Klimesch, W. (1996) Memory processes, brain oscillations and EEG synchronization. Int. J. Psychophysiol. 24: 61–100

Klimesch W (1997) EEG alpha rhythms and memory processes. In: Başar, E., Hari, R., Lopes da Silva, F.H., Schürmann, M. (eds.) Brain Alpha Activity – New Aspects and Functional Correlates. Int. J. Psychophysiol. 26: 319–340

Klimesch W, Pfurtscheller, G, Mohl W, Schimke H (1990) Event-related desynchronization, ERD-mapping and hemispheric differences for words and numbers. Int. J. Psychophysiol. 8: 297–308

Klimesch, W., Schimke, H., Pfurtscheller, G. (1993) Alpha frequency, cognitive load and memory performance. Brain Topogr. 5: 241–51

Knight, R.T., Hillyard, S.A., Woods, D.L., Neville, H.J., (1981) The effects of frontal cortex lesions on event-related potentials during auditory selective attention. Electroencephalogr. Clin. Neurophysiol. 52: 571–582

Koelle, B.G., Steiner E.C. (1956) The cerebral distributions of a tertiary and a quaternary anticholinesterase agent following intravenous and intraventricular injection. J. Pharmacol. Exp. Ther. 118: 420–434

Kojo, I., Liinasuo, M., Rovamo, J. (1993) Spatial and temporal properties of illusory figures. Vision Res., 33: 897–901

Kolev, V. Schürmann, M. (1992) Event-related prolongation of induced EEG rhythmicities in experiments with a cognitive task. Int. J. Neurosci. 67: 199–213

Kolev V, Daskalova M (1990) Recognition and analysis of single evoked potentials. In: Proceedings of the North Sea conference on Biomedical Engineering

Kopecz, K., Schöner, G., Spengler, F., Dinse, H.R. (1993) Dynamic properties of cortical evoked (10 Hz) oscillations: theory and experiment. Biol. Cybern. 69: 463–473

Kunze, H. (1917) Zur Topographie und Histologie des Centralnervensystems von *Helix pomatia* L. Inauguraldiss. Marburg

Kunze, H. (1921) Zur Topographie und Histologie des Zentralnervensystems von *Helix pomatia* L.Z. wiss. Zool. 118: 25–203

Kutas, M., McCarthy, G, Donchin, E. (1977) Augmenting mental chronometry: the P300 as a measure of stimulus evaluation time. Science 197: 792–795.

Lang, M., Lang, W., Diekmann, V., Kornhuber, H.H. (1987) The frontal theta rhythm indicating motor and cognitive learning. In: Johnson, R.Jr., Rohrbaugh, J.W., Parasuraman, R. (eds.), Current Trends in Event-Related Potential Research. Electroencephalogr. Clin. Neurophysiol. Suppl. 40

Lansing, R.W., Barlow, J.S. (1972) Rhythmic after-activity to flashes in relation to the background alpha which precedes and follows the photic stimuli. Electroencephalogr. Clin. Neurophysiol. 32: 149–160

Larson, J. Lynch, G. (1988) Role of N-methyl-D-aspartate receptors in the induction of synaptic potentiation by burst stimulation patterned after the hippocampal theta rhythm. Brain Res. 441: 111–118.

Lavin, A., Alcocer-Cuaron, C., Hernandez-Péon, R. (1959) Centrifugal arousal in the olfactory bulb. Science 129: 332–333.

Lehmann, D. (1989) From mapping to the analysis and interpretation of EEG/EP maps. In: Maurer, K. (ed.) Topographic Brain Mapping of EEG and Evoked Potentials. Springer, Berlin Heidelberg New York, pp. 53–75

Leiner, H.C., Leiner, A.L., Dow, R.S. (1989) Reappraising the Cerebellum: What Does the Hindbrain Contribute to the Forebrain? Behav. Neurosci. 103: 998–1008

Leung, L.S., Yim, C. (1986) Intracellular records of theta rhythm in hippocampal CA1 cells of the rat. Brain Res. 367: 323–327

Leung, L.S. (1992) Fast (beta) rhythms in the hippocampus: a review. Hippocampus 2: 93–98

Lieb, J., Sclabassi, R., Crandall, P., Buchness, R. (1974) Comparison of the action of diazepam and phenobarbital using EEG-derived power spectra obtained from temporal lobe epileptics. Neuropharmacol. 13: 769–784

Lissak, K., Grastyan, E. (1960) The changes of hippocampal electrical activity during conditioning. In: Jasper, E.E., Smirnov, G.D. (eds.) The Moscow Colloquium on Electroencephalography of Higher Nervous Activity. Electroencephalogr. Clin. Neurophysiol. Suppl. 13: 271–279

Lissmann, H. W. (1958) On the function and evolution of electric organs in fish. J. Exp. Biol. 35: 451–486

Llinás, R.R. (1988) The intrinsic electrophysiological properties of mammalian neurons: Insights into central nervous system function. Science, 242: 1654–1664

Llinás, R.R. (1992) Oscillations in CNS neurons: a possible role for cortical interneurons in the generation of 40 Hz oscillations. In: Başar, E., Bullock, T.H. (eds.) Induced Rhythms in the Brain. Birkhäuser, Boston, MA, pp. 269–283

Llinás, R.R. Ribary, U. (1992) Rostrocaudal scan in human brain: a global characteristic of the 40 Hz response during sensory input. In Başar, E. Bullock, T.H. (eds.) Induced Rhythms in the Brain. Birkhäuser, Boston, pp. 147–154

Llinás, R, Ribary U (1993) Coherent 40 Hz oscillation characterizes dream state in humans. Proc. Natl. Acad. Sci. USA 90: 2078–2081

Longo, V.G. (1962) Electroencephalographic Atlas for Pharmacological Research – Effect of Drugs on the Electrical Activity of the Rabbit Brain. Elsevier, Amsterdam New York

Lopes da Silva FH, Witter MP, Boeijinga PH, Lohman AH (1990) Anatomic organization and physiology of the limbic cortex. Physiol. Rev. 70: 453–511

Lopes da Silva, F.H. (1992) The rhythmic slow activity (theta) of the limbic cortex: an oscillation in search of a function. In: E. Başar, T. H. Bullock (eds.) Induced Rhythms in the Brain. Birkhäuser, Boston, pp. 269–283

Lopes da Silva, F.H., van Rotterdam, A., Storm van Leeuwen, W., Tielen, A.M. (1970a) Dynamic characteristics of visual evoked potentials in the dog. I. Cortical and subcortical potentials evoked by sine wave modulated light. Electroencephalogr. Clin. Neurophysiol. 29: 246–259

Lopes da Silva, F.H., van Rotterdam, A., Storm van Leeuwen, W., Tielen, A.M. (1970b) Dynamic characteristics of visual evoked potentials in the dog. II. Beta frequency selectivity in evoked potentials and background activity. Electroencephalogr. Clin. Neurophysiol. 29: 260–268

Lopes da Silva FH, Van Lierop THMT, Schrijier CF, Storm van Leeuwen W (1973a) Organization of thalamic and cortical alpha rhythms: spectra and coherences. Electroencephalogr. Clin. Neurophysiol. 35: 626–639

Lopes da Silva FH, Van Lierop THMT, Schrijier CF, Storm van Leeuwen W (1973b) Essential differences between alpha rhythms and barbiturate spindles: spectra and thalamo-cortical coherences. Electroencephalogr. Clin. Neurophysiol. 35: 641–645

Lopes da Silva, F.H. Arnolds, D.E.A.T. (1978) Physiology of the hippocampus and related structures. Annu. Rev. Physiol. 36: 291–301

Lopes da Silva, F.H., Arnolds, D.E.A.T., Neijt, H.C. (1984) A fuctional link between the limbic cortex and ventral striatum: physiology of the subiculum-accumbens pathway. Exp. Brain Res. 55: 205–214

Lopez L, Sannita WG (1997) Magnetically recorded oscillatory responses to luminance stimulation in man. Electroencephalogr. Clin. Neurophysiol. 104: 91–95

Lynch, J.C., Mountcastle, V.B., Talbot, W.H., Yin, T.C.H. (1977) Parietal lobe mechanisms for directed visual attention. J. Neurophysiol. 40: 362–389

Lutzenberger, W., Pulvermüller, F., Elbert, T., Birbaumer, N. (1995) Visual stimulation alters local 40 Hz responses in humans: an EEG study. Neurosci. Lett. 183: 1–4

Macchi, G. (1983) Old and new anatomo-functional criteria in the subdivision of the thalamic nuclei. In: Macchi, G., Rustioni, A., Spreafico, R. (eds.) Somatosensory Integration in the Thalamus. Elsevier, Amsterdam, pp. 3–16

Mangun, G.R. (1992) Human visual evoked potentials: induced rhythms or separable components. In: Başar, E., Bullock, T.H. (eds.) Induced Rhythms in the Brain. Birkhäuser, Boston, MA. pp. 217–231

Mansbach, R.S., Lorenz, D.N. (1983) Cholecystokinin (CCK-8) elicits prandial sleep in rats. Physiol. Behav. 30: 179–183

Mantyh, C.R., Mantyh, P.W. (1985) Differential localization of cholecystokinin-8 binding sites in the rat vs. the guinea pig brain. Europ. J. Pharmacol. 113: 137–139

Markowitsch, H.J., Emmans, D., Irle, E., Streicher, M., Preilowsky, B. (1985) Cortical and subcortical afferent connections of the primate's temporal pole: A study of rhesus monkeys, squirrel monkeys, and marmosets. J. Comp. Neurol. 242: 425–458

Mast, T.E., Watson, S.C. (1968) Attention and auditory evoked responses to low detectability signals. Percept. Psychophysics 4: 237–240

McCallum, W.C. (1980) Brain slow potential changes elicited by missing stimuli and by externally paced voluntary responses. Biol. Psychol. 11: 7–19

Meredith, M., Stein, B. (1983) Interactions among converging sensory inputs in the superior colliculus. Science 221: 389–391

Meredith, M., Stein, B. (1985) Descending efferent from the superior colliculus relay integrated multisensory information. Science 227: 657–659

Mesulam, M.-M. (1995) Cholinergic pathways and the ascending reticular activating system of the human brain. Ann. N. Y. Acad. Sci. 757: 169–179

Mesulam, M.-M. (1990) Large scale neurocognitive networks and distributed processing for attention, language, and memory. Ann. Neurol. 28: 597–613

Mesulam, M.-M. (1994) Neurocognitive networks and selectively distributed processing. Rev. Neurol. (Paris) 150: 564–569

Mesulam, M-M., Van Hoesen, G.W., Pandya, D.N., Geschwind, N. (1977) Limbic and sensory connections of the inferior parietal lobule (area PG) in the rhesus monkey: a study with a new method for horseradish peroxidase histochemistry. Brain Res., 136: 395–414

Michalewski J.J., Rosenberg, C., Starr, A. (1986) Event-related potentials in dementia. In: Cracco, R.Q., Bodis-Wollner, I. (eds.) Evoked Potentials. Liss, New York, pp. 521–528

Mikkelsen, K.B., Saermark, K., Lebech, J., Bak, C., Başar, E. (1989) Selective averaging in auditory magnetic field experiments. In: Williamson SJ, Hoke M, Stroink G, Kotani M, eds: Advances in Biomagnetism. Plenum, New York, pp. 93–96

Miller, R. (1991) Cortico-hippocampal Interplay and the Representation of Contexts in the Brain. Berlin Heidelberg: Springer

Mizuki, Y., Masotoshi, T., Isozaki, H., Nishijima, H., Inanaga, K. (1980) Periodic appearence of theta rhythm in the frontal midline area during performance of a mental task. Electroencephalogr. Clin. Neurophysiol., 49: 345–351

Mizuki, Y., Ushijima, I., Habu, K., Nakamura, K. Yamada, M. (1988) Effects of ceruletide on clinical symptoms and EEGs in schizophrenia. Progr. Neuropsychopharmacol. Biol. Psychiat. 12: 511–522

Montgomery, S.A. Green, M.D. (1988) The use of cholecystokinin in schizophrenia: A review. Psycholog. Med. 18: 593–604

Moran, M.A., Mufson, E.J., Mesulam, M.-M. (1987) Neural inputs into the temporal cortex of the rhesus monkeys. J. Comp. Neurol., 256: 88–103

Motter, B.C., Mountcastle, V.B. (1981) The functional properties of the light-sensitive neurons of the posterior prietal cortex studied in waking monkeys: foveal sparing and opponent vector organization. J. Neurosci. 1: 3–26

Mountcastle, V.B., Lynch, J.C., Georgopoulos, A., Sakata, H., Acuna, C. (1975) The posterior parietal association cortex of the monkey. Command functions for operations within extrapersonal space. J. Neurophysiol. 38: 871–908

Mountcastle, V.B. (1992). Preface. In: Başar, E., Bullock, T.H. (eds.) Induced Rhythms in the Brain. Birkhäuser, Boston, MA, pp. 217–231

Mountcastle, V.B., Motter, B.C., Steinmetz, M.A., Duffy, C.J. (1984) Looking and seeing: the visual functions of the parietal lobe. In: Edelman, G.M., Gall, W.E., Cowan, W.M. (eds.) Dynamic Aspects of Neocortical Function. Wiley, New York, pp. 159–193

Mountcastle, V.B., Andersen, R.A., Motter, B.C. (1981) The influence of attentive fixation upon the excitability of the light-sensitive neurons of the posterior parietal cortex. J. Neurosci. 1: 1218–1225

Mulholland, T. (1995) Human EEG, behavioral stillness and biofeedback. Int. J. Psychophysiol. 19:263–279

Munk, M.H., Roelfsema, P.R., König, P., Engel, A.K., Singer, W. (1996) Role of reticular activation in the modulation of intracortical synchronization. Science 272: 271–274

Näätänen, R. (1982) Processing negativity: An evoked-potential reflection of selective attention. Psych. Bull. 92: 605–640

Näätänen, R. (1988) Implications of ERP data for psychophysiological theories of attention. In: Renault, B., Kutas, M., Coles, M.G.H., Gaillard, A.W.K. (eds.) Event-related potential investigations of cognition. Elsevier, Amsterdam, pp. 117–163.

Näätänen, R. (1990) The role of attention in auditory information processing as revealed by event related potentials and other brain measures of cognitive function. Behav. Brain Sci. 13: 201–288.

Näätänen, R., Teder, W., Alho, K., Lavikainen, J. (1992) Auditory attention and selective input modulation: a topographical ERP study. Neuroreport 3: 493–496

Näätänen, R., Mitchie, P.T.: (1979) Different variants of endogenous negative brain potentials in performance situations. In: Callaway, E., Lehmann, D. (eds.) Human Evoked Potentials: Applications and Problems. Plenum, New York

Nair, N.P.V., Lal, S., Bloom, D.M. (1986) Cholecystokinin and schizophrenia. Progr. Brain Res. 65: 237–258

Nakamura, K., Toshima, T., Takemura, I. (1986) The comparative and developmental study of auditory information processing in autistic adults. J. Autism Dev. Disord. 16: 105–118

Narici, L., Pizzella, V., Romani, G.L., Torrioli, G., Traversa, R., Rossini, P.M. (1990) Evoked alpha and mu rhythm in humans: a neuro-magnetic study. Brain Res. 520: 222–231

Nauta, W.J.H. (1971) The problem of the frontal lobe. A reinterpretation. J. Psychiatr. Res. 8: 167–187

Nauta, W.J.H. (1972) Neural association of the frontal cortex. Acta Neurobiol. Exp. 32:125–140

Niehoff, D.L. (1989) Quantitative autoradiographic localization of cholecystokinin receptors in rat and guinea pig brain using 125I-Bolton-Hunter-CCK-8. Peptides 10: 265–274

Nogawa, T., K. Katayama, Y. Tabata, T. Ohshio, and T. Kawahra. (1976) Changes in amplitude of the EEG induced by a photic stimulus. Electroencephalogr. Clin. Neurophysiol. 40: 78–88

Nunez A, Garcia-Austt E, Buno W (1987) Intracellular theta rhythm generation in identified hippocampal pyramids. Brain Res. 416: 289–300

O'Connor, T.A., Starr, A. (1985) Intracranial potentials correlated with an event-related potential, P300, in the cat. Brain Res. 339: 27–38

O'Keefe, J. (1976) Place units in the hippocampus of the freely moving rat. Exp. Neurol. 51: 78–109

O'Keefe, J., Nadel, L. (1978) The hippocampus as a cognitive map. Clarendon Press, Oxford

O'Keefe, J. (1979) A review of the hippocampal place cells. Prog. Neurobiol. 13: 419–439

O'Keefe, J., Dostrovsky, J. (1971) The hippocampus as a spatial map. Preliminary evidence from unit activity in the freely-moving rat. Brain Res. 34: 171–175

Olds, J., Mink, W.D., Best, P.J. (1969) Single unit patterns during anticipatory behavior. Electroencephalogr. Clin. Neurophysiol. 26: 144–58

Ohman, A., Lader, M. (1977) Short-term changes of the human auditory evoked potentials during repetive stimulation. In Desmedt, J.E. (ed.) Progress in Clinical Neurophysiology, Vol.2. Karger, Basel

Olpe, H.R., Klebs, K., Küng, E., Campiche, P., Glatt, A., Ortmann, R., D'Amato, F., Pozza, M.F. Mondadori, C. (1987) Cholinomimetics induce theta-rhythm and reduce hippocampal pyramidal cell excitability. Eur. J. Pharmacol. 142: 275–283

Özesmi, C., Basar, E. (1974) Dynamics of potentials evoked in the auditory pathway and reticular formation of the cat. Studies during waking and sleep stages. Kybernetik 16: 27–35

Osaka M (1984) Peak alpha frequency of EEG during a mental task: task difficulty and hemispheric differences. Psychophysiol. 21: 101–105

Paller, K.A., Zola-Morgan, S., Squire, L.R., Hillyard, S.A. (1988) P3-like brain waves in normal monkeys and monkeys with medial temporal lesions. Behav. Neurosci. 102: 714–725

Pandya, D.N., Kuypers, H.G.J.M. (1969) Cortico-cortical association connections in the rhesus monkey. Brain Res., 13: 13–36

Pandya, D.N., Van Hoesen, G.W., Mesulam, M-M. (1981) Efferent connections of the cingulate gyrus in the rhesus monkey. Exp. Brain Res. 42: 319–330

Pantev, C., Elbert, T. (1994) The transient auditory evoked gamma-band field. In Pantev, C., Elbert, T., Lütkenhöner, B. (eds.) Oscillatory Event-Related Brain Dynamics. Plenum, New York, pp. 219–242

Pantev, C., Elbert, T., Lütkenhöner, B., eds (1994) Oscillatory Event-Related Brain Dynamics. Plenum, New York London

Pantev, C., Makeig, S., Hoke, M., Galambos, R., Gallen, C., Hampson, S. (1991) Human auditory evoked gamma band magnetic fields. Proc. Natl. Acad. Sci. USA, 88: 8996–9000

Parnefjord, R., Başar, E. (1995) Delta oscillations as a correlate to acoustical perception at the auditory threshold. Electroencephalogr. Clin. Neurophysiol. 94: 43P

Parnefjord, R. (1996) Untersuchungen zur Dynamik der evozierten Potentiale an der Hörschwelle. Med. Diss., Medizinische Universität zu Lübeck

Pavlov, I.P. (1927) Conditioned Reflexes: An Investigation of the Physiological Activity of the Cerebral Cortex. Oxford University Press, London

Petersen, S.E., Fox, P.T., Miezin, F.M., Raichle, M.E. (1988) Modulation of cortical visual responses by direction of spatial attention measured by PET. Assoc. Res. Vision Ophthal. p. 22

Petsche, H., Rappelsberger, P., Pockberger, H. (1987) EEG-Veränderungen beim Lesen. In: Weinmann, H.M. (ed.) Zugang zum Verständnis höherer Hirnfunktionen durch das EEG. Zuckschwerdt, München, pp. 59–74

Petsche, H., Stumpf, C., Gogolak, G. (1962). The significance of the rabbit's septum as a relay station between the midbrain and the hippocampus. 1. The control of hippocampal arousal activity by the septum cells. Electroencephalogr. Clin. Neurophysiol. 14: 202–211

Petsche, H., Rappelsberger, P. (1992) Is there any message hidden in the human EEG? In: Basar, E., Bullock, T.H. (eds.) Induced Rhythms in the Brain. Birkhäuser, Boston, MA, pp. 103–116

Pfurtscheller, G. (1988) Mapping of event-related desynchronization and type of derivation. Electroencephalogr. Clin. Neurophysiol. 70: 190–193

Pfurtscheller, G., Neuper, C. (1992) Simultaneous EEG 10 Hz desynchronization and 40 Hz synchronization during finger movements. NeuroReport 3: 1057–1060

Pfurtscheller, G., Neuper, C., Andrew, C., Edlinger, G. (1997). Foot and hand area mu rhythms. Int. J. Psychophysiol. 26: 121–135

Pfurtscheller G, Stancak A Jr., Neuper C (1996) Post-movement beta synchronization. A correlate of an idling motor area? Electroencephalogr. Clin. Neurophysiol. 98: 281–293

Picton, T.W. (1970) Amplitude of evoked responses to tones of high intensity. Acta. Otolaryngol. (Stockh.) 70: 77–82

Picton, T.W. (1992) The P300 wave of the human event-related potential. J. Clin. Neurophysiol. 9: 456–479

Picton, T.W., Woods, D., Proulx, G.B. (1978) Human auditory sustained potentials. II. Stimulus relationships. Electroencephalogr. Clin. Neurophysiol. 45: 198–210

Picton, T.W., Stuss, D.T. (1980) The component structure of the human event-related potentials. In: Kornhuber, H.H., Deecke, L. (eds.) Motivation, Motor and Sensory Processes of the Brain – Electrical Potentials, Behaviour and Clinical Use. Elsevier, Amsterdam, pp. 17–50

Pineda, J.A., Foote, S.L., Neville, H.J. (1987) Long latency event-related potentials in squirrel monkeys: further characterization of wave form morphology, topographic and functional properties. Electroencephalogr. Clin. Neurophysiol. 67: 77–90

Platt, C. J., Bullock, T. H., Czeh, G., Kovacevic, N., Konjevic, Dj. and Gojkovic, M. (1974) Comparison of electroreceptor, mechanoreceptor and optic evoked potentials in the brain of some rays and sharks. J. Comp. Physiol. 95: 323–355.

Polich, J., Kok, A. (1995) Cognitive and biological determinants of P300: An integrative review. Biol. Psychol. 41: 103–146

Porjesz, B., Begleiter, H., Samuelly, I. (1980) Cognitive deficits in chronic alcoholics and elderly subjects assessed by evoked brain potentials. Acta Psychiatr. Scand. 62: 15–29

Pritchard, W.S. (1981) Psychophysiology of P300. Psychol. Bull. 89: 506–540

Prosser, C.L. (1938) Evidence for chemical control of "spontaneous" activity of isolated ganglia. Am. J. Physiol. 123: 165.

Pulvermüller, F., Preissl, H., Eulitz, C., Pantev, C., Lutzenberger, W., Elbert, T., Birbaumer, N. (1994) Brain rhythms, cell assemblies and cognition: Evidence from the processing of words and pseudowords. psycoloquy.94.5.48.brain-rhythms.1

Pulvermüller, F., Lutzenberger, W., Preissl, H., Birbaumer, N. (1995) Spectral responses in the gamma band: physiological signs of higher cognitive processes? NeuroReport 6: 2059–2064

Radil-Weiss T, Zernicki B, Michalski A (1976) Hippocampal theta activity in the acute pretrigeminal cat. Acta Neurobiol. Exp. 36: 517–534

Radulovacki M, Adey WR (1965) The hippocampus and the orienting reflex. Exp. Neurol. 12: 68–83

Rahn E, Başar E (1993a) Prestimulus EEG-activity strongly influences the auditory evoked vertex responses: A new method for selective averaging. Int. J. Neurosci. 69: 207–220

Rahn E, Basşar E. (1993b) Enhancement of visual evoked potentials by stimulation during low prestimulus EEG stages. Int. J. Neurosci. 72: 123–136

Ranck, J.B. Jr. (1973) Studies on single neurons in dorsal hippocampal formation and septum in unrestrained rats. I. Behavioral correlates and firing repertoires. Exp. Neurol. 41: 461–531

Regan, D. (1966a) An effect of stimulus color on average steady-state potentials evoked in man. Nature 210: 1056

Regan, D. (1966b) Same characteristic of average steady-state and transient responses evoked by modulated light. Electroencephalogr. Clin. Neurophysiol. 20: 238–248

Regan, D. (1989) Human Brain Electrophysiology: Evoked Potentials and Evoked Magnetic Fields in Science and Medicine. Elsevier, Amsterdam

Ribary, U., Ioannides, A.A., Singh, K.D., Hasson, R., Bolten, J.P.R., Lado, F., Mogilner, A., Llinás, R.R. (1992) Magnetic field tomography of coherent thalamocortical 40 Hz oscillations in humans. Proc. Natl. Acad. Sci. USA 88: 11037–11041

Ritter, W., Vaughan, H.G., Costa, L.D. (1968) Orienting and habituation to auditory stimuli: a study of short term changes in averaged evoked responses. Electroencephalogr. Clin. Neurophysiol. 25: 550–556

Röschke, J. (1988) Untersuchungen zur elektrischen Aktivität des Zentralnervensystems von Invertebraten am Beispiel von *Helix pomatia*. Med. Diss. Medizinische Universität zu Lübeck

Röschke, J., Başar, E. (1988) The EEG is not a simple noise: strange attractors in intracranial structures. In: Başar, E. (ed.) Dynamics of Sensory and Cognitive Processing by the Brain. Springer, Berlin Heidelberg New York, pp. 203–216

Röschke, J., Başar, E. (1989) Correlation dimensions in various parts of cat and human brain in different states. In: Başar, E., Bullock, T.H. (eds.), Brain Dynamics – Progress and Perspectives. Springer, Berlin Heidelberg New York, pp. 131–148

Röschke, J. (1988) Untersuchung zur elektrischen Aktivität des Zentralnervensystems von Invertebraten am Beispiel von Helix pomatia. Med. Diss., Medizinische Universität zu Lübeck

Röschke, J., Mann, K., Riemann, D., Frank, C., Fell, J. (1995) Sequential analysis of the brain's transfer properties during consecutive REM episodes. Electroencephalogr. Clin. Neurophysiol. 96: 390–397

Rösler, D., Ruchkin, D., Siddle, D. (1984) Cognition and event-related potentials. II. The orienting reflex and P300. Ann. N. Y. Acad. Sci. 425: 39–57

Robinson, D.L., Goldberg, M.E., Staunton, G.B. (1978) Parietal association cortex in the primate: Sensory mechanisms and behavioral modulations. J. Neurophysiol. 41: 910–932

Rose, G.M., Dunwiddie, T.V. (1986) Induction of hippocampal long-term potentiation using physiologically patterned stimulation. Neurosci. Lett. 69: 244–248

Roth, W.T. (1973) Auditory evoked resonses to unpredictable stimuli. Psychophysiol. 10: 125–137.

Sachs, L. (1992) Angewandte Statistik. Berlin Heidelberg New York: Springer

Saermark, K., Mikkelsen, K.B., Başar, E. (1992) Magnetoencephalographic evidence for induced rhythms. In: Başar, E., Bullock, T.H. (eds.) Induced Rhythms in the Brain. Birkhäuser, Boston, MA, pp. 129–145

Sakowitz, O., Başar, E. (1996) Fronto-central theta response marks association of bimodally applied stimuli. In: Elsner, N., Schnitzler, H. (eds.) Göttingen Neurobiology Report 1996, Vol. 2. Thieme, Stuttgart New York, p. 777

Sato, S, ed. (1990) Magnetoencephalography. Raven Press, New York (Advances in Neurology, Vol. 54)

Samar, V.J., Swartz, K.P., Raghuveer, M.R. (1995) Multiresolution analysis of event-related potentials by wavelet decomposition. Brain Cogn. 27: 398–438

Satterfield, J.H. (1967) Evoked cortical response enhancement and attention in man. A study of responses to auditory and shock stimimuli. Electroencephalogr. Clin. Neurophysiol. 19: 470–475

Sayers, B.McA., Beagley, H.A. (1974a) Objective evaluation of auditory evoked EEG responses. Nature 251: 608–609

Sayers, B.McA., Beagley, H.A., Henshall, W.R. (1974b) The mechanism of auditory evoked EEG responses. Nature 274: 481–483

Sayers, B.McA., Beagley, H.A., Riha, J. (1979) Pattern analysis of auditory evoked EEG potentials. Audiology. 18: 1–16

Scheibel, M.E., Scheibel, A.B. (1966) Patterns of organization in specific and non-specific thalamic fields. In: Purpura, D.D., Yahr, M.D. (eds.) The Thalamus. Columbia University Press, New York

Schmalz, E. (1914) Zur Morphologie des Nervensystems von *Helix pomatia*. Z. wiss. Zool. 111: 506–568

Schoenberg, I. J. (1969) Cardinal Interpolation and Spline Functions. J. Approx. Theory, 2: 167–206

Schreiner, C.E. (1991) Functional topography in the primary auditory cortex of the cat. Acta Otolaryngol. (Stockh.) Suppl. 491: 7–16

Schultz (1981) A History of Modern Psychology. New York: Academic Press

Schürmann, M., Başar, E. (1994) Topography of alpha and theta oscillatory responses upon auditory and visual stimuli in humans. Biol. Cybern. 72: 161–174

Schürmann, M., Başar-Eroğlu, C., Rahn, E., Braasch, M., Dössel, O., Fuchs, M., Başar, E. (1992) A comparative study of alpha responses in human MEG temporo-parietal and occipital recordings and cat intracranial EEG recordings. Proceedings of IEEE Satellite Symposium on Neuroscience and Technology, pp. 132–137

Schürmann, M., Warecka, K., Basşar-Eroğlu, C., Basşar, E. (1993) Auditory evoked potentials in multiple sclerosis: alpha responses are reduced in amplitude, but theta responses are not altered. Int. J. Neurosci. 73: 259–276

Schürmann, M., Basşar-Eroğlu, C., Kolev, V., Basşar, E. (1995) A new metric for analyzing single-trial event-related potentials. Application to human visual P300 delta response. Neurosci. Lett. 197: 167–170

Schürmann M, Başar-Eroglu C, Başar E (1997) Gamma responses in the EEG: elementary signals with multiple functional correlates. NeuroReport 8: 1793–1796

Schütt, A., Başar, E. (1992) The effect of acetylcholine, dopamine and noradrenaline on the visceral ganglion of helix pomatia – II. Stimulus evoked field potentials. Comp. Biochem. Physiol. 102C: 169–176

Schütt, A., Başar, E., Bullock, T.H. (1992) The effect of acetylcholine, dopamine and noradrenaline on the visceral ganglion of helix pomatia – I. Ongoing compound field potentials of low frequencies. Comp. Biochem. Physiol. 102C: 159–168

Schütt A., Başar, E. (1993) Evoked field potentials in the visceral ganglion of *Helix pomatia*: serotonin (5-HT) increases 40 Hz response. 38. Jahrestagung der Deutschen EEG-Gesellschaft, Augsburg

Schweizer, J. (1986) Functional organization of the electroreceptive midbrain in an elasmobranch (*Platyrhinoidis triseriata*) a single unit study. J. Comp. Physiol. 158: 43–58.
Schwertfeger, W.K. (1979) Direct efferent and afferent connections of the hippocampus with the neocortex in the marmoset monkey. Am. J. Anat. 156: 77–82
Segal, M., Olds, J. (1972) Behavior of units in hippocampal circouit of the rat during learinig. J. Neurophysiol. 35: 680–690
Sekiguchi, R. Moroji, T. (1986) A comparative study on characterization and distribution of cholecystokinin binding sites among the rat, mouse and guinea pig brain. Brain Res. 399: 271–281
Shagass, C. (1972) Evoked potentials in psychiatry. Plenum, New York
Shallice, T. (1982) Specific impairments of planning. Philos. Trans. R. Soc. Lond. B. Biol. Sci. 298: 199–209
Shaw JC (1996) Intention as a component of the alpha-rhythm response to mental activity. Int. J. Psychophysiol. 24: 7–23
Sheer, D.E. (1976) Focused arousal and 40 Hz-EEG. In: Knight, R.M., Bakker, D.J. (eds.) The Neuropsychology of Learning Disorders. University Park Press, Baltimore, MD, pp. 71–87
Sheer, D.E. (1984) Focused arousal, 40 Hz EEG, and dysfunction. In: Elbert, T., Rockstroh, B., Lutzenberger, W., Birbaumer, N. (eds.) Self-Regulation of the Brain and Behaviour. Springer, Berlin Heidelberg New York, pp. 64–84
Sheer, D.E. (1989) Sensory and cognitive 40 Hz event-related potentials: behavioural correlates, brain function, and clinical application. In: Başar, E., Bullock, T.H. (eds.) Brain Dynamics. Springer, Berlin Heidelberg New York, pp. 339–374 (Springer Series in Brain Dynamics, Vol. 2)
Shepherd, G.M., (1988) Neurobiology. Oxford University Press
Sherrington, C. (1948) The Integrative Action of the Nervous System. Cambridge University Press
Silva, L.R., Amitai, Y., Connors, B.W. (1991) Intrinsic oscillations of neocortex generated by layer 5 pyramidal neurons. Science 251: 432–435
Simson, R., Vaughan, H.G., Ritter, W. (1976) The scalp topography of potentials associated with missing visual or auditory stimuli. Electroencephalogr. Clin. Neurophysiol. 40: 33–42
Sinton, C. (1988) Cholecystokinin and cholecystokinin antagonists enhance postsynaptic excitability in the dentate gyrus. Peptides 9: 1049–1053
Smith, E.M., E. Halgren, M. Sokolik, P. Baudena, A. Musolino, C. Liegeois-Chauvel, and P. Chauvel (1990) The intracranial topography of the P3 event related potential elicited during auditory oddball. Electroencephalogr. Clin. Neurophysiol. 76: 235–248
Smidt, H. (1900) Über die Darstellung der Begleit- und Gliazellen im Nervensystem von *Helix* mit der Golgi-Methode. Arch. mikr. Anat. 55: 300–313
Snider, R.S. Niemer, W.T. (1964) A Stereotaxic Atlas of the Cat Brain. University Chicago Press, Chicago
Sokolov, E.N. (1975) Neuronal mechanisms of the orienting reflex. In: Sokolov, E.N., Vinogradova, O.S. (eds.) Neuronal Mechanisms of the Orienting Reflex, pp. 217–238 (English translation: N. E. Weinberger, ed. Lawrence Erlbaum, Hillsdale, NJ and John Wiley, New York)
Spekreijse, H. (1966) Analysis of EEG responses in man evoked by sinewave-modulated light. Thesis, Univ. Amsterdam. Junk, The Hague
Squires, N.K., Squires, K.C., Hillyard, S.A. (1975) Two varieties of long-latency positive waves evoked by unpredictable auditory stimuli in man. Electroencephalogr. Clin. Neurophysiol. 38: 387–401

Squires, N.K., Donchin, E., Squires, K., Grossberg, S. (1977) Biosensory stimulation: inferring decision-related processes from P300 component. J. Exp. Psychol. 3: 299–315

Squire, L.R. (1992) Memory and the hippocampus: A synthesis from findings with rats, monkeys, and humans. Psychol. Rev. 99: 195–231

Stacher, G., Bauer, H., Steinringer, H. (1979) Cholecystokinin decreases appetite and activation evoked by stimuli arising from the preparation of a meal in man. Physiol. Behav. 23: 325–331

Stacher, G., Steinringer, H., Schmierer, G., Schneider, C. Winklehner, S. (1982a) Ceruletide decreases food intake in non-obese man. Peptides 3: 607–612

Stacher, G., Steinringer, H., Schmierer, G., Winklehner, S. Schneider, C. (1982b) Ceruletide increases threshold and tolerance to experimentally induced pain in healthy man. Peptides 3: 955–962

Stadler, M., Dieker, J. (1972) Untersuchungen zum Problem virtueller Konturen in der visuellen Wahrnehmung. Z. Exp. Angew. Psychol. 19: 325–350

Stampfer, H.G., Başar, E. (1985) Does frequency analysis lead to better understanding of human event related potentials. Int. J. Neurosci. 26: 181–196

Stampfer, H.G., Basar, E. (1988) An analysis of preparation and response activity in P300 experiments in humans. In: Basar, E. (ed.). Dynamics of Sensory and Cognitive Processing by the Brain. Springer, Berlin Heidelberg New York, pp. 275–286

Stancak, A. Jr., Pfurtscheller, G., (1996) The effects of handedness and type of movement on the contralateral preponderance of mu-rhythm desynchronisation. Electroencephalogr. Clin. Neurophysiol. 99: 174–182

Stein, B., Meredith, M., Wallace, M. (1993) The visually responsive neuron and beyond: multisensory integration in cat and monkey. Prog. Brain Res. 95: 79–90

Steriade, M., Gloor, P., Llinas, R.R., Lopes da Silva, F.H., Mesulam, M.-M. (1990) Basic mechanisms of cerebal rhythmic activities. Electroencephalogr. Clin. Neurophysiol. 76: 481–508

Stern L (1985) The Structures and Strategies of Human Memory. Belmont, CA: Dorsey Press

Sternberg, R.J., ed. (1985) Human Abilities: An Information Processing Approach. Freeman, New York.

Stowell, H. (1987) Cerebral slow waves and time parsing. Int. J. Neurosci. 32: 861–874

Stryker, M.P. (1989) Is grandmother an oscillation? Nature 338: 297–298

Sturbeck, K. (1988) Vergleichende Analyse evozierter Potentiale von Invertebraten und Vertebraten. Med. Diss., Medizinische Universität zu Lübeck

Sutton, S., Braren, M., Zubin, J., John, E.R. (1965) Evoked potential correlates of stimulus uncertainty. Science 150: 1187–1188

Sutton, S., Ruchkin, D.S. (1983) The late positive complex. Advances and new problems. Ann. N.Y. Acad. Sci. 425: 1–23

Sutton, S., Tueting, P., Zubin, J., John, E.R. (1967) Information delivery and the sensory evoked potential. Science 155: 1436–1439

Swanson, L.W. (1981) A direct projection from ammon's horn to the prefrontal cortex in the rat. Brain Res. 217: 150–154

Swanson, L. (1983)The Hippocampus and the concept of the limbic system. In: Seifert, W. (ed.) Neurobiology of the Hippocampus. Academic Press, London, pp. 1–19

Szenthágothai, J. (1983) The local neuronal apparatus of the cerebral cortex. In: Buser, P., Rougeul-Buser, A. (eds.) Cerebral Correlates of Conscious Experience. Elsevier, Amsterdam

Takasaka, Y. (1985) Expectancy-related cerebral potentials associated with voluntary time estimation and omitted stimulus. Folia Psychiatr. Neurol. Jpn. 39: 167–172

Tallon, C., Bertrand, O., Bouchet, P., Pernier, J. (1995) Gamma-range activity evoked by coherent visual stimuli in humans. Eur. J. Neurosci. 7: 1285–1291

Tamminga, C.A., Littman, R.L., Alphs, L.D., Chase, T.N., Thaker, G.K. Wagman, A.M. (1986) Neuronal cholecystokinin and schizophrenia: Pathogenic and therapeutic studies. Psychopharmacol. 88: 387–391

Thompson, R.F. (1967) Foundations of Physiological Psychology. Harper and Row, New York

Thompson, R.F. (1982) Training-induced increase in hippocampal unit activity. In: Swanson, L.W., Teyler, T.J., Thompson, R.F. (eds.) Hippocampal Long-Term Potentiation: Mechanisms and Implications for Memory. Neurosci. Res. Progr. Bull. 20: 710–722

Thompson, R.F., Berger, T.W., Berry, S.D., Hoehler, F.K., Kettner, R.E., Weisz, D.J. (1980) Hippocampal substrate of classical conditioning. Physiol. Psychol. 8: 262–279

Tiberghien, G. (1989) Models of human memory. Arch. Gerontol. Geriatr. Suppl. 1: 37–53

Tiihonen, J., Hari, R., Kajola, M., Ahlfors, S., Tissari, S. (1991) Magnetoencephalographic 10-Hz rhythm from the human auditory cortex. Neurosci. Lett. 129: 303–305

Tiitinen, H., Sinkkonen, J., Reinikainen, K., Alko, K., Lavikainen, J., Näätänen, R. (1993) Selective attention enhances the auditory 40-Hz transient response in humans. Nature 364: 59–60

Tömböl, T. (1967) Short neurons and their synaptic relations in the specific thalamic nuclei. Brain Res. 3: 307–326

Tononi, G., Sporns, O., Edelman, G.M. (1992) The problem of neural integration; Induced rhythms and short-term correlation. In: Basar, E., Bullock, T.H. (eds.) Induced Rhythms in the Brain. Birkhäuser, Boston, pp. 365–393

Trulson, M.E. Preussler, D.W. (1984) Dopamine-containing ventral tegmental area neurons in freely moving cats: Activity during the sleep-waking cycle and effects of stress. Exp. Neurol. 83: 367–377

Tsuchiya, T., Tami, S. Fukushima, H. (1979) Analysis of the dissociation between neocortical and hippocampal EEG activity induced by neuroleptics. Psychopharmacol. 63: 179–185

Tulving E (1983) Elements of Episodic Memory. Oxford: Clarendon Press

Twarog, B.M., Roeder, K. D. (1957) Pharmacological observations on the desheathed last abdominal ganglion of the cockroach. Ann. ent. Soc. Amer. 50: 231–237

Ursin, R. (1971) Differential effect of sleep deprivation on the two slow wave sleep stages in the cat. Acta Physiol. Scand. 83: 352–361

Van der Tweel, L.H., Verduyn Lunel, H.F.E. (1965) Human visual response to sinusoidally modulated light. Electroencephalogr. Clin. Neurophysiol. 18: 587–598

Van der Tweel, L.H., Spekreijse, H. (1969) Signal transport and rectification in the human evoked response system. Ann. N.Y. Acad. Sci. 156: 678–695

Van Dijk, A., Richards, J.G., Trzeciak, A., Gillesen, D. Möhler, H. (1984) Cholecystokinin receptors: biochemical demonstration and autoradiographical localization in rat brain and pancreas using [3H]cholecystokinin as radioligand. J. Neurosci. 4: 1021–1033

Vanderwolf, C.H. (1969) Hippocampal electrical activity and voluntary movement in the rat. Electroencephalogr. Clin. Neurophysiol. 26: 407–418

Vanderwolf, C.H. Robinson, T.E. (1981) Reticulo-cortical activity and behavior: A critique of the arousal theory and a new synthesis. Behav. Brain. Sci. 4: 459–514

Vanderwolf, C.H., Kramis, R., Gillespie, L.A., Bland, B.H. (1975) Hippocampal Rhythmic Slow Activity and Neocortical Low-Voltage Fast Activity: Relations to Behavior. In: Isaacson, R.L., Pribram, K.H. (eds.) The Hippocampus, Vol. 2. Plenum Press, New York, pp. 101–128

Vanderwolf, C.H., Kramis, R., Robinson, T.E. (1978) Hippocampal electrical activity during waking behaviour and sleep: analyses using centrally acting drugs. In: Functions of the Septo-Hippocampal System. Elsevier, Amsterdam, pp. 199–221 (CIBA Symposium No 58)

Vanderwolf, C.H., Leung, LWS (1983) Hippocampal rhythmic slow activity: a brief history and the effects of entorhinal lesions and phencyclidine. In: Seifert W (ed.) Neurobiology of the Hippocampus. Academic Press, London, pp. 275–302

Vaughan, H.G., Ritter, W. (1970) The sources of auditory evoked responses recorded from the human scalp. Electroencephalogr. Clin. Neurophysiol. 28: 360–67.

Vvedensky V, Hari R, Ilmoniemi R, Kajola M (1985) Physical basis of neuromagnetic fields. Biophys. (in Russian) 30: 154–158

Vertes RP (1982) Brain stem generation of the hippocampal EEG. Prog. Neurobiol. 19: 159–186

Vinogradova, O.S. (1975) The hippocampus and the orienting reflex. In: Sokolov, E.N., Vinogradova, O.S. (eds.) Neuronal Mechanisms of the Orienting Reflex, pp. 128–154 (English translation: N. E. Weinberger, ed. Lawrence Erlbaum, Hillsdale, NJ and John Wiley, New York)

Vinogradova OS (1995) Expression, control, and probable functional significance of the neuronal theta-rhythm. Prog. Neurobiol. 45: 523–583

Vinogradova, O.S., Dudaeva, K.I. (1972) Comparator function of the hippocampus. Dokl. Akad. Nauk SSSR, 202: 486–489

Vinogradova, O.S., Zolotukhina, L.I. (1972) Sensory characteristics of the neurons in the medial and lateral septal nuclei. Zh. Vyssh. Nerv. Deyat. 22: 1260–1269

Wallace, M., Meredith, M., Stein, B. (1992) Integration of multiple sensory modalities in cat cortex. Exp. Brain Res. 91: 484–488

Wallace, M.T., Stein, B.E. (1994) Cross-modal synthesis in the midbrain depends on input from cortex. J. Neurophysiol. 71: 429–432

Walter W.G. (1950) Normal rhythms: their development, distribution and significance. In: Hill, D., Parr, G. (eds.) Electroencephalography. McDonald, London

Walter, W.G. (1962). Oscillatory activity in the nervous system. In: Grenel, R.G. (ed.) Neural Physiopathology. Harper and Row, New York

Weinberg, H. (1991) Magnetoencephalography. In: Wong, P. (ed.) Introduction to Brain Topography. Plenum, New York, pp. 113–146

Weinberg H, Stroink G, Katila T, eds. (1985) Biomagnetism: Applications and Theory. Pergamon, New York

Westphal, K.P., Grözinger, B., Diekmann, V., Scherb, W., Reess, J., Leibing, U., Kornhuber, H.H. (1990) Slower theta activity over the midfrontal cortex in schizophrenic patients. Acta. Psychiatr. Scand., 81: 132–138

Wiener N (1955) Time and organization. In: Masani P, ed. (1985) Norbert Wiener: Collected Works, Vol. IV. MIT Press, Cambridge, MA

Wiener N (1956) Brain waves and interferometer. J. Phys. Soc. Jpn. 18: 499–507

Wiener N (1957) Rhythms in physiology with particular reference to electroencephalography. Proc. Rudolf Virchow Med. Soc. (NY) 16: 109–124

Williams, J.A., Gryson, K.A., McChesney, D.J. (1986) Brain CCK receptors: Species differences in regional distribution and selectivity. Peptides 7: 293–296

Williamson, SJ, Hoke, M., Stroink, G., Kotani, M., eds (1989) Advances in Biomagnetism. New York. Plenum Press

Williamson S.J., Kaufman, L. (1989) Advances in neuromagnetic instrumentation and studies of spontaneous brain activity. Brain Topogr. 2: 129–139

Wilder, M.B., Farley, G.R., Starr, A. (1981) Endogenous late positive component of the evoked potential in cats corresponding to P300 in humans. Science 211: 605–607

Wishaw, I.Q., Vanderwolf, C.H. (1973) Hippocampal EEG and behaviour: changes in amplitude and frequency of RSA (theta rhythm) associated with spontaneous and learned movement patterns in rats and cats. Behav. Biol. 8: 461–484

Woods, D.L. (1990) The physiological basis of selective attention: implication of event-related potential studies. In: Rohrbaugh, J.W., Parasuraman, R., Johnson R. Jr. (eds.) Event-related Brain Potentials – Basic Issues and Applications. Oxford University Press, Oxford, pp. 178–209

Wundt W (1874) Grundzüge der Physiologischen Psychologie. Engelmann, Leipzig

Wurtz, R.H., Richmond, B.J., Judge, S.J. (1980) Vision during saccadic eye movements. II. Visual interactions in monkey superior colliculus. J. Neurophysiol. 43: 1168–1181

Yin, T.C.T., Mountcastle, V.B. (1977) Visual input to the visuomotor mechanisms of the monkey's parietal lobe. Science 197: 1381–1383

Zarbin, M.A., Innis, R.B., Wamsley, J.K., Snyder, S.H. Kuhar, M.J. (1983) Autoradiographic localization of cholecystokinin receptors in rodent brain. J. Neurosci. 3: 877–906

Zetler, G. (1985a) Neuropharmacological profile of cholecystokinin-like peptides. Ann. N.Y. Acad. Sci. 448: 448–469

Zetler, G. (1985b) Caerulein and its analogues: Neuropharmacological properties. Peptides 6: Suppl. 3: 33–46

Zola-Morgan, S.M., Squire, L.R. (1990) The primate hippocampal formation: evidence for a time-limited role in memory storage. Science 250: 288–290

Author-Index

Aabus, L. 117
Adey, W.R. 278, 353, 356
Adler, G. 162
Adler, J. 162
Adrian, E.D. 61, 91, 337, 368
Aleksanov, S.N. 264
Allison, T. 142
Andersen, P. 333–337
Andersson, S.A. 333–337
Apostol, G. 357
Arellano, A. 160
Arieli, A. 435
Arnolds, D.E.A. 263

Babloyantz, A. 141, 318
Başar, E. 10, 12, 14, 15, 39, 51, 54, 55, 62–65, 69, 70, 72, 73, 75–79, 83–86, 95, 97, 99, 102, 105 106, 109, 130–142, 144–153, 172, 173, 180, 197, 209–215, 222, 226, 247, 265, 269–280, 310–314, 319, 324, 331–335, 338–339, 344, 348–350, 356, 360, 365–373, 390, 411–412, 422, 425,
Başar-Eroğlu, C. 36, 51, 55, 105, 118, 119, 144, 174, 197, 208, 229–239, 244–246, 263, 263, 284–289, 194, 339, 366, 369–372, 376, 378, 379, 425
Baddeley, A. 194–195, 203, 204
Ball, G.J. 174
Barlow H.B. 337, 395, 411, 434
Beagley, H.A. 172, 173
Berger, T.W. 251–254
Bishop, G.H. 337
Böhmig, L. 94
Bonnet, V. 337
Brandt, M.E. 349, 390
Brazier, M.A.B, 61, 331, 357
Bremer, F. 337
Bressler, S.L. 369
Buchwald, J. 174
Bullock, T.H. 61, 92–94, 97, 98, 109, 110, 114, 116 240, 249, 289, 331, 332, 367, 370, 414
Buszaki, G. 240
Butler, R.A. 162, 172
Buzsaki, G. 51

Cavada, C. 264
Chang, H.T. 337
Chapman, R.M. 197
Cohen, D. 177
Corwin, J.T. 110
Courchesne, E. 174
Creutzfeldt, O.D. 264, 282, 319, 348, 357, 421
Curio, G. 69, 73

Dafny, N. 118
Davis, H. 162
Demiralp, T, 39, 51, 55, 144, 160, 197, 256–262, 265, 270–280, 360, 365, 371
Desmedt, J.E. 195, 207, 269, 375, 425
Dietl, M.M. 118
Dinse, H.R. 3, 51, 333, 348, 367, 421, 434
Domesick, V.B. 124
Donald, N.W. 223, 225
Donchin, E, 172, 173, 195, 197, 223
Dostrovsky, J. 250
Douglas, R.J. 421
Dudaeva, K.I. 240
Duffy, F.H. 173

Eckhorn R. 3, 10, 52, 61, 105, 247, 369, 376, 421, 425, 431
Edelman, G.M. 378
Elazar, Z. 282, 356
Elbert, 370
Ellis, H.C. 194, 195
Engel, A.K. 376

Fairchild, M.D. 124
Fessard, A. 112, 430
Flohr, H. 387, 399, 404, 405
Fox, S.E. 359
Freeman, W.J. 61, 318, 368, 369, 397, 434
Friedman, 319
Fuster, J.M. 195, 204, 290, 390

Galambos, R. 37, 51 61, 195, 200, 289, 319, 331, 332, 369, 376–377
Gauderau, P. 118
Gauthier, 319
Gazzaniga, M.S. 382–383
Gelperin, A. 105
Gerschenfeld, H.M. 105
Giannitrapani, D. 319
Gille, H.G. 162
Goff, G.D. 142
Goldman-Rakic, P.S. 204, 378
Gönder, A, 70–73
Goodin, D.S. 216
Grandori, F. 177
Grastyan, E. 353
Gray, C.M. 3, 10, 53, 105, 247, 369–370, 376, 421, 425, 431
Green, R.L. 117
Gruzelier, J. 10

Haken, H. 332
Halgren, E. 174, 208
Haller, B. 93
Hari, R. 177, 178, 333
Harrison, J.B. 240
Hartline, P.H. 38, 55, 160
Hayashi, 359
Hebb, D.O. 363
Heiligenberg, W. 109, 112
Heinze, H.J. 414
Herkenham, M. 385
Hernandez-Peon, R. 347
Hillyard, S.A. 173, 195, 197, 200, 207, 318, 319
Hoke, M. 188
Horowitz, J.M. 355
Horridge, G.A. 91, 93, 94
Hunt, R.R. 194–195

Ishihara, T. 160

James, W. 196
Jansen, B.H. 349
Jarcho, L.V. 337
John, E.R. 38, 221, 240, 377, 397

Johnson, R. Jr. 208
Jokeit, H. 375
Joliot, M. 375

Kandel, E.R. 164
Kapas, L. 118
Karakaş, S. 69, 73 193, 203, 205 225
Kaukoranta, 177
Kerkut, G.A. 105
Kettner, R.E 251–254
Kirschfeld, 372
Klatzky, R.L. 194, 195
Klimesch, W. 10, 224, 289, 348, 421, 422
Knight, R.T. 208
Koelle, B.G. 118
Kojo, I. 376
Kok, A. 301
Kolev, V. 294, 338–341
Kopecz, K. 53
Kunze, H. 94
Kutas, M. 223

Lader, M. 173
Lang, M. 174, 264
Lange, M. 359
Larson, 263
Lavin, A. 368
Lehman, D. 320
Leiner, C.H. 391
Leung, L.S. 359, 368, 355, 372
Lieb, J. 357
Linas, R. 431
Lissman, H.W. 112
Llinás, R. 3, 333, 349, 375–377, 425
Longo, V.G. 118
Lopes da Silva, F. 51 197 263, 334–335, 355
Lopez, L. 69, 73
Lynch, J.C. 281

Machi, G. 385
Makeig, S. 61
Mangun, G.R. 331
Manthy, C.R. 118
Markowitsch, H.J. 264
McCallum, 319
McClune, M. 116
Meredith, M. 160
Mesulam, M.M, 15, 16, 264, 387
Mikkelsen, K.B. 177
Miller, R. 51, 160, 263, 264, 289, 345–346, 353, 354, 356, 366, 388
Mitchie, 195

Mizuki, Y. 118, 160, 358
Montgomery, S.A. 117
Moran, M.A. 264
Mountcastle, V.B. 198–200, 281, 387, 395, 434
Mulholland, T. 10

Näätänen, R. 173, 195, 208, 225, 269, 290, 355
Nair, N.P.V. 117
Nakamura, K. 319
Narici, L. 180
Nauta, W.J.H. 263
Neuper, C. 343, 378
Niehoff, D.L. 118
Niemer, W.T. 118
Nunez, P. 359

O'Keefe, J. 250, 355
Ohman, A. 173
Olpe, H. 118
Osaka, M. 348
Özesmi, Ç 70, 247, 339, 356, 368, 379

Paller, K.A. 208
Pandya, D.N. 264
Pantev, C. 53, 267, 369–370, 373, 376
Parnefjord, R. 161
Petersen, S.E. 281
Petsche, H. 289, 357–358
Pfurtscheller, G. 289, 341–344, 349, 378, 421
Picton, T.W. 269, 318, 414
Pineda, J.A. 240
Platt, C.J. 110
Polich, J. 301
Pribram, K. 207
Pulvermüller, F. 375

Rahn, E. 390
Ranck, J.B 250
Rappelsberger, P. 289
Regan, D. 173, 196, 208, 228, 395, 396
Ribary U. 52, 375–377
Ritter, W. 195, 197, 223
Robinson, D.L. 198, 263, 281
Roeder, K.D. 105
Röschke, J. 69, 89, 97, 99, 141, 318
Rose, G.N. 263
Rösler, F. 378
Roth, W.T. 197
Ruchkin, D.S. 207, 222

Salanki, J. 105
Salmelin, R. 177
Samar, V. 301
Sannita, W. 69
Sato, S. 177
Schreiner, C.E. 53
Schultz, 194
Schürmann, M. 53–55, 66, 130–140 144–153, 160, 294, 340–341, 372
Schütt, A. 95–97, 99, 102, 106, 344, 376
Schwab, R. 160
Schweizer, J. 110
Segal, M. 250
Shallice, T. 203
Shaw, J.C. 10
Sheer, D.E. 368
Shepherd, G.M, 52, 382–383
Silva, L.R. 348, 421, 434
Singer, W. 3, 53, 61, 105, 247, 369–370, 423, 425, 431, 434
Skarda, C.A. 318
Smidt, H. 93, 118, 208, 240
Sokolov, N.E. 240
Spekreije, H. 338, 368
Squire, L.R. 119, 282
Squires, N.K. 198, 223
Stacher, G. 118, 117
Stampfer, H. 209–210, 222, 339, 294
Stein, 160
Steiner, E.C. 118
Steriade, M. 378, 385–386
Stern, L. 195
Sternberg, R.J. 195
Stryker, M.P. 367, 414, 421
Sturbeck, K. 94, 102
Stuss, D.T. 269
Sutton, S. 222, 223, 269, 319
Swanson, L.W. 264, 345, 355
Szabo, T. 112
Szenthagothai 395, 398

Takasaka, Y. 319
Tallon, C. 376
Tamminga, C.A. 118
Thompson, R.F. 251–254
Tiberghien, G. 195
Tiihonen, J. 180
Tömberg 375, 425
Tulving, E. 203
Twarog, B,M. 105

Ullsperger, P. 162
Ungan, P. 70, 370, 368
Ursin, R. 75

Van der Tweel, L.H. 338, 368
Vanderwolf, C.H. 263, 372
Vaughan, H.G. 223
Verdun Lunel, H.F.E. 338
Vertes 355
Verzeano 434
Vinogrodova, O.J. 240, 250

Wallace, M. 160
Walter, G. 332, 333
Weinberg, H. 177
Wesdtphal, K.P. 174
Wiener, N. 349, 350
Williams, J.A. 118
Williamson 177
Woods, D.L. 174, 197, 208
Wundt, R.L. 194
Wurtz, R.H. 199

Yin, T.C.T. 199
Yoshii, N. 117, 124, 160

Subject Index

Acetylcholin 124
Adequate stimulation and inadequate stimulation 39–43, 147
AEP *see auditory evoked potential*
AFC *see Amplitude frequency characteristic*
Alpha System (selectively distributed alpha system) (or diffuse alpha system) *see also glossary* 5, 4, 143, 338–345, 363, 398, 400–401, 407, 420, 423, 431
Alphabet 9
Alphas *see also glossary* 331–353
Alpha activity 5, 6, 7,35, 40, 46, 415, 421
Alpha response 19, 41–47,53, 57, 59, 75–89, 110–111, 129–137, 143–153, 172–173, 180–181, 186 , 218, 223, 259, 280, 295-299, 331–353, 381, 397, 401–402, 417–426, 431
– secondary alpha response 341–349
– *see also Break of Alpha*
Ambiguous figures 379, 424
Amplitude frequency characteristics 23, 27,28 41–47, 58, 75–89, 42, 54-58, 61–67, 101–102, 106, 111–116, 120–125, 143–153, 157–158, 165, 170, 172, 181, 186–187, 216, 232 ,236, 238, 256, 273, 283–292, 410
Anamniotes 109–125
Anticipation 174, 390
Aplysia 91–108, 421
Association 197, 349, 355, 378, 381, 391, 418, 435
Association areas 152, 156, 387–388, 402, 418
Association cortex 229, 240, 346, 362–363, 381
Attention 161, 216, 323, 349, 353, 391, 394, 424
– focussed attention, 249–267, 283–292, 323, 406, 409, 419–420
– selective attention ,174, 193, 195–198, 202, 254, 265, 318, 424
Attractor, *strange attractor* 317–319
Auditory cortex 42, 46, 61–67, 117–126, 229, 232, 367, 369, 372, 400, 424
Auditory evoked potential 51, 58, 279
Auditory pathway 403, 404

BDRP *see Brain dynamics research program*
Beta response 35, 40, 88, 110, 289
Bimodal stimulation 59
Binding 367, 377, 423, 425
Brain alphabet 9, 325, 407–414, 432
Brain codes 196, 323, 359, 410, 423–424, 432
Brain dynamics research program 195, 224, 323, 393
Brain resonances *see resonances*
Brain response susceptibility 423, 432
Break of alpha 8, 51
Building blocks 93, 367, 379–380, 418, 425

CA3 233, 245–246, 390
CAP *see combined analysis procedure, see also glossary*
CE *see Cerebellum*
Cerebellum 19, 22–28, 69, 117–126, 350, 369, 421, 424, 431
– *see also Chap. 3 in Vol. I*
Ceruletide - CCK8 117–126
Chaos 392, 431
Coherence function 13, 29–32, 85–97, 336, 351, 356, 365–366, 423, 432
Compound EP 365–366, 407
Cross modality 6, 37, 39, 51, 53, 143–153, 185–186, 327, 388, 433

Decision making, 6, 55,174–175, 193, 197, 202, 216, 249–267, 323–325, 409
Delta oscillations 256, 264
Delta response *(evoked delta oscillations)* 41–47, 75–89, 161–175, 181, 223, 227, 228, 282–295, 324, 397, 405
Delta system *(selectively distributed)* 202, 398, 417–426

EEG 11, 433
EEG-modules 395–397
Encoding 202
Endogenous potential 207–228, 269, 375
Enhancement factor 143, 157–158, 173, *see also glossary*
EP *see evoked potentials*
ERP *see event related potentials*
Event related potential *see glossary* 207–228
Event related oscillations 16, 53, 118, 169, 207–228, 323–328, 393, 410, 417–426, 429, 431–434
Evolution 91
Evoked oscillations 5, 53, 124, 158, 171
Evoked potentials 410
Expectancy 249–267, 269–282, 389, 398

Filtering, filters
– adaptive 12, 42–49, 29, 96–123, 149, 183, 216, 258–275, 285–288, 368, 370, 409
– passband 107, 158
Frequency Stabilization Factor 426
Frontal cortex 160, 197, 240, 364–366, 308–392, 402, 418
– prefrontal cortex 263, 264, 269, 282–292, 358
Frontal recordings 129–147, 181, 324–326

Gamma activity 7, 101, 353, 411, 415, 431
Gamma response 19, 53, 62, 104–105, 110–116, 160, 223, 238, 240, 243–247, 256, 295–299, 332
Gamma system *(selectively distributed)* 4–5, 202, 367–381, 398, 417–426, *see also Glossary*
Gate function 346, 355, 418
GEA *see auditory cortex*
Going out of the system *see voyage*
Goldfish 109–125
Grandmother-cell 367, 395, 411, 415, 423

Habituation 225
Hebb's rule 363
Helix pomatia ganglia 91–108, 115, 372, 380, 421
Hippocampus 12, 19, 22–28, 54, 58, 61–67, 117–126, 229–230, 232, 233–236, 238, 240, 243–247, 249–267, 269, 333, 338, 346, 353, 358–359, 361, 363, 369, 372, 376, 389–392, 394, 418, 421, 423, 431, 433
Haloperidol 117–126

IC *see inferior colliculus*
Induced oscillations or induced rhythms 6, 160–175
Inferior colliculus 6, 9, 79, 253, 421
Integrative functions 378–381, 394, 415, 417–426, 429, 434
Integrative models 393–403
Integrative neurophysiology 4, 392, 429
Internal evoked potentials 432
Invariants 107, 115
Inverse relation between EEG and evoked potentials 349, 436
Invertebrates 8, 91–108, 430–432

Lateral geniculate nucleus 22, 28, 334
Learning 193, 218, 254, 363, 394, 409
LG *see lateral geniculate nucleus*
Limbic system 55, 353, 355, 359, 362, 381, 400, 406
Low-vertebrates brain 109–125

Magnetoencephalography 6, 9, 140, 150, 177–189, 372, 375, 410, 437-438
Major Operating Rhythms 399, 418, 420, 431–436
MEG *see magnetoencephalography*
MG *see medial geniculate nucleus*
Medial geniculate nucleus 13, 73, 253 , 333, 400
Memory 4, 124, 160, 175, 193, 197, 202–205, 228, 207–228, 282, 301, 344, 349, 363, 367, 376–377, 394, 421, 423
dynamic memory 318–327
– LTM 203, 250, 264, 267, 319
– STM 195, 204, 409, 418
– semantic memory 203, 205
– episodic memory 203
– working memory 204

MLR, middle latency response 369, 377
MMN 225, 327, 348, 424–426
MOR *see major operating rhythms*
Movement 198
MRI 9
Multiple functions 67, 348, 424–426
Multistability 375–379

N100 165, 197, 211–213, 211, 216, 274, 348
N200 233, 245
Natural frequency 356, 429, 431, 437
Neostigmine 117–126, 392, 431–4327
Neurons-Brain Doctrine 435–436
Noise 3, 10, 174, 336
Noradrenaline 106

OC *see visual cortex or occipital cortex, see also glossary*
Occipital recordings 129–147, 157, 186, 272, 282–292
Olactory bulb, 368–369
Omitted stimuli 231
Operator 351, 393, 415, 423
Orienting reflex 172, 195, 218, 240, 355, 357
Oscillatory responses 188, 324 *see also chapter 30*
– alpha 19, 41–47, 53, 57, 59, 75–89, 110–111, 129–137, 143–153, 172–173, 180–181, 186, 218, 223, 259, 280, 295–299, 331–353, 381, 397, 401–402, 417–426, 431
– beta 35, 40, 88, 110, 289
– delta 41–47, 75–89, 161–175, 181, 223, 227, 228, 282–295, 324, 397, 405
– gamma 19, 53, 62, 104–105, 110–116, 160, 223, 238, 240, 243–247, 256, 295–299, 332
– theta 4–5, 41–47, 51–53, 59, 110–116, 129–147, 159, 172, 181, 197, 202, 215–218, 220, 223, 227, 243–247, 280, 282–292, 295–299, 324–335, 353–366, 381, 390, 397, 405, 407, 420

P200 165, 211–215, 207–228, 245, 274
P300, *odd ball* 196–197, 223, 226, 230, 233, 234, 269, 272, 283–292, 293–302, 323, 326, 398, 408, 410–411, 418, 424, 433
P300-40 Hz 243–247, 355, 372
Parietal areas 55, 129–147, 197–199, 211, 264, 276, 281, 283–292, 324–326, 369–366, 376, 406, 407, 418
Parallel processing 369, 381, 436
Perception 4, 162, 174 ,368, 376, 395, 436
Pharmaco-EEG 117–126
Pharmacological agents 410, 421 426, 436
Phase-locking 375, 392, 424–426, *see also Glossary*
Phase spectra 30–32, 34, 381
Phase reordering 219–221
Post stimulus time histogram 198
Power spectra 95–97, 100, 103, 110–116
Preparation rhythms 219–222
Prestimulus EEG
Prolonged oscillation 51, 157, 217, 220, 324, 339, 365–366, 393, 409, 415, 426
PSTH *see Post stimulus time histogram*

Quantum mechanics 55

Ray 109–125
Regime change 314
Resonance *(in the brain)* 11, 51,69, 110, 112, 123–124, 142–143, 333, 349, 356, 358, 363, 411, 424–425, 437
Real response
– Brain's real response 51, 55, 432
Repertoire 394
Reticular formation 22–68, 61–67, 69, 72, 75–89, 117–126, 229, 232–233, 331, 338, 372, 400, 421, 423, 431
RF, *see reticular formation*

Schizophrenic 118
SEEG
Selective averaging 209, 323
Selectively distributed oscillatory (alpha, theta, delta, gamma) systems *see also glossary and Part IV*
– alpha 5, 4, 143, 338–345, 363, 398, 400–401, 407, 420, 423, 431
– delta 202, 398, 417–426
– gamma 4–5, 202, 367–381, 398, 417–426
– theta 143, 160, 174, 186, 262, 281, 353, 364–366, 390, 398, 401, 406, 407, 417–426, 431
Signal detection 162, 251, 292, 324, 379
Single trials 49, 52 ,141, 148, 156, 159, 170, 185, 209, 220, 244, 299–302,

323–325, 339, 372, 375, 409, 432 *see also single sweeps*
Sleep SWS 75–89, 124, 426
Storage 202
Superior colliculus 22–28, 38
Superposition 124, 163, 211, 213, 217, 395, 415–420, 436, 431
Surprise 6, 193, 202, 290, 325
Synchronization 12
Synchrony 196

Target, non-target 197, 211, 216–212, 243, 292, 293-302, 350, 409
Transition 394, 431
TRFC-method 28, 71–72, *see also glossary*
Thalamus 54, 334–335, 338, 346, 350, 367, 381–392, 419–420
Theta activity 6, 276
Theta response 4–5, 41–47, 51–53, 59, 110–116, 129–147, 159, 172, 181, 197, 202, 215–218, 220, 223, 227, 243–247, 280, 282–292, 295–299, 324–335, 353–366, 381, 390, 397, 405, 407, 420 *(see also evoked theta)*
– bisensory theta response 156–160
Theta system *(selectively distributed) see glossary*
Threshold (-hearing) 6, 161–175, 251, 292, 323–325, 327, 410, 424
Time-locking 69, 217, 327, 331, 339, 367, 369, 372, 380
Transfer function 411–412, 435, 429–430, *see also Chap. 4, Vol. I*

VEP *see visual evoked potential*
Visual cortex 42, 367, 368, 381, 400
Visual evoked potential 41–47, 51–58, 156, 267, 272
Visual pathway 403–404
Volume conduction 346
Voyage 8

Wavelet analysis 45–51, 293–302, *see also Chap. 5, Vol. I*

Printing: COLOR-DRUCK DORFI GmbH, Berlin
Binding: Buchbinderei Lüderitz & Bauer, Berlin